Error Control Systems for Digital Comunication and Storage

Stephen B. Wicker

School of Electrical and Computer Engineering
Georgia Institute of Technology

PRENTICE HALL, Upper Saddle River, New Jersey 07458

Library of Congress Cataloging-in-Publication Data

Wicker, Stephen B.
Error control systems for digital communication and storage /
Stephen B. Wicker.
p. cm.
Includes bibliographical references and index.
ISBN 0-13-200809-2
1. Digital communications. 2. Error-correcting codes (Information theory) 3. Digital modulation. I. Title.
TK5103.7.W53 1995
621.382'24--dc20 94-11649
CIP

Acquisitions Editor: *Don Fowley*
Managing Editor: *Linda Ratts*
Production Editor: *Joe Scordato*
Copy Editor: *Bob Lentz*
Cover Designer: *Wendy Alling Judy*
Buyers: *Dave Dickey/Phil Zolit*
Supplements Editor: *Alice Dworkin*
Editorial Assistants: *Jennifer Klein/Susan Handy*

Simon & Schuster / A Viacom Company
Upper Saddle River, New Jersey 07458

Printed in the United States of America

10 9 8 7 6 5 4

ISBN 0-13-200809-2

Prentice-Hall International (UK) Limited, *London*
Prentice-Hall of Australia Pty. Limited, *Sydney*
Prentice-Hall Canada Inc., *Toronto*
Prentice-Hall Hispanoamericana, S.A., *Mexico*
Prentice-Hall of India Private Limited, *New Delhi*
Prentice-Hall of Japan, Inc., *Tokyo*
Simon & Schuster Asia Pte. Ltd., *Singapore*
Editora Prentice-Hall do Brasil, Ltda., *Rio de Janeiro*

To Wendy

Contents

Preface

At this moment I have 25 books in my office that deal primarily with error control coding. I probably own more than that, but my students have a habit of wandering off with them. It is thus an act of sheer arrogance to add yet another volume to this wealth of literature. On the other hand, I have always felt that there remains a need for a particular type of text—one that includes all the following within its covers.

- A self-contained text for a one-year, comprehensive, graduate-level survey of error control systems.
- An up-to-date text, complete with bibliographical references, that will serve as a starting point for the student conducting doctoral research in error control coding.
- An applications-oriented text that provides the practicing engineer with the information necessary to design and implement error control subsystems for digital communication systems.
- And finally, a text that includes a good tutorial on trellis coded modulation and an up-to-date treatment of ARQ protocols.

I have thus made no attempt to match the mathematical rigor of Berlekamp or the encyclopedic depth of MacWilliams and Sloane. I have instead tried to direct my message toward the beginning graduate student and the practicing engineer, while providing sufficient direction for the advanced student into more detailed sources.

This book contains four basic parts: finite field theory, block codes, convolutional/trellis codes, and system design. These sections are organized as follows.

Chapters 2 and 3 provide an introduction to Galois fields and polynomials with coefficients over Galois fields. I have made every effort to minimize the amount of mathematical material in these chapters while still leaving enough to treat the structure of cyclic codes in general and BCH codes in particular in a reasonably rigorous manner.

Chapters 4 through 9 cover the various types of block error control codes that are currently being used or show promise of use in the future. I have emphasized BCH and Reed-Solomon codes because of their enormous power and popularity in current systems. I have also emphasized Reed-Muller codes, whose fast decoders are useful in high data rate (e.g., optical) applications.

Chapters 11 through 14 treat convolutional codes and their trellis coded progeny. The design and performance of the Viterbi and sequential decoding algorithms are presented here. Implementation issues are discussed in some detail, so that the practicing engineer can get some idea as to the scope of the encoder/decoder design problem. The design and use of rate compatible punctured convolutional codes are also discussed. A full chapter is devoted to trellis codes, a research/design/development area that is one of the most active in error control coding at the moment, and one that promises to bring us within sight of the promised land of Shannon's noisy channel coding theorem.

Finally, Chapter 1, 10, 15, and 16 discuss system level issues. Chapter 1 is designed to convey the basic idea of coding gain as well as some historical background. Chapter 10 discusses the various means for analyzing the performance of block codes over a variety of channels. In particular, the slowly fading channel is discussed in some detail. Chapter 15 discusses retransmission request systems that make use of the various block, convolutional, and trellis codes discussed earlier in the text. And finally Chapter 16 discusses some specific design applications, including the ubiquitous Compact Disc player and the magnetic recording channel.

This book is based on a series of class notes that I developed at Georgia Tech between 1987 and 1993. I estimate that some five hundred students have read through at least a portion of this book, perhaps even with some benefit (forgive me for not naming the students). When I joined the faculty at Georgia Tech in 1987 I designed a two-course sequence on error control coding and its applications. The basic outline was borrowed from the curriculum of the Communications faculty at the University of Southern California. The first course begins with a brief treatment of simple block codes (e.g., Hamming codes) and proceeds to develop convolutional and trellis codes. Particular emphasis is placed on the Viterbi decoding algorithm. This course does not require a great deal of mathematical background, and is thus ideal for the entry-level graduate student. The second course delves into Galois fields in some depth, and then proceeds to develop cyclic codes in general and BCH and Reed-Solomon codes in particular. It concludes with a treatment of the Berlekamp-Massey algorithm.

This book contains far more than can be comfortably taught in two quarters, or even in two semesters. Through appropriate selection of material, however, I think that it will serve as a suitable text in most situations. For a single-semester course in coding theory I suggest the following.

ONE-SEMESTER COURSE

Chapter 2
Chapter 3
Chapter 4
Chapter 5
Chapter 8 (partial)
Chapter 11
Chapter 12
Chapter 14 (partial)

Those who have the luxury of a two-course sequence may wish to use the format that I have used at Georgia Tech. The material marked with asterisks can be considered optional.

FIRST COURSE

Chapter 1
Chapter 2 (through vector spaces and finite fields of prime order)
Chapter 4
Chapter 7*
Chapter 11
Chapter 12
Chapter 13*
Chapter 14

SECOND COURSE

Chapter 2
Chapter 3
Chapter 5
Chapter 6*
Chapter 8
Chapter 9
Chapter 10
Chapter 15*
Chapter 16*

I would like to acknowledge all those who have played a major part in the preparation of this text. The editors at Prentice Hall and the reviewers, John J. Komo of Clemson University, and A. Brinton Cooper, III of the U.S. Army Research Laboratory, have been of enormous assistance. In the early stages of the writing, Ian Blake and Vijay Kumar kindly offered suggestions that made this book far better than I had originally intended. I am particularly grateful to Brint Cooper, who was a constant source of algebraic expertise and general wisdom.

I am grateful to my doctoral students, past and present, who contributed to this book in many ways. In particular, Thomas Tapp provided substantial assistance in weeding out errors during the production of the final text.

I would also like to thank those whose guidance over the past ten years has led me through a number of projects, of which this is but the latest. Vijay Bhargava, Ron Schafer, Gus Solomon and Lloyd Welch have always been willing to take time away from their own efforts to help me with mine. I am fortunate and grateful. I hope that this book reflects some small portion of their abilities as teachers, engineers, and mentors.

And finally, the most personal note. Thanks to the genius of Apple Computer, bits and pieces of this book have been written in five different countries on a dozen different machines. Chapter 5 was even written in a hospital room in Metz, France, while waiting for my son Alexander to be born. Given the various writing locales and the different technical resources used, one would expect this text to be a patchwork quilt. There are, however, two invisible threads that unite this three-year composition into a seamless whole: a single author and the constant love, support, and occasional sacrifice of his wife.

STEPHEN BRYANT WICKER
Atlanta, Georgia

1

Error Control Coding for Digital Communication Systems

This chapter is an overview of the key role played by error control coding in the design of digital communication systems. We begin with an overview of the elements of a communication system, focusing on the classical design trade-off between transmitted power, bandwidth, and data reliability. The chapter continues with a look at how Shannon and Hamming approached this problem, laying the foundations for the fields of information theory and error control coding. The chapter concludes with a brief outline of the contents of this text.

1.1 DIGITAL COMMUNICATION SYSTEMS

A digital communication system is a means of transporting information from one party (call him or her "User A") to another party ("User B"). The system is "digital" in that it uses a sequence of symbols from a finite alphabet to represent the information. The transmission of data in digital form allows for the use of a number of powerful signal processing techniques that would otherwise be unavailable, including, of course, error control coding. Figure 1-1 shows the basic elements of a digital communication system.

The "data source" block in Figure 1-1 may represent any of a number of sources of information. For example, in a digital cellular mobile telephone, the data source consists of a microphone and an analog-to-digital converter. In a remote sensing satellite, the source might consist of a two-dimensional grid of photodetectors, whose analog signals are read out by rows into analog to digital converters. In general, the data source provides a stream of symbols at some average rate R_S symbols per second that is passed along to the modulator.

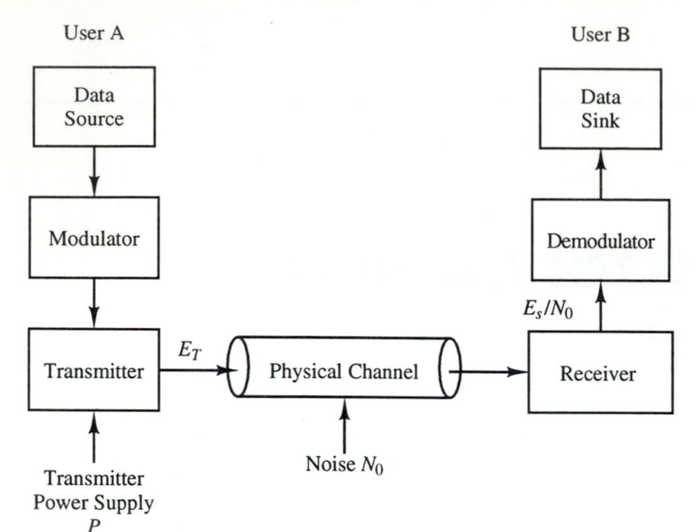

Figure 1-1. Basic elements of a digital communication system

The modulator maps the information symbols onto signals that can be efficiently transmitted over the communication channel. The selection of a modulation format can be a complex affair, for there are quite a few from which to choose, all providing a variety of different performance characteristics. In most cases, however, the selection of the modulation format reduces to a consideration of the power and bandwidth availability in the intended application. For example, in cellular telephony the principal design goal is the minimization of spectral occupancy by a single user (thus maximizing the number of paying customers in a given geographical area and a fixed bandwidth). Cellular telephone designers thus tend to select bandwidth-efficient modulation formats such as Gaussian minimum shift keying (GMSK) [Hir]. On the other hand, in deep space telecommunications the principal design goal is the minimization of transmitter power consumption (thus reducing battery weight and/or increasing the operational lifetime of the spacecraft). Satellite designers thus prefer power-efficient formats like binary phase-shift keying (BPSK, also written PSK) [Gag2].

A closer look at a modulation format may prove instructive, so consider the case of BPSK. BPSK maps the bits in a binary data steam onto one of two signals, which are typically represented as follows.

$$\begin{aligned} \text{data } 0 \leftrightarrow s_0(t) &= \sqrt{2P}\cos(2\pi f_c t) \qquad 0 \le t \le T \\ \text{data } 1 \leftrightarrow s_1(t) &= \sqrt{2P}\cos(2\pi f_c t + \pi) \qquad 0 \le t \le T \end{aligned} \tag{1-1}$$

The "phase-shift keying" portion of the name is readily apparent. The signal representing a binary 1 is π radians out of phase from the signal representing a binary 0. A fixed phase reference must thus be provided at the receiver in order to distinguish between the two signals.

The spectrum of the transmitted signal is centered about the frequency f_c. The selection of f_c is influenced by a number of factors, ranging from the channel characteristics to the dictates of national regulatory agencies. The signal period T is simply the inverse of the symbol transmission rate R_S. Given the average signal power P, the transmitted energy per bit is $E_T = P \cdot T = P/R_S$.

Power levels in communication systems are generally discussed in decibel units (dB), which are computed as follows.

$$P_{\text{dB}} = 10 \log_{10} P_{\text{watts}} \tag{1-2}$$

This convention is quite convenient, for the logarithmic scale not only compresses the range of numbers involved in the analysis, but also converts multiplicative effects such as channel attenuation into additive effects. Example problems are provided at the end of this chapter.

The physical channel attenuates the transmitted signal and introduces noise. The attenuation is generally caused by energy absorption and scattering in the propagation medium. In satellite communications this effect is often called the "free-space loss" and is typically measured as a dB power loss that is a function of distance. On a 14/12 GHz geosynchronous satellite channel the free-space loss can exceed 200 dB. In mobile radio applications the attenuation is not fixed, but tends to fluctuate with the speed of the vehicle and the geometry of the surrounding buildings and terrain. Buildings, mountains, and even trees can cause a signal moving from one point to another to take several different paths. As these paths may have different lengths, the various copies of the signal may constructively or destructively interfere with one another, depending on the location of the receiver. If the receiver is moving, it will see a series of energy peaks and troughs whose frequency of occurrence is a function of vehicle speed. Such communication channels are called "multipath fading channels"; their amplitude a is frequently modeled using Rayleigh or Rician probability density functions, as shown below.

$$\begin{aligned} p(a) &= ae^{-a^2/2}, & a \geq 0 \quad & \textbf{Rayleigh distribution} \\ p(a) &= ae^{-(a^2+B^2)/2} I_0(Ba), & a \geq 0 \quad & \textbf{Rician distribution} \end{aligned} \tag{1-3}$$

The function $I_0(a)$ is the zeroth-order modified Bessel function.

The noise introduced by the channel can also take on a number of different forms. The most common noise source is the ambient heat in the transmitter/receiver hardware and the propagation medium. Other sources include hardware-induced transients, co-channel and adjacent-channel interference from other communication systems, shot noise in photodetectors, and climatic phenomena. The most commonly assumed noise model is the additive white Gaussian noise (AWGN) model. The AWGN model is simple, easily analyzed in a variety of contexts, and can even provide an accurate model of what is actually happening in some communication systems. For example, many satellite channels and line-of-sight terrestrial channels can be accurately modeled as AWGN channels. AWGN is "additive" in the sense that the impact of the noise on the transmitted signal can be represented as a random variable n added to the demodulated signal [Woz]. The random variable n is assumed to be Gaussian with zero mean and variance σ^2, as shown in the Gaussian probability density function below.[1]

$$p(n) = \frac{1}{\sigma\sqrt{2\pi}} e^{-n^2/2\sigma^2} \qquad \textbf{Gaussian distribution} \tag{1-4}$$

The "white" portion of AWGN refers to the spectral density of the noise, which is assmed to have the constant one-sided value N_0 from frequency 0 to $+\infty$. This is clearly a physical impossibility, for it requires that the noise source have infinite energy; however, the assumption provides accurate results when applied to the analysis of bandlimited communication systems.

The transmitted signal arrives at the receiver with a symbol-energy to noise-spectral-density ratio E_s/N_0. The demodulator recovers the data from the received signal and passes it along to its final destination: User B's data sink. In telecommunication environments where noise is present, the demodulated data provided to the data sink contains errors. This is usually characterized in terms of a bit error rate (BER). The BER varies in proportion to the symbol rate R_S and in inverse proportion to E_s/N_0 and the transmitter power level P.

$$\text{BER} \propto R_S \propto \frac{1}{(E_s/N_0)} \propto \frac{1}{P}$$

If the BPSK modulation format discussed earlier is in use, then the BER is determined (assuming ideal demodulation) using the expression

$$\text{BER}_{BPSK} = Q\left(\sqrt{\frac{2E_b}{N_0}}\right), \qquad \text{where } Q(x) = \frac{1}{\sqrt{2\pi}} \int_x^{\infty} e^{-y^2/2}\, dy \tag{1-5}$$

Figure 1-2 contains a curve that shows how the BER varies with received signal energy for the BPSK modulation format (since the modulation format is binary, symbol energy E_s becomes bit energy E_b). This curve shows the ideal performance for a binary modulation format, for the two BPSK signals are maximally uncorrelated, or "antipodal" [Woz].

Unfortunately, as noted earlier, the BPSK receiver has to provide a coherent phase reference to demodulate the received signal. This phase reference is typically obtained by phase-locking to the received signal carrier [Gag1]. In some communication environments this is an extremely difficult task (e.g., when jamming is present). Some applications thus dictate the use of *noncoherent* modulation formats which do not require a phase reference. Figure 1-2 also contains a curve showing the performance of a binary format that consists of the two orthogonal signals listed below. These signals differ by a frequency shift and can thus be demodulated without a phase reference. This format is commonly called binary frequency-shift keying (BFSK or FSK).

$$\begin{aligned} \text{data } 0 \leftrightarrow s_0(t) &= \sqrt{2P}\cos(2\pi f_0 t), \qquad 0 \le t \le T \\ \text{data } 1 \leftrightarrow s_1(t) &= \sqrt{2P}\cos(2\pi f_1 t), \qquad 0 \le t \le T \end{aligned} \tag{1-6}$$

[1] One of the more interesting places one can find a Gaussian distribution is the German 10 Mark note that circulated in the early 1990s. It bears the likeness of Karl Friedrich Gauss (1777–1855) himself.

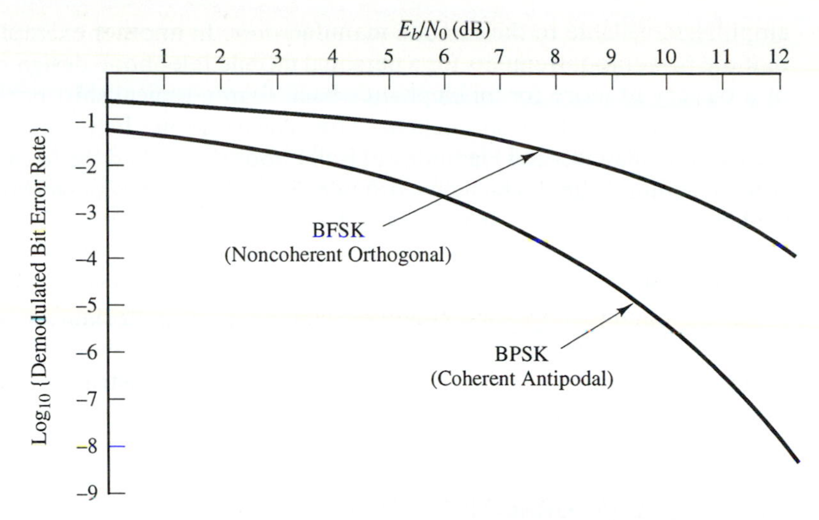

Figure 1-2. BER performance of ideal BPSK and BFSK

These two signals are orthogonal if they have an integral number of cycles on a symbol interval and $f_0 \neq f_1$. The bit error rate for noncoherently demodulated BFSK is determined using the expression [Gag1]

$$\text{BER}_{BFSK} = \frac{1}{2} e^{-(E_b/2N_0)} \tag{1-7}$$

BER is usually the most important performance parameter from the communication system customer's point of view. Throughput is also an important consideration: the customer generally wants to transmit information at some rate R_S symbols per second that is dictated by his or her application. Other design criteria include

- Complexity
- Cost
- Weight
- Heat dissipation
- Fault tolerance

This would seem to leave very little flexibility in the design process, for our simplistic development thus far seems to show that once the modulation format has been selected, the desired throughput and BER determine the required transmitter power level, which in turn determines battery weight, heat dissipation, etc. In some cases this analysis may indicate that the construction of the desired system is a physical impossibility. For example, in the design of a communication satellite, the power/throughput/BER equation may dictate a transmitter power level P that is higher than can be provided by the traveling-wave tube amplifiers or the solid state power amplifiers available to the satellite manufacturer. In another example, we may find that the power requirements for a personal mobile telephone design indicate the use of a battery fit more for an elephant's back than a typical shirt pocket. There was a time when these obstacles might have been insuperable. In the late 1940s, however, the work of Shannon and Hamming at Bell Laboratories laid the foundation for error control coding, a field which now provides us with a host of powerful techniques for achieving desired bit error rates at reduced transmitter power levels.

These two gentlemen attacked the problem of error control on noisy channels using radically different techniques. Shannon used a statistical/existential approach, while Hamming's approach could be labeled combinatorial/constructive [Ber2]. Consequently their results were also radically different: Shannon found the limits for ideal error control, while Hamming showed how to construct and analyze the first practical error control systems.

1.2 SHANNON AND INFORMATION THEORY

In 1948 Shannon published a paper entitled "A Mathematical Theory of Communication" in the *Bell Systems Technical Journal* [Sha1]. He published a second paper in this journal in 1949 under the title "Communication Theory of Secrecy Systems" [Sha2]. Both papers grew out of the cryptanalytic work he conducted during the Second World War. The first of these papers defines the way we think about digital communication systems today. It is the basis for the entire field of information theory. The second paper has had a similar impact within the more limited scope of cryptology. The reader who wants to pursue the elegant and rich field of information theory is referred first to Gallager [Gal1]. Though this text is somewhat dated (it contains no material on multiuser information theory), it contains, in this author's opinion, the clearest explanation of basic information theory in print. The best up-to-date text known to this author is that by Cover and Thomas (1991) [Cov].[2] Here we shall only make a few basic comments before moving on to the more practical side of error control coding.

In Shannon's 1948 paper he introduces a metric by which information can be quantified. This metric allows one to determine the minimum possible number of symbols necessary for the error-free representation of a given message. A longer message containing the same information is said to have *redundant* symbols. This basic start leads to the definition of three distinct types of codes.

- **Source codes.** Source codes are used to remove the *uncontrolled redundancy* that naturally occurs in some source information streams. Source coding reduces the symbol throughput requirement placed upon the transmitter. Source codes also include codes used to format the data for specialized modulator/transmitter pairs (e.g., Morse code in telegraphy).

[2] For those with a pragmatic bent, [Cov] includes a chapter on information theory and the stock market.

- **Secrecy codes.** Secrecy codes encrypt information so that the information cannot be understood by anyone except the intended recipient.
- **Error control codes.** Error control codes (also called channel codes) are used to format the transmitted information so as to increase its immunity to noise. This is accomplished by inserting *controlled redundancy* into the transmitted information stream (either as additional bits or as an expanded channel signal set), allowing the receiver to detect and possibly correct errors.

Figure 1-3 shows how these three types of codes are integrated into the basic communication system model shown in Figure 1-1. The order of the three codes (source encoding, then encryption, then channel encoding) is very important. Any rearrangement of this order would detract from the performance of one or more of the codes involved. The error control code has the inside position relative to the other codes. The error control encoder and decoder in conjunction with the physical channel create an error control channel that provides sufficiently reliable data for the effective operation of the secrecy and source codes.

Shannon used some very elegant statistical techniques to determine the limits of the performance of these three types of codes. For example, he showed that the "one-time pad" cryptosystem used by secret agents and the now obsolete "hot-line" between Washington, DC, and Moscow is unconditionally secure. In this text we

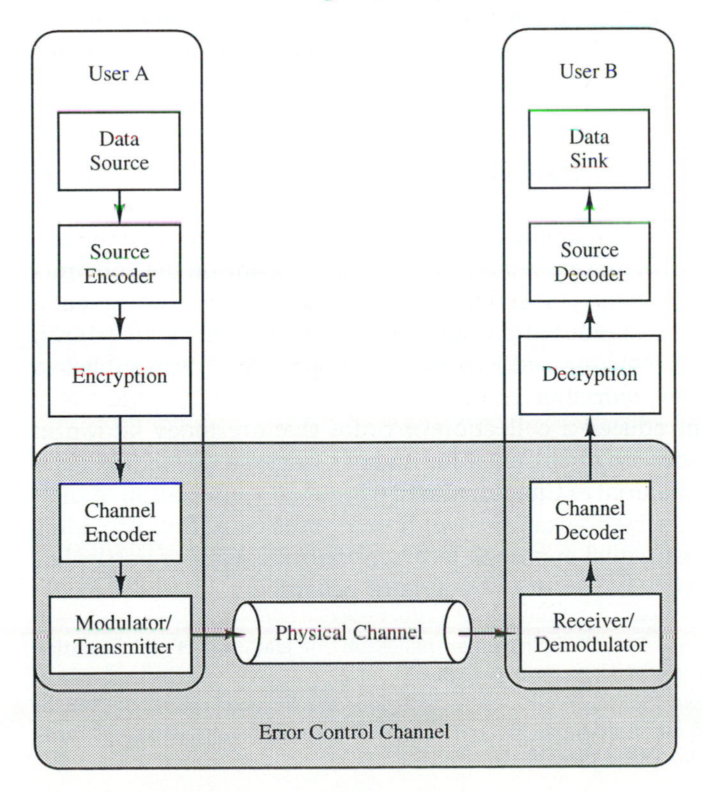

Figure 1-3. Digital communication system coding model

are interested solely in error control coding for noisy channels. Shannon's primary result in this area is called the **noisy channel coding theorem**, which can be paraphrased as follows

Theorem 1-1—Shannon's Noisy Channel Coding Theorem

> **With every channel we can associate a "channel capacity" *C*. There exist error control codes such that information can be transmitted across the channel at rates less than *C* with arbitrarily low bit error rate.**

The proof of this theorem is elegant but is not necessary for an understanding of the material in this book (the interested reader is, once again, referred to Gallager [Gal1]). The proof is not of immediate interest to us because it is an example of an "existential proof" as opposed to a "constructive proof." We are not actually given any codes that provide the desired "arbitrarily low bit error rate"; we are simply shown that such codes exist. This theorem thus describes the "promised land" for the error control coding theorist. Its publication initiated a search for good codes that has now lasted more than forty years.

Those interested in immediate practical applications must focus on what can be obtained from the codes that have already been discovered. Unfortunately all of these codes fall short of the goal established by the noisy channel coding theorem, but they do provide substantial performance improvement for digital communication systems. Henceforth the term "codes" shall refer to error control codes that not only exist but have actually been discovered.[3]

1.3 HAMMING AND ERROR CONTROL CODING

Hamming was a contemporary of Shannon at Bell Telephone Laboratories in the late 1940s. Hamming's work took a more pragmatic approach, focusing on the construction of error detecting and error correcting codes. Because of his work, he is considered by many to be the founder of the subject of error control coding [Ber2]. His seminal paper, "Error Detecting and Error Correcting Codes," was published in the *Bell System Technical Journal* in 1950.[4]

Hamming's paper introduced a collection of codes that are today known as Hamming codes. These codes are discussed at the end of Chapter 4, for they require some of the mathematical material in Chapters 2 and 4 for a full appreciation of their structure. However, it is possible to examine two special cases here, thus providing an idea of Hamming's constructive approach to error control coding in particular, and a feel for error detection and error correction in general.

[3] The author adheres to the Platonic/Pythagorean philosophy of mathematics: mathematical structures are discovered, not invented.

[4] Though Hamming discussed his single-error-correcting codes with Shannon in 1947 or 1948, he was apparently unable to publish his results until 1950 because of patent considerations [Ber2]. Shannon references Hamming's work in his 1948 paper [Sha1].

1.3.1 A Single-Error-Detecting Code

Take the output of a binary source and break it up into k-bit message blocks. We can represent such blocks using the notation

$$\mathbf{m} = (m_0, m_1, m_2, \ldots, m_{k-1})$$

At the end of every block we append a single redundant bit b. The value of b is selected so that the binary sum ($1 + 1 = 0, 1 + 1 + 1 = 1$, etc.) of all of the message bits and b in each block is zero.

$$\mathbf{c} = (m_0, m_1, m_2, \ldots, m_{k-1}, b), \qquad \text{where } b = \sum_{j=0}^{k-1} m_j \tag{1-8}$$

The receiver checks the received words by adding together the values in each coordinate position. If the sum is not zero, then we know that one or more of the coordinate values in the received word must have been corrupted by channel noise. It is clear that this simple system can detect all erroneous received words that contain an odd number of errors. The analysis of this and more complex error detecting systems is covered in Chapter 10.

This single-error-detecting code is commonly referred to as a "parity-check" code. Equation (1-8) ensures that all transmitted words have "even parity"—in other words, an even number of coordinates with the value 1.

1.3.2 A Single-Error-Correcting Code

We now take a look at one of Hamming's error-correcting codes. Let the message blocks discussed in the previous case be of length 4.

$$\mathbf{m} = (m_0, m_1, m_2, m_3)$$

Three redundant bits $\{b_0, b_1, b_2\}$ are appended according to the following formula.

$$\begin{aligned} b_0 &= m_1 + m_2 + m_3 \\ b_1 &= m_0 + m_1 + m_3 \\ b_2 &= m_0 + m_2 + m_3 \end{aligned} \tag{1-9}$$

The transmitted blocks of message bits and redundant bits have the form

$$\mathbf{c} = (c_0, c_1, c_2, c_3, c_4, c_5, c_6) = (m_0, m_1, m_2, m_3, b_0, b_1, b_2)$$

The received word $\mathbf{r}$ has the form $\mathbf{r} = (r_0, r_1, r_2, r_3, r_4, r_5, r_6)$. The receiver computes the following binary values for each received word.

$$\begin{aligned} s_0 &= r_1 + r_2 + r_3 + r_4 \\ s_1 &= r_0 + r_1 + r_3 + r_5 \\ s_2 &= r_0 + r_2 + r_3 + r_6 \end{aligned} \tag{1-10}$$

If $(s_0, s_1, s_2) = (0, 0, 0)$, then the received word is a valid word. The receiver then assumes that no errors have occurred and passes the data along to the data sink without modification. If $(s_0, s_1, s_2) \neq (0, 0, 0)$, then the receiver knows that an error has occurred. Each nonzero 3-tuple (s_0, s_1, s_2) corresponds to an error in a single location of the received word, as shown in Table 1-1 below.

After (s_0, s_1, s_2) has been computed, the receiver inverts the value in the corresponding location of the received word. The resulting word is a valid word, and is assumed to be the word that was transmitted. This code is guaranteed to correct any single error in a block of seven received bits. In general, there is a single-error-correcting code of this type for all lengths n of the form $(2^m - 1)$, where m is an integer greater than 1.

Example 1-1—Single-Error-Correction

Let us suppose that we want to transmit the message $\mathbf{m} = (1, 0, 0, 1)$. Using Eq. (1-9), we compute the values for the redundant bits, obtaining $(b_0, b_1, b_2) = (1, 0, 0)$. The transmitted word corresponding to the message is thus

$$\mathbf{c} = (1, 0, 0, 1, 1, 0, 0)$$

Let us now suppose that $\mathbf{c}$ is corrupted by noise during transmission such that the received word has an error in position r_1.

$$\mathbf{r} = (1, \underline{1}, 0, 1, 1, 0, 0)$$

The receiver computes the three values in Eq. (1-10), obtaining

$$(s_0, s_1, s_2) = (1, 1, 0)$$

According to Table 1-1, this corresponds to an error in received word location r_1. When we complement the value in this position, we correct the error, recovering the transmitted word $\mathbf{c}$. ■

This error correcting code typifies Hamming's combinatorial approach. At this point in the text, we are not yet in a position to appreciate all of the structure in his codes, but it should be clear that they are improving the reliability of the transmitted data. The next section describes how this improvement in reliability can be quantified in terms of a "coding gain."

TABLE 1-1 Hamming single-error-correcting code

(s_0, s_1, s_2)	Error location in $\mathbf{c}$
000	None
001	r_6
010	r_5
011	r_0
100	r_4
101	r_2
110	r_1
111	r_3

1.4 HOW ERROR CONTROL CODES IMPROVE THE PERFORMANCE OF COMMUNICATON SYSTEMS

As seen in the last section, error control codes insert redundancy into the transmitted data stream so that the receiver can detect and possibly correct errors that occur during transmission. The amount of inserted redundancy is usually expressed in terms of the **code rate** R. The code rate is the ratio of k, the number of data symbols transmitted per code word, to n, the total number of symbols transmitted per code word. For example, the single-error-correcting Hamming code in Section 1.3.2 had rate 4/7. Assuming that the data symbol transmission rate R_S is to remain constant, the added redundancy forces us to increase the overall symbol transmission rate to R_S/R. If the transmitter power level is constant, then the received energy per symbol is reduced from E_s to $R \cdot E_s$. The demodulated BER is thus *increased* with respect to its previous value! However, when the degraded, demodulated data symbols and redundant symbols are sent to the error control decoder, the redundancy is used to correct some of the errors, improving the reliability of the demodulated data. If the code is well selected, the BER at the output of the decoder is better than that at the output of the demodulator in the original, uncoded system. The amount of improvement in BER is usually discussed in terms of the additional transmitted power that is required to obtain the same performance without coding. This difference in power is called the **coding gain**.

Example 1-2—BER Improvement Through Error Control Coding

Suppose that BPSK modulation is in use over an uncoded AWGN channel. If the received E_b/N_0 is 10 dB, then the demodulated BER is 3.9×10^{-6} (see Eq. (1-5) or Figure 1-2). The (15, 11) single-error-correcting Hamming is discussed in some detail at the end of Chapter 4. For now a few basic facts concerning this code are presented for use in analyzing the performance of the example system.

The (15, 11) Hamming code contains 2^{11} code words of length 15 bits. Each code word thus represents 11 information bits. The code rate is $11/15 = 0.73$; the received E_b/N_0 thus drops to 8.75 dB, and the demodulated BER increases to 5.4×10^{-5}.

The (15, 11) Hamming code can correct a single error in every block of fifteen demodulated bits. Using the analysis discussed in Section 10.1, it can be shown that the information stream at the output of the decoder has a BER of 5.6×10^{-8}, which is better than the BER for the uncoded case. To achieve this same BER without error control coding, the power must be increased so that the received E_b/N_0 is 11.5 dB. The coding gain provided by the error control code in this example is thus $11.5 - 10.0 = 1.5$ dB. This is shown graphically in Figure 1-4. ■

The 1.5 dB coding gain provided by the (15, 11) Hamming code is small compared to that provided by some of the more complicated codes discussed in this book. Coding gains of 6 to 9 dB are readily achievable in standard error correcting systems, allowing for a reduction in transmitted power by a factor of up to 8. The use of error control coding thus introduces an important positive factor into the power budget used in the design of communication systems. Several problems based on this theme are provided at the end of this chapter.

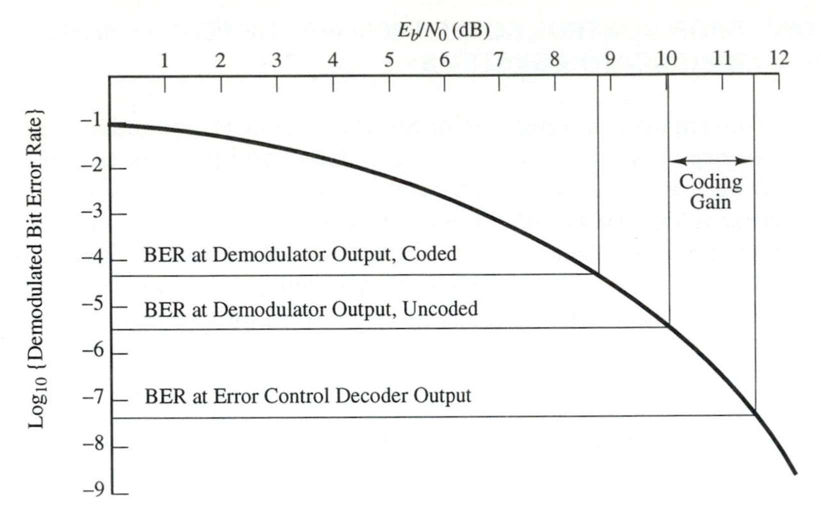

Figure 1-4. Coding gain provided by error control code

1.5 THE CONTENTS OF THIS BOOK

The following is a synopsis of the contents of this book. There are essentially three different types of chapters: mathematical background, discussions of specific codes, and system-level issues. We begin with the mathematics.

Galois fields. Mathematics is the language of coding theory. In fact, one could argue that each wave of innovations in coding theory has been caused by the application of increasingly powerful mathematical concepts (e.g., algebraic geometry in the early 1980s). It is not, however, necessary that the student or practicing engineer acquire a doctorate in mathematics in order to be conversant in the design and development of error control systems. In this, the second chapter of the text, we begin a two-chapter overview of the basic mathematical concepts necessary for an understanding of algebraic codes and their associated decoding techniques. The student is first introduced to several mathematical structures, including groups, rings, Galois fields, and vector spaces. Some of the basic properties of these structures are then discussed.

Polynomials over Galois fields. In the third chapter we examine in some detail the properties of polynomials with coefficients from Galois fields. This material is critical to the development of cyclic codes in general, and BCH and Reed-Solomon codes in particular. At the heart of these codes and their respective decoding algorithms lies the identification of code words with polynomials over Galois fields. In this chapter the basic structure of polynomial rings is examined, with an emphasis on Euclid's algorithm.

Linear block codes. The general structure of linear block codes is presented in Chapter 4. The concepts of error detection and error correction are defined and general decoding techniques discussed, including standard array and syndrome decoding. Several upper and lower bounds are introduced that demonstrate the most and the least that can be expected from error control codes. In this chapter we discuss the first class of error control codes to be discovered: the binary single-error-correcting Hamming codes. Hamming codes were originally developed in the late 1940s for error control in long-distance telephony. They have since continued to prove a useful basis for error control systems in a number of applications. Hamming codes are shown to be "perfect" in that they define optimal sphere packings.

Cyclic codes. In the fifth chapter the basic theory of cyclic codes is presented. Here we make use of the material in Chapter 3 in developing the generator polynomial description of cyclic codes. The structure of cyclic codes allows for the development of shift-register-based circuits for high-speed error detection and error correction. We also consider shortened cyclic codes, which include the popular CRC codes. CRC codes are error detecting codes that provide high-reliability data communication at the expense of an almost negligible reduction in throughput. The implementation and evaluation of the performance of CRC codes is considered in some detail.

Golay and Reed-Muller codes. In the sixth and seventh chapters we meet two families of codes developed during the 1950s. The binary Golay code, like the Hamming codes, defines a perfect sphere packing. The binary Golay code is somewhat more powerful in that it is able to decode three errors per code word. Among its many applications, the binary Golay code provided error control during the Jupiter fly-by portion of the *Voyager* mission. Both the binary and ternary Golay codes are considered by some to be the most beautiful in the world of coding theory because of their elegant structure. Though Golay's original presentation failed to fill one side of a sheet of paper ([Gol2]—it is perhaps the best short paper ever written), many a tree has since fallen in the efforts to describe the geometric structure of these codes. We follow Golay's example of brevity here, but do reference other works for the mathematically inclined. The Golay codes are introduced within the context of quadratic residue (QR) codes, so some information on this more general class of codes is provided.

Reed-Muller codes are not particularly strong, but they are very easy to decode. Reed-Muller codes were used extensively in the late 1960s and early 1970s in the *Mariner* spacecraft that explored Mars. They are now seeing renewed interest in optical communication because of their extremely high-speed decoding algorithms. The structure of Reed-Muller codes is considered in some detail, and an efficient and extremely fast decoding algorithm is presented.

BCH and Reed-Solomon codes. BCH and Reed-Solomon codes are arguably the most powerful in the family of block codes. Binary BCH codes have seen frequent application in military and commercial satellite programs and are generally considered the best of the binary block codes (though very long BCH codes are not as good as some other binary codes of the same length). The nonbinary Reed-Solomon codes are, quite simply, ubiquitous. A shortened pair of cross-interleaved Reed-Solomon codes provides error control for the digital audio disc. Another Reed-Solomon code was used for error control during the *Voyager* exploration of the outer solar system. In this chapter we examine the structure, design, and performance of BCH and Reed-Solomon codes. A brief discussion of the known weight distributions for BCH codes is provided. The fundamental theorems for Reed-Solomon codes are then presented and proved. It is shown that Reed-Solomon codes are optimal in that their distance properties are the best possible, given their length and dimension. Reed-Solomon codes are also easily tailored to meet the needs of a particular application. At the end of the chapter the various modification techniques (puncturing, shortening, etc.) are considered in depth. We also consider the Galois field Fourier transform interpretation of BCH and Reed-Solomon codes.

Decoding BCH and Reed-Solomon codes. The discovery of an efficient decoding algorithm for the Reed-Solomon codes converted a mathematical curiosity into one of the most powerful and popular error control systems known. In this chapter we begin by examining Peterson's direct solution of the binary BCH decoding problem. We then introduce Berlekamp's decoding algorithm for binary codes. Both algorithms are generalized for the nonbinary BCH and Reed-Solomon codes. In the nonbinary case we focus on Massey's shift-register synthesis interpretation of Berlekamp's algorithm. The nonbinary decoding algorithms are considered in detail for both error and erasure decoding.

The analysis of the performance of block codes. The performance of general binary and nonbinary block codes is considered here for the additive white Gaussian noise channel and the slowly fading Rayleigh channel. Particular emphasis is placed on the performance of BCH and Reed-Solomon codes.

Convolutional codes. Convolutional codes are radically different in their structure and analysis than their block-coded brethren. A convolutional encoder converts a data stream into one long code word, while the block encoder breaks the stream up into blocks of fixed length, assigning each block to a code word. In this chapter we must thus introduce a new set of terms and techniques that are used in the following three chapters. We begin with a discussion of the basic structure of convolutional codes. Graph-theoretic techniques are introduced to evaluate the performance of convolutional codes. A technique developed for finding the transfer functions of signal flow graphs is then used to obtain the weight distributions for these codes.

The Viterbi decoding algorithm. The Viterbi algorithm is an extremely powerful and flexible means for decoding convolutional codes. It is optimal in that it can be used to determine the maximum likelihood or maximum a posteriori code word associated with a given received word. The Viterbi algorithm also allows for soft-decision decoding, which uses channel reliability information to improve the performance of the error control system. This chapter provides a detailed discussion that includes performance analyses for various channels and a framework for decoder design and implementation. Rate-compatible punctured convolutional codes are introduced at the end of the chapter.

The sequential decoding algorithms. A sequential decoding algorithm was the first decoding technique discovered for convolutional codes. Sequential decoding is suboptimal in the same sense that Viterbi decoding is optimal, but it allows for the use of convolutional codes with a lot of memory. The complexity of the Viterbi decoder grows exponentially with code memory, while sequential decoder complexity grows linearly. In this chapter the Fano and Stack sequential decoding algorithms are presented. Both algorithms perform a search of the tree defined by a convolutional code, but differ somewhat in their respective approaches. Performance analyses and implementation issues are also discussed.

Trellis coded modulation. Trellis coded modulation was probably the single greatest development in error control coding during the 1980s. It allows for highly efficient, reliable data transmission in bandwidth-limited environments. In TCM the signals available to the modulator are divided into subsets such that the signals within each subset are well-separated. Some of the available information bits are convolutionally encoded, and the encoder output is used to select one of the subsets. The remaining information bits select a signal from within the designated subset. This seemingly simple idea has dramatically increased the rate at which information can be reliably transmitted across a telephone line (a twisted pair of copper wires). Where once it was believed that 2400 bits per second was the highest possible rate for reliable data communication over standard phone lines, 19,200-baud modems have now been commercially available for some time (*c.* 1994). In this chapter the basic theory of trellis coded modulation is introduced. The graph-theoretic techniques developed in Chapter 11 are used to evaluate the performance of TCM systems, and some of the standards thus far promulgated are listed.

Retransmission request systems. At this stage in the text we move to some of the system-related issues. Thus far we have considered codes as "forward error correcting" or one-way systems. If a feedback path is available to the decoder, retransmission requests become a possibility. Retransmission requests provide a powerful means of improving reliability performance at the cost of a reduction in throughput. In this chapter we start with the basic theory of automatic-repeat-request (ARQ) and hybrid-ARQ protocols. We then move to some of the more recently developed packet combining systems. These systems provide adaptive-rate error control by varying the number of packets used in the decoding process as the channel noise level varies. The graph-theoretic techniques developed in Chapter 11 are used here for system performance analysis.

Applications. In this, the final chapter, we discuss a series of specific examples that show how error control systems can be used to improve the performance of communication systems. The applications include digital audio, deep space satellite, and magnetic recording channels.

SUGGESTIONS FOR FURTHER READING

The author has found the following books to offer particularly good introductions to the stated areas. It is by no means a comprehensive list, but merely a suggested starting point for the interested student.

Communications theory

J. G. PROAKIS, *Digital Communications*, 2d ed., New York: McGraw-Hill, 1989.

J. M. WOZENCRAFT and I. M. JACOBS, *Principles of Communication Engineering*, New York: John Wiley & Sons, 1965.

Communications system design

R. M. GAGLIARDI, *Introduction to Communications Engineering*, 2nd ed., New York: John Wiley & Sons, 1988.

B. SKLAR, *Digital Communications, Fundamentals and Applications*, Englewood Cliffs, NJ: Prentice Hall, 1988.

S. G. WILSON, *Modulation and Coding*, Englewood Cliffs, NJ: Prentice Hall, 1994.

Satellite communications

V. K. BHARGAVA, D. HACCOUN, R. MATYAS, and P. NUSPL, *Digital Communications by Satellite*, New York: John Wiley & Sons, 1981.

R. M. GAGLIARDI, *Satellite Communications*, New York: John Wiley & Sons, 1988.

PROBLEMS

Satellite link budgets and coding gain. Satellite link budgets provide a very nice means for highlighting the impact of coding gain. Bhargava et al. provide an excellent discussion of satellite link budgets in the first chapter of [Bha], as does Gagliardi in [Gag1] and [Gag2]. The interested reader is invited to continue his or her investigations there.

If one assumes an isotropic transmitter antenna (equal energy propagation in all directions), then the amount of power lost due to signal propagation over some distance D meters in free space can be computed using the following expression [Bha]:

$$L_p = \frac{(4\pi D)^2}{\lambda^2}$$

where λ is the wavelength in meters of the transmitted carrier signal.

The received bit energy to noise spectral density ratio (E_b/N_0) for a satellite downlink can be computed in decibels as follows [Bha]. Note that the use of decibel units allows for a convenient additive expression.

$$\frac{E_b}{N_0} = EIRP - L_p + \frac{G}{T} - k - \frac{R}{r} \quad \text{(dB)}$$

EIRP is the transmitter effective isotropic radiated power, which takes into account the transmitter antenna gain. G/T is the receiver figure of merit, which takes into account the receiver antenna gain and the noise in the receiver front-end. k is Boltzmann's constant, R is the symbol transmission rate, and r is the rate of the error control code. Note that in this expression the receiver figure of merit and Boltzmann's constant account for the noise at the receiver's input.

1. Given that the distance between a satellite in geosynchronous orbit and an earth terminal may range between 36,000 and 41,000 kilometers, compute the smallest and greatest possible free-space losses (in dB) for a 12-GHz signal. It may help to recall that $\lambda = c/f$, where c is the speed of light (approximately 3×10^8 meters/second) and f is the transmission frequency.
2. Consider the following satellite system parameters.

   EIRP = 45.0 dBW $\quad R = 200 \times 10^6$ bps $\quad k = -228.6$ dBW/K°-Hz

   $G/T = 22.0$ dB/K° $\quad L_p = 206$ dB

   If coherently demodulated BPSK modulation is used, what is the BER at the receiver output if no error control coding is used (rate = 1)?
3. Using the system in the previous example, compute the coding gain that will be necessary if the BER is to be improved to 1×10^{-5}. Repeat the problem for a desired BER of 1×10^{-7}.

Coherent vs. noncoherent demodulation. In a coherent demodulator, the signal carrier phase is estimated using some form of phase-locked loop (PLL). If the channel is prone to shot noise and/or jamming, the PLL may slip cycles, causing burst errors in the demodulated data. In these situations noncoherent modulation formats are preferred, though they provide a higher BER at a given E_b/N_0 than comparable coherent schemes. This loss in performance can be recovered through error control coding.

4. Determine the coding gain required to maintain a BER of 1×10^{-4} when the received E_b/N_0 remains fixed and the modulation format is changed from coherent BPSK to noncoherent (BFSK). Repeat the problem for a desired BER of 1×10^{-6}. Figure 1-2 may be of some use.

M-ary signaling ($M > 2$) versus binary signaling. Nonbinary modulation formats frequently provide higher spectral efficiency than binary formats. For example, 8-ary phase-shift keying (8-PSK) uses one-third the bandwidth per bit per second that BPSK does. Naturally this increase in spectral efficiency is obtained at the expense of a decrease in power efficiency.

5. Determine the coding gain required to maintain a BER of 1×10^{-3} when the received E_b/N_0 remains fixed and the modulation format is changed from coherent BPSK to coherent 8-PSK. The BER for 8-PSK in an AWGN environment can be approximated as follows.

$$\text{BER}_{8\text{-}PSK} \approx \frac{2}{3} Q\left(\sin\left(\frac{\pi}{8}\right)\sqrt{\frac{6E_b}{N_0}}\right), \qquad \text{where } Q(x) = \frac{1}{\sqrt{2\pi}}\int_x^{\infty} e^{-y^2/2}\,dy$$

This expression and the BPSK performance curve are plotted in Figure 1-5.

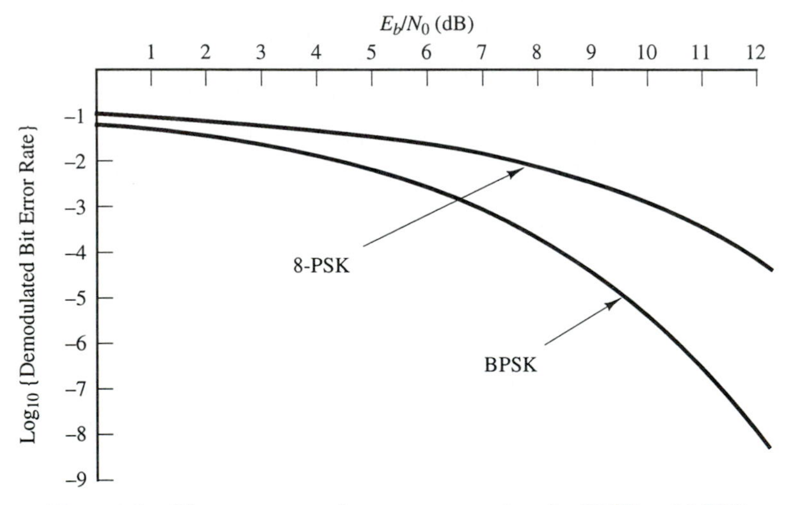

Figure 1-5. Bit error rate performance comparison for BPSK and 8-PSK

Error detecting codes

6. Show that the odd and even parity-check codes detect exactly the same error patterns. Is there any reason to prefer one over the other?
7. Let 0.1 be the probability that any given bit in a received word is incorrect. Compute the probability that a received word contains undetected errors given the following coding schemes.
   (a) No code, word length = 8
   (b) Even parity, word length = 4
   (c) Odd parity, word length = 9
   (d) Even parity, word length = n

Error correcting codes

8. Using the single-error-correcting code discussed in Section 1.3.2, encode the following messages.

(a) $\mathbf{m} = (0,0,0,1)$ **(b)** $\mathbf{m} = (0,1,1,0)$

(c) $\mathbf{m} = (0,1,0,1)$ **(d)** $\mathbf{m} = (1,1,1,1)$

9. Using Table 1-1, correct the errors in the following received words.

(a) $\mathbf{r} = (0,1,0,1,0,1,0)$ **(b)** $\mathbf{r} = (0,1,1,0,1,1,1)$

(c) $\mathbf{r} = (0,1,1,1,0,0,0)$ **(d)** $\mathbf{r} = (1,1,1,1,1,0,0)$

10. Prove that every binary 7-tuple differs in at most one coordinate from a valid word in the code described in Section 1.3.2.

2

Galois Fields

The word "algebra" comes from the old Arabic word "al-jabr," which means "the reuniting." The word may have been given its first mathematical application by the ninth-century Arabic mathematician Muhammad ibn Musa al-Khwarizmi. al-Khwarizmi wrote the first algebra textbook and is also credited with bringing "Arabic" numbers to the attention of the Western world.[1] Since then algebra has become the name for a large collection of mathematical topics that focus on the manipulation and analysis of discrete sets of objects. These topics include elementary abstract algebra, linear algebra, algebraic geometry, and algebraic topology.

The first generation of coding theorists used only a small portion of the available algebraic tools in the development of the first error control codes. For example, only a smattering of linear algebra and combinatorics is necessary for a functional understanding of the Hamming codes, as seen in the fourth chapter of this book. A bit more is necessary for the Reed-Muller codes. The independent development of Reed-Solomon and BCH codes, however, marked a significant increase in the depth of the algebraic techniques brought to bear on code design. It also marked a significant increase in the power of the resulting codes. This trend has continued into the 1990s with the application of algebraic geometry to the construction of good, long block codes.

In this chapter, the basic theory of groups, finite fields, and vector spaces is presented. This material provides sufficient background for an understanding of the general theory of block and convolutional codes (see Chapter 4). It also allows for

[1] Arabic numbers actually came from India; al-Khwarizmi translated several Sanskrit works on mathematics into Arabic. These works were later translated into Latin and thus made available in the West. al-Khwarizmi has since given his name to the word "algorithm."

an understanding of certain specific block codes (e.g., Hamming and Reed-Muller). Chapter 3 provides the more in-depth material needed for an understanding of the theory and implementation of Reed-Solomon and BCH error control systems.

2.1 GROUPS, FIELDS, AND VECTOR SPACES

A **set** is an arbitrary collection of objects, or elements, without any predefined operations between set elements. A set may be finite (e.g., the set of all cathedrals in France), countably infinite (e.g., the set of all positive integers), or uncountably infinite (e.g., the set of all real numbers). A set's primary characteristic is its **cardinality**, which is defined as the number of objects contained in the set.[2] The situation becomes more complicated mathematically when a binary operation is imposed on the set, along with a set of rules restricting the results of the operation. A binary operation operates on two set elements at a time, yielding a third (not necessarily distinct) element. The first major level of complexity achieved when a binary operation is imposed on a set is the group.

Definition 2-1—Groups

A **group** is a set of objects G on which a binary operation "$\cdot$" has been defined. The binary operation takes any two elements in G and generates as its result an element that is also in G (closure). The operation must satisfy the following requirements if G is to be a group.

1. Associativity: $(a \cdot b) \cdot c = a \cdot (b \cdot c)$ for all $a, b, c \in G$.
2. Identity: there exists $e \in G$ such that $a \cdot e = e \cdot a = a$ for all $a \in G$.
3. Inverse: for all $a \in G$ there exists a unique element $a^{-1} \in G$ such that $a \cdot a^{-1} = a^{-1} \cdot a = e$.

A group is said to be **commutative** (or **abelian**) if it also satisfies

4. Commutativity: for all $a, b \in G, a \cdot b = b \cdot a$.

The group operation for a commutative group is usually represented using the symbol "+", and the group is sometimes said to be "additive."

Example 2-1—Groups

- The set of integers forms an infinite commutative group under integer addition, but not under integer multiplication (the latter does not allow for the required multiplicative inverses).
- The set of $(n \times n)$ matrices with real elements forms a commutative group under matrix addition. ■

[2] Cardinality alone is a sufficient basis for an entire field of mathematics; see Cantor's works on set theory and transfinite numbers.

The **order of a group** is defined to be the cardinality of the group. It should be noted that the order of a group alone is not sufficient to completely specify the group unless we restrict ourselves to a particular operation.

In this text we are primarily interested in **finite** groups. The theory of finite groups is an extremely rich and fascinating field (see, for example, [Hun]). One of the more interesting and successful mathematical endeavors in recent history was the classification of all finite groups. Groups are classified by showing how they can be described, if possible, using smaller groups. One of the last of the previously unclassified groups was the "Monster," or "Friendly Giant," group; it is aptly named, for it contains approximately 8×10^{53} elements that describe a 196,884-dimensional space [Con1]. The Monster was slain in 1981 by R. L. Griess. This and several other highly structured and elegant groups have interesting connections to communications and coding theory. However, in our pursuit of the basic algebraic block codes and their applications we need make use of only a few of the more pedestrian groups.

One of the simplest methods for constructing finite groups lies in the application of modular arithmetic to the set of integers. Addition *modulo m* (or *mod m*) is often expressed in the following manner.

$$a + b \equiv c \text{ modulo } m \tag{2-1}$$

Equation (2-1) reads "*a* plus *b* is equivalent to *c modulo m*." The result *c* is obtained by summing *a* and *b* using standard integer addition and dividing the result by the modulus *m*; *c* is the positive remainder.

Example 2-2—Modular Addition

$$13 + 18 \equiv 11 \text{ modulo } 20 \qquad 3 + 8 \equiv 11 \text{ modulo } 20$$
$$9 + 5 \equiv 4 \text{ modulo } 10 \qquad 2 + 4 \equiv 0 \text{ modulo } 6$$
$$18 + 27 \equiv 0 \text{ modulo } 5 \qquad 15 + 100 \equiv 50 \text{ modulo } 65$$

■

Addition *modulo m* groups the infinite set of integers into *m* distinct **equivalence classes**. Two integers *a* and *b* are said to be in the same equivalence class *modulo m* if *a* can be written as $a = xm + b$ for some integer *x*. For example, if $m = 2$, we obtain two equivalence classes: one consisting of all even numbers, the other of all odd numbers. Elements in an equivalence class *modulo m* are "equivalent" in the sense that any element in a given class can be substituted for any other element in the same class without changing the outcome of a *modulo m* operation. Equivalence classes of integers are usually labeled with their smallest constituent nonnegative integer.

Example 2-3—Equivalence Classes of Integers under *Modulo* 5 Addition

Label ↔ Equivalence Class

$$\begin{aligned}
0 &\leftrightarrow \{\ldots, -20, -15, -10, -5, 0, 5, 10, 15, 20, \ldots\} \\
1 &\leftrightarrow \{\ldots, -19, -14, -9, -4, 1, 6, 11, 16, 21, \ldots\} \\
2 &\leftrightarrow \{\ldots, -18, -13, -8, -3, 2, 7, 12, 17, 22, \ldots\} \\
3 &\leftrightarrow \{\ldots, -17, -12, -7, -2, 3, 8, 13, 18, 23, \ldots\} \\
4 &\leftrightarrow \{\ldots, -16, -11, -6, -1, 4, 9, 14, 19, 24, \ldots\}
\end{aligned}$$

■

We shall henceforth refer to the equivalence classes only by their labels, though the reader is cautioned to remember that a large, often infinite collection of elements lies behind each label.

Given Definition 2-1 and the properties of the integers under integer addition, the following result can be readily obtained.

Theorem 2-1

The equivalence classes $\{0, 1, 2, 3, \ldots, m-1\}$ form a commutative group of order m under *modulo* m integer addition for any positive integer m.

Proof. Since the difference between any two elements in an equivalence class *modulo* m is zero in *modulo* m addition, we need only focus on the labels $G = \{0, 1, 2, 3, \ldots, m-1\}$. Since integer addition is associative and commutative, *modulo* m addition is associative and commutative as well. The identity element in G is clearly 0, while the additive inverse of an element x is the integer $m - x$. Closure is assured by the modularity of the additive operation and the result follows. **QED**

Example 2-4—The Group of Order 4 under *Modulo* 4 Addition

The following table completely defines the group operation.

+	0	1	2	3
0	0	1	2	3
1	1	2	3	0
2	2	3	0	1
3	3	0	1	2

The reader is invited to verify that the various requirements for a commutative group are satisfied. ■

Multiplication *modulo* m is performed over the integers in much the same manner as modular addition. The result of the integer multiplication operation is divided by the modulus m and the positive remainder retained as the result of the modular operation.

Example 2-5—Modular Multiplication

$$11 \cdot 5 \equiv 15 \textit{ modulo } 20 \qquad 6 \cdot 4 \equiv 6 \textit{ modulo } 18$$

$$9 \cdot 7 \equiv 13 \textit{ modulo } 50 \qquad 2 \cdot 5 \equiv 0 \textit{ modulo } 10$$

$$4 \cdot 17 \equiv 0 \textit{ modulo } 2 \qquad 6 \cdot 5 \equiv 0 \textit{ modulo } 6$$

■

The crucial difference between modular addition and multiplication is that the latter cannot be used to form a finite group using the integers and arbitrary moduli. Consider the set of elements $\{1, 2, 3, 4, 5, 6, 7\}$ under *modulo* 8 integer multiplication. Since $2 \cdot 4 \equiv 0$ *modulo* 8 and 0 is not in the set, the operation is not closed over the set and we therefore do not have a group. If 0 is included in the set, we still do not have a group because 0 does not have a multiplicative inverse (i.e., there does not exist x such that $x \cdot 0 \equiv 1$ *modulo* 8). The problem lies in the fact that under modulo 8 multiplication, the set $\{1, 2, 3, 4, 5, 6, 7\}$ contains several **zero divisors**. A zero divisor is any nonzero number a for which there exists nonzero b such that $a \cdot b \equiv 0$ *modulo m*. In general, if the modulus m has factors other than 1 in a given set, the set will have zero divisors under *modulo m* multiplication. To construct the multiplicative analog of Theorem 2-1, we must restrict our moduli to prime integers.

Theorem 2-2

The elements $S = \{1, 2, 3, \ldots, p - 1\}$ form a commutative group of order $(p - 1)$ under *modulo p* multiplication if and only if p is a prime integer.

Proof. The modular multiplication operation derives its associativity and commutativity from those of integer multiplication. The multiplicative identity is clearly 1. Closure and the existence of inverses for all elements is only assured, however, if p is prime. If p is not prime, then there exists $m, n \in S$ such that $1 < m, n < p$ and $mn \equiv 0$ *modulo p*; closure is not satisfied. If p is prime, then there can be no such pair of elements and closure is satisfied. Furthermore, given an element $x \in S$, the products $\{x \cdot 1, x \cdot 2, \ldots, x \cdot (p - 1)\}$ must be distinct, otherwise $xy = xz$ implies $x(y - z) \equiv 0$. Since the $(p - 1)$ products are distinct, one must equal the multiplicative identity, assuring the existence of inverses for all $x \in S$. **QED**

Example 2-6—The Group of Order 6 under *Modulo* 7 Multiplication

The following table completely defines the group operation.

·	1	2	3	4	5	6
1	1	2	3	4	5	6
2	2	4	6	1	3	5
3	3	6	2	5	1	4
4	4	1	5	2	6	3
5	5	3	1	6	4	2
6	6	5	4	3	2	1

■

Just as a group may be said to have an order, so may the individual elements in the group. Though the two concepts are related, they are by no means the same.

Definition 2-2—Order of a Group Element

Let g be an element in the group G with group operation "·". For convenience, let $g^2 = g \cdot g, g^3 = g \cdot g \cdot g$, etc. The order of g is the smallest positive integer ord (g) such that $g^{\mathrm{ord}(g)}$ is the group identity element.

Example 2-7—Orders of Group Elements

- The elements in the group in Example 2-6 can be seen to have the following orders.

ELEMENT	ORDER	ELEMENT	ORDER
1	1	4	3
2	3	5	6
3	6	6	2

■

Let S be a subset of the group G. If for all a and b in S, $c = a \cdot b^{-1}$ is also in S, then S is said to be a **subgroup** of G. This means that a subset of G is a subgroup if it exhibits closure and contains the necessary inverses. All other group properties are inherited from G. A subgroup is thus itself a group.

A subgroup S is said to be a **proper** subgroup of G if $S \subset G$, but $S \neq G$.

Example 2-8—Subgroups

- The group of integers under *modulo* 9 addition contains the proper subgroups $\{0\}$ and $\{0, 3, 6\}$.
- The group of integers under *modulo* 16 addition contains the proper subgroups $\{0\}$, $\{0, 8\}$, $\{0, 4, 8, 12\}$, and $\{0, 2, 4, 6, 8, 10, 12, 14\}$. ■

In the modular operations discussed so far, two integers were said to be equivalent if they differed by a multiple of an integer called the modulus. This concept can be generalized by allowing the modulus to be a subgroup S. Equivalence classes are thus formed whose constituent elements differ by some element in S. These equivalence classes are called **cosets**.

Definition 2-3—Left and Right Cosets

Let S be a subgroup of G with operation "+". A left coset of S in G is a subset of G whose elements can be expressed as $x + S = \{x + s, s \in S\}$. A right coset of S in G is a subset of G whose elements can be expressed as $S + x = \{s + x, s \in S\}$.

If G is commutative, every left coset $x + S$ is identical to every right coset $S + x$. Most of the groups encountered in this book are commutative, so no further distinction will be made between left and right cosets. They will simply be referred to as cosets.

Example 2-9—Cosets

- The group of integers G under *modulo* 9 addition contains the proper subgroup $S = \{0, 3, 6\}$. The distinct cosets of S in G are $\{0, 3, 6\}$, $\{1, 4, 7\}$, and $\{2, 5, 8\}$.
- The group of integers G under *modulo* 16 addition contains the proper subgroup $S = \{0, 4, 8, 12\}$. The distinct cosets of S in G are $\{0, 4, 8, 12\}$, $\{1, 5, 9, 13\}$, $\{2, 6, 10, 14\}$, and $\{3, 7, 11, 15\}$.

Theorem 2-3—Properties of Cosets

The distinct cosets of a subgroup S in a group G are disjoint.

Proof. Let a and b be in the same coset of S in G. This implies that $a = x + b$ for some $x \in S$. If b is also equivalent to c, then $b = y + c$ for some $y \in S$. It follows that a is equivalent to c, for $a = x + y + c$, and $x + y \in S$, since S is a subgroup. Thus if an element in coset A is equivalent to an element in coset B, then every element in A is equivalent to every element in B. It follows that B contains A. Since the reverse is also true, $A = B$. The distinct cosets must therefore be disjoint. **QED**

A comment on notation: Throughout the rest of the text the expression "$a \mid b$" is used to denote that an element a divides an element b without remainder.

Theorem 2-3 shows that a subgroup S of a group G defines a partitioning of G into distinct, disjoint cosets. This partitioning of G is called the *coset decomposition of G induced by S.*

Theorem 2-4—Lagrange's Theorem

If S is a subgroup of G, then ord $(S) \mid$ ord (G).

Proof. Let $a, b \in S, a \neq b$, and $x \in G - S$ (i.e., x is in G, but not in S). It is easy to show that $a + x \neq b + x$, for otherwise $a + x + (-x) = b + x + (-x)$ implies $a = b$, which contradicts the premise $a \neq b$. Given an element x in a coset A of S in G, it follows that all of the other elements y in the coset A are in a one-to-one relationship with elements $a \in S$ defined by $y = a + x$. All cosets of S in G thus have cardinality $|S|$ ($|S|$ is the number of elements in S). Since they are also disjoint by Theorem 2-3, the result follows. **QED**

Example 2-10—Subgroups and Cosets

- The group G of order 4 in Example 2-4 contains the subgroup $\{0, 2\}$. This subgroup induces the coset decomposition $G = \{0, 2\} \cup \{1, 3\}$.
- The group $G = \{0, 1, 2, 3, 4, 5, 6, 7, 8, 9, 10, 11\}$ under *modulo* 12 addition contains the subgroup $\{0, 4, 8\}$. This subgroup induces the coset decomposition $G = \{0, 4, 8\} \cup \{1, 5, 9\} \cup \{2, 6, 10\} \cup \{3, 7, 11\}$. ■

Groups were obtained by applying a single binary operation and a few restrictive conditions to a set of elements. We can further complicate matters by adding a second binary operation to the first. This brings us into the realm of rings and fields.

Definition 2-4—Rings

A **ring** is a collection of elements R with two binary operations "+" and "·" such that the following three properties hold:

1. R forms a commutative group under $+$. The additive identity element is labeled "0".
2. The operation $\cdot$ is associative: $(a \cdot b) \cdot c = a \cdot (b \cdot c)$ for all $a, b, c \in R$
3. The operation $\cdot$ distributes over $+$: $a \cdot (b + c) = (a \cdot b) + (a \cdot c)$

A ring is said to be a **commutative ring** if

4. The operation $\cdot$ commutes (i.e., $a \cdot b = b \cdot a$).

A ring is said to be a **ring with identity** if

5. The operation $\cdot$ has an identity element, which is labeled "1".

And, as one might imagine, a ring that satisfies both property 4 and property 5 is said to be a **commutative ring with identity**.

Example 2-11—Various Rings of Various Types

- Matrices with integer elements form a ring with identity under standard matrix addition and multiplication. The identity matrix provides the requisite multiplicative identity. Matrix multiplication is not, in general, commutative, so we do not have a commutative ring with identity.
- The integers under *modulo m* addition and multiplication form a commutative ring with identity.
- The set of all polynomials with binary coefficients forms a commutative ring with identity under standard polynomial addition and multiplication. This ring is usually denoted $F_2[x]$ or GF(2)[x]. ■

Rings of polynomials are of great interest in the study of algebraic codes. We will be particularly interested in polynomials whose coefficients are taken from finite fields. We must begin by defining just what is meant by the term "field." We return to the subject of polynomials over finite fields in the following chapter.

Definition 2-5—Fields

Let F be a set of objects on which two operations $+$ and $\cdot$ are defined. F is said to be a **field** if and only if

1. F forms a commutative group under $+$. The additive identity element is labeled "0".
2. $F - \{0\}$ (the set F with the additive identity removed) forms a commutative group under $\cdot$. The multiplicative identity element is labeled "1".
3. The operations $+$ and $\cdot$ distribute: $a \cdot (b + c) = (a \cdot b) + (a \cdot c)$

A field can also be defined as a commutative ring with identity in which every element has a multiplicative inverse.

Example 2-12—Various Infinite Fields

- The rational numbers form one of the "smallest" infinite fields.[3]
- The real numbers form an infinite field, as do the complex numbers.
- The integers do not form a field, for most integers do not have an integer multiplicative inverse. ■

Fields can be viewed as commutative rings with identity and multiplicative inverses, though to some this may not seem particularly enlightening. Perhaps the best approach is to remember the two-group structure of all fields: All of the field elements form an additive commutative group, while the nonzero elements form a multiplicative commutative group. This will be helpful in the construction of finite fields.

Fields of finite order (cardinality) are particularly interesting to coding theorists. Finite fields were discovered by Evariste Galois[4] and are thus known as **Galois fields**. A Galois field of order q is usually denoted GF(q).

The simplest of the Galois fields is GF(2). GF(2) can be represented by the two-element set {0, 1} under standard binary addition and multiplication, as shown below.

+	0	1
0	0	1
1	1	0

·	0	1
0	0	0
1	0	1

The multiplicative group within GF(2) is particularly simple, consisting only of the single multiplicative identity 1. The multiplication table includes the element 0 (which lies outside the multiplicative group) because the two field operations distribute.

Galois fields of size p, p a prime, can be constructed by taking advantage of Theorems 2-1 and 2-2. Consider the integers $\{0, 1, 2, \ldots, p - 1\}$. Under addition *modulo* p these elements form an additive commutative group (Theorem 2-1). Under multiplication *modulo* p the subset of elements $\{1, 2, \ldots, p - 1\}$ forms a multiplicative commutative group (Theorem 2-2). If the two operations are allowed to distribute as they do in integer arithmetic, then we have a field.

Theorem 2-5

The integers $\{0, 1, 2, \ldots, p - 1\}$, where p is a prime, form the field GF(p) under *modulo p* addition and multiplication.

[3] Yes, infinite sets do come in varying sizes. The smallest infinite number, $\aleph_0$, denotes the cardinality of countably-infinite sets (e.g., the set of integers and the set of rational numbers).

[4] The *Encyclopedia Britannica* has an excellent biography of this fascinating individual.

Example 2-13—Construction of GF(3)

Use the integers $\{0, 1, 2\}$ under *mod* 3 addition and multiplication. Distributivity is insured by setting $0 \times a = 0$ for all $a \in \text{GF}(3)$.

+	0	1	2
0	0	1	2
1	1	2	0
2	2	0	1

·	0	1	2
0	0	0	0
1	0	1	2
2	0	2	1

■

Finite fields of prime order are quite easy to construct. But this does leave a large range of possible field orders for which a construction has not yet been described. As we shall see, finite fields $\text{GF}(q)$ do not exist for all values of q. q must equal p^m, where p is a prime positive integer and m is a positive integer.

One cannot use modular arithmetic in the construction of $\text{GF}(q)$, q not a prime, for as noted earlier, the elements $\{1, 2, \ldots, q - 1\}$ do not form a group under multiplication *modulo* q. To construct the fields of size p^m, p a prime, $m > 1$, one must use methods that are somewhat more complex than simple modular arithmetic. We now show that finite fields of order p^m can be constructed as vector spaces over the prime order field $\text{GF}(p)$. We begin with a brief review of vector spaces.

Definition 2-6—Vector Spaces

Let V be a set of elements called **vectors** and F a field of elements called **scalars**. Two operations are introduced in addition to the two already defined between the field elements. Let "+" be a binary additive operation, henceforth called vector addition, that maps pairs of vectors $\mathbf{v}_1, \mathbf{v}_2 \in V$ onto a vector $\mathbf{v} = \mathbf{v}_1 + \mathbf{v}_2$ in V. Let "·" be a binary multiplicative operation, henceforth called scalar multiplication, that maps a scalar $a \in F$ and a vector $\mathbf{v} \in V$ onto a vector $\mathbf{w} = a \cdot \mathbf{v} \in V$. V forms a vector space over F if the following conditions are satisfied:

1. V forms a commutative group under the operation $+$.
2. For any element $a \in F$ and $\mathbf{v} \in V$, $a \cdot \mathbf{v} = \mathbf{u} \in V$.
3. The operations $+$ and $\cdot$ distribute: $a \cdot (\mathbf{u} + \mathbf{v}) = a \cdot \mathbf{u} + a \cdot \mathbf{v}$, and $(a + b) \cdot \mathbf{v} = a \cdot \mathbf{v} + b \cdot \mathbf{v}$. (Note that $(a + b)$ refers to the additive field operation, not the additive vector operation.)
4. Associativity: For all $a, b \in F$ and all $\mathbf{v} \in V$, $(a \cdot b) \cdot \mathbf{v} = a \cdot (b \cdot \mathbf{v})$.
5. The multiplicative identity 1 in F acts as a multiplicative identity in scalar multiplication: for all $\mathbf{v} \in V$, $1 \cdot \mathbf{v} = \mathbf{v}$.

F is commonly called the "scalar field" or the "ground field" of the vector space V.

The n-tuple $\mathbf{v} = (v_0, v_1, \ldots, v_{n-1})$ of elements $\{v_i\}$ from the ground field F is a type of vector. Such vectors allow for a convenient definition for vector addition and scalar multiplication. Let $\mathbf{v} = (v_0, v_1, \ldots, v_{n-1})$ and $\mathbf{u} = (u_0, u_1, \ldots, u_{n-1})$, with the $\{v_i\}$ and $\{u_i\} \subseteq F$.

vector addition: $\mathbf{u} + \mathbf{v} = (u_0 + v_0, u_1 + v_1, \ldots, u_{n-1} + v_{n-1})$

scalar multiplication: $a \cdot \mathbf{v} = (av_0, av_1, \ldots, av_{n-1})$

Example 2-14—Vector Spaces

- The set of binary n-tuples, designated V_n, forms a vector space over GF(2). Since each of the n coordinates in each vector can take on one of two values independent of the other coordinates, V_n has cardinality 2^n. For example, V_5 contains $2^5 = 32$ binary 5-tuples.
- The set of all $(n \times n)$ matrices with real elements forms a vector space of infinite cardinality over the field of real numbers. ■

The operations defined for vectors in vector spaces allow one to compute *linear combinations* of vectors. Let $\mathbf{v}_1, \mathbf{v}_2, \ldots, \mathbf{v}_n$ be vectors in V and let $a_1, a_2, \ldots, a_n$ be scalars in F. Since V forms a commutative group under $+$, the linear combination $\mathbf{v} = a_1 \cdot \mathbf{v}_1 + a_2 \cdot \mathbf{v}_2 + a_3 \cdot \mathbf{v}_3 + \cdots + a_n \cdot \mathbf{v}_n$ is a vector in V.

Definition 2-7—Spanning Sets

A collection of vectors $G = \{\mathbf{v}_1, \mathbf{v}_2, \ldots, \mathbf{v}_n\}$, the linear combinations of which include all vectors in a vector space V, is said to be a **spanning set** for V or to **span** V.

Example 2-16—Spanning Sets

- The vectors $\{(1000), (0110), (1100), (0011), (1001)\}$ can be shown to span the space V_4. ■

In the above example the reader may note that the five vectors in the spanning set are *linearly dependent.* A set of vectors is said to be linearly dependent when one or more of the vectors can be expressed as a linear combination of the others. In the example above, $(0110) + (1100) + (0011) = (1001)$. There is thus some redundancy in the spanning set, for the vector (1001) can be deleted from the set without reducing the span of the set.

Definition 2-8—Bases

A spanning set for V that has minimal cardinality is called a **basis** for V.

It follows immediately from the definition that the elements of a basis must be linearly independent; otherwise, one of the elements could be deleted, reducing the set's cardinality by one. Though a vector space may have several possible bases, all of the bases will have the same cardinality [Hun].

Example 2-17—Bases

- {(1000), (0100), (0010), (0001)} is a basis for V_4. This is frequently called the **canonical basis** *for* V_4.
- {(10101), (01011), (01101), (00111), (10001)} is a basis for V_5. ■

Definition 2-9—The Dimension of a Vector Space

If a basis for a vector space V has k elements, then the vector space is said to have **dimension** k, written $\dim(V) = k$.

Theorem 2-6

Let $\{\mathbf{v}_i\}$ be a basis for a vector space V. For every vector $\mathbf{v}$ in V there is a representation $\mathbf{v} = a_0 v_0 + \cdots + a_{k-1} v_{k-1}$. This representation is unique.

Proof. The first statement follows from the definition of bases. The second can be proven by contradiction. Suppose that there are two distinct representations for the same vector.

$$\mathbf{v} = a_0\mathbf{v}_0 + \cdots + a_{k-1}\mathbf{v}_{k-1} = b_0\mathbf{v}_0 + \cdots + b_{k-1}\mathbf{v}_{k-1}, \qquad \{a_i\} \neq \{b_i\}$$

$$\Rightarrow (a_0 - b_0)\mathbf{v}_0 + \cdots + (a_{k-1} - b_{k-1})\mathbf{v}_{k-1} = \mathbf{0}$$

$$\Rightarrow \text{The basis set vectors are not linearly independent}$$

If the basis vectors are not linearly independent, then one of them can be deleted. This contradicts the minimality of the cardinality of the basis! **QED**

Let a vector space V over the field F have dimension k. This implies that all vectors $\mathbf{v} \in V$ can be written as a linear combination $\mathbf{v} = a_0\mathbf{v}_0 + \cdots + a_{k-1}\mathbf{v}_{k-1}$ for some collection of scalars $\{a_i\} \subset F$. Each distinct vector $\mathbf{v} \in V$ is thus associated with a unique ordered set $(a_0, a_1, \ldots, a_{k-1})$ and vice versa. The number of vectors in the vector space thus equals the number of possible choices for the $\{a_i\}$: $|V| = |F|^k$. For example, the binary vector space V_n has cardinality 2^n, as noted earlier.

Vector spaces frequently contain smaller vector spaces within them. The contained spaces are called **vector subspaces**. The relationships between vector spaces and their subspaces is fundamental to the theory of linear block codes.

Theorem 2-7

Let $\mathbf{v}_1$ and $\mathbf{v}_2$ be an arbitrary pair of vectors in the subset S of the vector space V over F. S is a vector subspace of V if and only if any linear combination of $\mathbf{v}_1$ and $\mathbf{v}_2$ (i.e., $a \cdot \mathbf{v}_1 + b \cdot v_2, a,b \in F$) is also in S.

Proof. We begin by assuming that S is a vector subspace. It follows by definition of a vector space that S is closed under linear combinations, and the first half of the result follows.

Now assume that any linear combination $\mathbf{v} = a \cdot \mathbf{v}_1 + b \cdot \mathbf{v}_2, a, b \in F, \mathbf{v}_1, \mathbf{v}_2 \in S$, lies in S. It follows that the closure properties for vector addition and scalar multiplication are satisfied for S. Since S is closed under scalar multiplication, all additive inverses $(-1) \cdot \mathbf{v}$ for elements $\mathbf{v} \in S$ are also in S. It follows that additive identity must also be in S. The remainder of the vector-space properties follow by noting that since V is a vector space, the various properties for operations (associativity and commutativity) that hold in V must also hold in $S \subseteq V$. **QED**

Along with scalar multiplication and vector addition, a third operation may be imposed on vector spaces: the inner or "dot" product. The inner product is a binary operation that maps pairs of vectors in the vector space V over the field F onto scalars in F.

Definition 2-10—The Inner Product

Let $\mathbf{u} = (u_0, u_1, \ldots, u_{n-1})$ and $\mathbf{v} = (v_0, v_1, \ldots, v_{n-1})$ be vectors in the vector space V over the field F. The inner product $\mathbf{u} \bullet \mathbf{v}$ is defined as follows.

$$\mathbf{u} \bullet \mathbf{v} = \sum_{i=0}^{n-1} u_i \cdot v_i = u_0 \cdot v_0 + u_1 \cdot v_1 + \cdots + u_{n-1} \cdot v_{n-1}$$

Given the above definition, the following properties can be demonstrated.

1. Commutativity: $\mathbf{u} \bullet \mathbf{v} = \mathbf{v} \bullet \mathbf{u}$
2. Associativity with scalar multiplication: $a \cdot (\mathbf{u} \bullet \mathbf{v}) = (a \cdot \mathbf{u}) \bullet \mathbf{v}$
3. Distributivity with vector addition: $\mathbf{u} \bullet (\mathbf{v} + \mathbf{w}) = (\mathbf{u} \bullet \mathbf{v}) + (\mathbf{u} \bullet \mathbf{w})$

The inner product is used to characterize **dual spaces**.

Definition 2-11—Dual Spaces of Vector Spaces

Let S be a k-dimensional subspace of a vector space V. Let $S^\perp$ be the set of all vectors $\mathbf{v}$ in V such that for all $\mathbf{u} \in S$ and for all $\mathbf{v} \in S^\perp$, $\mathbf{u} \bullet \mathbf{v} = 0$. $S^\perp$ is said to be the **dual space** of S.

In some cases it is far easier to describe a vector space or its properties in terms of the dual space than in terms of the vector space itself. Many examples of this situation in block coding are seen in the following chapters.

Note that the space S and its dual $S^\perp$ are not disjoint, for $\mathbf{0}$ is in both spaces. In fact, we shall see in Chapter 6 that there exist binary vector subspaces that are their own duals (i.e., $S = S^\perp$).

The following two theorems describe the basic relationship between a vector space and its dual space.

Theorem 2-8

The dual space $S^\perp$ of a vector subspace $S \subsetneq V$ is itself a vector subspace of V.

Proof. Suppose that two vectors $\mathbf{v}$ and $\mathbf{w}$ lie in $S^\perp$

$\Rightarrow \mathbf{v} \cdot \mathbf{u} = 0$ and $\mathbf{w} \cdot \mathbf{u} = 0$ for all $\mathbf{u} \in S$ by definition of a dual space.

$\Rightarrow (\mathbf{v} + \mathbf{w}) \cdot \mathbf{u} = (\mathbf{v} \cdot \mathbf{u}) + (\mathbf{w} \cdot \mathbf{u}) = 0$ for all $\mathbf{u} \in S$.

$\Rightarrow (a \cdot \mathbf{w}) \cdot \mathbf{u} = a \cdot (\mathbf{w} \cdot \mathbf{u}) = 0$ for all $\mathbf{u} \in S$ and $a \in F$.

$\Rightarrow$ All linear combinations of elements in $S^\perp$ are elements in $S^\perp$.

$\Rightarrow S^\perp$ is a vector subspace by Theorem 2-7. **QED**

Theorem 2-9—The Dimension Theorem

Let S be a finite-dimensional vector subspace of V and let $S^\perp$ be the corresponding dual space. Then the dimension of S and the dimension of $S^\perp$ sum to the dimension of V.

$$\dim(S) + \dim(S^\perp) = \dim(V)$$

Proof. The proof to this theorem is of particular interest in that it brings out ideas that will be of use in our discussion of linear block codes in Chapter 4.

Let $\dim(V) = n$ and $\dim(S) = k$. Let $\{\mathbf{g}_1, \mathbf{g}_2, \ldots, \mathbf{g}_k\}$ be a basis for S. Let $\mathbf{G}$ be a $(k \times n)$ matrix, the rows of which are the vectors in a basis for S.

$$\mathbf{G} = \begin{bmatrix} \mathbf{g}_1 \\ \mathbf{g}_2 \\ \vdots \\ \mathbf{g}_k \end{bmatrix}$$

Since the rows of $\mathbf{G}$ span S, a vector $\mathbf{v}$ is in the dual space $S^\perp$ if and only if $\mathbf{G}\mathbf{v}^T = \mathbf{0}$. Since $S^\perp$ is a vector space by Theorem 2-8, we can also write down a basis $\{\mathbf{h}_1, \mathbf{h}_2, \ldots, \mathbf{h}_t\}$ for $S^\perp$. Furthermore, since $S^\perp \subset V$, the basis for $S^\perp$ can be extended to form a basis for V of the form

$$\{\mathbf{h}_1, \mathbf{h}_2, \ldots, \mathbf{h}_t, \mathbf{d}_1, \mathbf{d}_2, \ldots, \mathbf{d}_{n-t}\}.$$

The rows of $\mathbf{G}$ are independent by definition of a basis, so the row rank of $\mathbf{G}$ is k. The row and column ranks of a matrix are always equal [Hun], so the columns of $\mathbf{G}$ span a k-dimensional vector space.

By Definition 2-7, every vector $\mathbf{v}$ in the column space of $\mathbf{G}$ can be written in the (not necessarily unique) form $\mathbf{v} = \mathbf{G}\mathbf{x}^T$, where $\mathbf{x} \in V$. It follows that the column space of $\mathbf{G}$ is spanned by the vectors

$$\{\mathbf{G}\mathbf{h}_1^T, \mathbf{G}\mathbf{h}_2^T, \ldots, \mathbf{G}\mathbf{h}_t^T, \mathbf{G}\mathbf{d}_1^T, \mathbf{G}\mathbf{d}_2^T, \ldots, \mathbf{G}\mathbf{d}_{n-t}^T\}.$$

Since the vectors $\{\mathbf{h}_1, \mathbf{h}_2, \ldots, \mathbf{h}_t\}$ are in $S^\perp$, $\mathbf{G}\mathbf{h}_i^T = \mathbf{0}$ for $i = \{1, \ldots, t\}$. The vectors $\{\mathbf{G}\mathbf{d}_1^T, \mathbf{G}\mathbf{d}_2^T, \ldots, \mathbf{G}\mathbf{d}_{n-t}^T\}$ thus span the k-dimensional column space of $\mathbf{G}$. The vectors in this spanning set must also be independent, for if there exists a linear combination

$$a_1\mathbf{G}\mathbf{d}_1^T + a_2\mathbf{G}\mathbf{d}_2^T + \cdots + a_{n-t}\mathbf{G}\mathbf{d}_{n-t}^T = \mathbf{0}, \qquad \text{not all } a_i = 0$$

then

$$\mathbf{G}(a_1\mathbf{d}_1^T + a_2\mathbf{d}_2^T + \cdots + a_{n-t}\mathbf{d}_{n-t}^T) = \mathbf{0}, \qquad \text{not all } a_i = 0$$

Since the vectors $\{\mathbf{d}_1, \mathbf{d}_2, \ldots, \mathbf{d}_{n-t}\}$ are not in $S^\perp$, the linear combination $a_1\mathbf{d}_1^T + a_2\mathbf{d}_2^T + \cdots + a_{n-t}\mathbf{d}_{n-t}^T$ must sum to $\mathbf{0}$, contradicting the linear independence of the vectors in the basis for V above.

Since the vectors $\{\mathbf{G}\mathbf{d}_1^T, \mathbf{G}\mathbf{d}_2^T, \ldots, \mathbf{G}\mathbf{d}_{n-t}^T\}$ are linearly independent and span a k-dimensional space, it follows that $n - t = k$, and the dimension of $S^\perp$ is thus $t = n - k$. **QED**

Example 2-18—Dual Spaces and the Dimension Theorem

The vector space of binary 4-tuples contains the following vector subspace and its dual.

$$S = \{(0000), (0101), (0001), (0100)\} \qquad \text{(dimension} = 2)$$
$$S^\perp = \{(0000), (1010), (1000), (0010)\} \qquad \text{(dimension} = 2)$$

The sum of the dimensions is dim $(V_4) = 4$, as indicated by the dimension theorem.

■

2.2 ELEMENTARY PROPERTIES OF GALOIS FIELDS

In this section we examine some of the basic properties of Galois fields and their constituent elements. One of the first things to be noted is that, unlike a group, the order of a Galois field completely specifies the field. Thus two finite fields of the same size are always identical up to the labeling of their elements, regardless of how the fields are constructed. One may thus talk about *the* Galois field GF(q) without fear of ambiguity.[5]

Let β be an element in GF(q) and let 1 be the multiplicative identity. Consider the following sequence of elements.

$$1, \beta, \beta^2, \beta^3, \beta^4, \beta^5, \ldots$$

Since β is contained in GF(q), all of the successive powers of β must also be in GF(q) by closure under multiplication. Since GF(q) has only a finite number of elements, we must conclude that, at some point, the sequence begins to repeat values found earlier in the sequence.

It is a simple matter to show that the first element to repeat must be 1, the first

[5] To be rigorous, we would say that a finite field of order q is unique up to isomorphisms. A nice development of this result can be found in [McE1].

element in the sequence. The proof follows by contradiction. Assume that $\beta^x = \beta^y \neq 1$ is the first sequence element to repeat, where $x > y > 0$. It follows that

$$\beta^x = \beta^y \Rightarrow \beta^{x-y} = 1$$

and thus $\beta^{x-y} = 1$ is the first element to repeat, where $0 < x - y < x$. Since we assumed $\beta^x \neq 1$ to be the first repeating element, we have a contradiction.

Definition 2-12—The Order of a Galois Field Element

Let β be an element in GF(q). The **order of β** (written ord (β)) is the smallest positive integer m such that $\beta^m = 1$.

This definition is identical to that for the order of an element in a group. It should be noted, however, that for the case of the Galois field element, "order" is defined using the multiplicative operation and not the additive operation. As in a group, the nonzero elements in a Galois field have well-defined orders that meet some rather strict criteria. We first note that the order of an arbitrary element β in the Galois field GF(q) must be a divisor of $(q - 1)$.

Theorem 2-10

If $t = \text{ord}(\beta)$ for some $\beta \in \text{GF}(q)$, then $t \mid (q - 1)$.

Proof. If $t = \text{ord}(\beta)$ for some $\beta \in \text{GF}(q)$, then $\{\beta, \beta^2, \ldots, \beta^t = 1\}$ forms a subgroup of the nonzero elements in GF(q) under multiplication. The result follows from Lagrange's theorem (Theorem 2-4). **QED**

GCD (i, t) is defined to be the **greatest common divisor** of i and t; in other words, GCD (i, t) is the largest positive integer m such that $m \mid i$ and $m \mid t$.

Theorem 2-11

Let α and β be elements in GF(q) such that $\beta = \alpha^i$. If the ord $(\alpha) = t$, then ord $(\beta) = t/\text{GCD}(i, t)$.

Proof (adapted from [McE1]). We first note that if α has order t, then $\alpha^s = 1$ if and only if $t \mid s$. This can be proven quite readily. If $s = 0$, then $t \mid s$ trivially. If $0 < s < t$, then the premise contradicts the minimality of the order t of α. If $s > t$, then we can write $s = qt + r$, where $0 \leq r < t$. It follows that $\alpha^s = \alpha^{qt+r} = \alpha^r = 1$. The variable r must therefore be zero; otherwise, the minimality of the order is again contradicted.

Now let ord $(\beta) = x$. Since $i/\text{GCD}(i, t)$ is an integer by definition of GCD, we have $(\alpha^i)^{t/GCD(i,t)} = (\alpha^t)^{i/GCD(i,t)} = 1^{i/GCD(i,t)} = 1$, which implies that $x \mid t/\text{GCD}(i, t)$ by the result above. Similarly, since $(\alpha^i)^x = 1$, it follows that $t \mid ix$ and $t/\text{GCD}(i, t) \mid x$. Since $x \mid t/\text{GCD}(i, t)$ and $t/\text{GCD}(i, t) \mid x, x = \text{ord}(\beta)$ must equal $t/\text{GCD}(i, t)$. **QED**

Theorem 2-10 determines the possible orders a finite field element can display. For example, in GF(16) the elements can only have orders {1, 3, 5, 15}. Given a valid order, it is then possible to determine exactly how many elements in a finite field have that order. This information is obtained through the use of an extremely powerful function that appears throughout the theory of Galois fields and out into the broader reaches of number theory. It even lies at the heart of one of the most popular of the public key cryptosystems (RSA), but that is a tangential topic that must wait for another time and another book.

Definition 2-13—The Euler ϕ Function

The **Euler ϕ function** evaluated at an integer t, written $\phi(t)$, is the number of integers in the set $\{1, \ldots, t-1\}$ that are relatively prime to t (i.e., share no common divisors other than one).

$\phi(1)$ is defined to be 1. $\phi(t)$, $t > 1$, is computed in the following manner [McE1]:

$$\phi(t) = |\{1 \le i < t \,|\, \text{GCD}(i, t) = 1\}| = t \prod_{p|t} \left(1 - \frac{1}{p}\right) \tag{2-2}$$

The product expression in Eq. (2-2) is taken over all positive prime integers $p < t$ that divide t. The Euler ϕ function is called the **Euler totient function** in some mathematical texts.

Example 2-19—The Euler ϕ Function

- $\phi(56) = \phi(2 \cdot 2 \cdot 2 \cdot 7) = 56(1 - 1/2)(1 - 1/7) = 24$
- $\phi(256) = \phi(2^8) = 2^8(1 - 1/2) = 2^7(2 - 1) = 128$
- $\phi(30) = \phi(2 \cdot 3 \cdot 5) = 30(1 - 1/2)(1 - 1/3)(1 - 1/5) = (1)(2)(4) = 8$ ∎

The following useful properties of the Euler ϕ function follow from Eq. (2-2).

- If p is prime, then $\phi(p) = p - 1$, since all nonzero elements are relatively prime to a prime number.
- $\phi(p_1 \cdot p_2) = \phi(p_1) \cdot \phi(p_2) = (p_1 - 1)(p_2 - 1)$ if p_1 and p_2 are distinct primes.
- $\phi(p^m) = p^{m-1}(p - 1)$ for a prime p.
- $\phi(p^m r^n) = p^{m-1} r^{n-1}(p - 1)(r - 1)$ for distinct primes p and r.

Given that the integer t divides $(q - 1)$, it can be shown that the number of elements of order t in GF(q) is $\phi(t)$. This result and Theorem 2-10 combine to describe the multiplicative structure of all Galois fields.

Theorem 2-12—The Multiplicative Structure of Galois Fields

Consider the Galois field GF(q).

1. **If t does not divide $(q - 1)$, then there are no elements of order t in GF(q).**
2. **If $t \,|\, q - 1$, then there are $\phi(t)$ elements of order t in GF(q).**

Proof. Part 1 is a restatement of Theorem 2-10. Part 2 is proven as follows. If $t = \text{ord}(\alpha)$, then the set $\{\alpha, \alpha^2, \ldots, \alpha^t\}$ contains all of the solutions to the expression $x^t - 1 = 0$ (the expression has degree t and the set contains t distinct solutions). All of the elements of order t must thus be contained in this set. By Theorem 2-11 we see that only those elements of the form α^i, where $\text{GCD}(i, t) = 1$, have order t. By definition there are $\phi(t)$ such elements. **QED**

Definition 2-14—Primitive Elements in a Galois Field

An element with order $(q - 1)$ in GF(q) is called a **primitive element** in GF(q).

Corollary to Theorem 2-12

In every finite field GF(q) there are exactly $\phi(q - 1)$ primitive elements.

The above corollary takes on a great deal of significance when we note that $\phi(x)$ is always greater than zero for positive x. Every field GF(q) thus contains at least one primitive element α. Let α be a primitive element in GF(q) and consider the sequence

$$1, \alpha, \alpha^2, \alpha^3, \ldots, \alpha^{q-2}, \alpha^{q-1}, \alpha^q, \ldots$$

By the definition of a primitive element, α^{q-1} is the first positive power of α in the sequence to repeat the value 1. Using the same argument developed in the discussion on the order of a field element, it can be shown that 1 must be the first element to repeat. The first $q - 1$ elements in the above sequence are thus distinct. Since they are also nonzero, the first $q - 1$ sequence elements must comprise the $q - 1$ nonzero elements in GF(q). We reach the important conclusion that **all nonzero elements in GF(q) can be represented as ($q - 1$) consecutive powers of a primitive element α.**

Example 2-20—The Multiplicative Structure of GF(7)

As discussed in the previous section, since 7 is a prime we can construct GF(7) using the integers {0, 1, 2, 3, 4, 5, 6} and *modulo* 7 addition and multiplication. The following multiplication table results.

·	0	1	2	3	4	5	6
0	0	0	0	0	0	0	0
1	0	1	2	3	4	5	6
2	0	2	4	6	1	3	5
3	0	3	6	2	5	1	4
4	0	4	1	5	2	6	3
5	0	5	3	1	6	4	2
6	0	6	5	4	3	2	1

The Corollary to Theorem 2-12 indicates that there are $\phi(6) = 2$ primitive elements in GF(7). A brief investigation of the above table will show that the primitive elements are 3 and 5. Consider the consecutive powers of 5:

$$5^1 = 5,\ 5^2 = 4,\ 5^3 = 6,\ 5^4 = 2,\ 5^5 = 3,\ 5^6 = 1$$

All of the nonzero field elements are accounted for as powers of 5.

Using Theorem 2-11, the orders of the elements of GF(7) follow readily.

i	5^i	$\text{ord}(5^i) = 6/\text{GCD}(i, 6)$
0	1	1
1	5	6
2	4	3
3	6	2
4	2	3
5	3	6

We complete the example by noting that, as expected, the orders of the various elements are divisors of 6 and occur with the specified frequency.

ORDER i	ELEMENTS IN GF(7) WITH ORDER i	$\phi(i)$
1	{1}	1
2	{6}	1
3	{2, 4}	2
4	None: 4 does not divide 6	–
5	None: 5 does not divide 6	–
6	{3, 5}	2
		6 nonzero elements ■

The investigation of the additive structure of a Galois field begins in a manner similar to that used in the multiplicative case. All Galois fields contain a multiplicative identity element that is usually given the label "1". Consider the following sequence.

$$0, 1, 1 + 1, 1 + 1 + 1, 1 + 1 + 1 + 1, \ldots$$

Since the field is finite, this sequence must begin to repeat at some point. Let the notation $m(1)$ refer to the summation of m ones. If $j(1)$ is the first repeated element, being equal to $k(1)$ for $0 \leq k < j$, it follows that k must be zero; otherwise, $(j - k)(1) = 0$ is an earlier repetition than $j(1)$.

Definition 2-15

The **characteristic** of a Galois field GF(q) is the smallest positive integer m such that $m(1) = 0$.

Theorem 2-13

The characteristic of a Galois field is always a prime integer.

Proof (by contradiction). Consider the sequence 0, 1, 2(1), 3(1), . . . , $k(1)$, $(k + 1)(1), \ldots$. Suppose that the first repeated element is $k(1) = 0$, where k is not a prime. k is thus the characteristic of the field by definition. Since k is not prime, there exist positive integers $m, n > 1$ such that $m \cdot n = k$. It follows that $m(1) \cdot n(1) = k(1) = 0$. Since a field cannot contain zero divisors, either $m(1)$ or $n(1)$ must equal zero. Since $0 < m, n < k$, this is a contradiction of the minimality of the characteristic of the field. **QED**

Let GF(q) be a field of characteristic p. GF(q) thus contains a set of p distinct elements $Z_p = \{0, 1, 2(1), 3(1), \ldots, (p-1)(1)\}$. The preceding development has shown that Z_p is closed under both GF(q) addition and multiplication (the sum or product of sums of ones is still a sum of ones). The additive inverse of $j(1) \in Z_p$ is clearly $(p - j)(1) \in Z_p$. The multiplicative inverse of $j(1)$ ($j \neq 0$ or a multiple of p) is simply $k(1)$, where $j \cdot k \equiv 1 \; mod \; p$ (the proof that a solution always exists is left to the reader). The rest of the field requirements (associativity, distributivity, etc.) are satisfied by noting that Z_p is embedded in the field GF(q). Z_p is thus a Galois field of order p, a subfield of all fields GF(q) of characteristic p. Since the field of order p is unique up to isomorphisms (renaming of elements), Z_p must be the field of integers under *modulo p* addition and multiplication.

In a field GF(q) of characteristic p it follows that $p(\alpha) = 0$ for all $\alpha \in$ GF(q), for $p(\alpha) = p(\alpha \cdot 1) = \alpha \cdot p(1) = \alpha \cdot 0 = 0$.

Theorem 2-14

The order q of a Galois field GF(q) must be a power of a prime.

Proof. The proof to Theorem 2-13 and the discussion that followed showed that every finite field contains a prime-order subfield GF(p). Given that GF(q) forms an additive group, it may be viewed as a vector space over its subfield GF(p).

Let β_1 be a nonzero element in GF(q). There are p distinct elements of the form $\alpha_1\beta_1 \in$ GF(q), where α_1 ranges over all p of the elements in GF(p). (They must be distinct, for $\gamma\beta_1 = \alpha\beta_1$ implies $\gamma - \alpha = 0$.) If the field GF(q) contains no other elements, then the proof is complete. If there is an element β_2 that is not of the form $\alpha_1\beta_1, \alpha_1 \in$ GF(p), then there are p^2 distinct elements in GF(q) of the form $\alpha_1\beta_1 + \alpha_2\beta_2 \in$ GF(q), where $\alpha_1, \alpha_2 \in$ GF(p). This process continues until all elements in GF(q) can be represented in the form $\alpha_1\beta_1 + \alpha_2\beta_2 + \cdots + \alpha_m\beta_m$. Since each combination of coefficients $\{\alpha_1, \alpha_2, \ldots, \alpha_m\} \subset$ GF(p) corresponds by construction to a distinct element in GF(q), q must be of the form p^m. **QED**

2.3 PRIMITIVE POLYNOMIALS AND GALOIS FIELDS OF ORDER p^m

It has been shown that GF(q) can be represented using 0 and $(q-1)$ consecutive powers of a primitive field element $\alpha \in$ GF(q). Multiplication in a Galois field of nonprime order can thus be performed by representing the elements as powers of α and adding their exponents *modulo* $(q-1)$. It has also been shown that GF(q) contains a subfield of prime order whose additive operation is integer addition *modulo p*. In this section the primitive element representation is augmented to show the additive structure of the entire field. The technique presented is based on polynomials whose coefficients are taken from prime-order finite fields.

The notation GF(q)[x] is used here to denote the collection of all polynomials $a_0 + a_1x + a_2x^2 + \cdots + x^n$ of arbitrary degree with coefficients $\{a_i\}$ in the finite field GF(q). As noted earlier, such a collection of polynomials forms a commutative ring with identity. The additive and multiplicative operations are performed as one might expect.

$$(a_0 + a_1x + a_2x^2 + \cdots + a_nx^n) + (b_0 + b_1x + b_2x^2 + \cdots + b_nx^n)$$
$$= (a_0 + b_0) + (a_1 + b_1)x + (a_2 + b_2)x^2 + (a_3 + b_3)x^3 + \cdots + (a_n + b_n)x^n$$
$$(a_0 + a_1x + a_2x^2 + \cdots + a_nx^n) \cdot (b_0 + b_1x + b_2x^2 + \cdots + b_mx^m)$$
$$= (a_0 \cdot b_0) + [(a_1 \cdot b_0) + (a_0 \cdot b_1)]x + [(a_2 \cdot b_0) + (a_1 \cdot b_1) + (a_0 \cdot b_2)]x^2 + \cdots + a_nb_mx^{n+m}$$

The coefficient operations are performed using the operations for the field from which the coefficients were taken. For example, in GF(2)[x],

$$(x^5 + x^2) + (x^2 + x + 1) = x^5 + 2x^2 + x + 1 = x^5 + x + 1$$

Definition 2-16—Irreducible Polynomials

A polynomial $f(x)$ is **irreducible** in GF(q) if $f(x)$ cannot be factored into a product of lower-degree polynomials in GF(q)[x].

Example 2-21—Irreducible Polynomials

- $x^2 + x + 1$ is irreducible in GF(2)[x], but not in GF(4)[x].
- $x^2 + 1$ is irreducible in GF(3)[x].
- $x^{11} + x^2 + 1$ and $x^{21} + x^2 + 1$ are both irreducible in GF(2)[x]. ■

The first example brings out an important point. A polynomial may be irreducible in one ring of polynomials, but reducible in another. In fact *every* polynomial is reducible in *some* ring of polynomials. The term irreducible must thus be used only with respect to a specific ring of polynomials.

Definition 2-17—Primitive Polynomials

An irreducible polynomial $p(x) \in \mathrm{GF}(p)[x]$ of degree m is said to be **primitive** if the smallest positive integer n for which $p(x)$ divides $x^n - 1$ is $n = p^m - 1$.

It can be shown that *any* irreducible mth-degree polynomial $f(x) \in \mathrm{GF}(p)[x]$ must divide $x^{p^m-1} - 1$ [McE1].

Example 2-22—Primitive Polynomials

- $x^3 + x + 1$ is primitive in GF(2)[x], for the smallest polynomial of the form $x^n - 1$ for which it is a divisor is $x^7 - 1$ $(7 = 2^3 - 1)$.
- $x^4 + x + 1$ is primitive in GF(2)[x], for the smallest polynomial of the form $x^n - 1$ for which it is a divisor is $x^{15} - 1$ $(15 = 2^4 - 1)$.
- $x^{11} + x^2 + 1$ and $x^{21} + x^2 + 1$ are both primitive in GF(2)[x]. ■

There are $\phi(2^n - 1)/n$ binary primitive polynomials of degree n. An extensive list of binary primitive polynomials is provided in Appendix A at the end of the text.

A primitive polynomial $p(x) \in GF(p)[x]$ is always irreducible in $GF(p)[x]$, but irreducible polynomials are not always primitive. As an example, consider $x^4 + x^3 + x^2 + x + 1$, which is irreducible in GF(2)[x], but is a factor of $x^5 - 1$ and thus not primitive.

The roots of primitive polynomials have some interesting properties that follow directly from Definition 2-17.

Theorem 2-15

The roots $\{\alpha_j\}$ of an mth-degree primitive polynomial $p(x) \in GF(p)[x]$ have order $p^m - 1$.

Proof. We first state as given that all of the roots of an irreducible polynomial have the same order (this is the Corollary to Theorem 3-4). From this point, Theorem 2-15 can be proven quite easily.

Let α be an arbitrary root of an mth-degree primitive polynomial $p(x) \in GF(p)[x]$. This implies that α is also a root of the expression $x^{p^m-1} - 1 = 0$, since $p(x) \mid x^{p^m-1} - 1$ (Definition 2-17).

Claim: $p(x) \mid x^{p^m-1} - 1$ implies that $\text{ord}(\alpha) \mid p^m - 1$.

Proof: Since α is a root of $p(x)$, and $p(x) \mid x^{p^m-1} - 1$, α must be a $(p^m - 1)$st root of unity. We now proceed by contradiction.

If $\text{ord}(\alpha)$ does not divide $p^m - 1$, then $p^m - 1 = k[\text{ord}(\alpha)] + r$, where $0 < r < \text{ord}(\alpha)$.

$$\Rightarrow 1 = \alpha^{p^m-1} = \alpha^{k[\text{ord}(\alpha)]+r} = \alpha^r$$

This is a contradiction, for $\text{ord}(\alpha)$ is by definition the smallest positive integer x such that $\alpha^x = 1$.

This implies that all of the roots of $x^{\text{ord}(\alpha)} - 1 = 0$ are roots of $x^{p^m-1} - 1 = 0$, which in turn implies that $x^{\text{ord}(\alpha)} - 1 \mid x^{p^m-1} - 1$.

Since all of the roots of an irreducible polynomial have the same order, it follows that $p(x)$ divides $x^{\text{ord}(\alpha)} - 1$, which in turn divides $x^{p^m-1} - 1$. Finally, $\text{ord}(\alpha) = p^m - 1$ by Definition 2-17. **QED**

Given that α has order $(p^m - 1)$, the $(p^m - 1)$ consecutive powers of α form a multiplicative group of order $(p^m - 1)$. The multiplication operation is performed by adding the exponents of the powers of α *modulo* $(p^m - 1)$. These exponential representations can be reexpressed by reducing the sequence of powers of α *modulo* the primitive polynomial.

Let $p(x) = x^m + a_{m-1}x^{m-1} + \cdots + a_1x + a_0$ be primitive in $GF(p)[x]$. If α is a root of $p(x)$, it must satisfy $p(\alpha) = \alpha^m + a_{m-1}\alpha^{m-1} + \cdots + a_1\alpha + a_0 = 0$. It follows that

$$\alpha^m = -a_0 - a_1\alpha - a_2\alpha^2 - \cdots - a_{m-1}\alpha^{m-1}$$

The individual powers of α of degree greater than or equal to m can be reexpressed as polynomials in α of degree $(m - 1)$ or less. Since α has order $p^m - 1$, the distinct powers of α must have $p^m - 1$ distinct nonzero polynomial representations of the form $b_0 + b_1\alpha + b_2\alpha^2 + \cdots + b_{m-1}\alpha^{m-1}$. The coefficients $\{b_i\}$ are taken from GF(p), so there are exactly $p^m - 1$ distinct nonzero polynomial representations available. A one-to-one mapping is then defined between the distinct powers of α and the set of polynomials in α of degree less than or equal to $(m - 1)$ with coefficients in GF(p). These $p^m - 1$ polynomials and zero form an additive group using polynomial addition. The $p^m - 1$ consecutive powers of α can thus be shown to be the nonzero elements of the field GF(p^m). The roots of an mth-degree primitive polynomial in GF(p)[x] are primitive elements in GF(p^m).

Example 2-23—The Construction of GF(8)

- $p(x) = x^3 + x + 1$ is primitive in GF(2)[x]. Let α be a root of $p(x)$. This implies that $\alpha^3 + \alpha + 1 = 0$, or equivalently, $\alpha^3 = \alpha + 1$.

EXPONENTIAL REPRESENTATION		POLYNOMIAL REPRESENTATION
α^0	=	1
α^1	=	α
α^2	=	α^2
α^3	=	$\alpha + 1$
α^4	=	$\alpha^2 + \alpha$
α^5	=	$\alpha^3 + \alpha^2 = \alpha^2 + \alpha + 1$
α^6	=	$\alpha^2 + 1$
0	=	0

■

Using this type of construction for GF(p^m), addition is performed using the polynomial representation, as shown in the following example. To compute $\alpha^2 + \alpha^5$ in GF(8), one begins by substituting the polynomial representations for the exponential representations α^2 and α^5. The polynomials are then summed to obtain a third polynomial representation, which may then be reexpressed as a power of α.

$$\alpha^2 + \alpha^5 = (\alpha^2) + (\alpha^2 + \alpha + 1) = (\alpha + 1) = \alpha^3$$

As noted earlier, multiplication is most easily performed through the use of the exponential representations. The exponents of the two elements being multiplied together are added together *modulo* $(2^3 - 1)$ (in general *modulo* $(p^m - 1)$) to obtain the exponent of the product.

Multiplication can also be performed through the polynomial representation. If α^a and α^b have the polynomial representations $a_0 + a_1\alpha + a_2\alpha^2$ and $b_0 + b_1\alpha + b_2\alpha^2$, respectively, then $\alpha^{(a+b) mod 7}$ has polynomial representation

$$(a_0 + a_1\alpha + a_2\alpha^2)(b_0 + b_1\alpha + b_2\alpha^2)\ modulo\ (\alpha^3 + \alpha + 1)$$

In general, a finite field GF(p^m) constructed using an mth-degree primitive polynomial $p(x)$ has field elements represented by polynomials of the form $a_0 +$

$a_1\alpha + \cdots + a_{m-1}\alpha^{m-1}$, where α is a root of $p(x)$. Element multiplication can be performed as follows.

$$(a_0 + a_1\alpha + \cdots + a_{m-1}\alpha^{m-1})(b_0 + b_1\alpha + \cdots + b_{m-1}\alpha^{m-1})\ modulo\ p(\alpha) \quad (2\text{-}3)$$

The exponential and polynomial representations can be used to create standard operation tables, as in the following example.

Example 2-24—Operation Tables for GF(4)

- GF(4): $p(x) = x^2 + x + 1$ is primitive in GF(2)[x]. Let α be a root of $p(x)$. This implies that $\alpha^2 + \alpha + 1 = 0$.

EXPONENTIAL		POLYNOMIAL		LABEL
α^0	=	1	$\Leftrightarrow$	1
α^1	=	α	$\Leftrightarrow$	2
α^2	=	$\alpha + 1$	$\Leftrightarrow$	3
0	=	0	$\Leftrightarrow$	0

Using the labels, we can create the following tables.

+	0	1	2	3
0	0	1	2	3
1	1	0	3	2
2	2	3	0	1
3	3	2	1	0

·	0	1	2	3
0	0	0	0	0
1	0	1	2	3
2	0	2	3	1
3	0	3	1	2

■

In some cases the use of integer labels may be useful, but they present a subtle danger to the student. It is easy to forget that in general the labels do not represent integers, and that the associated addition and multiplication operations bear no resemblance to modular integer arithmetic. Integer labels also obscure the logarithmic properties of the field elements.

The polynomial representation for a finite field GF(p^m) has coefficients in the "ground field" GF(p). Clearly GF(p^m) can thus be interpreted as a vector space over GF(p).

Example 2-25–GF(8) as a Vector Space over GF(2)

Let α be a root of $x^3 + x + 1$. The set $\{1, \alpha, \alpha^2\}$ can be used as a basis for the vector-space representation of GF(8) in the following manner.

GF(8)			
	0	$\leftrightarrow$	$(0,0,0)$
	α	$\leftrightarrow$	$(0,1,0)$
	α^2	$\leftrightarrow$	$(0,0,1)$
	$\alpha^3 = \alpha + 1$	$\leftrightarrow$	$(1,1,0)$
	$\alpha^4 = \alpha^2 + \alpha$	$\leftrightarrow$	$(0,1,1)$
	$\alpha^5 = \alpha^2 + \alpha + 1$	$\leftrightarrow$	$(1,1,1)$
	$\alpha^6 = \alpha^2 + 1$	$\leftrightarrow$	$(1,0,1)$
	$\alpha^7 = 1$	$\leftrightarrow$	$(1,0,0)$

■

The vector representation makes software and hardware realization of finite-field mathematics easier, for GF(p^m) addition is reduced to vector addition over GF(p). The finite fields GF(8) through GF(1024) are listed in vector-space form in Appendix B. This Appendix also includes a C program that generates these representations.

It has now been seen that GF(p^m) contains the prime-order field GF(p) and can in fact be viewed as a construction over GF(p). For this reason fields of prime power order p^m are frequently called **extensions** of the field of order p. For example, in the literature one may see fields of order 2^m referred to as binary extension fields.

Just as groups may contain many subgroups, so may a field GF(p^m) contain subfields other than the base prime-order field GF(p). It can be shown that GF(p^m) contains all Galois fields of order p^b, where b divides m.

Example 2-26—Subfields of Galois Fields

- GF(64) = GF(2^6) contains GF(2^6), GF(2^3), GF(2^2), and GF(2^1) as subfields (all but the first being proper subfields).
- GF(59049) = GF(3^{10}) contains GF(3^{10}), GF(3^5), GF(3^2), and GF(3^1) as subfields. ■

The subfields of a given finite field can be represented in a graph that shows how the fields are contained within one another. The following graphs indicate the respective subfields in the examples above.

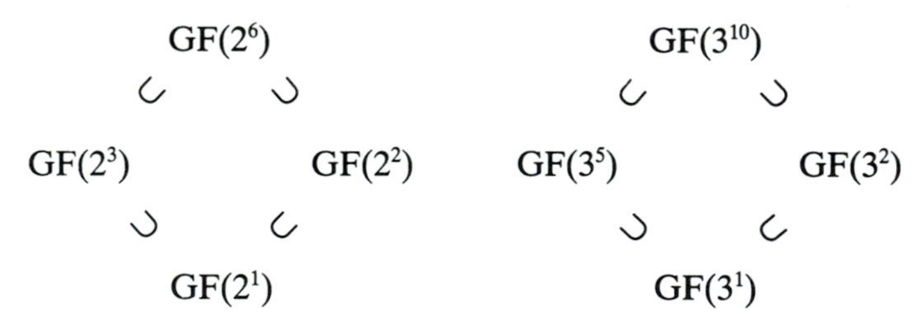

The following theorem is useful for identifying subfield elements within a larger field.

Theorem 2-16

An element β in GF(q^m) lies in the subfield GF(q) if and only if $\beta^q = \beta$.

Proof. Let $\beta \in GF(q) \subseteq GF(q^m)$.

$\Rightarrow$ ord $(\beta) \mid q - 1$ by Theorem 2-10 and thus $\beta^q = \beta$.

Now assume that $\beta^q = \beta$.

$\Rightarrow \beta$ is a root of $x^q - x = 0$.
$\Rightarrow$ The q elements of GF(q) comprise all q roots of $x^q - x = 0$ and the result follows. **QED**

Example 2-27—Subfield Elements in GF(64)

Let α be a primitive element in GF(64), and recall that GF(64) has GF(8), GF(4), and GF(2) as subfields. Since α is primitive in GF(64), all nonzero elements in GF(64) can be represented as α^j for some integer j. An element α^j is in the subfield GF(q) if and only if $j \cdot q \equiv j$ *modulo* 63.

$$GF(64) = \{0, \alpha^0 = 1, \alpha, \alpha^2, \alpha^3, \ldots, \alpha^{62}\} \qquad GF(4) = \{0, 1, \alpha^{21}, \alpha^{42}\}$$

$$GF(8) = \{0, 1, \alpha^9, \alpha^{18}, \alpha^{27}, \alpha^{36}, \alpha^{45}, \alpha^{54}\} \qquad GF(2) = \{0, 1\}$$

■

In the vector-space representation of finite fields it was noted that GF(p^m) could be represented as an m-dimensional subspace over GF(p), where p is prime. More generally, it is possible to represent GF(q^m) as an m-dimensional subspace over GF(q), where GF(q) is a subfield of GF(q^m) of prime power order.

2.4 ZECH'S LOGARITHMS

Except in the prime-order field case, GF(q) addition is not as easy to implement as multiplication. The simplest (though least efficient) approach is to construct a ($q \times q$) look-up table. A more efficient use of memory can be obtained through the use of Zech's logarithms, also known as "add-one tables." An add-one table has two columns: the first contains the logarithm of each element with respect to a primitive element α. The second column contains the logarithm to the base α of the corresponding element in the first column after it has been incremented by one.

Example 2-28—Add-One Table for GF(8)

Let α be a root of $x^3 + x + 1 = 0$. Using the representation for GF(8) in Example 2-23, the following add-one table can be constructed.

x	$\text{Log}_\alpha(x)$	$\text{Log}_\alpha(x+1)$
α	1	3
α^2	2	6
α^3	3	1
α^4	4	5
α^5	5	4
α^6	6	2
$\alpha^7 = 1$	0	*
0	*	0

■

The "$\text{Log}_\alpha(x)$" and the "$\text{Log}_\alpha(x + 1)$" columns of the add-one table are stored in a read-only-memory (ROM) look-up table. Addition in GF(p^m) is then performed using the following scheme:

1. Combine all terms that have the same exponent using modular addition of the exponents (i.e., GF(p) addition).

2. Arrange the resulting expression $\alpha^a + \alpha^b + \cdots + \alpha^z$ in order of decreasing exponents.
3. Factor the expression into the form $(\ldots(((\alpha^{a-b} + 1)\alpha^{b-c} + 1)\alpha^{c-d} + 1)\ldots)\alpha^z$. The summation can now be performed as a series of add-one operations and Galois field multiplications.

Example 2-29—GF(8) Addition Using the Add-One Table

- $\alpha^6 + \alpha^5 + \alpha^7 + \alpha^6 + \alpha^3 + \alpha^3 + \alpha^3 + \alpha + 1$

$$
\begin{aligned}
&= \alpha^5 + \alpha^7 + \alpha^3 + \alpha + 1 && \text{(Step 1)}\\
&= \alpha^7 + \alpha^5 + \alpha^3 + \alpha + 1 && \text{(Step 2)}\\
&= (((\alpha^2 + 1)\alpha^2 + 1)\alpha^2 + 1)\alpha + 1 && \text{(Step 3)}
\end{aligned}
$$

The computation is completed as follows:

$$
\begin{aligned}
(((\alpha^2 + 1)\alpha^2 + 1)\alpha^2 + 1)\alpha + 1 &= (((\alpha^6)\alpha^2 + 1)\alpha^2 + 1)\alpha + 1\\
&= ((\alpha + 1)\alpha^2 + 1)\alpha + 1\\
&= ((\alpha^3)\alpha^2 + 1)\alpha + 1\\
&= (\alpha^5 + 1)\alpha + 1\\
&= (\alpha^4)\alpha + 1\\
&= \alpha^5 + 1\\
&= \alpha^4
\end{aligned}
$$

■

Note that the storage requirement for an add-one table for GF(q) is of the order of $2q$, whereas that for a complete addition table for GF(q) is of the order of q^2. The difference can be substantial for algebraic codes over large fields (GF(256), GF(512), etc.).

Add-one tables for GF(8) through GF(64) are included in Appendix B.

PROBLEMS

Groups

1. Prove that the identity element e in a group G is unique.
2. Prove that the inverse a^{-1} of an element a in a group G is unique.
3. Show that all groups of order 3 are commutative.
4. Show that all groups of order 4 are commutative.
5. Construct the operation table for the group of order 6 under *modulo* 6 addition.
6. Show by counterexample that the integers in the set $\{0, 1, 2, 3, \ldots, 7\}$ do not form a group under *modulo* 8 multiplication.
7. Find the orders of each of the elements in the group of order 8 under *modulo* 8 addition.
8. Show that $S = \{0, 4, 8, 12\}$ forms a subgroup of the group of integers $G = \{0, 1, 2, 3, \ldots, 15\}$ under *modulo* 16 addition. Decompose G into cosets *modulo* S.

Rings

9. Let R be a ring. A nonempty subset $I \subseteq R$ is said to be an **ideal** if it satisfies the following:
(a) I forms a group under the addition operation in R.
(b) $a \cdot r = b \in I$ for all $a \in I$ and for all $r \in R$.
Show that $\{0, 4, 8, 12\}$ forms an ideal in the ring of integers $R = \{0, 1, 2, 3, \ldots, 15\}$ under *modulo* 16 addition and multiplication.

10. Any element in a ring R that has its multiplicative inverse in R is called a **unit**. What is (are) the unit(s) in the ring R discussed in Problem 9?

11. Show that if an ideal I contained in a ring R contains a unit, then $I = R$.

Fields

12. Show that a field F contains only two ideals: $\{0\}$ and F.

13. Construct the operation tables for GF(5) and GF(7).

Vector Spaces

14. Show that the set of binary polynomials (i.e., polynomials with binary coefficients) with degree less than or equal to r forms a vector space over GF(2) with dimension $(r + 1)$.

15. Find a basis for the vector space discussed in Problem 14.

16. What is the dimension of the vector space spanned by the set of vectors {(001010), (101000), (001100), (100010), (011111)} over GF(2)? What is the dimension if the ground field is changed to GF(4)? GF(31)? Support your answer.

17. Find a basis for the dual space to the vector space spanned by {(11100), (01110), (00111)} over GF(2).

18. Find a basis for the dual space to the vector space spanned by {(12322), (14312), (41233)} over GF(5).

19. V_5 is the vector space of binary 5-tuples. The canonical vector space for V_5 is {(10000), (01000), (00100), (00010), (00001)}. Express the vector (11010) as a linear combination of the vectors in the canonical basis for V_5.

20. S = {(11010), (00111), (10101), (10001), (00010)} also forms a basis for V_5. Express the vector (10111) as a linear combination of the vectors in S.

21. Let $S = \{\mathbf{v}_1, \mathbf{v}_2, \ldots, \mathbf{v}_n\}$ be an arbitrary basis for the vector space V_n and let $\mathbf{v}$ be an arbitrary vector in V_n. Derive an expression for computing the coefficients $\{a_i\}$ for the representation $\mathbf{v} = a_1\mathbf{v}_1 + a_2\mathbf{v}_2 + \cdots + a_n\mathbf{v}_n$.

Galois Fields

22. List the possible orders taken on by the elements in the field GF(125) and determine the number of elements in the field that display each allowed order.

23. Prove that $\phi(p_1 \cdot p_2) = \phi(p_1) \cdot \phi(p_2) = (p_1 - 1)(p_2 - 1)$ for distinct primes p_1 and p_2.

24. Prove that $\phi(p^m) = p^{m-1}(p - 1)$ for a prime p.

25. Prove that $\phi(p^m r^n) = p^{m-1} r^{n-1}(p - 1)(r - 1)$ for distinct primes p and r.

26. Prove that $\sum_{x|n} \phi(x) = n$.

27. Compute $\phi(193)$, $\phi(284)$, $\phi(440)$, $\phi(699)$, $\phi(788)$, and $\phi(1024)$.

28. Express all of the nonzero elements of GF(11) as powers of a single element in GF(11).

29. Express 7 as a power of 3 in the field GF(17). Express 3 as a power of 17 in GF(23). Can you think of a fast way to solve such problems? Taking logarithms in finite fields is a difficult task. In fact, the first suggested public key cryptosystem was based on this problem. Diffie and Hellman [Dif] feel that a cryptosystem whose successful cryptoanalysis requires the computation of logarithms in GF(p), where p is a prime with one to two hundred decimal digits, is secure. Having dealt with two-digit primes, you are probably inclined to agree.

30. Determine whether each of the following polynomials from GF(2)[x] is irreducible.

(a) $x^2 + x + 1$ **(b)** $x^3 + x^2 + 1$
(c) $x^4 + x^2 + 1$ **(d)** $x^4 + x^3 + x^2 + x + 1$
(e) $x^5 + x^4 + 1$ **(f)** $x^5 + x^4 + x^3 + x^2 + x + 1$
(g) $x^6 + x^5 + x^2 + x + 1$ **(h)** $x^6 + x^3 + 1$
(i) $x^7 + x + 1$ **(j)** $x^{128} + x^{90} + 1$

31. Determine whether each of the following polynomials from GF(3)[x] is irreducible.

(a) $x^2 + 2x + 1$ **(b)** $x^3 + x^2 + x + 1$
(c) $x^4 + x^2 + 1$ **(d)** $x^4 + 2x^3 + 2x^2 + x + 1$
(e) $x^5 + x^4 + x^3 + 1$ **(f)** $x^5 + x^3 + x^2 + x + 1$
(g) $x^6 + x^5 + x^2 + x + 1$ **(h)** $x^6 + 2x^3 + x + 1$
(i) $x^7 + 2x + 1$ **(j)** $x^{129} + x^{90} + 1$

32. Determine whether each of the following is primitive in GF(2)[x].

(a) $x^2 + x + 1$ **(b)** $x^3 + x^2 + 1$
(c) $x^6 + x^5 + x^4 + x^3 + x^2 + x + 1$ **(d)** $x^4 + x + 1$
(e) $x^5 + x^2 + 1$

33. Show a polynomial and exponential representation of the elements in GF(16) and GF(9).

34. Using the representation for GF(8) in Example 2-24, compute the following.

(a) $\alpha^4 + \alpha^2 + \alpha + 1$ **(b)** $(\alpha^3 + \alpha^2)\cdot(\alpha^6 + \alpha^3)$
(c) $\alpha^6(\alpha^5 + \alpha^4) + \alpha^2$ **(d)** $(\alpha^3x^2 + \alpha x + 1)(\alpha^5x^3 + x + \alpha^2)$

35. Construct GF(32) as a vector space over GF(2).

36. Let α be a primitive element in GF(1024). List the elements in the subfields GF(4) and GF(32) as powers of α.

37. What are the subfields of GF(4096)?

38. Using the results from Problem 35, construct an add-one table for GF(32) and use it to compute the following.

(a) $\alpha^{27} + \alpha^{13} + \alpha + 1$ **(b)** $(\alpha^{23} + \alpha^{21})\cdot(\alpha^{16} + \alpha^3)$
(c) $\alpha^{16}(\alpha^5 + \alpha^{24}) + \alpha^{30}$ **(d)** $(\alpha^{13}x^{23} + \alpha x + \alpha^{18})(\alpha^{15}x^3 + x + \alpha^{12})$

3

Polynomials over Galois Fields

In this chapter we continue the study of the theory of Galois fields by examining the properties of collections of polynomials whose coefficients are taken from Galois fields. In the last chapter the importance of polynomials over Galois fields to the overall theory of Galois fields became apparent. It was shown that a primitive polynomial of degree m in $GF(p)[x]$ can be used to obtain representations of the elements in $GF(p^m)$. In this chapter it is shown that, through the definition of a metric and the introduction of the cancellation property, rings of polynomials form Euclidean domains. Euclidean domains allow for the use of the division algorithm and Euclid's algorithm. Euclid's algorithm and the domains in which it can be applied are named after Euclid, the Greek mathematician and father of geometry who was born around 300 B.C. Euclid's algorithm lies at the heart of the most popular decoding algorithms for the BCH and Reed-Solomon codes. It is thus interesting to note that there is a direct link between this mathematician/philosopher of the ancient world (mathematics and philosophy were inseparable then) and the digital audio disc player of the late twentieth century. The relationship between conjugate elements and irreducible polynomials is also developed in this chapter, and the properties of minimal polynomials are investigated. This chapter culminates in the examination of the factorization of the polynomial $x^n - 1$, which is central to the development of all cyclic codes.

3.1 EUCLIDEAN DOMAINS AND EUCLID'S ALGORITHM

As in the previous chapter, $GF(q)[x]$ is taken to denote all polynomials with coefficients taken from the field $GF(q)$. It was noted in the last chapter that $GF(q)[x]$ forms a commutative ring with identity. A bit more structure can be introduced to this ring by noting that there is a simple method for comparing the "size" of two or more selected polynomials. With each polynomial $f(x) \in GF(q)[x]$ we associate the integer a that is the degree of $f(x)$. For example, we associate 3 with $x^3 + x + 1 \in GF(2)[x]$. We can thus say that $x^3 + x + 1$ is "larger" in some sense than $x^2 + x + 1$, but "smaller" than $x^4 + x^3 + x^2 + x + 1$. This association is called a **metric**, and its application to a commutative ring with identity makes the ring a **Euclidean domain**.

Definition 3-1—Euclidean Domains

A **Euclidean domain** is a set D with two binary operations "+" and "·" that satisfy the following:

1. D forms an additive commutative ring with identity.
2. Cancellation: if $ab = bc, b \neq 0$, then $a = c$.
3. Every element $a \in D$ has an associated metric $g(a)$ such that
   a. $g(a) \leq g(a \cdot b)$ for all nonzero $b \in D$.
   b. For all nonzero $a, b \in D, g(a) > g(b)$, there exist q and r such that $a = qb + r$ with $r = 0$ **or** $g(r) < g(b)$. q is called the **quotient** and r the **remainder**.

The metric for the additive identity element, $g(0)$, is generally taken to be undefined, though a value of $-\infty$ can be assigned if desired.

Example 3-1—Euclidean Domains

- The ring of integers under integer addition and multiplication with metric $g(n) = |n|$ (absolute value).
- The ring of polynomials over a finite field with metric $g(f(x)) = \text{degree}(f(x))$. ■

Euclidean domains allow for the definition of "division" in a manner that makes intuitive sense. Let a and b be two elements in a Euclidean domain D. a is said to be a divisor of b (written $a \mid b$) if there exists $c \in D$ such that $a \cdot c = b$.

Definition 3-2—Common Divisors

An element a is said to be a **common divisor** of a collection of elements $\{b_1, b_2, \ldots, b_n\}$ if $a \mid b_i$ for $i = 1, \ldots, n$.

Definition 3-3—Greatest Common Divisors

If d is a common divisor of the $\{b_i\}$ and all other common divisors are less than d, then d is called the **greatest common divisor (GCD)** of the $\{b_i\}$.

Euclid's algorithm is a very fast method for finding the GCDs of sets of elements in Euclidean domains. It works in the following manner.

Euclid's Algorithm

1. Let a, b be a pair of elements contained in a Euclidean domain D, where $g(a) > g(b)$.
2. Let the indexed variable r_i take on the initial values $r_{-1} = a$ and $r_0 = b$. Proceed by using the following recursion formula.
3. If $r_{i-1} \neq 0$, then define r_i using $r_{i-2} - q_i r_{i-1} = r_i$, where $g(r_i) < g(r_{i-1})$.

   Repeat until $r_i = 0$. Note that with each iteration of the recursion formula, the size of the remainder r_i gets smaller. It can be shown that, in a Euclidean domain, the remainder r_i will always take on the value zero after a finite number of steps.[1]

   If $r_i = 0$, then $r_{i-1} = \text{GCD}(a, b)$.

For a proof showing that r_{i-1} is indeed the GCD when r_i first takes on the value zero, see [McE1].

Example 3-2—Euclid's Algorithm

- Find GCD (168, 166).

$$r_{-1} = 168, \quad r_0 = 166$$

$$168 = (166) \cdot 1 + 2 \qquad q_1 = 1, \quad r_1 = 2$$

$$166 = (2) \cdot 83 + 0 \qquad q_2 = 83, \quad r_2 = 0$$

$$\Rightarrow \text{GCD}(168, 166) = r_1 = 2$$

- Find GCD (336, 54).

$$r_{-1} = 336, \quad r_0 = 54$$

$$336 = (54) \cdot 6 + 12 \qquad q_1 = 6, \quad r_1 = 12$$

$$54 = (12) \cdot 4 + 6 \qquad q_2 = 4, \quad r_2 = 6$$

$$12 = (6) \cdot 2 + 0 \qquad q_3 = 2, \quad r_3 = 0$$

$$\Rightarrow \text{GCD}(336, 54) = r_2 = 6$$

  To find the GCD of three or more numbers, one should make use of the fact (easily proven) that $\text{GCD}(a, b, c) = \text{GCD}(\text{GCD}(a, b), c)$.

- Find GCD (336, 108, 42).

  1. Find GCD (336, 108).

$$r_{-1} = 336, \quad r_0 = 108$$

$$336 = (108) \cdot 3 + 12 \qquad q_1 = 3, \quad r_1 = 12$$

$$108 = (12) \cdot 9 + 0 \qquad q_2 = 9, \quad r_2 = 0$$

$$\Rightarrow \text{GCD}(336, 108) = r_1 = 12$$

[1] The worst case: Euclid's algorithm requires a maximal number of steps to complete when a and b are consecutive Fibonacci numbers.

2. Find GCD (42, 12).

$r_{-1} = 42, \quad r_0 = 12$

$42 = (12) \cdot 3 + 6 \qquad q_1 = 3, \quad r_1 = 6$

$12 = (6) \cdot 2 + 0 \qquad q_2 = 2, \quad r_2 = 0$

$\Rightarrow \text{GCD}(42, 12) = r_1 = 6$

$\Rightarrow \text{GCD}(336, 108, 42) = 6$

Since polynomials over finite fields form Euclidean domains, Euclid's algorithm can be readily applied. Consider the following example using polynomials from GF(2)[x] (binary polynomials).

- GCD $(x^5 + x^3 + x + 1, \quad x^4 + x^2 + x + 1)$.

$$r_{-1} = x^5 + x^3 + x + 1, \quad r_0 = x^4 + x^2 + x + 1$$

$$x^5 + x^3 + x + 1 = (x^4 + x^2 + x + 1) \cdot x + (x^2 + 1)$$

$$q_1 = x, \quad r_1 = (x^2 + 1)$$

$$x^4 + x^2 + x + 1 = (x^2 + 1) \cdot x^2 + (x + 1)$$

$$q_2 = x^2, \quad r_2 = (x + 1)$$

$$x^2 + 1 = (x + 1) \cdot (x + 1) + 0$$

$$q_3 = (x + 1), \quad r_3 = 0$$

$$\Rightarrow \text{GCD}(x^5 + x^3 + x + 1, \quad x^4 + x^2 + x + 1) = r_2 = (x + 1)$$ ■

A careful examination of the computations in the above example shows that, in all cases, GCD (a, b) can be written as a linear combination of a and b. Working backward through the computations in the second example yields the following.

$$\begin{aligned} \text{GCD}(336, 54) &= 6 \\ &= 54 - 12 \cdot (4) \\ &= 54 - 4 \cdot (336 - 6 \cdot (54)) \\ &= 54 \cdot (1 + 24) - 336 \cdot (4) \\ &= 54 \cdot (25) + 336 \cdot (-4) \end{aligned}$$

In the last example the following is obtained.

$$\begin{aligned} &\text{GCD}(x^5 + x^3 + x + 1, \quad x^4 + x^2 + x + 1) \\ &\quad = (x + 1) \\ &\quad = (x^4 + x^2 + x + 1) + (x^2 + 1) \cdot x^2 \\ &\quad = (x^4 + x^2 + x + 1) + (x^5 + x^3 + x + 1 + (x^4 + x^2 + x + 1) \cdot x) \cdot x^2 \\ &\quad = (x^4 + x^2 + x + 1)(1 + x^3) + (x^5 + x^3 + x + 1) \cdot x^2 \end{aligned}$$

It is always possible to represent GCD (a, b) as a linear combination of a and b. The following is the general result.

Theorem 3-1

If $B = \{b_1, b_2, \ldots, b_n\}$ is any finite subset of elements from a Euclidean domain D, then B has a GCD d which can be expressed as a linear combination $\Sigma\,\lambda_k b_k$, where the coefficients $\{\lambda_i\} \subset D$.

Proof [McE1]. Let S be the set of all linear combinations of the form $\Sigma\,\lambda_k b_k$, where the coefficients $\{\lambda_i\} \subset D$. Let d be the element in S with the smallest metric $(g(d))$. By definition d is a linear combination of the elements in the set B. We shall now prove that d is the GCD of the elements in B. If d does not divide some element $b_i \in B$, then we can write $b_i = qd + r$, where $g(r) < g(d)$. But $r = b_i - qd$ must be in S, since b_i and d are in S. This contradicts the minimality of the metric of d in S. Thus d is a common divisor of all of the elements in B.

Now let e be any other common divisor of the elements in B. We can then write $b_i = q_i' e$ for each $b_i \in B$. Since $d \in S$, we can write d as a linear combination of the elements in B: $d = \Sigma\,\lambda_i b_i = \Sigma\,\lambda_i q_i' e = e\,\Sigma\,\lambda_i q_i'$. So d is a multiple of every common divisor and thus the GCD of all of the elements in B. **QED**

It is not necessary to work backward through the various expressions to obtain the coefficients of the linear combination. They can be obtained in a more direct manner using the extended version of Euclid's algorithm.

The Extended Version of Euclid's Algorithm

1. We wish to find s and t such that $\text{GCD}(a,b) = sa + tb$. A set of indexed variables $\{r_i, s_i, t_i\}$ is given the following initial conditions: $r_{-1} = a$, $r_0 = b$, $s_{-1} = 1$, $s_0 = 0$, $t_{-1} = 0$, $t_0 = 1$.
2. If $r_{i-1} \neq 0$, then define r_i using $r_i = r_{i-2} - q_i r_{i-1}$, $g(r_i) < g(r_{i-1})$.
3. Compute s_i using $s_i = s_{i-2} - q_i s_{i-1}$, where the q_i is from step 2.
4. Compute t_i using $t_i = t_{i-2} - q_i t_{i-1}$.
5. Repeat steps 2 through 4 until $r_i = 0$. At this point $r_{i-1} = \text{GCD}(a,b)$ and $s_{i-1}a + t_{i-1}b = r_{i-1}$. The sequences $\{s_i\}$ and $\{t_i\}$ provide the desired linear combination, since for all j, $s_j a + t_j b = r_j$.

Example 3-3—Extended Euclid's Algorithm

GDC(256, 108):

i	r_i	q_i	s_i	t_i
−1	256	–	1	0
0	108	–	0	1
1	40	2	1	−2
2	28	2	−2	5
3	12	1	3	−7
4	4	2	−8	19
5	0	3	27	−64

$\text{GCD} = r_3 = 4;\qquad -8 \cdot (256) + 19 \cdot (108) = 4$ ■

3.2 MINIMAL POLYNOMIALS AND CONJUGATE ELEMENTS

In Chapter 5 we shall see that the design of algebraic codes focuses on the selection of polynomials over finite fields that have a required set of roots selected from a given finite field GF(q^m). It is further required that the polynomials have coefficients in the subfield GF(q) and that the polynomials have minimal degree. To meet these requirements, we must understand the relationship between the subfield from which the coefficients of a polynomial are taken and the roots of the polynomial. We begin by looking at polynomials that contain a specified root α.

Definition 3-4—Minimal Polynomials

Let α be an element in the field GF(q^m). The minimal polynomial of α with respect to GF(q) is the smallest-degree nonzero polynomial $p(x)$ in GF(q)[x] such that $p(\alpha) = 0$.

Some of the properties of minimal polynomials are presented in the following theorem.

Theorem 3-2

For each element α in GF(q^m) there exists a unique monic (and thus nonzero) polynomial $p(x)$ of minimal degree in GF(q)[x] such that the following are true.

1. **$p(\alpha) = 0$.**
2. **The degree of $p(x)$ is less than or equal to m.**
3. **$f(\alpha) = 0$ implies that $f(x)$ is a multiple of $p(x)$.**
4. **$p(x)$ is irreducible in GF(q)[x].**

Proof. 1 and 2. Consider the $(m + 1)$ field elements $1, \alpha, \alpha^2, \ldots, \alpha^m \in$ GF(q^m). Since GF(q^m) is a vector space of dimension m over GF(q), the $(m + 1)$ elements must be linearly dependent over GF(q). This means that there must be a polynomial of the form $f(x) = a_0 + a_1 x + a_2 x^2 + \cdots + a_m x^m$ in GF(q)[x] such that $f(\alpha) = 0$.

Uniqueness. Given that we know that at least one polynomial $f(x)$ exists such that $f(\alpha) = 0$, there must be at least one such polynomial $p(x)$ of minimal degree, for the degrees of all polynomials in GF(q)[x] are bounded below by 0, while the degree of $p(x)$ is bounded above by m.

The uniqueness of the monic minimal polynomial of α is readily proven by contradiction. Let $f(x)$ and $g(x)$ be monic minimal polynomials of α, where $f(x) \neq g(x)$. Since $f(x) = g(x) + r(x)$, it follows that there is a nonzero polynomial $r(x)$, where $\deg(r(x)) < \deg(f(x))$, such that $r(\alpha) = 0$. This contradicts the minimality of the degree of the minimal polynomial.

3. Suppose that there exists another polynomial $f(x)$ such that $f(\alpha) = 0$. Since GF(q)[x] is a Euclidean domain, $f(x) = p(x)q(x) + r(x)$, where the

degree of $r(x)$ is less than the degree of $p(x)$ or $r(x)$ is identically zero.[2] Since $f(\alpha) = p(\alpha) = 0$, it follows that $r(\alpha) = 0$. $r(x)$ must thus be identically zero by the minimality of the degree of $p(x)$. This implies that $f(x)$ is a multiple of $p(x)$.

4. If $p(x)$ factors into a product of lower-degree polynomials $f(x) \cdot g(x)$, where $f(x), g(x) \in GF(q)[x]$, then either $f(\alpha) = 0$ or $g(\alpha) = 0$ (otherwise $f(\alpha)$ and $g(\alpha)$ would be zero divisors). The minimality of the degree of $p(x)$ is contradicted. **QED**

We can now provide another definition for primitive polynomials. Primitive polynomials are the minimal polynomials for primitive elements in a Galois field.

The factorization of minimal polynomials in higher-order fields is important to the understanding of cyclic codes. The principal design question is as follows: If we want a polynomial $p(x)$ with coefficients in $GF(q)$ to have a root α from $GF(q^m)$, what other roots must the polynomial have? The answer (and its proof) begins with a definition.

Definition 3-5—Conjugates of Field Elements

Let β be an element in the Galois field $GF(q^m)$. The conjugates of β with respect to the subfield $GF(q)$ are the elements $\beta, \beta^q, \beta^{q^2}, \beta^{q^3}, \ldots$.

The conjugates of α with respect to $GF(q)$ form a set called the **conjugacy class of α with respect to GF(q)**. Henceforth the modifying phrase "with respect to $GF(q)$" is assumed and not always written out. It should be noted, however, that the term "conjugate" is meaningless unless used with respect to a subfield.

Theorem 3-3

The conjugacy class of $\alpha \in GF(q^m)$ with respect to $GF(q)$ contains d elements, where $\alpha^{q^d} = \alpha$ and $d \mid m$.

Proof. Consider the sequence of elements

$$\alpha, \alpha^q, \alpha^{q^2}, \alpha^{q^3}, \ldots$$

We proceed by using an argument familiar to the faithful reader of Chapter 2. Since there are only a finite number of elements in $GF(q^m)$, there can only be a finite number of elements in the conjugacy class of α. Not only must this sequence begin to repeat elements at some point, but it can be shown that the first element to repeat must be α. The proof of this claim proceeds as follows.

Suppose that α^{q^d} is the first element to repeat, taking on the same value as α^{q^s}, where $0 < s < d$. This implies that $(\alpha^{q^d})/(\alpha^{q^s}) = \alpha^{(q^d - q^s)} = \alpha^{q^s[q^{(d-s)}-1]} = 1$, and thus $\text{ord}(\alpha) \mid q^s[q^{(d-s)} - 1]$. But according to Theorem 2-10,

[2] Nonzero field elements are considered to be zero-degree polynomials. The zero element, however, is not considered a polynomial at all, because most metrics used with Euclidean rings of polynomials are undefined for the zero element.

$\text{ord}(\alpha) \mid (q^m - 1)$. Since the greatest common divisor of q^s and $(q^m - 1)$ must be one (recall that q is a prime or the power of a prime), $\text{ord}(\alpha) \mid [q^{(d-s)} - 1]$. It follows that $\alpha^{q^{(d-s)}} = \alpha$, contradicting the supposition that $\alpha^{q^d} = \alpha^{q^s} \neq \alpha$ was the first element in the sequence to repeat. The sequence of conjugates of α thus progresses as follows:

$$\alpha, \alpha^q, \alpha^{q^2}, \alpha^{q^3}, \ldots, \alpha^{q^d} = \alpha, \alpha^q, \alpha^{q^2}, \alpha^{q^3}, \ldots$$

We now show that d must be a divisor of m. The proof proceeds as follows. Since α is an element in GF(q^m), $\alpha^{(q^m-1)} = 1$ and $\alpha^{q^m} = \alpha$ by Theorem 2-10. It follows that $a^{q^{\text{GCD}(m,d)}} = \alpha$. Since α^{q^d} is the first repetition of the element α, d must equal GCD (m, d) and is therefore a divisor of m. **QED**

Example 3-4—Conjugacy Classes

- Let α be an element of order 3 in GF(16). The conjugates of α with respect to GF(2) are $\alpha, \alpha^2, \alpha^{2^2} = \alpha^{3+1} = \alpha$, etc. The conjugacy class with respect to GF(2) of α is thus $\{\alpha, \alpha^2\}$.
- The conjugacy class with respect to GF(5) of α, an element of order 6 in GF(25), is $\{\alpha, \alpha^5\}$.
- The conjugacy class with respect to GF(4) of α, an element of order 63 in GF(64), is $\{\alpha, \alpha^4, \alpha^{16}\}$. ■

Conjugacy classes are the key to the factorization of minimal polynomials in GF(q^m)[x].

Theorem 3-4

Let α be an element in GF(q^m). Let $p(x)$ be the minimal polynomial of α with respect to GF(q). The roots of $p(x)$ are exactly the conjugates of α with respect to GF(q).

Proof. This proof is a long one, but worth the effort, for it makes use of a number of the more important results developed in this and the previous chapter. We begin by proving a pair of lemmas.

Lemma 3-1

$p \left| \binom{p}{k} \right.$ for all $k \in \{1, 2, 3, \ldots, p - 1\}$ and for all prime integers p.

Proof. $\binom{p}{k} = \frac{p(p-1)(p-2)\cdots(p-k+1)}{k(k-1)(k-2)\cdots(2)(1)}$ is always an integer. Since p is prime, none of the integers $k, (k-1), \ldots, 3, 2$ are divisors of p. $\binom{p}{k}$ is thus a multiple of p. **QED (Lemma)**

Lemma 3-2

Let $\alpha_1, \alpha_2, \ldots, \alpha_t$ be elements in the field GF(p^m). Then

$$(\alpha_1 + \alpha_2 + \cdots + \alpha_t)^{p^r} = \alpha_1^{p^r} + \alpha_2^{p^r} + \cdots + \alpha_t^{p^r} \quad \text{for } r = 1, 2, 3, \ldots.$$

Proof (for the $t = 2$ case—the rest follows by induction). We wish to show that $(\alpha + \beta)^{p^r} = \alpha^{p^r} + \beta^{p^r}$.

$$(\alpha + \beta)^p = \alpha^p + \binom{p}{1}\alpha^{p-1}\beta + \binom{p}{2}\alpha^{p-2}\beta^2 + \cdots + \beta^p$$

The previous lemma shows that $p \,\Big|\, \binom{p}{i}$ for all $i \in \{1, 2, \ldots, p-1\}$. It follows that

$$\binom{p}{1} \equiv \binom{p}{2} \equiv \cdots \equiv \binom{p}{p-1} \equiv 0 \; mod\, p$$

Since GF(p^m) has characteristic p, we see that

$$(\alpha + \beta)^p = \alpha^p + (0)\alpha^{p-1}\beta + (0)\alpha^{p-2}\beta^2 + \cdots + (0)\alpha\beta^{p-1} + \beta^p = \alpha^p + \beta^p$$

which implies that $(\alpha + \beta)^{p^2} = [(\alpha + \beta)^p]^p = (\alpha^p + \beta^p)^p = \alpha^{p^2} + \beta^{p^2}$. We conclude by induction on r, the power of p in the exponent. **QED (Lemma)**

We now show that if $p(x)$ is the minimal polynomial of α with respect to GF(q), then the conjugates of α with respect to GF(q) must also be roots of $p(x)$. By definition of minimal polynomial with respect to GF(q), $p(\alpha) = p_0 + p_1\alpha + p_2\alpha^2 + \cdots + p_w\alpha^w = 0$, where the coefficients $\{p_i\}$ lie in the field GF(q). It follows from Lemma 3-2 that

$$0 = \left(\sum_{k=0}^{w} p_k \alpha^k\right)^q = \sum_{k=0}^{w} p_k \alpha^{qk} = p(\alpha^{qk}) = 0$$

which implies that if α is a root of $p(x)$, then so are $\{a, a^q, a^{q^2}, \ldots\}$.

We can now complete the proof of Theorem 3-4. Let α be an element in GF(q^m) with d distinct conjugates with respect to GF(q). Define $f_\alpha(x)$ to be the polynomial whose roots are the d conjugates of α. It follows that $f_\alpha(x) = (x - \alpha)(x - \alpha^q)(x - \alpha^{q^2}) \cdots (x - \alpha^{q^{d-1}})$.

We can rewrite this in the form

$$(x - \alpha)(x - \alpha^q) \cdots (x - \alpha^{q^{d-1}}) = A_d x^d + A_{d-1} x^{d-1} + \cdots + A_1 x + A_0$$

Using Lemma 3-2 and rearranging the order of multiplication, we then have

$$\begin{aligned}
\{(x - \alpha)(x - \alpha^q) \cdots (x - \alpha^{q^{d-1}})\}^q &= A_d^q x^{qd} + A_{d-1}^q x^{q(d-1)} + \cdots + A_1^q x^q + A_0^q \\
&= (x^q - \alpha^q)(x^q - \alpha^{q^2}) \cdots (x^q - \alpha^{q^d}) \\
&= (x^q - \alpha)(x^q - \alpha^q) \cdots (x^q - \alpha^{q^{d-1}}) \\
&= f_\alpha(x^q) = A_d x^{qd} + A_{d-1} x^{q(d-1)} + \cdots + A_1 x^q + A_0
\end{aligned}$$

This implies that

$$A_d x^{qd} + A_{d-1}x^{q(d-1)} + \cdots + A_1 x^q + A_0 = A_d^q x^{qd} + A_{d-1}^q x^{q(d-1)} + \cdots + A_1^q x^q + A_0^q$$

which finally shows that

$$A_d = A_d^q, \quad A_{d-1} = A_{d-1}^q, \quad A_{d-2} = A_{d-2}^q, \quad \ldots, \quad A_0 = A_0^q$$

The coefficients of $f_\alpha(x)$ thus satisfy the expression $\beta^q = \beta$ and are thus in the subfield GF(q) by Theorem 2-16. We have shown that the minimal polynomial of $\alpha \in \mathrm{GF}(q^m)$ with respect to GF(q) must contain all of the conjugates of α with respect to GF(q) as roots, and that those roots alone are sufficient to ensure that the coefficients of the resulting polynomial are in the subfield GF(q). **QED**

Corollary to Theorem 3-4

All of the roots of an irreducible polynomial have the same order.

Proof. Let GF(q^m) be the smallest field containing all of the roots of an irreducible polynomial $p(x) \in \mathrm{GF}(q)[x]$. Our previous results tell us the following.

Theorem 2-12 shows that all of these roots have orders that divide $(q^m - 1)$.

Theorem 3-4 shows that the roots of $p(x)$ are conjugates with respect to GF(q), and are thus of the form $\{\beta, \beta^q, \beta^{q^2}, \ldots\}$.

Since q is the order of a finite field, it must be a power of a prime. q and its powers are thus relatively prime to $(q^m - 1)$ and all divisors of $(q^m - 1)$.

Theorem 2-11 provides the final conclusion that

$$\mathrm{ord}\,(\beta^{q^k}) = \frac{\mathrm{ord}\,(\beta)}{\mathrm{GCD}(q^k, \mathrm{ord}\,(\beta))} = \mathrm{ord}\,(\beta)$$ **QED**

Example 3-5—The Minimal Polynomials of the Elements in GF(8) with Respect to GF(2)

Recall the construction of GF(8) in Chapter 2 using α, a root of the primitive polynomial $x^3 + x + 1$.

EXPONENTIAL REPRESENTATION		POLYNOMIAL REPRESENTATION
α^0	=	1
α^1	=	α
α^2	=	α^2
α^3	=	$\alpha + 1$
α^4	=	$\alpha^2 + \alpha$
α^5	=	$\alpha^2 + \alpha + 1$
α^6	=	$\alpha^2 + 1$
0	=	0

The eight elements in GF(8) are arranged in conjugacy classes and their minimal polynomials computed as follows:

CONJUGACY CLASS	ASSOCIATED MINIMAL POLYNOMIAL
$\{0\}$	$M_*(x) = (x - 0) = x$
$\{\alpha^0 = 1\}$	$M_0(x) = (x - 1) = x + 1$
$\{\alpha, \alpha^2, \alpha^4\}$	$M_1(x) = (x - \alpha)(x - \alpha^2)(x - \alpha^4) = x^3 + x + 1$
$\{\alpha^3, \alpha^6, \alpha^5\}$	$M_3(x) = (x - \alpha^3)(x - \alpha^6)(x - \alpha^5) = x^3 + x^2 + 1$

(Recall that in fields with characteristic two, addition and subtraction are identical. $(x - 1)$ and $(x + 1)$ are thus the same element in any ring of the form $GF(2^m)[x]$.) The minimal polynomials are indexed by the smallest logarithm to the base α of the roots of the polynomial. ■

Example 3-6—The Minimal Polynomials of the Elements in GF(16) with Respect to GF(4)

The first step in obtaining the minimal polynomials of the elements in GF(16) with respect to GF(4) is the construction of a representation of GF(16). Let α be a root of the primitive polynomial $x^4 + x + 1$. The following exponential and polynomial representations are obtained.

0	$\alpha^7 = \alpha^3 + \alpha + 1$
1	$\alpha^8 = \alpha^2 + 1$
α	$\alpha^9 = \alpha^3 + \alpha$
α^2	$\alpha^{10} = \alpha^2 + \alpha + 1$
α^3	$\alpha^{11} = \alpha^3 + \alpha^2 + \alpha$
$\alpha^4 = \alpha + 1$	$\alpha^{12} = \alpha^3 + \alpha^2 + \alpha + 1$
$\alpha^5 = \alpha^2 + \alpha$	$\alpha^{13} = \alpha^3 + \alpha^2 + 1$
$\alpha^6 = \alpha^3 + \alpha^2$	$\alpha^{14} = \alpha^3 + 1$

The desired polynomials will have coefficients exclusively in GF(4), a subfield of GF(16). We can identify the subfield elements in the list above by noting that elements in GF(4) satisfy $\beta^4 = \beta$. In the above representation the subfield $GF(4) = \{0, 1, \alpha^5, \alpha^{10}\}$.

CONJUGACY CLASS	ASSOCIATED MINIMAL POLYNOMIAL
$\{0\}$	$M*(x) = x$
$\{1\}$	$M_0(x) = x + 1$
$\{\alpha, \alpha^4\}$	$M_1(x) = (x + \alpha)(x + \alpha^4) = x^2 + x + \alpha^5$
$\{\alpha^2, \alpha^8\}$	$M_2(x) = (x + \alpha^2)(x + \alpha^8) = x^2 + x + \alpha^{10}$
$\{\alpha^3, \alpha^{12}\}$	$M_3(x) = (x + \alpha^3)(x + \alpha^{12}) = x^2 + \alpha^{10}x + 1$
$\{\alpha^5\}$	$M_5(x) = x + \alpha^5$
$\{\alpha^6, \alpha^9\}$	$M_6(x) = (x + \alpha^6)(x + \alpha^9) = x^2 + \alpha^5 x + 1$
$\{\alpha^7, \alpha^{13}\}$	$M_7(x) = (x + \alpha^7)(x + \alpha^{13}) = x^2 + \alpha^5 x + \alpha^5$
$\{\alpha^{10}\}$	$M_{10}(x) = x + \alpha^{10}$
$\{\alpha^{11}, \alpha^{14}\}$	$M_{11}(x) = (x + \alpha^{11})(x + \alpha^{14}) = x^2 + \alpha^{10}x + \alpha^{10}$ ■

Appendix D contains a listing of all of the minimal polynomials with respect to GF(2) of the elements in the binary extension fields GF(4) through GF(1024).

3.3 FACTORING $X^n - 1$

In Theorem 2-10 it was noted that any element α in the field $GF(q^m)$ has an order ord(α) that divides $(q^m - 1)$. It follows that the nonzero elements in the field $GF(q^m)$ are all roots of the expression $x^{q^m-1} - 1 = 0$, or equivalently, the elements of $GF(q^m)$ are $(q^m - 1)$st roots of unity. Since the expression $x^{q^m-1} - 1 = 0$ is of degree $(q^m - 1)$, it can only have $(q^m - 1)$ roots. Since there are $(q^m - 1)$ nonzero elements in $GF(q^m)$, they must comprise the complete set of roots for $x^{q^m-1} - 1 = 0$.

Theorem 3-5

> **The set of nonzero elements in $GF(q^m)$ form the complete set of roots of the expression $x^{(q^m-1)} - 1 = 0$.**

The minimal polynomials with respect to $GF(q)$ of the nonzero elements in a given field $GF(q^m)$ thus provide the complete factorization of $[x^{(q^m-1)} - 1]$ into irreducible polynomials in the ring $GF(q)[x]$.

Example 3-7—The Factorization of $x^{q^m-1} - 1$

- The factorization of $x^7 - 1$ in $GF(2)[x]$ (using the results in Example 3-5):

$$x^7 - 1 = (x + 1)(x^3 + x + 1)(x^3 + x^2 + 1)$$

- The factorization of $x^{15} - 1$ in $GF(4)[x]$ (using the results in Example 3-6):

$$x^{15} - 1 = (x + 1)(x^2 + x + \alpha^5)(x^2 + x + \alpha^{10})(x^2 + \alpha^{10}x + 1)(x + \alpha^5)$$
$$\cdot (x^2 + \alpha^5 x + 1)(x^2 + \alpha^5 x + \alpha^5)(x + \alpha^{10})(x^2 + \alpha^{10}x + \alpha^{10}) \quad \blacksquare$$

The use of conjugacy classes in factoring polynomials of the form $x^{(q^m-1)} - 1$ can be extended to the more general form $x^n - 1$ in a simple manner. All of the roots of $x^n - 1$ are nth roots of unity. We need only identify the field where we can find all of these roots, separate the roots into conjugacy classes, and compute the minimal polynomials of the nth roots of unity.

We begin by assuming the existence of an element β with order n in some field $GF(p^m)$. β and all powers of β satisfy $x^n - 1 = 0$ by definition of order. It also follows that the elements $1, \beta, \beta^2, \beta^3, \ldots, \beta^{n-1}$ must be distinct. The n roots of $x^n - 1 = 0$ are thus generated by computing n consecutive powers of β. For this reason elements of order n are often called **primitive *n*th roots of unity**. We must now show that β exists in the first place and determine where to find it.

Theorem 2-12 showed that if n is a divisor of $(p^m - 1)$, then there are $\phi(n)$ elements of order n in $GF(p^m)$. Since $\phi(n)$ is always positive for positive n, it follows that we are guaranteed the existence of a primitive nth root of unity in an extension field of $GF(p)$ so long as we can find a positive integer m such that $n \mid (p^m - 1)$.

Definition 3-6—The Order of q *modulo* n

The **order of q *modulo* n** is the smallest positive integer m such that n divides $(q^m - 1)$.

If m is the order of q *modulo* n, then $GF(q^m)$ is the smallest extension field of $GF(q)$ in which one may find primitive nth roots of unity.

Example 3-8—Primitive *n*th Roots of Unity

- Since $5 | (2^4 - 1)$ but does not divide $(2^3 - 1)$, $(2^2 - 1)$, or $(2 - 1)$, GF(16) is the smallest binary extension field in which one may find primitive fifth roots of unity.
- GF(27) is the smallest extension field of GF(3) in which one may find primitive thirteenth roots of unity.
- GF(125) is the smallest extension field of GF(5) in which one may find primitive thirty-first roots of unity. ■

Once the desired primitive root has been located, the factorization of $x^n - 1$ can be completed by forming the conjugacy classes and computing the associated minimal polynomials.

Example 3-9—Factoring $x^5 - 1$ in GF(4)[x]

We note from the previous example that primitive fifth roots of unity may be found in GF(16). Consider the construction of GF(16) provided in Example 3-6. α is a primitive element in GF(16) and thus has order 15. So α^3 must have order 5 and is thus a primitive fifth root of unity. Let $\beta = \alpha^3$. Note that the minimal polynomials below are indexed with respect to α.

CONJUGACY CLASS	ASSOCIATED MINIMAL POLYNOMIAL
$\{1\}$	$M_0(x) = x + 1$
$\{\beta, \beta^4\} = \{\alpha^3, \alpha^{12}\}$	$M_3(x) = x^2 + \alpha^{10}x + 1$
$\{\beta^2, \beta^3\} = \{\alpha^6, \alpha^9\}$	$M_6(x) = x^2 + \alpha^5 x + 1$

$x^5 + 1 = (x + 1)(x^2 + \alpha^{10}x + 1)(x^2 + \alpha^5 x + 1)$ in the ring GF(4)[x] ■

In many cases it is sufficient to know the number and degree of the factors of $x^n - 1$ without actually computing them. In these cases we need only construct the conjugacy classes containing the nth roots of unity. The number of classes equals the number of factors. The cardinality of the conjugacy class dictates the degree of the associated minimal polynomial.

Example 3-10—Characterizing the Factors of $x^n - 1$ in GF(q)[x]

- Consider the factorization of $x^{25} - 1$ in GF(2)[x].

  Let β be a primitive 25th root of unity. Such elements can be found in $GF(2^{20})$. The 25 roots of $x^{25} - 1$ can be grouped into the following conjugacy classes with respect to GF(2).

  $\{1\}$

  $\{\beta, \beta^2, \beta^4, \beta^8, \beta^{16}, \beta^7, \beta^{14}, \beta^3, \beta^6, \beta^{12}, \beta^{24}, \beta^{23}, \beta^{21}, \beta^{17}, \beta^9, \beta^{18}, \beta^{11}, \beta^{22}, \beta^{19}, \beta^{13}\}$

  $\{\beta^5, \beta^{10}, \beta^{20}, \beta^{15}\}$

$x^{25} - 1$ thus factors into three irreducible binary polynomials: one of degree one, one of degree four, and one of degree twenty. This seemingly obscure bit of knowledge is sufficient to indicate that the selection of binary cyclic codes of length 25 is extremely limited! The knowledge necessary for this dramatic leap will be provided in Chapter 5.

- Consider the factorization of $x^{15} - 1$ in GF(7)[x].

  Let γ be a primitive 15th root of unity, which can be found in GF(7^4). The conjugacy classes with respect to GF(7) are as follows.

$$\{1\}$$
$$\{\gamma, \gamma^7, \gamma^4, \gamma^{13}\}$$
$$\{\gamma^2, \gamma^{14}, \gamma^8, \gamma^{11}\}$$
$$\{\gamma^3, \gamma^6, \gamma^{12}, \gamma^9\}$$
$$\{\gamma^5\}$$
$$\{\gamma^{10}\}$$

$x^{15} - 1$ thus factors into six irreducible polynomials in GF(7)[x]: three of degree one and three of degree four. ■

In Theorem 3-3 it was shown that the number of elements in the conjugacy class with respect to GF(q) of $\alpha \in$ GF(q^m) had to be a divisor of m. This is clearly seen in the factorization information derived in the previous example. For example, the primitive 25th root of unity was found in GF(2^{20}). The conjugacy classes could thus be of cardinality 1, 4, 5, or 20. In the second case the conjugates with respect to GF(7) of an element in GF(7^4) were examined. The conjugacy classes are thus of cardinality 1, 2, or 4.

In many cases the common base of the elements in a conjugacy class is obvious within the context of the discussion, and may be assumed without actually being written down. When the base element is deleted, the resulting partition of the powers of the base element form cyclotomic cosets.

Definition 3-7—Cyclotomic Cosets

The cyclotomic cosets *modulo n* with respect to GF(q) are a partitioning of the integers $\{0, 1, \ldots, n-1\}$ into sets of the form

$$\{a, aq, aq^2, aq^3, \ldots, aq^{d-1}\}$$

The cyclotomic cosets *modulo n* with respect to GF(q) thus contain the exponents of the n distinct powers of a primitive nth root of unity with respect to GF(q), each coset corresponding to a conjugacy class. Cyclotomic cosets provide a convenient means for summarizing the factorization of polynomials of the form $x^n - 1$. Appendix C contains a listing of the cyclotomic cosets *modulo* ($2^m - 1$), for $m = 2$ to 10, with respect to GF(2).

Example 3-11—Cyclotomic Cosets

The conjugacy classes formed by the powers of γ, a primitive fifteenth root of unity, with respect to GF(7), and the corresponding cyclotomic cosets are as follows.

CONJUGACY CLASS		CYCLOTOMIC COSETS
$\{1\}$	$\leftrightarrow$	$\{0\}$
$\{\gamma, \gamma^7, \gamma^4, \gamma^{13}\}$	$\leftrightarrow$	$\{1, 4, 7, 13\}$
$\{\gamma^2, \gamma^{14}, \gamma^8, \gamma^{11}\}$	$\leftrightarrow$	$\{2, 8, 11, 14\}$
$\{\gamma^3, \gamma^6, \gamma^{12}, \gamma^9\}$	$\leftrightarrow$	$\{3, 6, 9, 12\}$
$\{\gamma^5\}$	$\leftrightarrow$	$\{5\}$
$\{\gamma^{10}\}$	$\leftrightarrow$	$\{10\}$

■

3.4 IDEALS IN THE RING GF(q)[x]/(x^n − 1)

In Chapter 2 it was demonstrated that a number of interesting structures could be formed by reducing the ring of integers *modulo m*, m an integer. If m is prime, the resulting structure is a field; otherwise it is a commutative ring with identity. When Euclidean domains of polynomials are reduced *modulo* $f(x)$, $f(x)$ a polynomial, analogous results are obtained.

The ring of polynomials GF(q)[x] *modulo* $f(x)$ is usually denoted GF(q)[x]/$f(x)$. In many cases this ring has a number of very interesting properties; for example, the results in Chapter 2 concerning finite fields of prime power order can be restated in the following form.

Theorem 3-6

If $p(x)$ is an irreducible polynomial in GF(q)[x], then GF(q)[x]/$p(x)$ is a field.

The ring GF(q)[x]/(x^n − 1) also has some interesting properties which will be of great use in the construction of cyclic codes. This ring contains a series of equivalence classes that each contain a single polynomial $f(x)$ of degree less than n or the zero element.

Example 3-12—GF(2)[x]/(x^3 − 1)

The equivalence classes contained in GF(2)[x]/(x^3 + 1) are as follows:

EQUIVALENCE CLASS

$\{0, x^3 + 1, x^4 + x, x^5 + x^2, \ldots\}$

$\{1, x^3, x^4 + x + 1, x^5 + x^2 + 1, \ldots\}$

$\{x, x^3 + x + 1, x^4, x^5 + x^2 + x, \ldots\}$

$\{x + 1, x^3 + x, x^4 + 1, x^5 + x^2 + x + 1, \ldots\}$

$\{x^2, x^3 + x^2 + 1, x^4 + x^2 + x, x^5, \ldots\}$

$\{x^2 + 1, x^3 + x^2, x^4 + x^2 + x + 1, x^5 + 1, \ldots\}$

$\{x^2 + x, x^3 + x^2 + x + 1, x^4 + x^2, x^5 + x, \ldots\}$

$\{x^2 + x + 1, x^3 + x^2 + x, x^4 + x^2 + 1, x^5 + x + 1, \ldots\}$

■

Let $R_n = GF(2)[x]/x^n + 1$. Each equivalence class in R_n is labeled with the smallest-degree element that it contains (a polynomial of degree $< n$) or, if it contains zero, it is labeled with zero.

$GF(q)[x]/(x^n - 1)$ is highly structured with respect to its constituent ideals. Ideals were treated briefly in Problems 9, 10, and 11 at the end of Chapter 2. The definition is repeated here for convenience.

Definition 3-8—Ideals

Let R be a ring. A nonempty subset $I \subseteq R$ is said to be an **ideal** if it satisfies the following:

1. I forms a group under the addition operation in R.
2. $a \cdot r = b \in I$ for all $a \in I$ and for all $r \in R$.

Example 3-13—Ideals

- $\{0\}$ and R form the trivial ideals in any ring R.
- $\{0, x^4 + x^3 + x^2 + x + 1\}$ forms an ideal in R_5.
- All multiples of $x^4 + x + 1$ in R_{15} form a nontrivial ideal in R_{15}. ■

Definition 3-9—Principal Ideals

An ideal I contained in a ring R is said to be **principal** if there exists $g \in I$ such that every element $c \in I$ can be expressed as the product $m \cdot g$ for some $m \in R$.

The element g used to represent all of the elements in a principal ideal is commonly called the **generator element**. The ideal generated by g is denoted $\langle g \rangle$.

In Chapter 5 it will be shown that the ideals in $GF(q)[x]/x^n - 1$ define linear cyclic codes. The following theorem will prove extremely useful in characterizing these codes.

Theorem 3-7—Ideals in $GF(q)[x]/x^n - 1$

Let I be an ideal in $GF(q)[x]/x^n - 1$. The following is true.

1. **There exists a unique monic polynomial $g(x) \in I$ of minimal degree.**
2. **I is principal with generator $g(x)$.**
3. **$g(x)$ divides $x^n - 1$ in $GF(q)[x]$.**

Proof. 1. Since there is at least one nonempty ideal in $GF(q)[x]/x^n - 1$ (the entire ring) and the degree of the polynomials is bounded below by zero, there must be at least one polynomial of minimal degree. These polynomials can be made monic by dividing through by the leading nonzero coefficient.

We now proceed with a proof by contradiction. Let $g(x)$ and $f(x)$ be monic polynomials in I of minimal degree, where $g(x) \neq f(x)$. Since I forms an additive group, $h(x) = g(x) - f(x)$ must also be in I. Since $g(x)$ and $h(x)$ are monic, $h(x)$ must be of lower degree, contradicting the minimality of the degree of $g(x)$.

2. (By contradiction) Let there be $f(x) \in I$ such that $f(x)$ is not a multiple of $g(x)$. Since GF$(q)[x]$ forms a Euclidean domain and $\deg(f(x)) > \deg(g(x))$, $f(x)$ can be expressed as $f(x) = m(x)g(x) + r(x)$, where $\deg(r(x)) < \deg(g(x))$. $m(x)g(x) \in I$, since $g(x) \in I$ and I is an ideal. Since I forms an additive group, $r(x) = f(x) - m(x)g(x)$ is also in I, contradicting the minimality of the degree of $g(x)$.

3. (By contradiction) Suppose $g(x)$ does not divide $x^n - 1$ in GF$(q)[x]$. Since GF$(q)[x]$ is a Euclidean domain, $x^n - 1$ can be expressed as $h(x)g(x) + r(x)$, where $\deg(r(x)) < \deg(g(x))$. Since $r(x) = (x^n - 1) - h(x)g(x)$, it is the additive inverse of $h(x)g(x) \in \langle g(x) \rangle \subseteq \text{GF}(q)[x]/x^n - 1$, contradicting the minimality of the degree of $g(x)$. **QED**

Using the material developed in the last section concerning the factorization of polynomials of the form $x^n - 1$, it is possible to characterize all of the ideals in the ring GF$(q)[x]/x^n - 1$.

Example 3-14—Principal Ideals in GF$(q)[x]/x^n - 1$

- Using information from Example 3-7, one sees that GF$(2)[x]/(x^7 - 1)$ contains the ideals $\langle x + 1\rangle$, $\langle x^3 + x + 1\rangle$, $\langle x^3 + x^2 + 1\rangle$, $\langle (x + 1)(x^3 + x^2 + 1)\rangle$, $\langle (x^3 + x + 1)(x^3 + x^2 + 1)\rangle$, $\langle (x + 1)(x^3 + x + 1)\rangle$, and the two trivial ideals.
- Using information from Example 3-9, one sees that GF$(4)[x]/(x^5 - 1)$ contains the ideals $\langle x + 1\rangle$, $\langle x^2 + \alpha^{10}x + 1\rangle$, $\langle x^2 + \alpha^5 x + 1\rangle$, $\langle (x + 1)(x^2 + \alpha^{10}x + 1)\rangle$, $\langle (x + 1)(x^2 + \alpha^5 x + 1)\rangle$, $\langle (x^2 + \alpha^{10}x + 1)(x^2 + \alpha^5 x + 1)\rangle$ and the two trivial ideals.
- Using information from Example 3-10, one sees that GF$(2)[x]/(x^{25} - 1)$ contains eight ideals, two of which are trivial. ■

SUGGESTIONS FOR FURTHER READING

The scope of this text precludes any but the most superficial treatment of the elegant theory of finite fields. The reader who wishes to pursue a more detailed knowledge of finite fields is referred to the following.

R. J. McEliece, *Finite Fields for Computer Scientists and Engineers*, Boston: Kluwer Academic Publishers, 1987.

T. W. Hungerford, *Algebra*, New York: Springer-Verlag, 1974.

R. Lidl and H. Niederreiter, *Finite Fields*, Reading, MA: Addison-Wesley, 1983.

PROBLEMS

Euclidean domains

1. Let D be a Euclidean domain. Suppose that for $a, b, c \in D$, a divides bc, but a and b are relatively prime. Show that a divides c.
2. Show that all finite Euclidean domains are fields.

Polynomial division and Euclid's algorithm

3. Express the following pairs $(f(x), g(x))$ of binary polynomials in the form $f(x) = m(x)g(x) + r(x)$.
   (a) $(x^8 + 1, x^4 + 1)$
   (b) $(x^8 + 1, x^4 + x)$
   (c) $(x + 1, x^{10} + x^9 + x^5 + x^4)$
   (d) $(x^5 + x^3 + 1, x^2 + x + 1)$
   (e) $(x^4 + x^2 + x + 1, x^2 + x + 1)$
   (f) $(x^{10} + 1, x + 1)$
   (g) $(x^5 + 1, x^4 + x^3 + x^2 + x + 1)$
   (h) $(x^{10} + x^9 + x + 1, x^3 + x^2 + 1)$
4. Find the greatest common divisor (GCD) for the following pairs of binary polynomials.
   (a) $(x^6 + x^5 + x + 1, x^4 + x^3)$
   (b) $(x^8 + 1, x^4 + x)$
   (c) $(x^{10} + x^9 + x^5 + x^4, x^{18} + x^8)$
   (d) $(x^{23} + x^{16} + 1, x^{19} + 1)$
   (e) $(x^4 + x^2 + 1, x^2 + x + 1)$
   (f) $(x^{10} + 1, x^8 + x^6 + x^4 + x^2 + 1)$
   (g) $(x^5 + 1, x^4 + x^3 + x^2 + x + 1)$
   (h) $(x^{10} + x^9 + x + 1, x^3 + x^2 + x + 1)$
5. Find the greatest common divisor for the following sets of binary polynomials.
   (a) $(x^{23} + x^{22} + x + 1, x^{10} + x^9 + x^7 + x^6 + x^3 + x^2, x^{10} + x^8 + x^2 + 1)$
   (b) $(x^{10} + x^9 + x + 1, x^{18} + 1, x^4 + x^3 + x + 1)$
6. Find the greatest common divisor for the following sets of binary polynomials and express it as a linear combination of the given polynomials.
   (a) $(x^6 + x^5 + x + 1, x^4 + x^3)$
   (b) $(x^4 + x^2 + 1, x^2 + x + 1)$
   (c) $(x^{10} + x^9 + x^5 + x^4, x^{18} + x^8)$
   (d) $(x^5 + 1, x^4 + x^3 + x^2 + 1)$
   (e) $(x^{23} + x^{22} + x + 1, x^{10} + x^9 + x^7 + x^6 + x^3 + x^2, x^{10} + x^8 + x^2 + 1)$

Conjugacy classes and minimal polynomials

7. Let α be an element of order 1023 in GF(1024). Find the conjugates of α with respect to GF(2), GF(4) and GF(32).
8. Let α be an element of order 65535 in GF(65536). Find the conjugates of α with respect to GF(2), GF(4), GF(16), and GF(256).
9. Construct a representation for GF(9) and determine the minimal polynomials with respect to GF(3) of all of the elements in GF(9).
10. Construct a representation for GF(32) and determine the minimal polynomials with respect to GF(2) of all of the elements in GF(32).
11. Determine the degree of the minimal polynomial with respect to GF(2) for field elements with the following orders.
    (a) 3
    (b) 5
    (c) 7
    (d) 9
    (e) 13
    (f) 15
    (g) 21
    (h) 1023

Factoring $x^n - 1$

12. Express the following as products of binary irreducible polynomials.
    (a) $x^3 + 1$
    (b) $x^5 + 1$
    (c) $x^7 + 1$
    (d) $x^{15} + 1$
    (e) $x^{21} + 1$
    (f) $x^{31} + 1$

13. Determine the number of binary irreducible polynomials in the factorization of the following.

(a) $x^9 + 1$ **(b)** $x^{11} + 1$
(c) $x^{13} + 1$ **(d)** $x^{17} + 1$
(e) $x^{19} + 1$ **(f)** $x^{29} + 1$
(g) $x^{127} + 1$ **(h)** $x^{341} + 1$

14. Determine the number of irreducible polynomials in the factorization of the following in GF(3)[x].

(a) $x^9 - 1$ **(b)** $x^{11} - 1$
(c) $x^{13} - 1$ **(d)** $x^{17} - 1$
(e) $x^{19} - 1$ **(f)** $x^{29} - 1$
(g) $x^{127} - 1$ **(h)** $x^{341} - 1$

Ideals in GF(2)[x]/(x^n − 1)

15. List all of the distinct ideals contained in GF(2)[x]/($x^{15} - 1$) and GF(2)[x]/($x^{31} - 1$) by their generator elements.

16. Let $\langle f(x), h(x) \rangle$ be the ideal $I \subseteq R_n$ formed by all linear combinations of the form $a(x)f(x) + b(x)h(x)$, where $a(x), b(x) \in R_n$. Since I is principal, it has a single generator $g(x)$. Express $g(x)$ in terms of $f(x)$ and $h(x)$ and support your conclusion.

Linear Block Codes

In this chapter we leave the rarefied atmosphere of abstract algebra and turn toward the practical world of error control systems. We begin by considering block coding as a general concept. Block codes introduce controlled amounts of redundancy into a transmitted data stream, providing the receiver with the ability to detect and possibly correct errors caused by noise on the communication channel. In the first section of this chapter, various general properties of block codes are discussed. The concept of minimum distance is developed and used to characterize a code's ability to detect and correct errors.

Linear block codes, the subject of the second section, are the most easily implemented and therefore most widely used of the block codes. By definition they form vector subspaces over finite fields. All of the material in the previous chapters can thus be brought to bear in an examination of the properties of linear block codes. One of the results emerging from this examination is the representation of linear codes using generator and parity check matrices.

In the third section the standard array and syndrome table decoding algorithms for linear block codes are discussed. These decoding algorithms become impracticable as the size of the code increases, but they are useful for small codes and serve to highlight some interesting properties of linear codes and their ability to detect/correct errors. The chapter continues with a quick look at weight distributions and concludes with an introduction to Hamming codes. Hamming codes provide the first step in complexity beyond the simple parity check codes. They thus provide a good subject for the text's first development and analysis of an error control code.

4.1 BLOCK ERROR CONTROL CODES

A block error control code **C** consists of a set of M code words $\{\mathbf{c}_0, \mathbf{c}_1, \mathbf{c}_2, \ldots, \mathbf{c}_{M-1}\}$. Each code word is of the form $\mathbf{c} = (c_0, c_1, \ldots, c_{n-1})$; if the individual coordinates take on values from the Galois field GF(q), then the code **C** is said to be ***q*-ary**. The encoding process consists of breaking up the data stream into blocks, and mapping these blocks onto code words in **C** (see Figure 4-1). This mapping is usually one-to-one, ensuring that the encoding process can be reversed at the receiver and the original data block recovered.

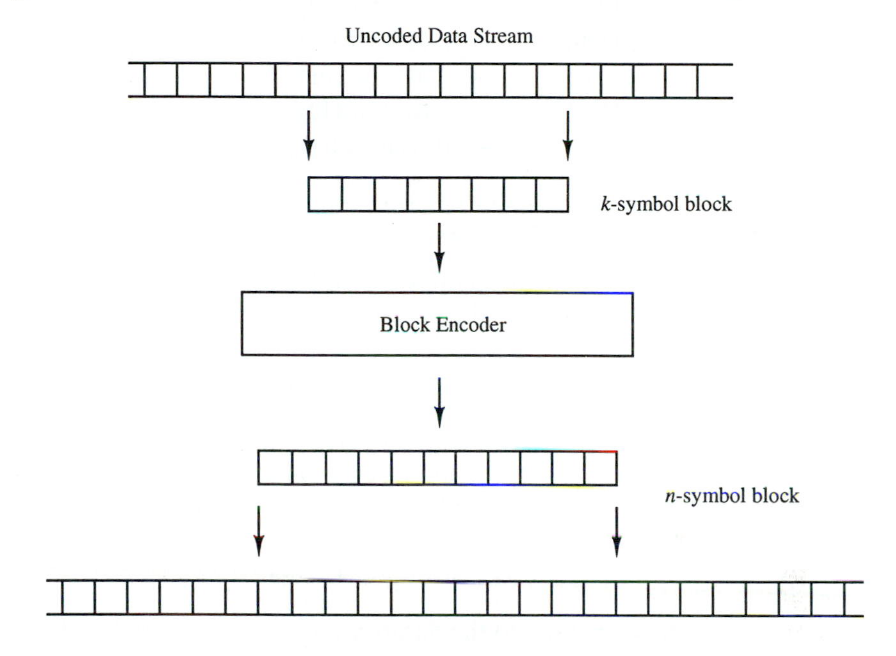

Figure 4-1. Block encoding

If the symbols in the data stream can take on any value in GF(q), then the collection of all possible k-tuples $\mathbf{m} = (m_0, m_1, \ldots, m_{k-1})$ forms a vector space over GF(q). A simple counting exercise shows that there are q^k possible k-symbol data vectors. If $M = q^k$, then the encoder breaks the data stream up into k-symbol blocks. If M is not of this form, then the encoder must work with messages of varying length, as shown in Example 4-1.

Example 4-1—The Binary Hadamard Code $\mathcal{A}_{12}$

The binary Hadamard code $\mathcal{A}_{12}$ consists of 12 code words of length 11 (Hadamard codes are discussed in Chapter 6), so the information blocks must be of varying length. For example, we can use the following mapping for a binary message stream.

MESSAGE BLOCKS		CODE WORDS
(000)	→	(00000000000)
(001)	→	(10100011101)
(010)	→	(11010001110)
(011)	→	(01101000111)
(100)	→	(10110100011)
(101)	→	(11011010001)
(1100)	→	(11101101000)
(1101)	→	(01110110100)
(11100)	→	(00111011010)
(11101)	→	(00011101101)
(11110)	→	(10001110110)
(11111)	→	(01000111011)

The message blocks are selected so that any binary message stream can be broken up into a sequence of blocks without creating ambiguity at the decoder output. ■

Though there are many good q-ary codes for which M is not of the form q^k, as is evident in the example, they are harder to implement than codes for which $M = q^k$.

The collection of all possible n-tuples over GF(q) forms a vector space over GF(q) containing q^n vectors. There are thus $(q^n - M)$ n-symbol patterns which are not associated with data blocks and are thus not valid code words. The code **C** is said to contain **redundancy**, for the number of valid code words is smaller than the number of possible n-symbol blocks. The redundancy r is expressed in logarithmic form.

$$r = n - \log_q M$$

If $M = q^k$, then r simplifies to the difference $r = (n - k)$ between the length of the code words and the length of the data blocks. The redundancy is also frequently expressed in terms of the **code rate**.

Definition 4-1—The Rate of a Block Code

Let M be the number of code words, each of length n, in a code **C**. The rate of **C** is

$$R = \frac{\log_q M}{n}$$

The code rate simplifies to $R = k/n$ for those cases in which $M = q^k$.

The corruption of a code word by channel noise is often modeled as an additive process, as shown in Figure 4-2. Figure 4-2 is a **baseband model** in that it suppresses all of the modulation and demodulation functions, though they are, of course, present in most real communication systems. The modulation format, transmitter

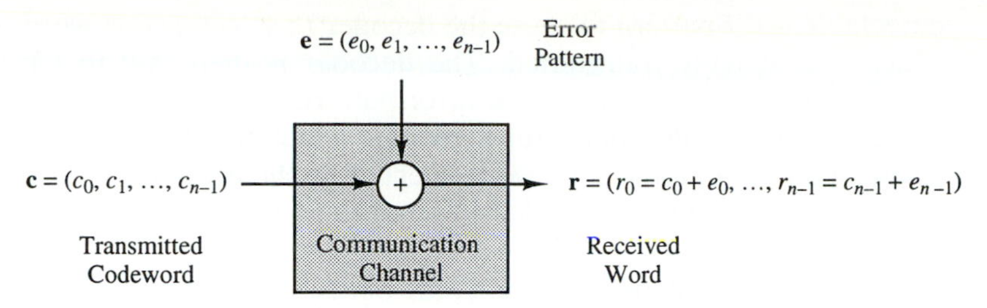

Figure 4-2. Baseband model for an additive noisy channel

power level, and the amount of noise on the channel determine the likelihood that each of the q^n possible error patterns **e** occurs. This model is particularly useful with binary channels, in which the error pattern addition is performed *modulo* 2.

The error control decoder in the receiver performs several functions. In almost all cases, the decoder's first task is to examine the received word and to determine whether or not it is a code word. If the received word is found to be invalid, then it is assumed that the channel has caused one or more symbol errors. The determination of whether errors are present in a received word is **error detection**.

An error pattern is **undetectable** if and only if it causes the received word to be a valid code word other than that which was transmitted. Given a transmitted code word **c**, there are $(M - 1)$ code words other than **c** that may arrive at the receiver and thus $(M - 1)$ undetectable error patterns.

The decoder can react to a detected error with one of the following three responses:

- Request a retransmission of the word.
- Tag the word as being incorrect and pass it along to the data sink.
- Attempt to correct the errors in the received word.

Retransmission requests are used in a very reliable set of error control strategies collectively referred to as **automatic repeat request (ARQ)** protocols. These are examined in Chapter 15. For now we shall note in passing that this approach is very important for data transfer applications in which a premium is placed on data reliability.

The second option listed above is commonly referred to as **muting**. It is typical of applications in which delay constraints do not allow for the retransmission of the transmitted word and it is deemed better to set the received word to a predetermined muted value than to attempt to correct the errors (e.g., voice communication and digital audio).

The final option is referred to as **forward error correction (FEC)**. In FEC systems the arithmetic or algebraic structure of the code is used to determine which of the valid code words is most likely to have been sent, given the erroneous received word.

If the decoder is performing forward error correction, it is possible for a detectable error pattern to cause the decoder to select a code word other than that which was actually transmitted. The decoder is then said to have committed a **decoder error**. Though the initial error pattern that causes the decoder error may be detectable, the decoder error is itself undetectable. In FEC systems the receiver always assumes that the code word selected by the decoder is correct. The number of error patterns that cause decoder errors is greater than the number of undetectable error patterns.

If the code is chosen carefully, the nonzero error patterns that are most likely to occur will cause the received word to be an invalid word. On many communication channels the error patterns with a smaller number of nonzero coordinates are more likely to occur than error patterns with a larger number of nonzero coordinates.

Definition 4-2—Code Word and Error Pattern Weight

The **weight** of a code word or error pattern is the number of nonzero coordinates in the code word or error pattern.

The weight of the code word $\mathbf{c}$ is commonly written $w(\mathbf{c})$.

Example 4-2—Weights

- $w(1,0,0,1,1,0,1,1,1,1) = 7$
- $w(\alpha^3,\alpha^{29},0,0,1,\alpha^6,0) = 4$ ■

Consider a pair of n-symbol blocks $\mathbf{v} = (v_0, v_1, \ldots, v_{n-1})$ and $\mathbf{w} = (w_0, w_1, \ldots, w_{n-1})$. There are a variety of ways in which we can discuss the "distance" $d(\mathbf{v}, \mathbf{w})$ between these two blocks. In the familiar Euclidean geometry one encounters the following definition for distance.

$$d_{\text{Euclidean}}(\mathbf{v}, \mathbf{w}) = \sqrt{(v_0 - w_0)^2 + (v_1 - w_1)^2 + \cdots + (v_{n-1} - w_{n-1})^2} \tag{4-1}$$

Euclidean distance will be used extensively in the analysis of convolutional and trellis codes in Chapters 11 through 14. In block coding, however, we are more frequently concerned with **Hamming distance**.

Definition 4-3—Hamming Distance

The Hamming distance between two blocks $\mathbf{v}$ and $\mathbf{w}$ is the number of coordinates in which the two blocks differ.

$$d_{\text{Hamming}}(\mathbf{v}, \mathbf{w}) = d(\mathbf{v}, \mathbf{w}) = |\{i \mid v_i \neq w_i, i = 0, 1, \ldots, n-1\}| \tag{4-2}$$

Hamming distance allows for a useful characterization of the error detection and error correction capabilities of a block code as a function of the code's **minimum distance**.

Definition 4-4—The Minimum Distance of a Block Code

The **minimum distance** of a block code $\mathbf{C}$ is the minimum Hamming distance between all distinct pairs of code words in $\mathbf{C}$.

Recall that the only undetectable error patterns are those that cause the transmitted code word to look like another code word. Let $d_{\min}$ be the minimum distance of the code in use. A transmitted code word is thus guaranteed to differ in at least $d_{\min}$ coordinates from any other code word. For an error pattern to be undetectable, it must change the symbol values in the transmitted code word in at least $d_{\min}$ coordinates.

A code with minimum distance $d_{\min}$ can thus detect all error patterns of weight less than or equal to ($d_{\min} - 1$).

A given code can detect a large number of error patterns of weight $w \geq d_{\min}$. The above statement simply provides a limit on the weight for which we can detect *all* error patterns. The above limit does, however, provide a convenient means for computing bounds on the probability of undetected error for specific systems in conjunction with certain communication channels, as will be seen later in this chapter and in Chapter 10.

In FEC systems the goal is the minimization of the probability of decoder error given a received word $\mathbf{r}$. Suppose that the code words $\{\mathbf{c}_i\}$ are transmitted according to some probability mass function $p_C(\mathbf{c}_i)$, while the received words $\{\mathbf{r}_i\}$ arrive at the decoder according to the probability mass function $p_r(\mathbf{r}_i)$. The detailed characteristics of the communication system and channel (e.g., transmitter power level, channel noise level, modulation format) can be used to derive the probability $p(\mathbf{c}\,|\,\mathbf{r})$ that code word $\mathbf{c}$ is transmitted conditioned on the receipt of the word $\mathbf{r}$. The **maximum a posteriori decoder** identifies the code word $\mathbf{c}_i$ that maximizes $p(\mathbf{c} = \mathbf{c}_i\,|\,\mathbf{r})$ [Woz]. The **maximum likelihood decoder**, on the other hand, identifies the code word $\mathbf{c}_i$ that maximizes $p(\mathbf{r}\,|\,\mathbf{c} = \mathbf{c}_i)$. $p(\mathbf{c}\,|\,\mathbf{r})$ can be related to $p(\mathbf{r}\,|\,\mathbf{c})$ by Bayes' rule [Pap].

$$p(\mathbf{c}\,|\,\mathbf{r}) = \frac{p_C(\mathbf{c})p(\mathbf{r}\,|\,\mathbf{c})}{p_R(\mathbf{r})} \tag{4-3}$$

If the code words are not equally likely to be transmitted, then the maximum likelihood decoder does not provide the minimum probability of decoder error. In many cases, however, the probability of transmission of the various code words is not known or can be assumed to be uniformly distributed (in which case the two decoders are identical). The maximum likelihood decoder is the focus of the following discussion.

The conditional probability $p(\mathbf{r}\,|\,\mathbf{c})$ is equal to the probability of occurrence of the error pattern $\mathbf{e} = (\mathbf{r} - \mathbf{c})$ given that $\mathbf{c}$ has been transmitted. If it is assumed that lower-weight error patterns are more likely to occur than higher-weight error patterns, the code word $\mathbf{c}_i$ that maximizes $p(\mathbf{r}\,|\,\mathbf{c}_i)$ is the code word that minimizes $d(\mathbf{r}, \mathbf{c}_i) = w(\mathbf{r} - \mathbf{c})$. The maximum likelihood transmitted code word is thus the code word that is closest in Hamming distance to the received word $\mathbf{r}$. A decoder operating on this principle commits a **decoder error** whenever the received word is closer to an incorrect code word than to the correct code word. By definition, incorrect code words are at least a distance $d_{\min}$ away from the transmitted code word. Decoder errors are thus a possibility only if the weight of the error pattern induced by the channel is greater than or equal to $d_{min}/2$. This can be stated in an equivalent fashion as follows.

A code with minimum distance d_{min} can correct all error patterns of weight less than or equal to $\lfloor (d_{min} - 1)/2 \rfloor$.[1]

$\lfloor (d_{min} - 1)/2 \rfloor$ is the upper bound on the weight for which one can correct *all* error patterns. It is sometimes possible to correct ($\lfloor (d_{min} - 1)/2 \rfloor + 1$) or more errors in certain received blocks.

Definition 4-5—Complete Decoders

A **complete error correcting decoder** is a decoder that, given a received word $\mathbf{r}$, selects as the transmitted code word $\mathbf{c}$ a code word that minimizes $d(\mathbf{r}, \mathbf{c})$.

The complete decoder is thus, for most channels, the maximum likelihood decoder. It selects the code word that is closest to the received word regardless of the distance between the two. It is possible that for a given received block $\mathbf{r}$ there is more than one code word $\mathbf{c}$ that minimizes $d(\mathbf{r}, \mathbf{c})$. In this case the complete decoder chooses randomly from among the closest code words (assuming that there is no additional information to help distinguish between them).

For many codes there has not yet been found an efficient complete decoding algorithm (e.g., Reed-Solomon codes). If the error correction requirements are relaxed, however, efficient algorithms are more readily available.

Definition 4-6—Bounded-Distance Decoders

Given a receiver word $\mathbf{r}$, a ***t*-error correcting bounded-distance decoder** selects the code word $\mathbf{c}$ that minimizes $d(\mathbf{r}, \mathbf{c})$ if and only if there exists $\mathbf{c}$ such that $d(\mathbf{r}, \mathbf{c}) \leq t$. If no such $\mathbf{c}$ exists, then a **decoder failure** is declared.

Clearly the error correction capability t for a bounded distance decoder must satisfy $t \leq \lfloor (d_{min} - 1)/2 \rfloor$.

There is an important distinction to be made between decoder errors and decoder failures in a bounded-distance decoder: The latter are detectable while the former are not. Decoder failures can be resolved through retransmission requests.

Example 4-3—Binary Repetition Codes

The length-four binary repetition code has two code words: (1111) and (0000). Since it has a minimum distance of 4, it is guaranteed to correct all error patterns of weight one or zero. The complete decoder and the single error correcting bounded-distance decoder operate as follows:

[1] The function $f(x) = \lfloor x \rfloor$ is commonly called the "floor function." It is the largest integer less than or equal to x. Its counterpart, the "ceiling function" $g(x) = \lceil x \rceil$, is the smallest integer greater than or equal to x.

Received word	Selected code word	Received word	Selected code word
(0000)	(0000)	(1000)	(0000)
(0001)	(0000)	(1001)	(0000) or (1111)*
(0010)	(0000)	(1010)	(0000) or (1111)*
(0011)	(0000) or (1111)*	(1011)	(1111)
(0100)	(0000)	(1100)	(0000) or (1111)*
(0101)	(0000) or (1111)*	(1101)	(1111)
(0110)	(0000) or (1111)*	(1110)	(1111)
(0111)	(1111)	(1111)	(1111)

* Decoder failure declared by the bounded-distance decoder.

The length-three binary repetition code also has two code words: (000) and (111). Since it has a minimum distance of 3, it can correct all error patterns of weight one or zero. In this case the bounded-distance and the complete decoder are identical.

Received word	Selected code word	Received word	Selected code word
(000)	(000)	(100)	(000)
(001)	(000)	(101)	(111)
(010)	(000)	(110)	(111)
(011)	(111)	(111)	(111)

■

The geometric interpretation of error correction can offer some useful insights. A **Hamming sphere** of radius t contains all possible received vectors that are at a Hamming distance $\leq t$ from a code word. $V_q(n,t)$ is the volume, or number of vectors in a Hamming sphere with radius t in an n-dimensional vector space over GF(q). A quick exercise in combinatorics yields the following result.

$$V_q(n,t) = \sum_{j=0}^{t} \binom{n}{j} (q-1)^j \tag{4-4}$$

If a received vector **r** falls within a Hamming sphere, then the decoder selects the code word **c** at the center of the sphere as that which was transmitted. If the received vector falls into the region between the various decoding spheres, then a decoder failure is declared (bounded-distance decoding) or one of the nearest code words is selected (complete decoding).

There is a very interesting and unsolved problem that naturally follows from the geometric approach. Given a fixed code word length n and minimum distance $d_{\min}$, what is the maximum number of code words $A_q(n, d_{\min})$ that can be selected from within an n-dimensional vector space over GF(q)? This problem can be restated as a question of redundancy r: What is the minimum redundancy required for a t-error correcting q-ary code of length n? The following results bound the answer to this question.

Theorem 4-1—The Hamming Bound

A t-error correcting q-ary code of length n must have redundancy r that satisfies

$$r \geq \log_q V_q(n, t)$$

Proof. Each of the M code words in **C** is associated with a Hamming sphere of radius t. The spheres do not overlap, and the total volume of the spheres associated with **C** cannot exceed the total number of vectors in the space of all possible received words. It follows that $MV_q(n,t) \leq q^n$, and thus $q^n/M \geq V_q(n,t)$. This in turn implies that $r = n - \log_q M \geq \log_q V_q(n,t)$. **QED**

Theorem 4-2—The Gilbert Bound [Gil]

There exists a t-error correcting q-ary code of length n and redundancy r that satisfies

$$r \leq \log_q V_q(n, 2t)$$

Proof. A t-error correcting code **C** is to be constructed by randomly selecting vectors one at a time from the vector space of n-tuples over GF(q). When a code word is selected, all surrounding vectors that are Hamming distance less than or equal to $2t$ away from the selected code word are deleted from further consideration. This ensures that the resulting code has minimum distance $(2t + 1)$ and is thus t-error correcting. The selection of each code word results in the deletion of at most $V_q(n, 2t)$ vectors from the space of n-tuples. It follows that at least M code words are selected, where

$$M = \left\lceil \frac{q^n}{V_q(n, 2t)} \right\rceil \geq \frac{q^n}{V_q(n, 2t)}$$

The result follows. **QED**

Figure 4-3 compares the two bounds presented in Theorems 4-1 and 4-2 for the case of binary single-error-correcting block codes. Code word length is allowed to vary between 10 and 1000. Clearly the bounds are not very tight!

There are a few codes that satisfy the Hamming bound with equality. Such codes are called **perfect codes**.

Definition 4-7—Perfect Codes

A block code is **perfect** if it satisfies the Hamming bound (Theorem 4-1) with equality.

Perfection often proves to be a limiting constraint. For example, in the following theorem it is shown that the number of code words in a perfect q-ary code must be of the form q^k.

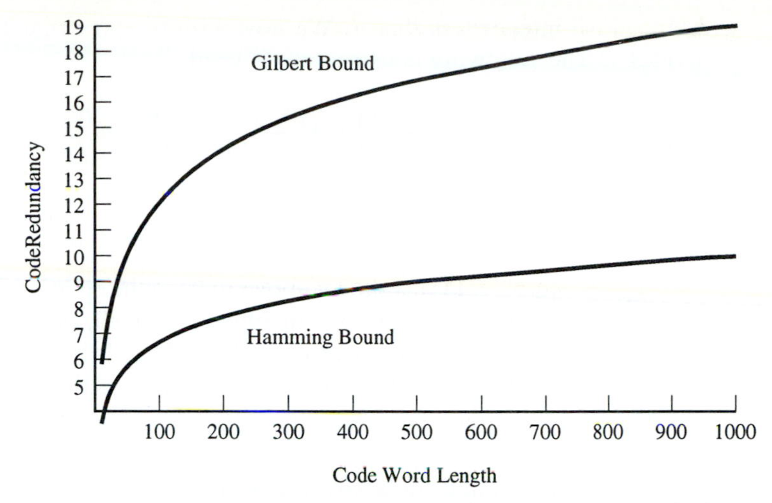

Figure 4-3. A comparison of the Hamming and Gilbert bounds on required redundancy for binary single-error-correcting codes

Theorem 4-3

The number of code words in a perfect q-ary code must be of the form $M = q^k$ for some positive integer k.

Proof [Mac]. By definition the code must satisfy the Hamming bound with equality. By Theorem 2-14, the order q of the code symbol field must be of the form p^r, where p is prime and r is a positive integer. We then have

$$M\sum_{i=0}^{t}\binom{n}{i}(q-1)^i = q^n = p^{nr}$$

It follows that $M = p^j$ for some integer j, and

$$\sum_{i=0}^{t}\binom{n}{i}(q-1)^i = \sum_{i=0}^{t}\binom{n}{i}(p^r-1)^i = p^{nr-j}$$

We now show that $r\,|\,j$. Subtracting one from both sides, we get

$$\begin{aligned}\sum_{i=0}^{t}\binom{n}{i}(p^r-1)^i - 1 &= \sum_{i=1}^{t}\binom{n}{i}(p^r-1)^i \\ &= (p^r-1)\sum_{i=1}^{t}\binom{n}{i}(p^r-1)^{i-1} \\ &= p^{nr-j}-1\end{aligned}$$

and thus $(p^r - 1)$ divides $(p^{nr-j} - 1)$. Let $s = (nr - j)$. Since r and s are integers, we can write $s = qr + t$, where q is a positive integer and t is a nonnegative integer less than r. We now wish to show that $t = 0$, and that r divides s. The following construction is used.

$$\left(\frac{p^s - 1}{p^r - 1}\right) = \left(\frac{p^{qr+t} - p^t + p^t - 1}{p^r - 1}\right)$$

$$= p^t\left(\frac{p^{qr} - 1}{p^r - 1}\right) + \left(\frac{p^t - 1}{p^r - 1}\right)$$

$(p^s - 1)/(p^r - 1)$ was shown above to be an integer, and it can be shown that $(p^{qr} - 1)/(p^r - 1)$ is always an integer as well. Since $t < r$, $(p^t - 1)/(p^r - 1)$ must be either a fraction or zero. It follows that $t = 0$ and $r|s$. Since $s = (nr - j)$, r must also divide j. M is thus of the form $q^{j/r} = q^k$, k an integer. **QED**

By the Hamming bound and the above theorem, a perfect t-error-correcting code with q^k code words of length n must have parameters $\{q, n, k, t\}$ that satisfy the following relation.

$$\sum_{j=0}^{t} \binom{n}{j} (q - 1)^j = q^{n-k} \tag{4-5}$$

The solutions to Eq. (4-5) for which codes exist have all been found. They can be grouped into the following five classes.

- $\{q, n, k = n, t = 0\}$. This is a trivial case. The "codes" in this category consist of the q^n n-tuples over GF(q). Since these codes have no redundancy, they cannot detect or correct any errors.
- $\{q = 2, n \text{ odd}, k = 1, t = (n - 1)/2\}$. The codes corresponding to this set of solutions are the odd-length binary repetition codes. These codes contain two code words: an all-zero word and an all-one word.
- $\{q = 2, n = 2^m - 1, k = 2^m - m - 1, t = 1\}$, m a positive integer. The only linear codes corresponding to this set of solutions are the Hamming codes, which are discussed at the end of this chapter. There are several nonlinear codes with these parameters.
- $\{q = 2, n = 23, k = 12, t = 3\}$. The only code, linear or nonlinear, that has these parameters is the binary Golay code $\mathcal{G}_{23}$.
- $\{q = 3, n = 11, k = 6, t = 2\}$. The only code, linear or nonlinear, that has these parameters is the ternary Golay code $\mathcal{G}_{11}$. Both $\mathcal{G}_{11}$ and $\mathcal{G}_{23}$ are discussed in Chapter 6.

There are solutions to Eq. (4-5) other than the above. For example, Golay noted that $\{q = 2, n = 90, k = 78, t = 2\}$ is a valid solution, but he showed that a corresponding code cannot exist [Gol2]. In 1973 Tietäväinen showed that no perfect codes exist other than those with parameters from the above list. The proof is rather long, so we shall simply state the result.

Theorem 4-4

Any nontrivial perfect code must have the same length n, symbol alphabet GF(q), and cardinality $M = q^k$ as a Hamming, Golay, or repetition code.

Proof. See [Tie] or [Mac].

There is some danger in taking the term "perfect" too literally. This does not mean that perfect codes are the best of all possible error control codes. It merely means that they define nonoverlapping Hamming spheres that "perfectly" fill an n-dimensional vector space. The sphere packing problem and the error control problem are not entirely equivalent (see [Con1]). Many codes that are highly imperfect (e.g., Reed-Solomon codes) provide for substantially more powerful error control systems than the perfect codes.

We now consider the performance of error control codes as their length is allowed to increase without bound. Suppose that each bit transmitted over a noisy communication channel is received in error with probability p. If the noise process is independent from bit to bit, then we expect an average of np out of every n bits received to be in error. If the length n of the error control code for this channel is allowed to increase, it follows that the minimum distance $d_{\min}$ of the code must increase accordingly if reliability is to be maintained. Let δ be the ratio of the minimum distance to the length.

$$\delta = \frac{d_{\min}}{n}$$

Let $A_q(n, d_{\min})$ be the maximum possible number of code words for a q-ary code of length n and minimum distance $d_{\min}$. The asymptotic rate for a code with maximal cardinality is defined to be

$$a(\delta) = \limsup_{n \to \infty} \left[\frac{\log_q A_q(n, \lfloor \delta n \rfloor)}{n} \right]$$

$a(\delta)$ is the maximum possible code rate that a code can have if it is to maintain a minimum distance/length ratio δ as its length increases without bound.

In 1957 Varsharmov extended the Gilbert bound to obtain a lower bound on $a(\delta)$. We begin the development of the Gilbert-Varsharmov bound with a definition.

Definition 4-8—The Entropy Function

$H_q(x) = x \log_q (q - 1) - x \log_q x - (1 - x) \log_q (1 - x)$,
for $0 < x \leq (q - 1)/q$

The entropy function plays an extremely important role as a metric in information theory [Sha1]. It is used here in a somewhat different, though related context.

Lemma 4-5

Let $0 \le \delta \le (q-1)/q$. Then

$$\lim_{n\to\infty}\left[\frac{\log_q V_q(n, \lfloor \delta n \rfloor)}{n}\right] = H_q(\delta)$$

Proof [VanL]. Let $m = \lfloor \delta n \rfloor$. From Eq. (4-4) we have

$$V_q(n,m) = \sum_{j=0}^{m} \binom{n}{j}(q-1)^j$$

The last term in this summation is the largest when $\delta \le 1/2$. As $n \to \infty$, $m \to \delta n + O(1)$.

$$\binom{n}{m}(q-1)^m \le V_q(n,m) \le (1+m)\binom{n}{m}(q-1)^m$$

We now take the logarithm of all three terms and divide by n.

$$\begin{aligned}\frac{1}{n}\log_q\binom{n}{m} + \frac{m}{n}\log_q(q-1) &\\ &\le \frac{1}{n}\log_q V_q(n,m)\\ &\le \frac{1}{n}\log_q(1+m) + \frac{1}{n}\log_q\binom{n}{m} + \frac{m}{n}\log_q(q-1)\end{aligned}$$

This implies that

$$\begin{aligned}\frac{1}{n}\log_q\binom{n}{m} + \delta\log_q(q-1) &\le \frac{1}{n}\log_q V_q(n,m)\\ &\le \frac{1}{n}\log_q(1+m) + \frac{1}{n}\log_q\binom{n}{m} + \delta\log_q(q-1)\end{aligned}$$

As $n \to \infty$, $1/n \log_q(1+m) \to 1/n \log_q n \to 0$. It follows that

$$\lim_{n\to\infty}\frac{1}{n}\log_q V_q(n,m) = \frac{1}{n}\log_q\binom{n}{m} + \delta\log_q(q-1)$$

Stirling's formula provides the following useful identity.

$$\lim_{n\to\infty}\log(n!) = n\log n - n + O(\log n)$$

Applying this to the binomial coefficient $\binom{n}{m}$ we obtain

$$\begin{aligned}\lim_{n\to\infty}\frac{1}{n}\log_q V_q(n,m) &= \frac{1}{n}\{n\log_q n - m\log_q m - (n-m)\log_q(n-m) + o(n)\}\\ &\quad + \delta\log_q(q-1)\\ &= \log_q n - \delta\log_q m - (1-\delta)\log_q(n-m) + \delta\log_q(q-1)\\ &= \delta\log_q n - \delta\log_q m - (1-\delta)\log_q(1-\delta) + \delta\log_q(q-1)\\ &= -\delta\log_q\delta - (1-\delta)\log_q(1-\delta) + \delta\log_q(q-1)\\ &= H_q(\delta)\end{aligned}$$

QED

The Gilbert-Varsharmov bound follows from the Lemma.

Theorem 4-6—The Gilbert-Varsharmov Bound

If $0 \le \delta \le (q-1)/q$, then $a(\delta) \ge 1 - H_q(\delta)$.

Proof. Given the error correction capability $t = \left[\frac{(\delta n - 1)}{2}\right]$, it follows that $V_q(n, 2t) \le V_q(n, [\delta n])$. We then apply the Gilbert bound and Lemma 4-5.

$$\begin{aligned} a(\delta) &= \lim_{n\to\infty} \sup\left[\frac{\log_q A_q(n, \lfloor \delta n \rfloor)}{n}\right] \\ &\ge \lim_{n\to\infty} \frac{1}{n} \log_q\left(\frac{q^n}{V_q(n, \lfloor \delta n \rfloor)}\right) \\ &= \lim_{n\to\infty}\left[1 - \frac{\log_q V_q(n, \lfloor \delta n \rfloor)}{n}\right] \\ &= 1 - H_q(\delta) \end{aligned}$$

QED

Theorem 4-6 shows that there exist good codes of arbitrarily long length. It was not until 1982, however, that Tsfasman, Vladut, and Zink were able to construct a sequence of nonbinary codes whose asymptotic performance exceeded the Gilbert-Varsharmov bound [Tsf]. Their construction technique was taken from Goppa's work on codes over algebraic curves [Gop], [Wic1], and makes heavy use of algebraic geometry.

The best known upper bound on $a(\delta)$ for binary codes was derived by McEliece, Rodemich, Rumsey, and Welch. We state a simplified form without proof.

Theorem 4-7—The McEliece-Rodemich-Rumsey-Welch Bound for Binary Codes

$$a(\delta) \le H_2\left[\frac{1}{2} - \sqrt{\delta(1-\delta)}\right]$$

Proof. See [McE3].

The Gilbert-Varsharmov bound and the McEliece-Rodemich-Rumsey-Welch bound are plotted in Figure 4-4.

4.2 LINEAR BLOCK CODES

In this section the various properties of linear codes are discussed. The structure inherent in linear codes makes them particularly easy to implement and analyze.

Definition 4-9—q-ary Linear Codes

Consider a block code **C** consisting of n-tuples $\{(c_0, c_1, \ldots, c_{n-1})\}$ of symbols from GF(q). **C** is a ***q*-ary linear code** if and only if **C** forms a vector subspace over GF(q).

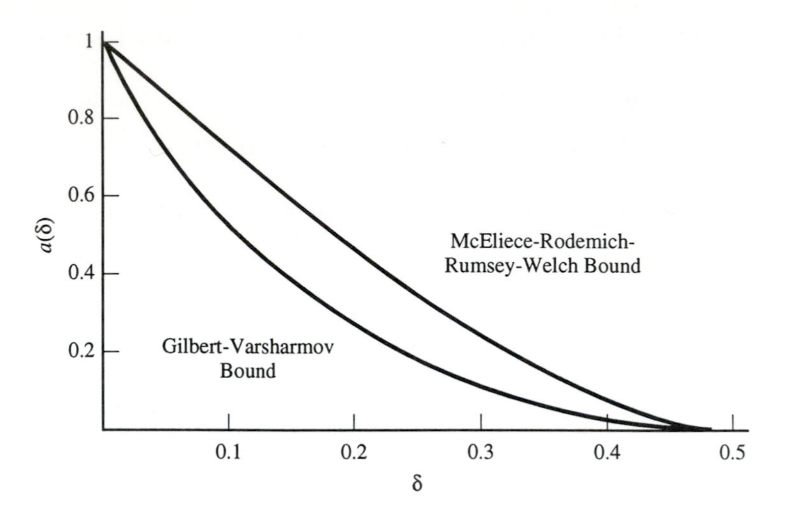

Figure 4-4. Upper and lower bounds for asymptotic binary code performance

Definition 4-10—The Dimension of a Linear Code

The **dimension** of a linear code is the dimension of the corresponding vector space.

A notational convenience is commonly used to refer to linear codes: A linear code of length n and dimension k is called an (n, k) code. An (n, k) code with symbols in GF(q) (henceforth a q-ary code) will thus have a total of q^k code words of length n.

Linear codes have a number of interesting properties that can be derived directly from the properties of vector spaces discussed in Chapter 2.

Property One

The linear combination of any set of code words is a code word. One consequence of this is that linear codes always contain the all-zero vector.

Proof. This follows directly from Theorem 2-7. **QED**

Property Two

The minimum distance of a linear code is equal to the weight of the lowest-weight nonzero code word.

Proof. Minimum distance is defined as $d_{\min} = \min_{\mathbf{c},\mathbf{c}'\in\mathbf{C},\mathbf{c}\neq\mathbf{c}'} d(\mathbf{c},\mathbf{c}')$, which can be re-expressed as $d_{\min} = \min_{\mathbf{c},\mathbf{c}'\in\mathbf{C},\mathbf{c}\neq\mathbf{c}'} w(\mathbf{c}-\mathbf{c}')$. Since the code is linear, $\mathbf{c}'' = (\mathbf{c}-\mathbf{c}')$ is a code word and $d_{\min} = \min_{\mathbf{c}''\in\mathbf{C},\mathbf{c}''\neq\mathbf{0}} w(\mathbf{c}'')$. **QED**

Property Two implies that the determination of the minimum distance (and hence the error detection and correction capabilities) of a linear code is far easier than that for a general block code. At worst, a computer can be used to determine the weight of all of the nonzero code words (numbering $q^k - 1$) instead of having to examine all distinct pairs of code words (numbering $q^k(q^k - 1)/2$).

Property Three

The undetectable error patterns for a linear code are independent of the code word transmitted and always consist of the set of all nonzero code words.

Proof. Let $\mathbf{c}$ be the transmitted code word and $\mathbf{c}'$ the incorrectly received code word. The corresponding undetectable error pattern $\mathbf{e} = \mathbf{c} - \mathbf{c}'$ must be a code word by Property One. **QED**

In Theorem 2-6 it was shown that, given a particular basis $\{\mathbf{v}_i\}$ for a vector space V, there is a one-to-one relationship between the linear combinations of the basis elements and the vectors in V. This result can be used to obtain an efficient representation for linear block codes.

Let $\{\mathbf{g}_0, \mathbf{g}_1, \ldots, \mathbf{g}_{k-1}\}$ be a basis of code words for the (n, k) q-ary code $\mathbf{C}$. By Theorem 2-6 there exists a unique representation $\mathbf{c} = a_0\mathbf{g}_0 + a_1\mathbf{g}_1 + \cdots + a_{k-1}\mathbf{g}_{k-1}$ for every code word $c \in \mathbf{C}$. Since every linear combination of the basis elements must also be a code word, there is one-to-one mapping between the set of k-symbol blocks $(a_0, a_1, \ldots, a_{k-1})$ over GF(q) and the code words in $\mathbf{C}$. A matrix $\mathbf{G}$ is constructed by taking as its rows the vectors in the basis.

$$\mathbf{G} = \begin{bmatrix} \mathbf{g}_0 \\ \mathbf{g}_1 \\ \vdots \\ \mathbf{g}_{k-1} \end{bmatrix} = \begin{bmatrix} g_{0,0} & g_{0,1} & \cdots & g_{0,n-1} \\ g_{1,0} & g_{1,1} & \cdots & g_{1,n-1} \\ \vdots & \vdots & \ddots & \vdots \\ g_{k-1,0} & g_{k-1,1} & \cdots & g_{k-1,n-1} \end{bmatrix} \tag{4-6}$$

This matrix is a **generator matrix** for the code $\mathbf{C}$. It can be used to directly encode k-symbol data blocks in the following manner. Let $\mathbf{m} = (m_0, m_1, \ldots, m_{k-1})$ be a q-ary block of uncoded data.

$$\mathbf{mG} = (m_0, m_1, \ldots, m_{k-1})\begin{bmatrix} \mathbf{g}_0 \\ \mathbf{g}_1 \\ \vdots \\ \mathbf{g}_{k-1} \end{bmatrix} = m_0\mathbf{g}_0 + m_1\mathbf{g}_1 + \cdots + m_{k-1}\mathbf{g}_{k-1} = \mathbf{c} \tag{4-7}$$

A q-ary code $\mathbf{C}$ of length n is a vector subspace V embedded within the space of all n-tuples over GF(q). Given this context, we can talk meaningfully about the dual space of $\mathbf{C}$ within V. The dual space of a linear code $\mathbf{C}$ is called the **dual code** of $\mathbf{C}$ and is denoted here by $\mathbf{C}^\perp$. By Theorems 2-8 and 2-9, $\mathbf{C}^\perp$ is a vector space of dimension $(n - k)$. It follows that a basis $\{\mathbf{h}_0, \mathbf{h}_1, \ldots, \mathbf{h}_{n-k-1}\}$ for $\mathbf{C}^\perp$ can be found and used to construct a **parity-check matrix H**.

$$\mathbf{H} = \begin{bmatrix} \mathbf{h}_0 \\ \mathbf{h}_1 \\ \vdots \\ \mathbf{h}_{n-k-1} \end{bmatrix} = \begin{bmatrix} h_{0,0} & h_{0,1} & \cdots & h_{0,n-1} \\ h_{1,0} & h_{1,1} & \cdots & h_{1,n-1} \\ \vdots & \vdots & \ddots & \vdots \\ h_{n-k-1,0} & h_{n-k-1,1} & \cdots & h_{n-k-1,n-1} \end{bmatrix} \tag{4-8}$$

Theorem 4-8—The Parity-Check Theorem

A vector c is a code word in C if and only if $\mathbf{c}\mathbf{H}^T = \mathbf{0}$.

Proof. Given a vector $\mathbf{c} \in \mathbf{C}$, $\mathbf{c} \cdot \mathbf{h} = \mathbf{0}$ for all $\mathbf{h}$ in $\mathbf{C}^\perp$ by the definition of dual spaces. It follows that $\mathbf{c}\mathbf{H}^T = \mathbf{0}$, for this matrix operation is simply $(n - k)$ inner products of the form $\mathbf{c} \cdot \mathbf{h}_i$, where the $\{\mathbf{h}_i\} \subseteq \mathbf{C}^\perp$.

Given a vector $\mathbf{c}$ such that $\mathbf{c}\mathbf{H}^T = \mathbf{0}$, it follows that $\mathbf{c} \cdot \mathbf{h} = \mathbf{0}$ for all $\mathbf{h}$ in $\mathbf{C}^\perp$ (the rows of $\mathbf{H}$ span $\mathbf{C}^\perp$). If $\mathbf{c} \notin \mathbf{C}$, then $\dim(\mathbf{C}^\perp)^\perp \geq \dim(\mathbf{C})$. This contradicts the dimension theorem (Theorem 2-9). **QED**

The parity-check matrix for a code also offers a convenient means for determining the minimum distance of the code.

Theorem 4-9

Let C have parity check matrix H. The minimum distance of C is equal to the minimum nonzero number of columns in H for which a nontrivial linear combination sums to zero.

Proof. Let the column vectors of $\mathbf{H}$ be $\{\mathbf{d}_0, \mathbf{d}_1, \ldots, \mathbf{d}_{n-1}\}$. The matrix operation $\mathbf{c}\mathbf{H}^T$ can be reexpressed as follows.

$$\begin{aligned}\mathbf{c}\mathbf{H}^T &= (c_0, c_1, \ldots, c_{n-1})[\mathbf{d}_0 \quad \mathbf{d}_1 \quad \cdots \quad \mathbf{d}_{n-1}]^T \\ &= c_0\,\mathbf{d}_0 + c_1\,\mathbf{d}_1 + \cdots + c_{n-1}\,\mathbf{d}_{n-1}\end{aligned}$$

If $\mathbf{c}$ is a weight-w code word, then $\mathbf{c}\mathbf{H}^T$ is a linear combination of w columns of $\mathbf{H}$. The above expression defines a one-to-one mapping between weight-w code words and linear combinations of w columns of $\mathbf{H}$. The result follows. **QED**

Theorem 4-9 can be used to bound the possible minimum distance for an (n, k) code.

Theorem 4-10—Singleton Bound [Sin]

The minimum distance $d_{\min}$ for an (n, k) code is bounded by

$$d_{\min} \leq n - k + 1$$

Proof. An (n, k) code has parity-check matrices $\mathbf{H}$ containing $(n - k)$ linearly independent rows. The row rank and hence the column rank of any $\mathbf{H}$ is thus $(n - k)$. Any collection of $(n - k + 1)$ columns of $\mathbf{H}$ must thus be linearly dependent. The result then follows from Theorem 4-9. **QED**

The generator and parity-check matrices for linear block codes greatly simplify encoding at the transmitter and error detection at the receiver. Both operations are reduced to simple matrix multiplication, obviating the need for look-up tables stored in memory. It is thus possible to efficiently implement encoders and decoders for codes with a very large number of code words. Since, as we shall see, a (64, 56)

Reed-Solomon code contains 1.4×10^{101} code words, such developments are highly significant.[2]

Continuing on the subject of implementation, the problem of recovering the data block from a code word can be greatly simplified through the use of **systematic encoding**. Consider a linear code **C** with generator matrix **G**. Using Gaussian elimination and column reordering, it is always possible to obtain a generator matrix of the form in Figure 4-5. (This can be proved by noting that the rows of a generator matrix are linearly independent and that the column rank of a matrix is equal to the row rank.)

$$\mathbf{G} = [\mathbf{P} | \mathbf{I}_k] = \left[\begin{array}{cccc|ccccc} p_{0,0} & p_{0,1} & \cdots & p_{0,n-k-1} & 1 & 0 & 0 & \cdots & 0 \\ p_{1,0} & p_{1,1} & \cdots & p_{1,n-k-1} & 0 & 1 & 0 & \cdots & 0 \\ p_{2,0} & p_{2,1} & \cdots & p_{2,n-k-1} & 0 & 0 & 1 & \cdots & 0 \\ \vdots & \vdots & \ddots & \vdots & \vdots & \vdots & \vdots & \ddots & \vdots \\ p_{k-1,0} & p_{k-1,0} & \cdots & p_{k-1,n-k-1} & 0 & 0 & 0 & \cdots & 1 \end{array}\right]$$

Figure 4-5. A Systematic Generator Matrix for an (n, k) Code

When a data block is encoded using a systematic generator matrix, the data block is embedded without modification in the last k coordinates of the resulting code word.

$$\begin{aligned} \mathbf{c} &= \mathbf{mG} \\ &= [m_0 \quad m_1 \quad \cdots \quad m_{k-1}][\mathbf{P} | \mathbf{I}_k] \\ &= [c_0 \quad c_1 \quad \cdots \quad c_{n-k-1} | m_0 \quad m_1 \quad \cdots \quad m_{k-1}] \end{aligned} \tag{4-9}$$

After decoding, the last k symbols are removed from the selected code word and passed along to the data sink.

The performance of Gaussian elimination operations on a generator matrix does not alter the code word set for the associated code. Column reordering, on the other hand, may generate code words that are not in the original code. If a given application requires that a particular code word set be used and thus does not allow for column reordering, it is always possible to use some set of coordinates other than the last k for the message positions. Unfortunately this can slightly complicate certain encoder and decoder designs, particularly those discussed in the next chapter.

Given a systematic generator matrix of the form in Figure 4-5, a corresponding parity check matrix can be obtained as shown in Figure 4-6.

$$\mathbf{H} = [\mathbf{I}_{n-k} | -\mathbf{P}^T] = \left[\begin{array}{ccccc|ccccc} 1 & 0 & 0 & \cdots & 0 & -p_{0,0} & -p_{1,0} & -p_{2,0} & \cdots & -p_{k-1,0} \\ 0 & 1 & 0 & \cdots & 0 & -p_{0,1} & -p_{1,1} & -p_{2,1} & \cdots & -p_{k-1,1} \\ 0 & 0 & 1 & \cdots & 0 & -p_{0,2} & -p_{1,2} & -p_{2,2} & \cdots & -p_{k-1,2} \\ \vdots & \vdots & \vdots & \ddots & \vdots & \vdots & \vdots & \vdots & \ddots & \vdots \\ 0 & 0 & 0 & \cdots & 1 & -p_{0,n-k-1} & -p_{1,n-k-1} & -p_{2,n-k-1} & \cdots & -p_{k-1,n-k-1} \end{array}\right]$$

Figure 4-6. A Systematic Parity-Check Matrix for an (n, k) Code

[2] If 1.4×10^{101} code words does not seem terribly impressive, please note that it is currently believed that there are only around 1×10^{79} atomic particles in the entire universe, creating a serious storage problem for the look-up table for the (64, 56) code.

4.3 STANDARD ARRAY AND SYNDROME TABLE DECODING

In Figure 4-2 the received vector $\mathbf{r}$ was modeled by the summation $\mathbf{r} = \mathbf{c} + \mathbf{e}$, where $\mathbf{c}$ is the transmitted code word and $\mathbf{e}$ is the error pattern induced by the channel noise. If lower-weight error patterns are more likely to occur than those with higher weight, the maximum likelihood decoder picks a code word $\mathbf{c}'$ such that $\mathbf{r} = \mathbf{c}' + \mathbf{e}'$, where $\mathbf{e}'$ has the smallest possible weight. In this section a simple look-up table called a **standard array decoder** is used to implement this process.

A standard array decoder is constructed for an (n, k) linear q-ary code $\mathbf{C}$ in the following manner. Let the set V_n^q consist of all of the n-tuples over GF(q).

1. Remove from V_n^q (without replacement) all of the code words in the code $\mathbf{C}$. Write down the code words in a single row, starting with the all-zero code word.
2. Select from the remaining n-tuples (without replacement) one of the patterns of lowest weight. Write this pattern down in the column under the all-zero code word. Now add this pattern to each of the other code words, writing each sum down under the respective code word and removing the sum from the set of remaining n-tuples.
3. Repeat step 2 until V_n^q is exhausted. The standard array is now complete.

A received word $\mathbf{r}$ is decoded by looking up the word in the table. The code word at the top of the column in which $\mathbf{r}$ is located is the maximum likelihood code word $\mathbf{c}$ associated with $\mathbf{r}$. The error pattern at the far left of the row on which $\mathbf{r}$ is located is the error pattern most likely to have corrupted $\mathbf{r}$. The construction technique ensures that this difference vector $\mathbf{e} = \mathbf{r} - \mathbf{c}$ has the minimum possible weight.

In algebraic terms, each row of the standard array is a coset of $\mathbf{C}$ in V_n^q (see Definition 2-3).

Example 4-4—Standard Array Decoder for a (7, 3) Code

Let $\mathbf{C}$ be the $(7, 3)$ binary linear code with basis {(1000111), (0101011), (0011101)} (this is the dual of a $(7, 4)$ Hamming code). One possible generator matrix for this code is thus

$$\mathbf{G} = \begin{bmatrix} 1 & 0 & 0 & 0 & 1 & 1 & 1 \\ 0 & 1 & 0 & 1 & 0 & 1 & 1 \\ 0 & 0 & 1 & 1 & 1 & 0 & 1 \end{bmatrix}$$

A standard array decoding table for $\mathbf{C}$ is shown in Table 4-1.

Suppose that the received vector is $\mathbf{r} = 0011011$. This is the underlined vector in the table. The column header 0011101 is the maximum likelihood code word. The row header 0000110 is the error pattern that is most likely to have corrupted the code word during transmission.

TABLE 4-1. Standard Array for a (7, 3) Linear Block Code

0000000	1000111	0101011	0011101	1101100	1011010	0110110	1110001
0000001	1000110	0101010	0011100	1101101	1011011	0110111	1110000
0000010	1000101	0101001	0011111	1101110	1011000	0110100	1110011
0000100	1000011	0101111	0011001	1101000	1011110	0110010	1110101
0001000	1001111	0100011	0010101	1100100	1010010	0111110	1111001
0010000	1010111	0111011	0001101	1111100	1001010	0100110	1100001
0100000	1100111	0001011	0111101	1001100	1111010	0010110	1010001
1000000	0000111	1101011	1011101	0101100	0011010	1110110	0110001
0000011	1000100	0101000	0011110	1101111	1011001	0110101	1110010
0000110	1000001	0101101	**0011011**	1101010	1011100	0110000	1110111
0001100	1001011	0100111	0010001	1100000	1010110	0111010	1111101
0011000	1011111	0110011	0000101	1110100	1000010	0101110	1101001
0001010	1001101	0100001	0010111	1100110	1010000	0111100	1111011
0010100	1010011	0111111	0001001	1111000	1001110	0100010	1100101
0010010	1010101	0111001	0001111	1111110	1001000	0100100	1100011
0111000	1111111	0010011	0100101	1010100	1100010	0001110	1001001

The reader may have noted that the construction of the standard array table in the previous example involves a certain number of arbitrary choices. All of the weight-one error patterns were used as row headers, but this was not the case with the weight-two patterns. In selecting weight-two patterns as headers we placed weight-two patterns within the body of the table, making those patterns unavailable for later selection as row headers. We are thus not able to guarantee correction of all weight-two error patterns. A different set of weight-two patterns could just as easily have been selected as row headers. Perfect codes are structured such that this cannot happen; the row headers consist of all patterns of weight t or less and no others.

Example 4-5—Standard Array Decoder for a Perfect Code

The (5, 1) repetition code has the following standard array.

TABLE 4-2. Standard Array for (5, 1) Repetition Code

00000	11111	00000	11111
00001	11110	00110	11001
00010	11101	01100	10011
00100	11011	11000	00111
01000	10111	00101	11010
10000	01111	01010	10101
00011	11100	10100	01011
01001	10110	10010	01101
10001	01110		

Note that the row headers consist of all weight-zero, one, and two error patterns and no others. ■

The problem with the standard array approach to decoding is probably clear at this point. The table for a q-ary table has q^n entries, all of which must be stored in memory. This is fine for small codes like those in the examples, but becomes rapidly impractical as the number of code words in the code increases. The physical impossibility of the use of this approach with most Reed-Solomon codes is an obvious extension of the comment in footnote 2 of this chapter. In later chapters more practical approaches to the problem of decoding certain specific algebraic codes are presented. For the moment, however, we must be content with a method for reducing the size of the standard array for an (n, k) q-ary code to a q^{n-k} entry table.

In Theorem 4-8 it was shown that given a parity-check matrix $\mathbf{H}$ for a code $\mathbf{C}$, a vector $\mathbf{c}$ in the received space is a code word if and only if $\mathbf{cH}^T = \mathbf{0}$. The parity-check matrix $\mathbf{H}$ can thus be thought of as a linear transformation whose null space is $\mathbf{C}$. Consider a received vector $\mathbf{r} = \mathbf{c} + \mathbf{e}$, where $\mathbf{c}$ is a valid code word and $\mathbf{e}$ an error pattern induced by the channel. The matrix product $\mathbf{rH}^T$ is the **syndrome vector s** for the received vector $\mathbf{r}$. The following is a result of the linearity of matrix multiplication by $\mathbf{H}^T$.

$$\begin{aligned} \mathbf{s} &= \mathbf{rH}^T \\ &= (\mathbf{c} + \mathbf{e})\mathbf{H}^T \\ &= \mathbf{cH}^T + \mathbf{eH}^T \\ &= \mathbf{0} + \mathbf{eH}^T \\ &= \mathbf{eH}^T \end{aligned} \tag{4-10}$$

The syndrome vector $\mathbf{s}$ is thus solely a function of the error pattern $\mathbf{e}$ and is independent of the transmitted code word $\mathbf{c}$. It follows that all of the vectors in a given row of a standard array must have the same syndrome, for the difference between any pair of vectors in a row is a code word. The syndromes for vectors in different rows of the standard array, however, cannot be equal. This can be proven as follows. If two error vectors $\mathbf{e}$ and $\mathbf{e}'$ have the same syndrome, then they must differ by a nonzero code word:

$$\begin{aligned} &\mathbf{s} = \mathbf{eH}^T = \mathbf{e}'\mathbf{H}^T \\ &\Rightarrow \mathbf{eH}^T - \mathbf{e}'\mathbf{H}^T = (\mathbf{e} - \mathbf{e}')\mathbf{H}^T = 0 \\ &\Rightarrow (\mathbf{e} - \mathbf{e}') = \mathbf{c} \in \mathbf{C} \end{aligned} \tag{4-11}$$

$\mathbf{e}$ and $\mathbf{e}'$ are thus in the same row of the standard array. Since every vector appears exactly once in the standard array, the result follows.

The result can also be proved by noting that the column rank of $\mathbf{H}$ is $(n - k)$. The linear transformation $\mathbf{r} \rightarrow \mathbf{rH}^T$ is thus surjective ("onto") on the space of q^{n-k} $(n - k)$-tuples over GF(q). There are q^{n-k} rows in the standard array, and all of the vectors in a row have the same syndrome. The result follows.

This result shows that one need only store the row headers of the standard array and their syndromes in memory to be able to perform maximum likelihood decoding.

Decoding is performed by computing the syndrome of a received vector, looking up the corresponding error pattern, and subtracting the error pattern from the received word.

Example 4-6—Syndrome Table Decoding

The following can be readily shown to be a valid parity-check matrix for the (7, 3) code used in Example 4-4.

$$\mathbf{H} = \begin{bmatrix} 0 & 1 & 1 & 1 & 0 & 0 & 0 \\ 1 & 0 & 1 & 0 & 1 & 0 & 0 \\ 1 & 1 & 0 & 0 & 0 & 1 & 0 \\ 1 & 1 & 1 & 0 & 0 & 0 & 1 \end{bmatrix}$$

Using **H**, the syndrome table in Table 4-3 is constructed.

TABLE 4-3. Syndrome Decoding Table for a (7, 3) Binary Code

Error Pattern	Syndrome
0000000	0000
0000001	0001
0000010	0010
0000100	0100
0001000	1000
0010000	1101
0100000	1011
1000000	0111
0000011	0011
0000110	0110
0001100	1100
0011000	0101
0001010	1010
0010100	1001
0010010	1111
0111000	1110

Suppose that the received vector is (0011011). The decoder first computes the syndrome $\mathbf{s} = (0011011)\mathbf{H}^T = (0110)$. Looking up this syndrome in the table, we see that the corresponding error pattern is $\mathbf{e} = (0000110)$. The maximum likelihood transmitted code word is thus $\mathbf{r} - \mathbf{e} = (0011101)$, the same value obtained in Example 4-4. ∎

4.4 THE WEIGHT DISTRIBUTION OF BLOCK CODES

One of the keys to obtaining an exact expression for the error detection and error correction performance of a block code is the weight distribution of the code.

Definition 4-11—The Weight Distribution of a Block Code

The weight distribution of an (n, k) code **C** is a series of coefficients $A_0, A_1, A_2, \ldots, A_n$, where A_i is the number of code words in **C** of weight i. ■

The weight distribution $\{A_0, A_1, A_2, \ldots, A_n\}$ for a code is often written as a polynomial $A(x) = A_0 + A_1 x + A_2 x^2 + \cdots + A_n x^n$. This representation is often called the **weight enumerator**.

There are a large number of codes for which a weight distribution has not yet been found. Fortunately there are several classes of codes (including the Reed-Solomon codes) for which the weight distribution is known.

In many cases the weight enumerator for the dual of a code is easier to work with than that of the code itself. The following extremely powerful result is thus offered without proof (though interesting, it is a bit long; see [Mac] or [Van]). This result will be quite useful in dealing with certain BCH codes.

Theorem 4-11—The MacWilliams Identity[3]

Let $A(x)$ and $B(x)$ be the weight enumerators for an (n, k) code C and its $(n, n-k)$ dual code $C^{\perp}$, respectively. $A(x)$ and $B(x)$ are related by the following identity.

$$B(x) = 2^{-k}(1 + x)^n A\left[\frac{(1-x)}{(1+x)}\right]$$

Both [Mac] and [Van] contain detailed discussions of the weight distributions of block codes.

4.5 THE HAMMING CODES

Hamming codes were the first major class of linear binary codes designed for error correction (they are discussed in a publication dated April 1950 [Ham], though the work was well underway in 1947 or 1948). Their first application was in error control for long-distance telephony. In Chapter 1 Hamming codes were briefly introduced from a statistical viewpoint. In this section we shall use the algebraic machinery from this chapter and Chapters 2 and 3 to more fully develop the fundamental structure of these interesting codes.

The performance parameters for the family of binary Hamming codes are typically expressed as a function of a single integer $m \geq 2$.

Code length:	$n = 2^m - 1$
Number of information symbols:	$k = 2^m - m - 1$
Number of parity symbols:	$n - k = m$
Error correcting capability:	$t = 1$

[3] Florence MacWilliams (1917–1990) was one of the pioneers in the field of error control coding.

The parity-check matrices for binary Hamming codes are quite easy to construct. For a Hamming code of length $(2^m - 1)$, construct a matrix whose columns consist of all nonzero binary m-tuples. For example, a $(15, 11)$ Hamming code is defined by the parity-check matrix in Figure 4-7. The ordering of the columns is arbitrary; another arrangement would still define a $(15, 11)$ Hamming code, albeit a different $(15, 11)$ Hamming code than the one associated with the matrix shown in the figure. The parity-check matrix in Figure 4-7 was specifically designed to describe a systematic code. The corresponding generator matrix in Figure 4-8 clearly shows that the message bits always occupy the last 11 coordinates of each code word.

$$\mathbf{H} = \begin{bmatrix} 1&0&0&0&0&0&0&0&1&1&1&1&1&1&1 \\ 0&1&0&0&1&1&1&0&0&0&0&1&1&1&1 \\ 0&0&1&0&0&1&1&1&0&1&1&0&0&1&1 \\ 0&0&0&1&1&0&1&1&1&0&1&0&1&0&1 \end{bmatrix}$$

Figure 4-7. Systematic Parity-Check Matrix for a (15, 11) Hamming Code

$$\mathbf{G} = \begin{bmatrix} 0&1&0&1&1&0&0&0&0&0&0&0&0&0&0 \\ 0&1&1&0&0&1&0&0&0&0&0&0&0&0&0 \\ 0&1&1&1&0&0&1&0&0&0&0&0&0&0&0 \\ 0&0&1&1&0&0&0&1&0&0&0&0&0&0&0 \\ 1&0&0&1&0&0&0&0&1&0&0&0&0&0&0 \\ 1&0&1&0&0&0&0&0&0&1&0&0&0&0&0 \\ 1&0&1&1&0&0&0&0&0&0&1&0&0&0&0 \\ 1&1&0&0&0&0&0&0&0&0&0&1&0&0&0 \\ 1&1&0&1&0&0&0&0&0&0&0&0&1&0&0 \\ 1&1&1&0&0&0&0&0&0&0&0&0&0&1&0 \\ 1&1&1&1&0&0&0&0&0&0&0&0&0&0&1 \end{bmatrix}$$

Figure 4-8. Systematic Generator Matrix for a (15, 11) Hamming Code

An examination of the parity-check matrix in Figure 4-7 quickly shows that the $(15, 11)$ Hamming code is single-error-correcting. In general, note that the smallest number of distinct nonzero binary m-tuples that can sum to zero is always three. Hamming codes thus have minimum distance three and are single-error-correcting by Theorem 4-9.

Hamming codes can be shown to be perfect codes. A $(2^m - 1, 2^m - m - 1)$ single-error-correcting binary code divides the received space up into 2^{2^m-m-1} Hamming spheres of volume $V_2(2^m - 1, 1) = 2^m$ (see Eq. (4-4)). The decoding spheres thus take up $(2^{(2^m-m-1)})(2^m) = 2^{2^m-1}$ vectors in the received space. Since the length of the code is 2^{m-1}, there are exactly 2^{2^m-1} vectors in the received space and the Hamming bound is satisfied with equality.

Hamming codes can be decoded through the use of a syndrome table. Since Hamming codes are single-error-correcting, the syndrome table has a simple relationship to the parity-check matrix used in the computation of the syndrome. Let the received vector $\mathbf{r}$ be a code word $\mathbf{c}$ corrupted by $\mathbf{e}$, an error pattern with a single one in the jth coordinate position. Let $\{\mathbf{d}_0, \mathbf{d}_1, \ldots, \mathbf{d}_{n-1}\}$ be the set of columns of the parity-check matrix $\mathbf{H}$. When the syndrome is computed, we obtain the transposition of the jth column of $\mathbf{H}$.

$$\mathbf{s} = \mathbf{r}\mathbf{H}^T = \mathbf{e}\mathbf{H}^T = (0, \ldots, 0, 1, 0, \ldots, 0)\begin{bmatrix} \mathbf{d}_0^T \\ \mathbf{d}_1^T \\ \vdots \\ \mathbf{d}_{n-1}^T \end{bmatrix} = \mathbf{d}_j^T$$

Hamming codes can thus be decoded as follows.

Hamming Code Decoding Algorithm

1. Compute the syndrome $\mathbf{s}$ for the received word. If $\mathbf{s} = \mathbf{0}$, then go to step 4.
2. Determine the position j of the column of $\mathbf{H}$ that is the transposition of the syndrome.
3. Complement the jth bit in the received word.
4. Output the resulting code word and STOP.

Example 4-7—Decoding the (15, 11) Hamming Code

Consider the (15, 11) Hamming code described by the parity-check matrix in Figure 4-7. Three vectors are received:

$$\mathbf{r}_1 = (111000100000010)$$

$$\mathbf{r}_2 = (100100001000001)$$

$$\mathbf{r}_3 = (000100000000011)$$

Using the parity-check matrix in Figure 4-7, the following syndromes $\mathbf{s} = \mathbf{r}\mathbf{H}^T$ are obtained.

$$\mathbf{s}_1 = (0111), \quad \mathbf{s}_2 = (1111), \quad \mathbf{s}_3 = (0000)$$

The first syndrome is the transposition of the sixth column of $\mathbf{H}$. Inverting the sixth coordinate of $\mathbf{r}_1$, we obtain the following code word (the ninth row of the generator matrix in Figure 4-8):

$$\mathbf{c}_1 = (111000000000010)$$

$\mathbf{s}_2$ is the transposition of the last column of $\mathbf{H}$. Inverting the last coordinate of $\mathbf{r}_2$, we obtain

$$\mathbf{c}_2 = (100100001000000)$$

Since $\mathbf{s}_3 = \mathbf{0}$, $\mathbf{r}_3$ is a valid code word. ■

The weight enumerator for (n, k) binary Hamming codes can be shown to be

$$A(x) = \frac{1}{n+1}\{(1+x)^n + n(1-x)(1-x^2)^{(n-1)/2}\} \tag{4-12}$$

where the coefficient A_i for the term x^i is the number of code words of weight i [Lin1].

The weight enumerator for a (15, 11) Hamming code is thus

$$A(x) = \frac{1}{16}\{(1 + x)^{15} + 15(1 - x)(1 - x^2)^7\}$$

$$= 1 + 35x^3 + 105x^4 + 168x^5 + 280x^6 + 435x^7 \qquad (4\text{-}13)$$

$$+ 435x^8 + 280x^9 + 168x^{10} + 105x^{11} + 35x^{12} + x^{15}$$

Using the weight distribution, we can obtain exact performance curves for the probability of undetected error and decoder error as a function of the binary symmetric channel (BSC) crossover probability (see Chapter 10). Figure 4-9 shows the error detection and correction performance of a (15, 11) and a (63, 57) binary Hamming code.

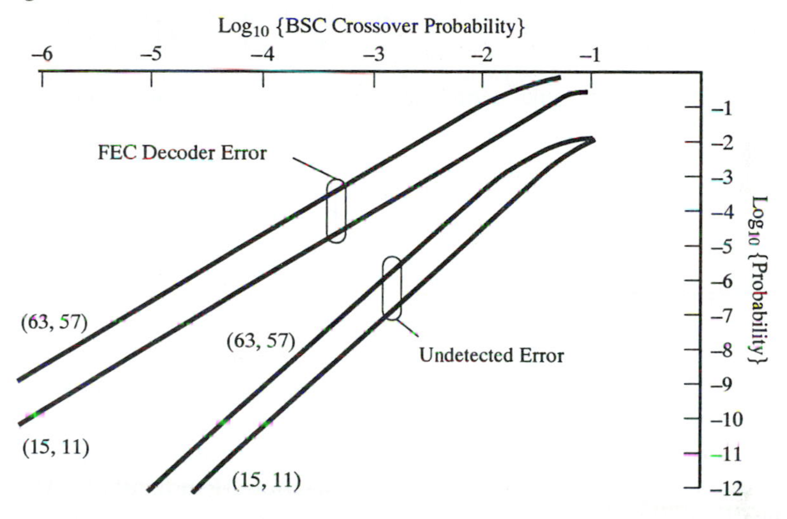

Figure 4-9. Error detection and error correction performance of two Hamming codes

It is also possible to construct nonbinary Hamming codes using the same basic approach. There are exactly $(q^m - 1)$ distinct nonzero q-ary m-tuples, but not all pairs of these m-tuples are linearly independent. For each q-ary m-tuple there are $(q - 1)$ distinct nonzero m-tuples that are multiples of that m-tuple, pairs of which are clearly linearly dependent. The q-ary Hamming code parity-check matrix **H** is constructed by selecting exactly one m-tuple from each set of multiples. This can be done by selecting as columns of **H** all distinct q-ary m-tuples for which the uppermost nonzero element is 1. **H** thus has $(q^m - 1)/(q - 1)$ columns and defines a q-ary Hamming code with the following properties.

Code length:	$n = (q^m - 1)/(q - 1)$
Number of information symbols:	$k = (q^m - 1)/(q - 1) - m$
Number of parity symbols:	$n - k = m$
Error correcting capability:	$t = 1$

Example 4-8—The (5, 3) 4-ary Hamming Code

The integer labeling for GF(4) shown in Example 2-24 is used in this example. Recall that the truth tables for this field are as follows.

+	0	1	2	3
0	0	1	2	3
1	1	0	3	2
2	2	3	0	1
3	3	2	1	0

·	0	1	2	3
0	0	0	0	0
1	0	1	2	3
2	0	2	3	1
3	0	3	1	2

The columns in the parity-check matrix can be written out as follows.

$$\mathbf{H} = \begin{bmatrix} 1 & 0 & 1 & 1 & 1 \\ 0 & 1 & 1 & 2 & 3 \end{bmatrix}$$

Since this particular parity-check matrix has a systematic form, the generator matrix is easily obtained.

$$\mathbf{G} = \begin{bmatrix} 1 & 1 & 1 & 0 & 0 \\ 1 & 2 & 0 & 1 & 0 \\ 1 & 3 & 0 & 0 & 1 \end{bmatrix}$$ ■

4.6 MODIFIED LINEAR CODES

In many applications there are external constraints unrelated to error control that determine the allowed length of the error control code. For example, computer applications have a propensity for word lengths that are a multiple of 8. When the "natural" length of the code we wish to use is unsuitable, the code's length can be changed by **puncturing**, **extending**, **shortening**, **lengthening**, **expurgating**, or **augmenting** [Ber1], [Mac], [Wic8]. Figure 4-10 provides a graphic depiction of these terms. It is suggested that the reader refer to Figure 4-10 while reading the following definitions.

Definition 4-12—Puncturing

A code is punctured by deleting one of its parity coordinates. An (n, k) code thus becomes a $(n - 1, k)$ code.

Definition 4-13—Shortening

A code is shortened by deleting a message coordinate from the encoding process. An (n, k) code thus becomes an $(n - 1, k - 1)$ code.

Definition 4-14—Expurgating

A code is expurgated by deleting some of its code words. If $(q - 1)/q$ of the code words are deleted in a manner such that the remainder form a linear subcode, then a q-ary (n, k) code becomes a q-ary $(n, k - 1)$ code.

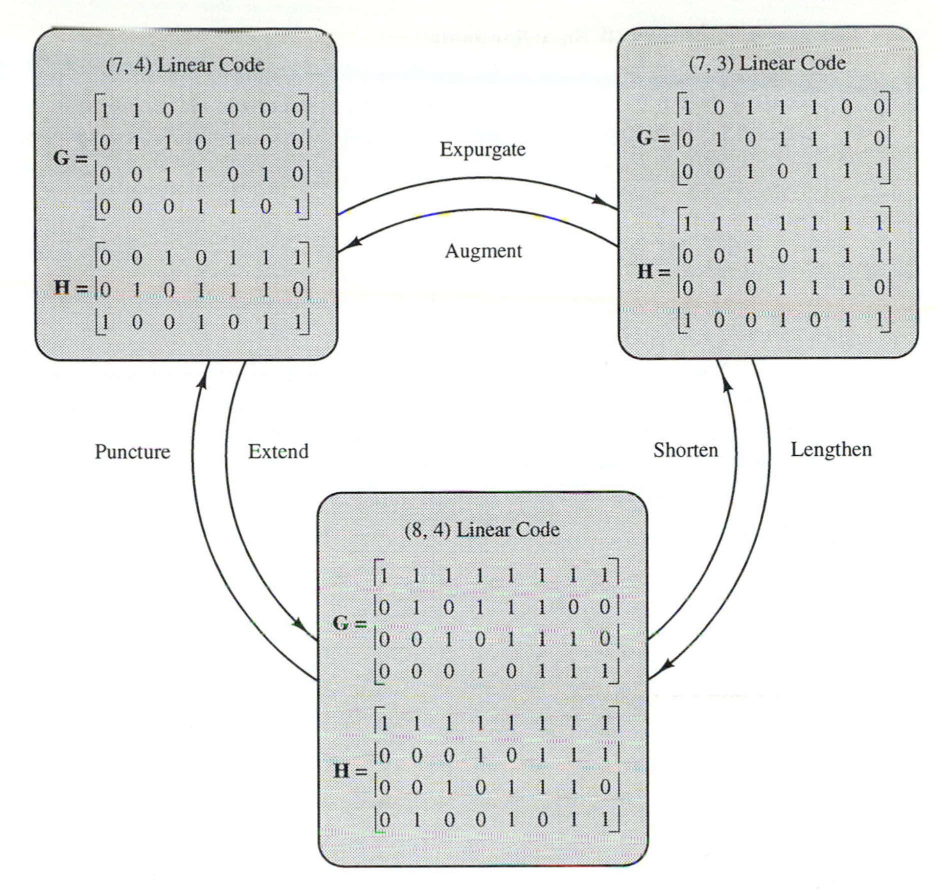

Figure 4-10. Methods for Modifying Linear Block Codes[4]

Definition 4-15—Extending

A code is extended by adding an additional redundant coordinate. An (n, k) code thus becomes an $(n + 1, k)$ code.

Definition 4-16—Augmenting

A code is augmented by adding new code words. If the number of code words is increased by the factor q such that the resulting code is linear, then a q-ary (n, k) code becomes an $(n, k + 1)$ code.

Definition 4-17—Lengthening

A code is lengthened by adding message coordinates. An (n, k) code thus becomes an $(n + 1, k + 1)$ code.

[4] The codes and the basic pictorial concept of this figure are due to Berlekamp [Ber1].

Example 4-9—Extended Hamming Codes

The extended Hamming codes are formed by adding a row of ones and the column vector $[0, 0, 0, \ldots, 0, 1]^T$ to the parity-check matrix for a Hamming code. For example, the (15, 11) parity-check matrix in Figure 4-7 becomes the following when extended.

$$\mathbf{H}' = \left[\begin{array}{ccccccccccccccc|c} 1 & 0 & 0 & 0 & 0 & 0 & 0 & 0 & 1 & 1 & 1 & 1 & 1 & 1 & 1 & 0 \\ 0 & 1 & 0 & 0 & 1 & 1 & 1 & 0 & 0 & 0 & 1 & 1 & 1 & 1 & 1 & 0 \\ 0 & 0 & 1 & 0 & 0 & 1 & 1 & 1 & 0 & 1 & 1 & 0 & 0 & 1 & 1 & 0 \\ 0 & 0 & 0 & 1 & 1 & 0 & 1 & 1 & 1 & 0 & 1 & 0 & 1 & 0 & 1 & 0 \\ \hline 1 & 1 & 1 & 1 & 1 & 1 & 1 & 1 & 1 & 1 & 1 & 1 & 1 & 1 & 1 & 1 \end{array}\right]$$

The bottom row of **H** insures that all code words must have even weight. The submatrix formed by the original Hamming code insures that all nonzero code words must have a weight of at least three. The extended parity-check matrix above thus defines a code with minimum distance four.

The 4-ary (5, 3) parity-check matrix in Example 4-8 becomes the following under extension.

$$\mathbf{H}' = \left[\begin{array}{ccccc|c} 1 & 0 & 1 & 1 & 1 & 0 \\ 0 & 1 & 1 & 2 & 3 & 0 \\ \hline 1 & 1 & 1 & 1 & 1 & 1 \end{array}\right]$$

In general, the extension of an (n, k) single-error-correcting Hamming code creates an $(n + 1, k)$ single-error-correcting, double-error-detecting extended Hamming code. ■

PROBLEMS

General block codes

1. Consider the binary code **C** composed of the following four code words.

$$\mathbf{C} = \{(00100), (10010), (01001), (11111)\}$$

   (a) What is the minimum distance of this code?
   (b) What is the maximum weight for which the detection of all error patterns is guaranteed?
   (c) What is the maximum weight for which the correction of all error patterns is guaranteed?
   (d) Is this code linear? Prove your answer.
   (e) Is this code perfect? Prove your answer.

2. For the code **C** in Problem 1, find the maximum likelihood code word associated with each of the following received vectors. The maximum likelihood decoder in this case minimizes $d(\mathbf{r}, \mathbf{c})$, where **r** is the received word and **c** is a code word.
   (a) (00000) **(b)** (00111) **(c)** (01101) **(d)** (10110) **(e)** (01010)

3. Determine the following.
   (a) $V_2(n, 1)$ **(b)** $V_q(n, 1)$ **(c)** $V_q(n, n - 1)$

4. Find the lower bound on required redundancy for the following codes.
   (a) A single-error-correcting binary code of length 7.
   (b) A single-error-correcting binary code of length 15.
   (c) A triple-error-correcting binary code of length 23.
   (d) A triple-error-correcting 4-ary code of length 23.
   (e) A triple-error-correcting 16-ary code of length 23.
5. Find the upper bound on required redundancy for the codes in Problem 4.

Repetition codes

6. Prove that all odd-length binary repetition codes are perfect.
7. Prove that no even-length binary repetition code can be perfect.

Linear codes

8. Find the length, dimension, and minimum distance for the linear codes defined by the following parity-check matrices.

(a) $\mathbf{H} = \begin{bmatrix} 1 & 0 & 0 & 0 & 1 & 0 & 1 \\ 0 & 1 & 0 & 0 & 1 & 1 & 1 \\ 0 & 0 & 1 & 0 & 1 & 1 & 1 \\ 0 & 0 & 0 & 1 & 0 & 1 & 1 \end{bmatrix}$ (binary)

(b) $\mathbf{H} = \begin{bmatrix} 1 & 0 & 1 & 0 & 0 & 0 & 1 \\ 1 & 0 & 0 & 1 & 1 & 0 & 0 \\ 0 & 0 & 0 & 1 & 0 & 1 & 1 \\ 1 & 1 & 0 & 1 & 0 & 0 & 1 \end{bmatrix}$ (binary)

(c) $\mathbf{H} = \begin{bmatrix} 0 & 1 & 1 & 1 \\ 1 & \alpha & \alpha^2 & 0 \end{bmatrix}$, α primitive in GF(4) (4-ary)

9. Consider the 4-ary code **C** defined by the following parity-check matrix. α is primitive in GF(4).

$$\mathbf{H} = \begin{bmatrix} \alpha & \alpha^2 & 1 & 1 \\ \alpha^2 & \alpha & 1 & 0 \end{bmatrix}$$

   (a) Write out the 16 code words in **C**.
   (b) Does **C** achieve the Gilbert bound on redundancy?
   (c) Does **C** exceed the Hamming bound on redundancy?
   (d) Does **C** achieve the Singleton bound?
10. Find a systematic generator matrix for the code **C** in Problem 9.

Binary Hamming codes

11. Construct parity-check and generator matrices for a (7, 4) Hamming code.
12. Construct a standard array decoding table for a (7, 4) Hamming code.
13. Construct a syndrome decoding table for a (7, 4) Hamming code.
14. Compute the weight distribution for a (7, 4) Hamming code.
15. Compute the weight distribution for the dual code to a (7, 4) Hamming code.

16. Decode the following received vectors using the (15, 11) parity-check matrix in Figure 4-7.
 (a) $\mathbf{r} = (010000000000000)$
 (b) $\mathbf{r} = (001111100000000)$
 (c) $\mathbf{r} = (001100011100000)$
 (d) $\mathbf{r} = (000110000001100)$
 (e) $\mathbf{r} = (010000110000000)$
 (f) $\mathbf{r} = (000100000000011)$
 (g) $\mathbf{r} = (110000000001100)$
 (h) $\mathbf{r} = (111100001111010)$

Nonbinary Hamming codes

17. Construct parity-check matrix and generator matrices for a (13, 10) 3-ary Hamming code (i.e., with code symbols taken from GF(3)).
18. Construct a parity-check matrix for a (21, 18) 4-ary Hamming code.
19. Construct a syndrome decoding table for a (21, 18) 4-ary Hamming code.

Extended Hamming codes

20. Determine the length, dimension, and minimum distance of the family of extended Hamming codes.
21. Write out the code words in the (8, 4) extended Hamming code.
22. Express the weight distribution for an extended Hamming code in terms of the weight distribution for the regular Hamming code from which it was derived.

5

Cyclic Codes

Cyclic codes were first discussed in a series of technical notes and reports published from 1957 to 1959 by E. Prange at the Air Force Cambridge Research Labs [Pra1], [Pra2], [Pra3]. Prange identified cyclic codes with ideals, an idea which led to the development of BCH codes and a reinterpretation of Reed-Solomon codes a few years later. Cyclic codes have been and will probably continue to be a constant focus of interest to both mathematicians and engineers. Cyclic codes are important practical error control codes for a variety of reasons. The general class of cyclic codes can be implemented using high-speed shift-register-based encoders and decoders. This is of great interest in applications involving fiber optics, where the lack of bandwidth limitations allows for the use of gigabit-per-second data rates. Within the family of cyclic codes there are certain special families of codes that are extremely powerful. These include the Golay, BCH, and Reed-Solomon codes, which will be discussed in detail in the chapters that follow.

In this chapter we investigate the general theory of cyclic codes and consider a few examples. The shift-register-based methods for encoding and decoding are discussed, and the high-speed CRC error detecting codes presented in some detail. It should be noted that a code need not be linear to be cyclic, but the nonlinear cyclic codes have less useful structure and see only a tiny fraction of the application of their linear counterparts. In this chapter the focus is thus placed solely on the linear cyclic codes.

5.1 THE GENERAL THEORY OF LINEAR CYCLIC CODES

The definition of linear cyclic block codes is quite simple. It is thus utterly fascinating to discover the wealth of properties, restrictions, and overall structure that emerges.

Definition 5-1—Linear Cyclic Block Codes

An (n, k) linear block code $\mathbf{C}$ is said to be **cyclic** if for every code word $\mathbf{c} = (c_0, c_1, \ldots, c_{n-2}, c_{n-1}) \in \mathbf{C}$, there is also a code word $\mathbf{c}' = (c_{n-1}, c_0, c_1, \ldots, c_{n-2}) \in \mathbf{C}$.

In computer parlance, the code word $\mathbf{c}'$ is a right cyclic shift of the code word $\mathbf{c}$. Since $\mathbf{c}$ has been arbitrarily selected from among the code words in $\mathbf{C}$, it follows that all n of the distinct cyclic shifts of $\mathbf{c}$ must also be code words in $\mathbf{C}$ (replace $\mathbf{c}$ with $\mathbf{c}'$ and apply the definition again).

The key to the underlying structure of cyclic codes lies in the association of a **code polynomial** $c(x) = c_0 + c_1x + c_2x^2 + \cdots + c_{n-1}x^{n-1}$ with every code word $\mathbf{c} = (c_0, c_1, \ldots, c_{n-2}, c_{n-1}) \in \mathbf{C}$. If $\mathbf{C}$ is a q-ary (n, k) code, then the collection of code words in $\mathbf{C}$ forms a vector subspace of dimension k within the space of all n-tuples over GF(q) (Definition 4-9). It follows that the code polynomials associated with $\mathbf{C}$ also form a vector subspace, this time within $\mathrm{GF}(q)[x]/(x^n - 1)$. The terms code word and code polynomial are henceforth used interchangeably.

If the code word $\mathbf{c}'$ is the right cyclic shift of the code word $\mathbf{c} \in \mathbf{C}$, then $c'(x) = x \cdot c(x)$ *modulo* $(x^n - 1) \in \mathbf{C}$. This can be seen as follows.

$$\begin{aligned} x \cdot c(x) &\equiv (c_0x + c_1x^2 + \cdots + c_{n-1}x^n) \ mod\ (x^n - 1) \\ &\equiv (c_{n-1} + c_0x + \cdots + c_{n-2}x^{n-1}) \ mod\ (x^n - 1) \\ &\equiv c'(x) \ mod\ (x^n - 1) \end{aligned} \tag{5-1}$$

Two right cyclic shifts of a code word are thus equivalent to the multiplication *modulo* $(x^n - 1)$ of the associated code polynomial by x^2; three right cyclic shifts are equivalent to multiplication *modulo* $(x^n - 1)$ by x^3; etc. Let the cyclic shifts of $\mathbf{c}$ and the associated polynomials be represented as follows.

$$\begin{aligned} \mathbf{c} &= (c_0, c_1, \ldots, c_{n-1}) \leftrightarrow c(x) = c_0 + c_1x + \cdots + c_{n-1}x^{n-1} \\ \mathbf{c}' &= (c_{n-1}, c_0, \ldots, c_{n-2}) \leftrightarrow c'(x) = c_{n-1} + c_0x + \cdots + c_{n-2}x^{n-1} \\ \mathbf{c}'' &= (c_{n-2}, c_{n-1}, \ldots, c_{n-3}) \leftrightarrow c''(x) = c_{n-2} + c_{n-1}x + \cdots + c_{n-3}x^{n-1} \\ &\vdots \\ \mathbf{c}^{(n-1)} &= (c_1, c_2, \ldots, c_0) \leftrightarrow c^{(n-1)}(x) = c_1 + c_2x + \cdots + c_0x^{n-1} \end{aligned} \tag{5-2}$$

Let $a(x) = a_0 + a_1x + a_2x^2 + \cdots + a_{n-1}x^{n-1}$ be an arbitrary polynomial in $\mathrm{GF}(q)[x]/(x^n - 1)$. The product $a(x)c(x)$ is a linear combination of cyclic shifts of $\mathbf{c}$. Since $\mathbf{C}$ forms a vector space, $a(x)c(x)$ must be a valid code polynomial.

$$a(x)c(x) \in \mathbf{C} \qquad \text{for all } a(x) \in \mathrm{GF}(q)[x]/(x^n - 1), c(x) \in \mathbf{C} \tag{5-3}$$

A cyclic code is an ideal (see Definition 3-8) within $\mathrm{GF}(q)[x]/(x^n - 1)$.

Theorem 5-1

C is a q-ary linear cyclic code of length n if and only if the code polynomials in C form an ideal in $GF(q)[x]/(x^n - 1)$.

The following theorem is simply a translation of Theorem 3-7 into the language of cyclic codes.

Theorem 5-2—The Basic Properties of Cyclic Codes

Let C be a q-ary (n, k) linear cyclic code.

1. **Within the set of code polynomials in C there is a unique monic polynomial $g(x)$ with minimal degree $r < n$. $g(x)$ is called the generator polynomial of C.**
2. **Every code polynomial $c(x)$ in C can be expressed uniquely as $c(x) = m(x)g(x)$, where $g(x)$ is the generator polynomial of C and $m(x)$ is a polynomial of degree less than $(n - r)$ in $GF(q)[x]$.**
3. **The generator polynomial $g(x)$ of C is a factor of $x^n - 1$ in $GF(q)[x]$.**

Since $g(x)$ is monic, it takes the form $g(x) = g_0 + g_1x + \cdots + g_{r-1}x^{r-1} + x^r$. We can also claim that $g_0 \neq 0$, for if this were not the case, we could obtain a lower-degree code polynomial by cyclically shifting the code word associated with $g(x)$ one place to the left.

Property 3 in Theorem 5-2 limits the selection of generator polynomials for cyclic codes of length n to divisors of $(x^n - 1)$. In Section 3.3 the factorization of $(x^n - 1)$ into irreducible polynomials in $GF(q)[x]$ was discussed in some detail. Using the results in that section, it is possible to catalog by dimension all q-ary cyclic codes of a given length. For example, consider the set of binary cyclic codes of length 15. The generator polynomial for one of these codes must be a divisor of $(x^{15} - 1)$. $(x^{15} - 1)$ is the product of the distinct minimal polynomials of the nonzero elements in GF(16) with respect to GF(2). We do not have to actually derive these polynomials to catalog the available cyclic codes of length 15. We need only consider the conjugacy classes formed by the powers of α, an element of order 15 (and thus a primitive element) in GF(16).

$$\{1\}$$
$$\{\alpha, \alpha^2, \alpha^4, \alpha^8\}$$
$$\{\alpha^3, \alpha^6, \alpha^9, \alpha^{12}\}$$
$$\{\alpha^5, \alpha^{10}\}$$
$$\{\alpha^7, \alpha^{14}, \alpha^{13}, \alpha^{11}\}$$

$(x^{15} - 1)$ factors into a degree-one binary polynomial, a degree-two binary polynomial, and three degree-four polynomials. It is thus possible to construct a generator polynomial of any degree between 1 and 15 for a binary cyclic code of length 15. On the other hand, consider the binary cyclic codes of length 25. A primitive 25th root of unity β may be found in GF(2^{20}). The powers of β form the following conjugacy classes with respect to GF(2).

$\{1\}$

$\{\beta, \beta^2, \beta^4, \beta^8, \beta^{16}, \beta^7, \beta^{14}, \beta^3, \beta^6, \beta^{12}, \beta^{24}, \beta^{23}, \beta^{21}, \beta^{17}, \beta^9, \beta^{18}, \beta^{11}, \beta^{22}, \beta^{19}, \beta^{13}\}$

$\{\beta^5, \beta^{10}, \beta^{20}, \beta^{15}\}$

$(x^{25} - 1)$ factors into a degree-one binary polynomial, a degree-four binary polynomial, and a degree-twenty binary polynomial. The selection of binary cyclic codes of length 25 is correspondingly limited to the following.

$$\{(25,1), (25,4), (25,5), (25,20), (25,21), (25,24), (25,25)\}$$

Property 2 in Theorem 5-2 suggests a convenient method for mapping data blocks onto code words in a cyclic code. Let $g(x)$ be the rth-degree generator polynomial for an (n, k) q-ary cyclic code $\mathbf{C}$ (all of the cyclic codes discussed in the remainder of this chapter are linear). An $(n - r)$-symbol data block $(m_0, m_1, \ldots, m_{n-r-1})$ can be associated with a **message polynomial** $m_0 + m_1 x + \cdots + m_{n-r-1}x^{n-r-1}$ and encoded through multiplication by the generator polynomial, as shown below. It follows that $\mathbf{C}$ has dimension $k = (n - r)$ and contains $q^{(n-r)}$ code words.

$$\begin{aligned} \mathbf{m} &= (m_0, m_1, \ldots, m_{n-r-1}) \leftrightarrow m(x) = m_0 + m_1 x + \cdots + m_{n-r-1}x^{n-r-1} \\ \mathbf{c}_m &= (c_0, c_1, \ldots, c_{n-1}) \leftrightarrow c_m(x) = m(x)g(x) = c_0 + c_1 x + \cdots + c_{n-1}x^{n-1} \end{aligned} \tag{5-4}$$

The polynomial multiplication in Eq. (5-4) can be reexpressed using matrix multiplication, as shown in Eq. (5-5).

$$\begin{aligned} c_m(x) &= m(x)g(x) = (m_0 + m_1 x + \cdots + m_{n-r-1}x^{n-r-1})g(x) \\ &= m_0 g(x) + m_1 x g(x) + \cdots + m_{n-r-1}x^{n-r-1}g(x) \\ &= [m_0 \quad m_1 \quad \cdots \quad m_{n-r-1}] \begin{bmatrix} g(x) \\ x \cdot g(x) \\ \vdots \\ x^{n-r-1} \cdot g(x) \end{bmatrix} \end{aligned} \tag{5-5}$$

This provides a convenient general form for generator matrices for cyclic codes.

$$\mathbf{c}_m = \mathbf{m} \begin{bmatrix} g_0 & g_1 & \cdots & g_r & & & & \mathbf{0} \\ & g_0 & g_1 & \cdots & g_r & & & \\ & & \ddots & \ddots & \ddots & \ddots & & \\ & & & g_0 & g_1 & \cdots & g_r & \\ \mathbf{0} & & & & g_0 & g_1 & \cdots & g_r \end{bmatrix} = \mathbf{mG} \tag{5-6}$$

A general form for parity check matrices in terms of the coefficients of the generator polynomial can be obtained in a similar manner. It follows from Property 3 in Theorem 5-2 that for every generator polynomial $g(x)$ there exists a **parity**

polynomial $h(x)$ of degree $k = (n - r)$ such that $g(x)h(x) = (x^n - 1)$. Since a polynomial $c(x)$ is a code polynomial if and only if it is a multiple of $g(x)$, it follows that $c(x)$ is a code polynomial if and only if $c(x)h(x) \equiv 0$ *modulo* $(x^n - 1)$.

The polynomial product $c(x)h(x)$ *modulo* $(x^n - 1)$ is a polynomial of the form $s(x) = s_0 + s_1 x + \cdots + s_{n-1}x^{n-1} \in GF(q)[x]/(x^n - 1)$. If $s(x)$ is identically zero, then the n coefficients $\{s_j\}$ must be zero for $j = 0, 1, \ldots, n - 1$, providing us with n parity-check equations.

$$c(x) = \sum_{i=0}^{n-1} c_i x^i, \qquad h(x) = \sum_{j=0}^{n-1} h_j x^j$$

$$\Rightarrow s(x) = \sum_{t=0}^{n-1} s_t x^t \equiv c(x)h(x) \equiv \left(\sum_{i=0}^{n-1} c_i x^i\right)\left(\sum_{j=0}^{n-1} h_j x^j\right) \equiv 0 \; mod \; (x^n - 1) \tag{5-7}$$

$$\Rightarrow s_t = \sum_{i=0}^{n-1} c_i h_{(t-i) \, mod \, n} = 0, \qquad t = 0, 1, \ldots, n - 1$$

The last $(n - k)$ of these parity-check equations can be expressed in matrix form.

$$\mathbf{s} = \begin{bmatrix} s_k \\ s_{k+1} \\ \vdots \\ s_{n-1} \end{bmatrix}^T = \begin{bmatrix} \sum_{i=0}^{n-1} c_i h_{(k-i) \, mod \, n} \\ \sum_{i=0}^{n-1} c_i h_{(k+1-i) \, mod \, n} \\ \vdots \\ \sum_{i=0}^{n-1} c_i h_{(n-1-i) \, mod \, n} \end{bmatrix}^T$$

$$= [c_0 \quad c_1 \quad \cdots \quad c_{n-1}] \begin{bmatrix} h_k & \cdots & h_1 & h_0 & & & & \mathbf{0} \\ & h_k & \cdots & h_1 & h_0 & & & \\ & & \ddots & \ddots & \ddots & \ddots & & \\ & & & h_k & \cdots & h_1 & h_0 & \\ \mathbf{0} & & & & h_k & \cdots & h_1 & h_0 \end{bmatrix}^T = \mathbf{cH}^T \tag{5-8}$$

It has been shown so far that if $\mathbf{c}$ is a code word, then $\mathbf{cH}^T = \mathbf{0}$. The rows of $\mathbf{H}$ are thus vectors in $\mathbf{C}^\perp$. Since $h(x)$ is monic, it follows that the $(n - k)$ rows of $\mathbf{H}$ are linearly independent, spanning a subspace of dimension $(n - k)$. The dimension theorem (Theorem 2-9) then implies that the rows of $\mathbf{H}$ span $\mathbf{C}^\perp$, and $\mathbf{H}$ is thus a valid parity-check matrix.

Theorem 5-3

Let C be an (n, k) cyclic code with generator polynomial $g(x)$. $C^\perp$ is then an $(n, n - k)$ cyclic code with generator polynomial $h^*(x)$, the reciprocal[1] of the parity-check polynomial for C.

[1] The reciprocal of an nth-degree polynomial $f(x) = f_0 + f_1 x + f_2 x^2 + \cdots + f_n x^n$ is $f^*(x) = x^n f(x^{-1}) = f_n + f_{n-1}x + f_{n-2}x^2 + \cdots + f_0 x^n$.

Proof. The parity-check matrix in Eq. (5-8) has the same structure as the generator matrix in Eq. (5-6).

Example 5-1—Binary Cyclic Codes of Length 7

To design a binary cyclic code of length 7, we need to obtain a generator polynomial that is a factor of $x^7 - 1$ in GF(2)[x]. Let α be a root of the primitive polynomial $x^3 + x + 1 = 0$. The conjugacy classes of the powers of α with respect to GF(2) and their associated minimal polynomials are as follows.

$$\begin{array}{lcl} \{1\} & \leftrightarrow & x+1 \\ \{\alpha, \alpha^2, \alpha^4\} & \leftrightarrow & x^3+x+1 \\ \{\alpha^3, \alpha^6, \alpha^5\} & \leftrightarrow & x^3+x^2+1 \end{array}$$

The generator polynomials for binary cyclic codes of length 7 must have one or more of the above minimal polynomials as factors. Consider the case $g(x) = (x^3 + x + 1)(x + 1) = x^4 + x^3 + x^2 + 1$. The corresponding parity polynomial is $h(x) = (x^7 + 1)/g(x) = x^3 + x^2 + 1$. The message polynomials consist of all binary polynomials of degree less than or equal to 2. The corresponding code is thus a (7, 3) code with the following eight code words.

$m(x)g(x)$ FACTORIZATION				**CODE POLYNOMIAL**		**CODE WORD**
$0 \cdot g(x)$	$=$	$c_0(x)$	$=$	0	$\leftrightarrow$	(0000000)
$1 \cdot g(x)$	$=$	$c_1(x)$	$=$	$1 + x^2 + x^3 + x^4$	$\leftrightarrow$	(1011100)
$x \cdot g(x)$	$=$	$c_2(x)$	$=$	$x + x^3 + x^4 + x^5$	$\leftrightarrow$	(0101110)
$x^2 \cdot g(x)$	$=$	$c_3(x)$	$=$	$x^2 + x^4 + x^5 + x^6$	$\leftrightarrow$	(0010111)
$(x^2 + 1) \cdot g(x)$	$=$	$c_4(x)$	$=$	$1 + x^3 + x^5 + x^6$	$\leftrightarrow$	(1001011)
$(x^2 + x + 1) \cdot g(x)$	$=$	$c_5(x)$	$=$	$1 + x + x^4 + x^6$	$\leftrightarrow$	(1100101)
$(x + 1) \cdot g(x)$	$=$	$c_6(x)$	$=$	$1 + x + x^2 + x^5$	$\leftrightarrow$	(1110010)
$(x^2 + x) \cdot g(x)$	$=$	$c_7(x)$	$=$	$x + x^2 + x^3 + x^6$	$\leftrightarrow$	(0111001)

The following generator and parity-check matrices are derived from the generator and parity-check polynomials using the general forms developed in Eqs. (5-4) and (5-6), respectively.

$$\mathbf{G} = \begin{bmatrix} 1 & 0 & 1 & 1 & 1 & 0 & 0 \\ 0 & 1 & 0 & 1 & 1 & 1 & 0 \\ 0 & 0 & 1 & 0 & 1 & 1 & 1 \end{bmatrix}$$

$$\mathbf{H} = \begin{bmatrix} 1 & 1 & 0 & 1 & 0 & 0 & 0 \\ 0 & 1 & 1 & 0 & 1 & 0 & 0 \\ 0 & 0 & 1 & 1 & 0 & 1 & 0 \\ 0 & 0 & 0 & 1 & 1 & 0 & 1 \end{bmatrix}$$ ■

Example 5-2—4-ary Cyclic Codes of Length 5

This example is somewhat more complicated than the previous one in that it involves a nonbinary code whose length is not of the form $p^m - 1$, where p is a prime. The first step is to factor the polynomial $x^5 - 1$ in the ring GF(4)[x]. We use the techniques developed in Chapter 3.

The roots of $x^5 - 1$ are the five distinct powers of a primitive fifth root of unity. The smallest extension field of GF(4) in which a fifth root of unity can be found is GF(16). In Example 3-6 the following exponential and polynomial representations were obtained for GF(16) using α, a root of the primitive polynomial $x^4 + x + 1$.

0	$\alpha^7 = \alpha^3 + \alpha + 1$
1	$\alpha^8 = \alpha^2 + 1$
α	$\alpha^9 = \alpha^3 + \alpha$
α^2	$\alpha^{10} = \alpha^2 + \alpha + 1$
α^3	$\alpha^{11} = \alpha^3 + \alpha^2 + \alpha$
$\alpha^4 = \alpha + 1$	$\alpha^{12} = \alpha^3 + \alpha^2 + \alpha + 1$
$\alpha^5 = \alpha^2 + \alpha$	$\alpha^{13} = \alpha^3 + \alpha^2 + 1$
$\alpha^6 = \alpha^3 + \alpha^2$	$\alpha^{14} = \alpha^3 + 1$

Since α has order 15, α^3 must have order 5 and is thus a primitive fifth root of unity. The conjugacy classes with respect to GF(4) formed by the five distinct powers of α^3 and the corresponding minimal polynomials are as follows.

CONJUGACY CLASS	**ASSOCIATED MINIMAL POLYNOMIAL**
$\{1\}$	$M_0(x) = (x + 1)$
$\{\alpha^3, \alpha^{12}\}$	$M_3(x) = (x + \alpha^3)(x + \alpha^{12}) = x^2 + \alpha^{10}x + 1$
$\{\alpha^6, \alpha^9\}$	$M_6(x) = (x + \alpha^6)(x + \alpha^9) = x^2 + \alpha^5 x + 1$

The generator polynomials for 4-ary cyclic codes of length 5 must factor into one or more of the above polynomials in GF(4)[x]. Having found these polynomials, all of the rest of the computations in this example can be performed in GF(4). To keep our computations from becoming too bogged down in notation, we rename the subfield elements to be $\{0, 1, \alpha^5 = \beta, \alpha^{10} = \delta\}$. The minimal polynomials with respect to GF(4) for the fifth roots of unity are then:

$$M_0(x) = x + 1$$

$$M_3(x) = x^2 + \delta x + 1$$

$$M_6(x) = x^2 + \beta x + 1$$

A set of truth tables was created for GF(4) in Example 2-24. Taking into account the change in labels (we shall avoid the use of integer labels, as suggested), we obtain

$+$	0	1	β	δ
0	0	1	β	δ
1	1	0	δ	β
β	β	δ	0	1
δ	δ	β	1	0

$\cdot$	0	1	β	δ
0	0	0	0	0
1	0	1	β	δ
β	0	β	δ	1
δ	0	δ	1	β

As an example generator polynomial, let $g(x) = M_0(x)M_3(x) = 1 + \beta x + \beta x^2 + x^3$. This generator polynomial defines a (5, 2) 4-ary cyclic code with the following code words. The message polynomials consist of all polynomials of degree one or zero in GF(4)[x]. The corresponding parity polynomial is $h(x) = M_6(x) = x^2 + \beta x + 1$.

m(x)g(x)	**CODE POLYNOMIAL**	**CODE WORD**
$0 \cdot g(x)$	$= c_0(x) = 0$	$\leftrightarrow (00000)$
$1 \cdot g(x)$	$= c_1(x) = 1 + \beta x + \beta x^2 + x^3$	$\leftrightarrow (1\beta\beta 10)$
$\beta \cdot g(x)$	$= c_2(x) = \beta + \delta x + \delta x^2 + \beta x^3$	$\leftrightarrow (\beta\delta\delta\beta 0)$
$\delta \cdot g(x)$	$= c_3(x) = \delta + x + x^2 + \delta x^3$	$\leftrightarrow (\delta 11\delta 0)$
$x \cdot g(x)$	$= c_4(x) = x + \beta x^2 + \beta x^3 + x^4$	$\leftrightarrow (01\beta\beta 1)$
$\beta x \cdot g(x)$	$= c_5(x) = \beta x + \delta x^2 + \delta x^3 + \beta x^4$	$\leftrightarrow (0\beta\delta\delta\beta)$
$\delta x \cdot g(x)$	$= c_6(x) = \delta x + \beta x^2 + \beta x^3 + \delta x^4$	$\leftrightarrow (0\delta 11\delta)$
$(x + 1) \cdot g(x)$	$= c_7(x) = 1 + \delta x + \delta x^3 + x^4$	$\leftrightarrow (1\delta 0\delta 1)$
$(x + \beta) \cdot g(x)$	$= c_8(x) = \beta + \beta x + x^2 + x^4$	$\leftrightarrow (\beta\beta 101)$
$(x + \delta) \cdot g(x)$	$= c_9(x) = \delta + \delta x^2 + x^3 + x^4$	$\leftrightarrow (\delta 0\delta 11)$
$(\beta x + 1) \cdot g(x)$	$= c_{10}(x) = 1 + x^2 + \beta x^3 + \beta x^4$	$\leftrightarrow (101\beta\beta)$
$(\beta x + \beta) \cdot g(x)$	$= c_{11}(x) = 1 + x^2 + \beta x^3 + \beta x^4$	$\leftrightarrow (\beta 101\beta)$
$(\beta x + \delta) \cdot g(x)$	$= c_{12}(x) = \delta + \delta x + \beta x^2 + \beta x^4$	$\leftrightarrow (\delta\delta\beta 0\beta)$
$(\delta x + 1) \cdot g(x)$	$= c_{13}(x) = 1 + x + \delta x^2 + \delta x^4$	$\leftrightarrow (11\delta 0\delta)$
$(\delta x + \beta) \cdot g(x)$	$= c_{14}(x) = \beta + \beta x^2 + \delta x^3 + \delta x^4$	$\leftrightarrow (\beta 0\beta\delta\delta)$
$(\delta x + \delta) \cdot g(x)$	$= c_{15}(x) = \delta + \beta x + \beta x^3 + \delta x^4$	$\leftrightarrow (\delta\beta 0\beta\delta)$

This code has minimum distance 4 and can thus correct all weight-one error patterns while simultaneously detecting all weight-two error patterns.

The following are generator and parity-check matrices for the code.

$$\mathbf{G} = \begin{bmatrix} 1 & \beta & \beta & 1 & 0 \\ 0 & 1 & \beta & \beta & 1 \end{bmatrix}$$

$$\mathbf{H} = \begin{bmatrix} 1 & \beta & 1 & 0 & 0 \\ 0 & 1 & \beta & 1 & 0 \\ 0 & 0 & 1 & \beta & 1 \end{bmatrix}$$ ■

The polynomial multiplication encoding technique has the advantage of simplicity, but in most cases it is not systematic. In the following section it is shown that there is a substantial design benefit to be obtained through having the message bits occupy the last k positions of the code word. This systematic mapping between message blocks and code words can be obtained through a procedure that is only slightly more complicated than polynomial multiplication by the generator polynomial.

Consider an (n, k) cyclic code $\mathbf{C}$ with generator polynomial $g(x)$. We wish to systematically encode a k-symbol message block $\mathbf{m} = (m_0, m_1, \ldots, m_{k-1})$. When the corresponding message polynomial $m(x)$ is multiplied by x^{n-k}, the resulting polynomial $x^{n-k}m(x) = m_0x^{n-k} + m_1x^{n-k+1} + \cdots + m_{k-1}x^{n-1}$ is associated with an n-symbol block $(0, 0, \ldots, 0, m_0, m_1, \ldots, m_{k-1})$ whose first $(n - k)$ coordinates are zero. Now divide $x^{n-k}m(x)$ by $g(x)$ to obtain the following expression.

$$x^{n-k}m(x) = q(x)g(x) + d(x) \tag{5-9}$$

Since $c(x) = [x^{n-k}m(x) - d(x)] = q(x)g(x)$ is a multiple of $g(x)$, it must be a valid code polynomial by Theorem 5-2. The remainder $d(x)$ has degree less than $(n - k)$, the degree of the generator polynomial $g(x)$. $-d(x)$ can thus be associated with an n-symbol block whose last k coordinates are zero: $-d(x) \leftrightarrow (-d_0, -d_1, \ldots, -d_{n-k-1}, 0, 0, \ldots, 0)$. The code word associated with the code polynomial $c(x) = [x^{n-k}m(x) - d(x)]$ thus has the form

$$c(x) = [x^{n-k}m(x) - d(x)] \leftrightarrow (-d_0, -d_1, \ldots, -d_{n-k-1}, m_0, m_1, \ldots, m_{k-1}).$$

$m(x)$ has been systematically mapped to a code word. The encoding algorithm is summarized below.

Systematic encoding algorithm for an (n, k) cyclic code C

Step 1. Multiply the message polynomial $m(x)$ by x^{n-k}.

Step 2. Divide the result of Step 1 by the generator polynomial $g(x)$. Let $d(x)$ be the remainder.

Step 3. Set $c(x) = x^{n-k}m(x) - d(x)$.

Example 5-3—Systematic Encoding of the (7, 3) Binary Code in Example 5-1

The binary $(7, 3)$ code in Example 5-1 has generator polynomial $g(x) = x^4 + x^3 + x^2 + 1$. Consider the encoding of the message block (101).

Step 1. $x^{n-k}m(x) = x^4(x^2 + 1) = x^6 + x^4$

Step 2.

$$\begin{array}{r}
x^2 + x + 1 = q(x) \\
x^4 + x^3 + x^2 + 1 \overline{) x^6 + x^4 \qquad\qquad\qquad} \\
\underline{x^6 + x^5 + x^4 + x^2 \qquad\quad} \\
x^5 + x^2 \qquad\quad \\
\underline{x^5 + x^4 + x^3 + x \quad} \\
x^4 + x^3 + x^2 + x \\
\underline{x^4 + x^3 + x^2 + 1} \\
x + 1 = d(x)
\end{array}$$

Step 3. $c_m(x) = x^{n-k}m(x) - d(x) = 1 + x + x^4 + x^6 \leftrightarrow c_m = (1100101)$

The following table shows the systematic mapping for all of the code words in the $(7, 4)$ code.

$m(x)$	CODE POLYNOMIAL	CODE WORD
0	$= c_0(x) = 0$	$\leftrightarrow (0000000)$
1	$= c_1(x) = 1 + x^2 + x^3 + x^4$	$\leftrightarrow (1011100)$
x	$= c_2(x) = 1 + x + x^2 + x^5$	$\leftrightarrow (1110010)$
x^2	$= c_3(x) = x + x^2 + x^3 + x^6$	$\leftrightarrow (0111001)$
$1 + x^2$	$= c_4(x) = 1 + x + x^4 + x^6$	$\leftrightarrow (1100101)$
$1 + x + x^2$	$= c_5(x) = x^2 + x^4 + x^5 + x^6$	$\leftrightarrow (0010111)$
$1 + x$	$= c_6(x) = x + x^3 + x^4 + x^5$	$\leftrightarrow (0101110)$
$x + x^2$	$= c_7(x) = 1 + x^3 + x^5 + x^6$	$\leftrightarrow (1001011)$

The systematic generator matrix for this code is obtained by selecting as rows those code words associated with the message blocks (100), (010), and (001). The corresponding parity-check matrix is obtained using the form in Figure 4-6.

$$\mathbf{G} = \left[\begin{array}{cccc|ccc} 1 & 0 & 1 & 1 & 1 & 0 & 0 \\ 1 & 1 & 1 & 0 & 0 & 1 & 0 \\ 0 & 1 & 1 & 1 & 0 & 0 & 1 \end{array}\right]$$

$$\mathbf{H} = \left[\begin{array}{cccc|ccc} 1 & 0 & 0 & 0 & 1 & 1 & 0 \\ 0 & 1 & 0 & 0 & 0 & 1 & 1 \\ 0 & 0 & 1 & 0 & 1 & 1 & 1 \\ 0 & 0 & 0 & 1 & 1 & 0 & 1 \end{array}\right]$$

■

Example 5-4—Systematic Encoding of the (5, 2) 4-ary Code in Example 5-2

The 4-ary (5, 2) code in Example 5-2 has generator polynomial $g(x) = 1 + \beta x + \beta x^2 + x^3$. Consider the encoding of the message block (1δ).

Step 1. $x^{n-k}m(x) = x^3(1 + \delta x) = x^3 + \delta x^4$

Step 2.

$$\begin{array}{r} \delta x = q(x) \\ x^3 + \beta x^2 + \beta x + 1 \overline{)\delta x^4 + x^3 \qquad\qquad} \\ \underline{\delta x^4 + x^3 + x^2 + \delta x} \\ x^2 + \delta x = d(x) \end{array}$$

Step 3. $c_m(x) = x^{n-k}m(x) - d(x) = \delta x + x^2 + x^3 + \delta x^4 \leftrightarrow c_m = (0\delta 11\delta)$

The following table shows the systematic mapping for all of the code words in the (5, 2) code.

$m(x)$	CODE POLYNOMIAL	CODE WORD
0	$c_0(x) = 0$	$\leftrightarrow (00000)$
1	$c_1(x) = 1 + \beta x + \beta x^2 + x^3$	$\leftrightarrow (1\beta\beta 10)$
β	$c_2(x) = \beta + \delta x + \delta x^2 + \beta x^3$	$\leftrightarrow (\beta\delta\delta\beta 0)$
δ	$c_3(x) = \delta + x + x^2 + \delta x^3$	$\leftrightarrow (\delta 11\delta 0)$
x	$c_4(x) = \beta + \beta x + x^2 + x^4$	$\leftrightarrow (\beta\beta 101)$
βx	$c_5(x) = \delta + \delta x + \beta x^2 + \beta x^4$	$\leftrightarrow (\delta\delta\beta 0\beta)$
δx	$c_6(x) = 1 + x + \delta x^2 + \delta x^4$	$\leftrightarrow (11\delta 0\delta)$
$(x + 1)$	$c_7(x) = \delta + \delta x^2 + x^3 + x^4$	$\leftrightarrow (\delta 0\delta 11)$
$(x + \beta)$	$c_8(x) = x + \beta x^2 + \beta x^3 + x^4$	$\leftrightarrow (01\beta\beta 1)$
$(x + \delta)$	$c_9(x) = 1 + \delta x + \delta x^3 + x^4$	$\leftrightarrow (1\delta 0\delta 1)$
$(\beta x + 1)$	$c_{10}(x) = 1 + x^2 + \beta x^3 + \beta x^4$	$\leftrightarrow (\beta 101\beta)$
$(\beta x + \beta)$	$c_{11}(x) = 1 + x^2 + \beta x^3 + \beta x^4$	$\leftrightarrow (101\beta\beta)$
$(\beta x + \delta)$	$c_{12}(x) = \beta x + \delta x^2 + \delta x^3 + \beta x^4$	$\leftrightarrow (0\beta\delta\delta\beta)$
$(\delta x + 1)$	$c_{13}(x) = \delta x + \beta x^2 + \beta x^3 + \delta x^4$	$\leftrightarrow (0\delta 11\delta)$
$(\delta x + \beta)$	$c_{14}(x) = \delta + \beta x + \beta x^3 + \delta x^4$	$\leftrightarrow (\delta\beta 0\beta\delta)$
$(\delta x + \delta)$	$c_{15}(x) = \beta + \beta x^2 + \delta x^3 + \delta x^4$	$\leftrightarrow (\beta 0\beta\delta\delta)$

The systematic generator matrix for this code is obtained by selecting as rows those code words associated with the message blocks (100), (010), and (001). The corresponding parity-check matrix is obtained using the form in Figure 4-6.

$$\mathbf{G} = \left[\begin{array}{ccc|cc} 1 & \beta & \beta & 1 & 0 \\ \beta & \beta & 1 & 0 & 1 \end{array}\right]$$

$$\mathbf{H} = \left[\begin{array}{ccc|cc} 1 & 0 & 0 & 1 & \beta \\ 0 & 1 & 0 & \beta & \beta \\ 0 & 0 & 1 & \beta & 1 \end{array}\right]$$

■

5.2 SHIFT-REGISTER ENCODERS AND DECODERS FOR CYCLIC CODES

Data rates in the hundreds or even thousands of megabits per second are common in many applications. Unfortunately such data rates severely limit the device technologies that can be used to implement error control systems and, within a given technology, limit also the complexity of the circuits. It is thus extremely important to note that encoders and decoders for cyclic codes can be implemented using simple exclusive-OR gates, switches, shift registers, and, in the case of nonbinary encoders and decoders, finite field adder and multiplier circuits. Shift registers are among the simplest of digital circuits, consisting of a collection of flip-flops connected in series. They are thus operable at speeds quite close to the maximum speed possible for a single gate using a given device technology. In this section a series of shift-register encoders and decoders for cyclic codes are examined. We begin with nonsystematic and systematic encoders and then proceed to error detecting and error correcting decoders.

5.2.1 Binary and Nonbinary Shift-Register Operational Elements

Figures 5-1 and 5-2 show the symbology that is used in this and subsequent chapters to describe the operational elements used in shift-register (SR) circuits. We first consider the simple binary elements and then show how they can be used to construct the nonbinary elements.

The binary operational elements are extremely simple: the half-adder (no carry) is an exclusive-OR gate, the SR cell is a flip-flop, and the fixed multiplier is implemented through either the existence of a connection (multiplication by one) or by the absence of a connection (multiplication by zero).

A nonbinary element performs addition or multiplication in the field $GF(p^m)$ or stores a value from $GF(p^m)$. These elements are slightly more complex than the binary elements, for they are generally constructed using several of the binary elements. We shall assume here that $GF(p^m)$ is a binary extension field and thus of the form $GF(2^m)$. The following can be extended to fields of the form $GF(p^m)$, where $p \neq 2$, but the circuits are substantially more complicated.

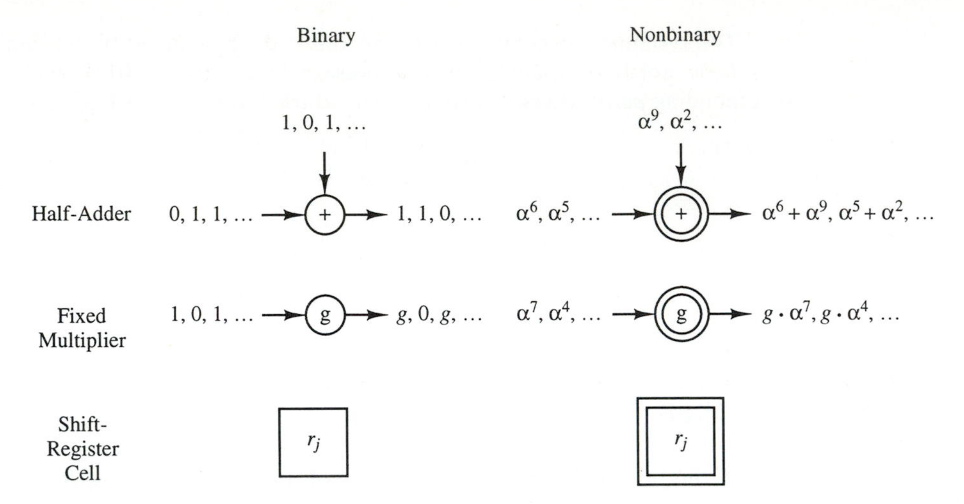

Figure 5-1. Shift-Register Operational Elements

It was shown in Chapter 2 that a pair of elements α and β in $GF(2^m)$ can be represented as m-tuples $(a_0, a_1, \ldots, a_{m-1})$ and $(b_0, b_1, \ldots, b_{m-1})$, respectively. $GF(2^m)$ addition consists of the coordinate-by-coordinate binary addition of the two m-tuples. The corresponding hardware is shown in Figure 5-2.

Multiplication by a fixed finite field element is achieved through a similar circuit. Consider the multiplication of an arbitrary value $\beta = b_0 + b_1\alpha + b_2\alpha^2 + b_3\alpha^3$ by a fixed value $g = 1 + \alpha$, both elements in $GF(2^4)$ with α a root of the

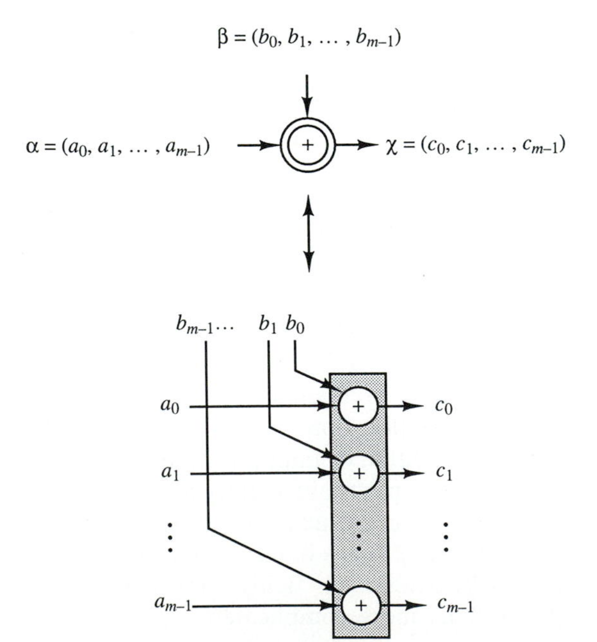

Figure 5-2. Nonbinary Addition Circuit

primitive polynomial $x^4 + x + 1$. The product $\chi = \beta \cdot g$ is obtained by multiplying together the polynomial representations for β and g and reducing the result *modulo* $p(x)$.

$$\begin{aligned}\chi &= (c_0 + c_1\alpha + c_2\alpha^2 + c_3\alpha^3) \\ &= (b_0 + b_1\alpha + b_2\alpha^2 + b_3\alpha^3)(1 + \alpha) \\ &= b_0 + (b_0 + b_1)\alpha + (b_1 + b_2)\alpha^2 + (b_2 + b_3)\alpha^3 + b_3\alpha^4 \qquad (5\text{-}10) \\ &= b_0 + (b_0 + b_1)\alpha + (b_1 + b_2)\alpha^2 + (b_2 + b_3)\alpha^3 + b_3(\alpha + 1) \\ &= [(b_0 + b_3) + (b_0 + b_1 + b_3)\alpha + (b_1 + b_2)\alpha^2 + (b_2 + b_3)]\alpha^3\end{aligned}$$

Figure 5-3 shows how the last expression in Eq. (5-10) is used to realize a circuit that multiplies elements in GF(4) by $g = (1 + \alpha)$.

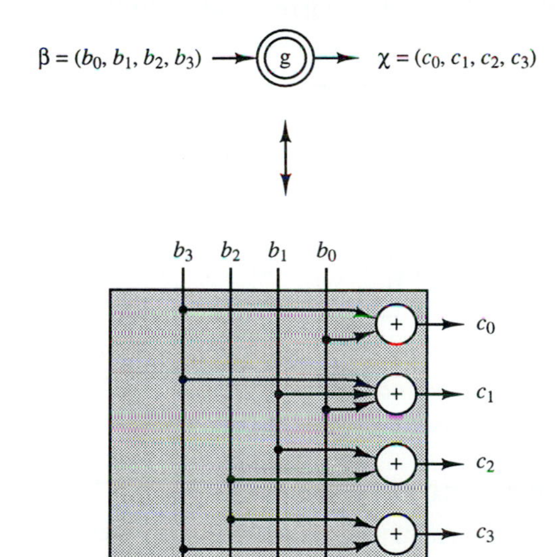

Figure 5-3. Nonbinary Fixed Multiplier Circuit

The nonbinary shift-register cells are implemented as one might expect: a flip-flop is dedicated to the storage of each coordinate in the m-tuple representation of the field element, as shown in Figure 5-4 for $\beta = (b_0, b_1, \ldots, b_{m-1})$.

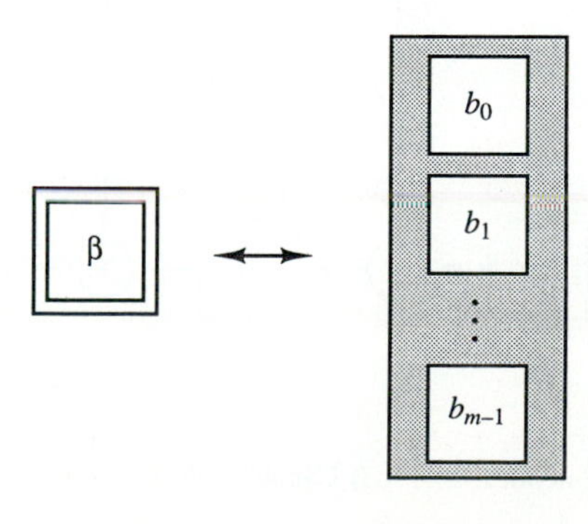

Figure 5-4. Nonbinary Shift-Register Cell

5.2.2 Nonsystematic and Systematic Encoders

In Section 5.1 a nonsystematic encoding technique suggested by Property 2 of Theorem 5-2 was discussed. In this approach, a k-symbol message block $\mathbf{m} = (m_0, m_1, \ldots, m_{k-1})$ is associated with a polynomial $m(x) = m_0 + m_1 x + \cdots + m_{k-1}x^{k-1}$. The message polynomial is multiplied by the code's generator polynomial $g(x)$ to obtain a code polynomial $c_m(x) = c_0 + c_1 + \cdots + c_{n-1}x^{n-1}$. The associated n-symbol block $(c_0, c_1, \ldots, c_{n-1})$ is the desired code word.

The product $m(x)g(x)$ can be computed as a weighted sum of cyclic shifts of $g(x)$. The corresponding SR circuit is shown in Figure 5-5. The coefficients of $m(x)$ are fed into the SR in descending order of index. Each time a new coefficient is placed on the input line, the SR clock is pulsed, and the contents of the SR cells shifted one cell to the right. When the final coefficient (m_0) has been fed into the SR, the SR cells contain the code word $\mathbf{c}_m$ corresponding to the message block $\mathbf{m}$.

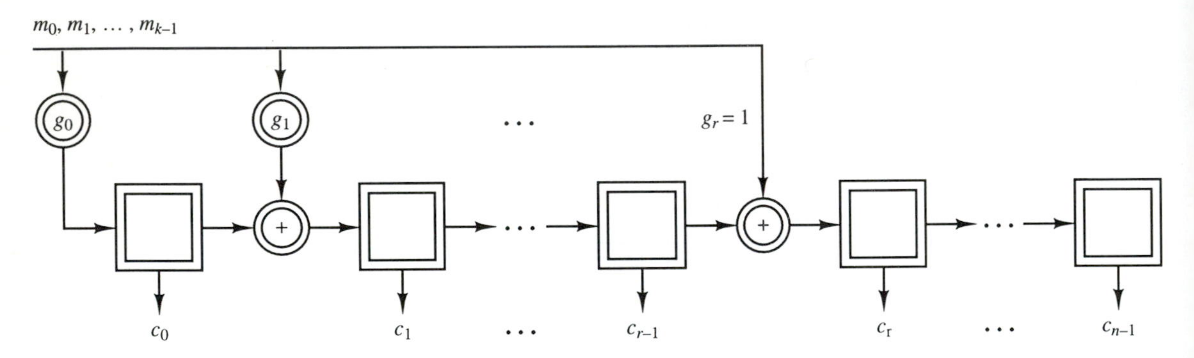

Figure 5-5. Shift-Register Encoder (Nonsystematic)

Example 5-6—Shift-Register Polynomial Multiplication

Figure 5-6 shows the polynomial multiplier encoder for the (7, 3) binary code discussed in Example 5-1. In Figure 5-7 the contents of the SR cells in this encoder are traced during the encoding of the message block $\mathbf{m} = (101)$. At the end of the encoding process the SR cells contain the code word $\mathbf{c}_4 = (1001011)$.

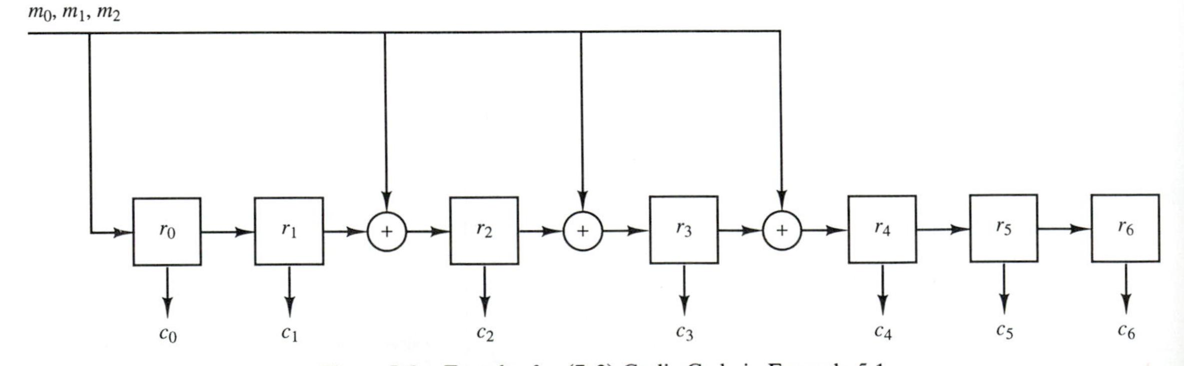

Figure 5-6. Encoder for (7, 3) Cyclic Code in Example 5-1

SR cells	r_0	r_1	r_2	r_3	r_4	r_5	r_6
Initial state	0	0	0	0	0	0	0
Input $m_2 = 1$	1	0	1	1	1	0	0
Input $m_1 = 0$	0	1	0	1	1	1	0
Input $m_0 = 1$	1	0	0	1	0	1	1
Final state $= \mathbf{c}_4$	1	0	0	1	0	1	1

Figure 5-7. Shift-Register Cell Contents During Encoding of $m(x) = x^2 + 1$ ■

The systematic encoding procedure discussed at the end of the last section also has an SR implementation, though it is slightly more complex than the nonsystematic SR encoder. The three steps in the encoding algorithm have distinct SR operations associated with them.

Step 1. Multiply the message polynomial $m(x)$ ***by*** x^{n-k}. Polynomial multiplication by SR was demonstrated during the discussion of the nonsystematic encoder. Applying the same techniques here, multiplication by x^{n-k} is performed as shown in Figure 5-8.

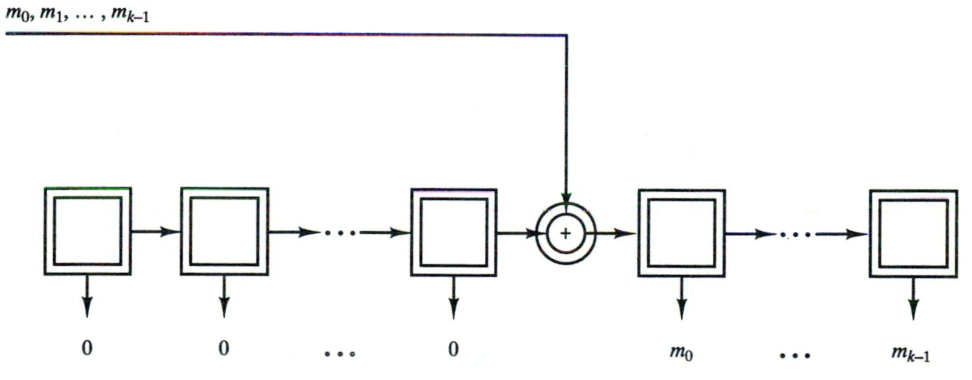

Figure 5-8. Shift-Register Multiplication of $m(x)$ by x^{n-k}

Step 2. Divide the result of Step 1 by the generator polynomial $g(x)$***. Let*** $d(x)$ ***be the remainder.*** Polynomial division is performed through the use of a linear feedback shift register (LFSR). The following circuit divides $a(x) = a_0 + a_1x + \cdots + a_{n-1}x^{n-1}$ by $g(x) = g_0 + g_1x + \cdots + g_{r-1}x^{r-1} + x^r$ and retains the remainder $d(x) = d_0 + d_1x + \cdots + d_{r-1}x^{r-1}$. The symbols $a_0, a_1, \ldots, a_{n-2}, a_{n-1}$ are fed into the shift register one at a time in order of decreasing index. When the last symbol (a_0) has been fed into the rightmost SR cell, the SR cells will contain the coefficients of the remainder polynomial.

The SR divider in Figure 5-9 operates in the same manner in which one performs long division. The following example demonstrates this for the binary case.

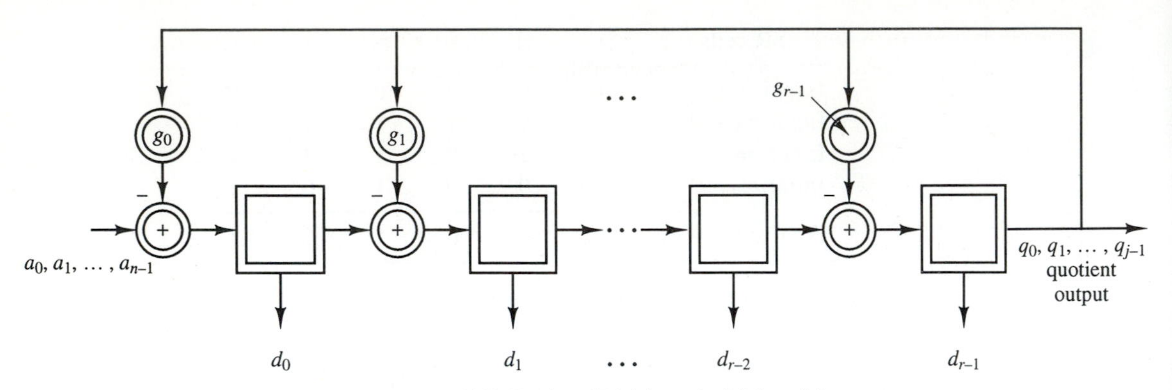

Figure 5-9. Shift-Register Division of $a(x)$ by $g(x)$

Example 5-7—Shift-Register Polynomial Division

Consider the division of the binary polynomial $x^6 + x^4$ by $x^4 + x^3 + x^2 + 1$. The circuit in Figure 5-10 is obtained using the general form in Figure 5-9. The contents of the SR cells at every step of the operation are shown in Figure 5-11.

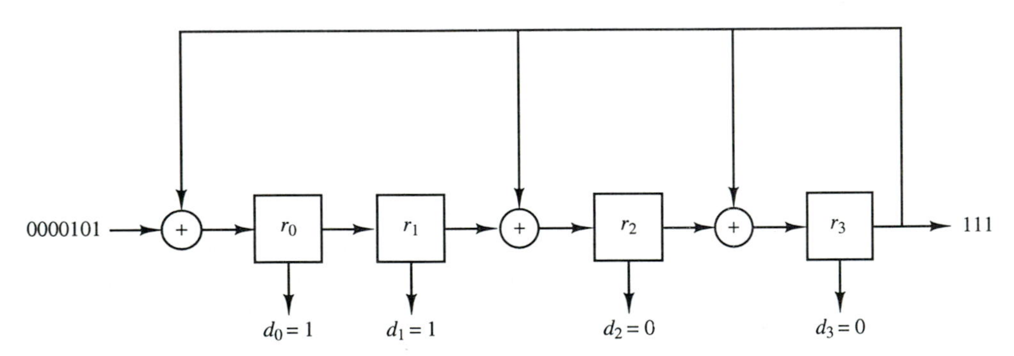

Figure 5-10. Shift-Register Division of $x^6 + x^4$ by $x^4 + x^3 + x^2 + 1$

SR cells	r_0	r_1	r_2	r_3	
Initial state	0	0	0	0	
Input $a_6 = 1$	1	0	0	0	
Input $a_5 = 0$	0	1	0	0	
Input $a_4 = 1$	1	0	1	0	
Input $a_3 = 0$	0	1	0	1	
Input $a_2 = 0$	1	0	0	1	
Input $a_1 = 0$	1	1	1	1	
Input $a_0 = 0$	1	1	0	0	
Final state $= r$	1	1	0	0	$\Leftrightarrow d(x) = x + 1$

Figure 5-11. Shift-Register Cell Contents During Division of $x^6 + x^4$ by $x^4 + x^3 + x^2 + 1$

Step 3. Set $c(x) = x^{n-k}m(x) - d(x)$. The final step in the design of the systematic encoder is achieved by combining the two SR circuits in Figures 5-8 and 5-9, as shown in Figure 5-12.

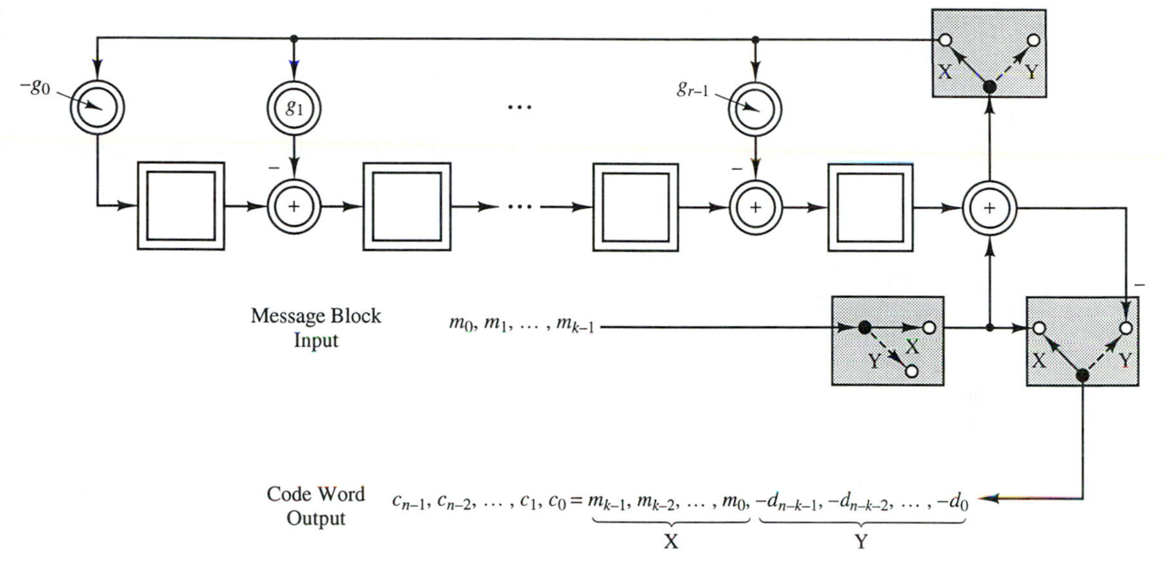

Figure 5-12. Systematic Encoder for Cyclic Codes

During the first step of the encoding operation the three switches are placed in position X and the k message symbols are fed into the encoder in order of decreasing index. The k message symbols are simultaneously sent to the transmitter, for they comprise the last k coordinates of the systematic code word. This, by the way, is the primary rationale for placing the message symbols at the end of a systematic code word. After the kth message symbol has been fed into the SR, the switches are moved to position Y. At this point the SR cells contain the remainder generated by the division operation. These symbols are then shifted out of the SR and to the transmitter, where they comprise the remaining systematic code word coordinates.

It is also possible to design a systematic encoder based on the coefficients of the code's parity polynomial $h(x)$. In Eq. (5-7) the following parity equations were derived for an (n, k) cyclic code **C** with parity polynomial $h(x)$.

$$s_t = \sum_{i=0}^{n-1} c_i h_{(t-i) \bmod n} = 0, \qquad t = 0, 1, \ldots, n-1 \tag{5-11}$$

We now change the counting variable from i to $j = t - i$ and note that $h_x = 0$ for all x greater than k.

$$s_t = \sum_{j=0}^{k} c_{(t-j) \bmod n} h_j = 0, \qquad t = 0, 1, \ldots, n-1 \tag{5-12}$$

Since h_k must be 1 (recall that the generator polynomial is monic), the expressions in Eq. (5-12) can be solved for the code word coordinate values c_{t-k}.

$$c_{(t-k)\ mod\ n} = -\sum_{j=0}^{k-1} c_{(t-j)\ mod\ n} h_j, \qquad t = 0, 1, \ldots, n-1 \tag{5-13}$$

Equation (5-13) is a recursive equation. If we know k consecutive values in a sequence $\{c_i\}$, we can generate a $(k + 1)$st value; using the new value we can generate a $(k + 2)$nd value, and so on. This is precisely the position we find ourselves in when encoding data systematically. The values for the code word coordinates $c_{n-k}, c_{n-k+1}, \ldots, c_{n-1}$ are provided by the data. The value for the coordinate c_{n-k-1} is then computed recursively using Eq. (5-13) with $t = n - 1$. The new value is then used in the computation of c_{n-k-2} $(t = n - 2)$, and so on until c_0 has been computed $(t = k)$. Since Eq. (5-13) is little more than polynomial multiplication, the SR multiplier in Figure 5-5 can be adapted to create a systematic encoder based on the parity polynomial, as shown in Figure 5-13.

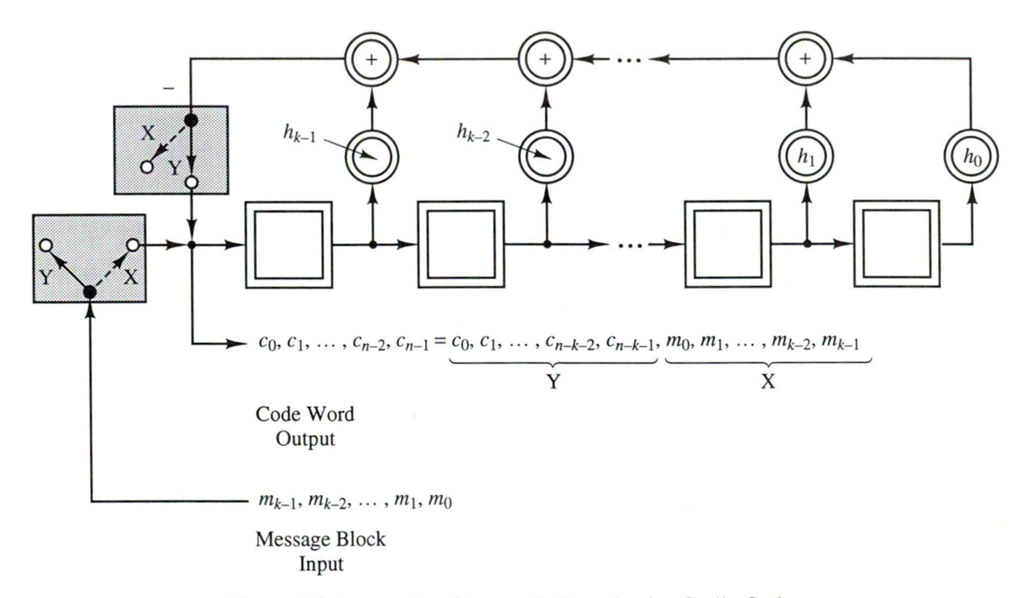

Figure 5-13. Another Systematic Encoder for Cyclic Codes

When the switches are in position X, the message block is simultaneously shifted into the SR cells and sent to the transmitter. The switches are then placed in the Y position, allowing the SR to recursively generate the remaining coordinate values.

The complexity of the encoder in Figure 5-12 is a function of the degree of the generator polynomial $(n - k)$, while that of the encoder in Figure 5-13 is a function of the degree of the parity polynomial (k). The Figure 5-12 encoder is thus a natural selection for high rate coding applications (code dimension $\geq$ code length/2), while the Figure 5-13 encoder is a better choice for low rate applications.

5.2.3 Shift-Register Syndrome Computation and Error Detection

Cyclic codes allow for a number of highly convenient techniques for detecting errors using shift-register circuits. We first consider the systematic case. Recall that a systematic encoding is obtained by dividing the generator polynomial $g(x)$ into the weighted message polynomial $x^{n-r}m(x)$. The code polynomial $c(x)$ is equal to $x^{n-r}m(x)$ minus the remainder $d(x)$ obtained from the division operation. The transmitted code word thus has the following form.

$$\mathbf{c} = (c_0, c_1, \ldots, c_{n-1}) = (\underbrace{-d_0, -d_1, \ldots, -d_{n-k-1}}_{\text{remainder block } \mathbf{d}}, \underbrace{m_0, m_1, \ldots, m_{k-1}}_{\text{message block } \mathbf{m}})$$

Error detection is performed on a received vector $\mathbf{r}$ as follows:

1. Construct an estimated message block $\mathbf{m}'$ and remainder block $\mathbf{d}'$ using the values in the message and parity positions of the received word $\mathbf{r}$.

$$\mathbf{r} = (r_0, r_1, \ldots, r_{n-1}) = (\underbrace{-d_0', -d_1', \ldots, -d_{n-k-1}'}_{\text{estimated remainder block } \mathbf{d}'}, \underbrace{m_0', m_1', \ldots, m_{k-1}'}_{\text{estimated message block } \mathbf{m}'})$$

2. Encode $\mathbf{m}'$ using an encoder identical to that used by the transmitter (e.g., see Figure 5-12) and obtain an estimated remainder block $\mathbf{d}''$.
3. Compare $\mathbf{d}'$ to $\mathbf{d}''$. If they are not the same, then $\mathbf{r}$ is not a valid code word, indicating the presence of errors in the received word.

This approach to error detection has a significant advantage. The encoder and error detection circuits are essentially identical, and the design process is correspondingly simplified.

Recall from Chapter 4 that the syndrome is defined by the matrix product $\mathbf{r}\mathbf{H}^T$. Using the systematic parity-check matrix in Figure 4-5, it can be shown that the syndrome $\mathbf{s}$ for a received word $\mathbf{r}$ is simply the difference $\mathbf{s} = \mathbf{d}' - \mathbf{d}''$.

Syndromes for nonsystematic cyclic codes can also be computed through the use of shift-register techniques. Equations (5-7) and (5-8) showed that the syndrome $\mathbf{s} = (s_0, s_1, \ldots, s_{n-k-1}) = \mathbf{r}\mathbf{H}^T$ is the set of coefficients of the following polynomial product

$$s(x) = s_0 + s_1 x + \cdots + s_{n-k-1}x^{n-k-1} = r(x)h(x) \text{ } modulo \text{ } (x^n - 1) \quad (5\text{-}14)$$

Since $g(x)h(x) = x^n - 1$, $s(x)$ can also be computed through division by the generator polynomial.

$$r(x) = a(x)g(x) + s(x), \qquad \text{where degree } (s(x)) < \text{degree } (g(x)) \quad (5\text{-}15)$$

Both Equations (5-14) and (5-15) can be implemented using the polynomial multiplication and division circuits discussed earlier. For example, Eq. (5-15) can be implemented using the SR circuit in Figure 5-9. The received vector **r** is fed into the SR circuit one symbol at a time in order of decreasing index. When the last symbol has entered the SR, the SR cells will contain the desired syndrome. For error detection, we need only add a circuit that tests the SR cells for nonzero values to complete the design.

5.2.4 Shift-Register Syndrome Computation and Error Correction

In Chapter 4 the syndrome values were used in a table look-up scheme to implement error correction using an arbitrary linear code **C**. The basic idea is that every syndrome value corresponds to a coset of **C** in V_n^q. The error pattern with the lowest weight within a given coset is the most likely to occur, and is thus selected as the coset leader. Maximum likelihood error correction is performed by computing the syndrome for a received vector, looking up the corresponding coset leader, and subtracting the coset leader from the received vector. The primary drawback to this approach, it will be recalled, was the size of the syndrome look-up table. For an (n, k) q-ary code the syndrome table must contain q^{n-k} n-tuples, with each n-tuple coordinate taking on one of q different values. Cyclic codes have an interesting property that allows us to cut the size of the syndrome table to $(1/n)$th its original size.

Theorem 5-3

Let $s(x)$ be the syndrome polynomial corresponding to a received polynomial $r(x)$. Let $r^{(1)}(x)$ be the polynomial obtained by cyclically shifting the coefficients of $r(x)$ once to the right. Then the remainder obtained when dividing $xs(x)$ by $g(x)$ is the syndrome $s^{(1)}(x)$ corresponding to $r^{(1)}(x)$.

Proof. The polynomial $r^{(1)}(x)$ can be expressed as follows.

$$r^{(1)}(x) = xr(x) - r_{n-1}(x^n - 1)$$

$r^{(1)}(x)$ and $r(x)$ are then expressed as multiples of $g(x)$ added to a remainder:

$$r^{(1)}(x) = x[a(x)g(x) + s(x)] - r_{n-1}[g(x)h(x)] = [b(x)g(x) + d(x)]$$

$d(x)$ is the remainder resulting from the division of $r^{(1)}(x)$ by $g(x)$ and is thus the syndrome $s^{(1)}(x)$ associated with $r^{(1)}(x)$. The division of $xs(x)$ by $g(x)$ is then seen to also have the remainder $d(x) = s^{(1)}(x)$, proving the theorem.

$$xs(x) = [b(x)g(x) + d(x)] - x[a(x)g(x)] + r_{n-1}[g(x)h(x)]$$
$$= [b(x) - xa(x) + r_{n-1}h(x)]g(x) + d(x) \qquad \textbf{QED}$$

This theorem greatly reduces the size of syndrome look-up table decoders for cyclic codes. Given a received vector **r**, the corresponding syndrome **s** is obtained by entering **r** into an SR division circuit. As before, when the last symbol of **r** has been shifted into the circuit, the SR cells contain the syndrome. The input to the circuit of an additional zero at this point is equivalent to multiplying $s(x)$ by x and dividing by $g(x)$. The remainder $s^{(1)}(x)$ is now in the SR cells. According to Theorem 5-3, $s^{(1)}(x)$ is the syndrome for $r^{(1)}(x)$. This process can be repeated n times, bringing us back to the starting point with the original syndrome in the SR cells. We thus need only store one syndrome **s** for an error pattern **e** and all cyclic shifts of **e**. A syndrome decoder with a reduced-size look-up table is used as follows.

Syndrome decoder for cyclic codes

1. Set a counting variable j to 0.
2. Compute the syndrome **s** for a received vector **r**.
3. Look for the error pattern **e** corresponding to **s** in the syndrome look-up table. If there is such a value, go to step 8.
4. Increment the counter j by 1 and enter a zero into the SR circuit at the input, computing $\mathbf{s}^{(j)}$.
5. Look for the error pattern $\mathbf{e}^{(j)}$ corresponding to $\mathbf{s}^{(j)}$ in the syndrome look-up table. If there is such a value, go to step 7.
6. Go to step 4.
7. Determine the error pattern **e** corresponding to **s** by cyclically shifting $\mathbf{e}^{(j)}$ j times to the left.
8. Subtract **e** from **r**, obtaining the code word **c**.

The following example shows how the above approach can be used to realize a fairly simple single-error-correcting decoder. For a more detailed exposition on decoders for the general class of cyclic codes, the reader is referred to Lin and Costello, Section 4.5 [Lin1]. It should be noted that the complexity of decoders for general cyclic codes increases exponentially with code length and with the number of errors to be corrected per received word. In the following chapters we will examine specific cyclic codes whose inherent algebraic structure allows for simple, yet extremely powerful, decoder designs.

Example 5-8—A Shift-Register Error Correction Circuit

Consider the (7, 4) binary cyclic code generated by $g(x) = x^3 + x + 1$. The parity-check polynomial is $(x^7 + 1)/(x^3 + x + 1) = x^4 + x^2 + x + 1$. Using the general form in Eq. (5-8), we obtain the following parity-check matrix.

$$\mathbf{H} = \begin{bmatrix} 1 & 0 & 1 & 1 & 1 & 0 & 0 \\ 0 & 1 & 0 & 1 & 1 & 1 & 0 \\ 0 & 0 & 1 & 0 & 1 & 1 & 1 \end{bmatrix}$$

This code is a Hamming code (the columns of the parity-check matrix consist of all nonzero binary 3-tuples) and is thus a perfect single-error-correcting code. The correctable error patterns and their corresponding syndromes are as follows.

ERROR PATTERN	ERROR POLYNOMIAL	SYNDROME	SYNDROME POLYNOMIAL
(0000000)	0	(000)	0
(1000000)	1	(100)	1
(0100000)	x	(010)	x
(0010000)	x^2	(001)	x^2
(0001000)	x^3	(110)	$1 + x$
(0000100)	x^4	(011)	$x + x^2$
(0000010)	x^5	(111)	$1 + x + x^2$
(0000001)	x^6	(101)	$1 + x^2$

Since the error patterns consist of all of the cyclic shifts of (0000001), our decoder need only be able to recognize one of the seven nonzero syndromes to be able to correct all of the nonzero error patterns. The syndrome corresponding to (0000001) is the best choice, for it allows us to release corrected code-word bits before the error location has actually been identified.

Figure 5-14 shows the complete decoder. Decoding begins by first setting all of the SR cells in the syndrome computation circuit to zero. Copies of the received bits are then shifted into the received-word buffer and the syndrome computation circuit simultaneously. Once the received word has been completely shifted into the buffer, the SR cells in the syndrome computation circuit contain the syndrome for the received word. As we continue to shift the contents of the received-word buffer and the syndrome computation circuit, the syndrome computation circuit computes the syndromes for the cyclically shifted versions of the received word and corrected bits leave the decoder at the output. If at any point the computed syndrome is $\mathbf{s} = 101$, it is "recognized" by the AND gate, whose output goes to 1. This value is then used to complement the rightmost bit in the buffer as it leaves the buffer.

Consider the case in which the received word is 1101011. The following chart tracks the computation of the syndrome and the decoder output during the decoding process.

RECEIVED WORD BUFFER	SYNDROME	DECODER OUTPUT
1101011	010	1
-110101	001	11
--11010	110	011
---1101	011	1011
----110	111	01011
-----11	101	001011
------1	100	1001011

The sixth syndrome computed is 101 and thus is recognized by the decoder. Since it is the sixth syndrome computed, the corresponding error pattern must be shifted cyclically five times to the left before it is added to the received word. This is handled by shifting the received-word buffer synchronously with the syndrome computation circuit. The erroneous bit (sixth from the right) is automatically in position to be corrected as it leaves the received-word buffer. ■

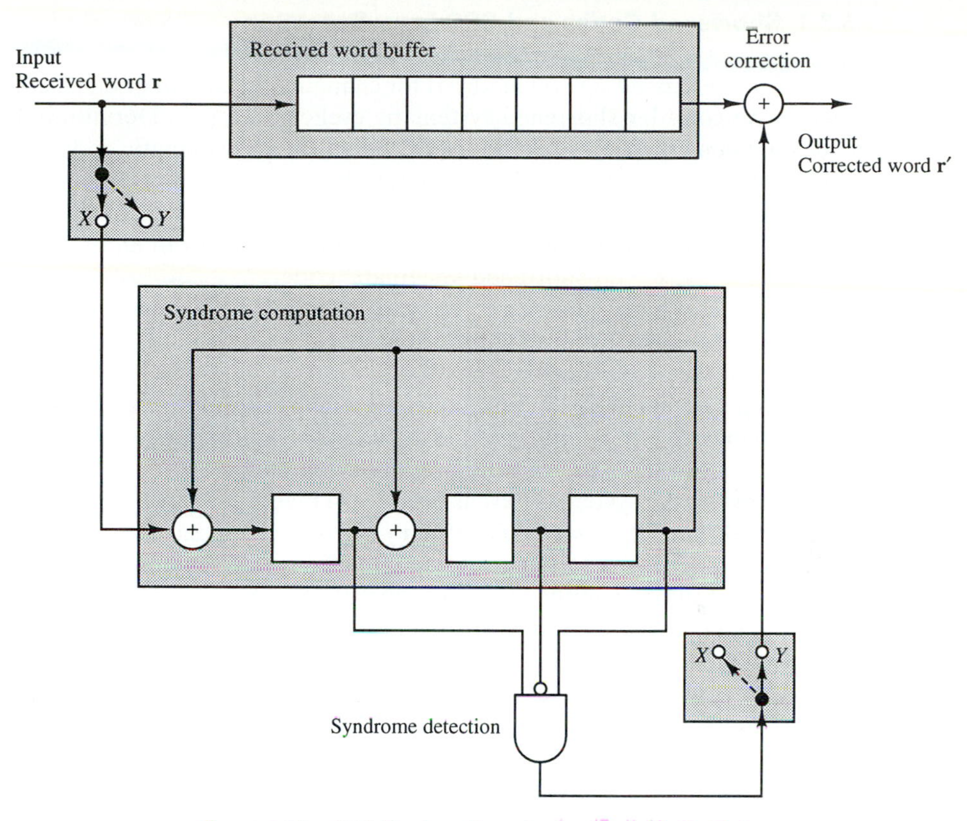

Figure 5-14. Shift-Register Decoder for (7, 4) Cyclic Code

5.3 SHORTENED CYCLIC CODES AND CRC ERROR DETECTION

A good case can be made for the claim that the most frequently used error control technique over the past 50 years has been the simple one-bit parity check. The second most frequently used would be the set of CRC (cyclic redundancy check) error detecting codes. Both techniques owe their predominance to the computer industry and the ubiquity of computer communication networks. Beyond that, parity bits are so simple in concept and implementation as to be almost trivial (not to imply that they do not offer a moderate amount of error control through retransmission requests). On the other hand, CRC codes are not so easily understood conceptually; they owe their popularity to extremely simple (and fast) encoder and decoder implementations and very good error detection performance. In this section CRC error detection is briefly described within the context of shortened cyclic codes. This discussion is followed by a series of results that describe the error detection performance of cyclic codes in general and the CRC codes in particular.

5.3.1 Shortened Codes and CRC Error Detection

Section 4.6 discussed several methods for changing the length of block codes. In this section we consider shortened systematic cyclic codes (see Definition 4-13). Such codes are constructed as follows. Let **C** be an (n, k) systematic code. Let **S** be the subset of code words in **C** whose j high-order information coordinates (i.e., the j rightmost coordinates in the code word) have the value zero. Let **C**′ be the set of words obtained by deleting the j rightmost coordinates from all of the words in **S**. **C**′ is an $(n - j, k - j)$ shortened systematic code.

Since shortening decreases the rate of a code, shortened codes have error detection and correction capabilities that are at least as good as those of the code from which they were derived.

Codes obtained from cyclic codes through shortening are almost always non-cyclic.[2] However, it is still possible to use the same shift-register encoders and decoders as can be used with the original cyclic code.

Consider the systematic cyclic encoder in Figure 5-12. Before encoding begins, the SR cells are set to zero. If the first j high-order message symbols are equal to zero, their insertion into the encoding circuit does not change the state of the SR cells, and results only in the output of j zeroes. The deletion of j high-order symbols from the message block thus has no impact on the encoding process except for the deletion of j zeroes in the corresponding message positions of the code word.

Using this same reasoning, it can be seen that the SR syndrome computation circuit for the original cyclic code (see Figure 5-9) can be used to compute the syndrome for the shortened code as well.

Shortened cyclic codes have seen extensive use, particularly in computer communications, under the label **cyclic redundancy check (CRC)** codes. CRC codes are generally not cyclic, as noted above, but they are derived from cyclic codes and hence the name. CRC codes are primarily useful in error detection systems, for they take advantage of the considerable burst-error detection capability provided by cyclic codes (this capability is quantified in the subsection that follows). A given system's response to detected errors (e.g., a retransmission request) is dictated by the individual application and need not concern us here.

CRC codes are designated by a generator polynomial $g(x)$ and may be said to have arbitrary length up to a limit N, the length of the original cyclic code defined by $g(x)$. The literature on CRC codes discusses a variety of "rules of thumb" for selecting CRC generator polynomials. Gallager and Bertsekis [Gal2] note that it is common practice to select $g(x) = (x + 1)b(x)$, where $b(x)$ is primitive. The $(x + 1)$ factor ensures that all odd-weight error patterns are detectable. Detailed examinations of various polynomials of various degree are provided in [Cas], [Mer], and [Wit]. The list below contains CRC generator polynomials of varying degree that offer optimal or near-optimal performance. Three of these polynomials are now international standards.

[2] Shortened cyclic codes are frequently called **polynomial codes** (e.g., [Cla]), because the generator polynomial for the original cyclic code generates the shortened code as well.

CRC CODE	GENERATOR POLYNOMIAL
CRC-4	$g_4(x) = x^4 + x^3 + x^2 + x + 1$
CRC-7	$g_7(x) = x^7 + x^6 + x^4 + 1 = (x^4 + x^3 + 1)(x^2 + x + 1)(x + 1)$
CRC-8	$g_8(x) = (x^5 + x^4 + x^3 + x^2 + 1)(x^2 + x + 1)(x + 1)$
CRC-12	$g_{12}(x) = x^{12} + x^{11} + x^3 + x^2 + x + 1 = (x^{11} + x^2 + 1)(x + 1)$
CRC-ANSI[3]	$g_{ANSI}(x) = x^{16} + x^{15} + x^2 + 1 = (x^{15} + x + 1)(x + 1)$
CRC-CCITT[4]	$g_{CCITT}(x) = x^{16} + x^{12} + x^5 + 1$ $= (x^{15} + x^{14} + x^{13} + x^{12} + x^4 + x^3 + x^2 + x + 1)(x + 1)$
CRC-SDLC[5]	$g_{SDLC}(x) = x^{16} + x^{15} + x^{13} + x^7 + x^4 + x^2 + x + 1$ $= (x^{14} + x^{13} + x^{12} + x^{10} + x^8 + x^6 + x^5 + x^4 + x^3 + x + 1)$ $\cdot (x + 1)^2$
CRC-24	$g_{24}(x) = x^{24} + x^{23} + x^{14} + x^{12} + x^8 + 1$ $= (x^{10} + x^8 + x^7 + x^6 + x^5 + x^4 + x^3 + x + 1)$ $\cdot (x^{10} + x^9 + x^6 + x^4 + 1)(x^3 + x^2 + 1)(x + 1)$
CRC-32$_A$ [Mer]	$x^{32} + x^{30} + x^{22} + x^{15} + x^{12} + x^{11} + x^7 + x^6 + x^5 + x$ $= (x^{10} + x^9 + x^8 + x^6 + x^2 + x + 1)(x^{10} + x^7 + x^6 + x^3 + 1)$ $\cdot (x^{10} + x^8 + x^5 + x^4 + 1)(x + 1)(x)$
CRC-32$_B$ [Gal2]	$x^{32} + x^{26} + x^{23} + x^{22} + x^{16} + x^{12} + x^{11} + x^{10} + x^8 + x^7 + x^5$ $+ x^4 + x^2 + x + 1$

Let us examine one of the CRC codes a bit more closely. CRC-12 has a generator polynomial $g_{12}(x)$ that divides $(x^{2047} - 1) = [x^{(2^{11}-1)} - 1]$, but nothing of smaller degree that has the form $(x^m - 1)$. $g_{12}(x)$ thus defines a cyclic code of length 2047 and dimension $2047 - 12 = 2035$. We may thus encode up to 2035 message bits at a time, generating 12 bits of redundancy that are appended to each message block by an encoder based on the design shown in Figure 5-12.

5.3.2 Error Detection Performance Analysis for Cyclic and Shortened Cyclic (CRC) Codes

Unfortunately the weight distribution and even the minimum distance for many cyclic codes are not known. Given the generator polynomial $g(x)$ for a cyclic code **C**, however, we can make meaningful comments about **C**'s ability to detect errors in a variety of situations. We consider three situations below that are generally considered in the design of computer communication systems.

Error pattern coverage. Many transmitter and receiver hardware faults can result in the total corruption of the transmitted code word. This is particularly true in high data rate systems. In such cases the received word is frequently completely

[3] American National Standards Institute.

[4] Comité Consultatif International de Télegraphique et Téléphonique.

[5] IBM Synchronous Data Link Control.

independent of the transmitted code word. It is thus desirable to use a code that ensures that the probability that a random word is a valid code word stays below some preset threshold. Let **coverage λ** be the ratio of the number of invalid n-tuples over GF(q) to the number of n-tuples over GF(q).

$$\lambda = \frac{q^n - q^k}{q^n} = 1 - q^{k-n} \tag{5-16}$$

Coverage is thus solely a function of the number of redundant symbols in the transmitted code words.

Example 5-9—CRC Coverage Performance

The error detection coverage provided by several binary CRC codes, including those listed in the previous section, is as follows.

CRC CODE	ERROR DETECTION COVERAGE
CRC-7	$\lambda = 1 - 2^{-7} = 0.992188$
CRC-12	$\lambda = 1 - 2^{-12} = 0.999756$
CRC-CCITT, CRC-ANSI	$\lambda = 1 - 2^{-16} = 0.999985$
CRC-32	$\lambda = 1 - 2^{-32} = 0.99999999977$

■

Burst error detection. Transient or intermittent hardware faults in computer circuitry can introduce errors over several consecutive transmitted symbols. This creates, from the decoder's perspective, **burst errors** in the received words. Burst errors are also prevalent in mobile communication systems, where the communication channel suffers from multipath fading. A burst-error pattern of length b begins and ends with a nonzero value; the intervening symbols in the pattern may take on any value, including zero. Suppose that the transmission alphabet is GF(q). Let N be a nonzero symbol and X be any symbol, zero or otherwise. An error pattern containing a single burst of length 7 looks like this.

. . . 00000NXXXXXN00000 . . .

It follows that for a given length b, there are $[(q-1)^2 \cdot q^{b-2}]$ different burst-error patterns.

Cyclic codes are quite good at performing burst-error detection. In a low-noise environment that suffers from occasional bursts of noise, a CRC error detection system coupled with retransmission requests is frequently an excellent means of controlling errors (if an error burst affects the first transmission, it is hoped that the retransmission reaches the receiver error free—see Chapter 15). Using the burst-error pattern model presented above, some useful results concerning the detection capability of cyclic codes can be obtained.

Let **C** be an (n, k) cyclic code or shortened cyclic code with generator polynomial $g(x)$ of degree r. (If **C** is a cyclic code, then $r = (n - k)$.) A cyclic code's burst-error detection capability is best expressed in terms of the ratio γ_b of the number of detectable burst-error patterns of length b to the total number of burst-error patterns of length b. For the case $b = r$, we proceed as follows. An error burst $e(x)$ is undetectable if and only if $e(x)$ is a valid code polynomial. An undetectable pattern $e(x)$ must thus be of degree greater than or equal to r, otherwise the minimality of the degree of $g(x)$ is contradicted. Since **C** (or the code from which it is obtained) is cyclic, $e(x)$ cannot be a shifted version of a polynomial of degree less than r for the same reason. A degree-r polynomial has $(r + 1)$ coefficients, and thus $\gamma_b = 1.0$ for all b less than or equal to r.

Theorem 5-4

> **A q-ary cyclic or shortened cyclic code with generator polynomial $g(x)$ of degree r can detect all burst-error patterns of length r or less.**

The only valid code polynomials of degree r are scalar multiples of the generator polynomial (otherwise we would be able to generate a nonzero code polynomial of degree less than r through subtraction). The only undetectable burst-error patterns of length $(r + 1)$ are thus scalar multiples of $g(x)$ and their $(n - r - 1)$ *non*cyclic shifts to the right. γ_{r+1} can thus be derived as follows. Note that the terms corresponding to shifts of the error patterns cancel out in the analysis.

$$\gamma_{r+1} = \frac{\left|\left\{\begin{matrix}\text{detectable error patterns}\\ \text{of length } r+1\end{matrix}\right\}\right|}{\left|\left\{\begin{matrix}\text{error patterns}\\ \text{of length } r+1\end{matrix}\right\}\right|}$$

$$= 1 - \frac{(n-r)(q-1)}{(n-r)(q-1)^2 q^{r-1}} \tag{5-17}$$

$$= 1 - \frac{q^{1-r}}{q-1}$$

Theorem 5-5

> **A q-ary cyclic or shortened cyclic code with generator polynomial $g(x)$ of degree r can detect the fraction $1 - \dfrac{q^{1-r}}{q-1}$ of all burst-error patterns of length $(r + 1)$.**

Any code polynomial $c(x)$ in a cyclic or shortened cyclic code can be uniquely expressed in the product form $c(x) = m(x)g(x)$, where the degree of $m(x)$ is less than or equal to $(n - r)$. If the code polynomial is to be associated with an error burst, the coefficients of the degree-b and the degree-0 terms must be nonzero. Since the degree-r and the degree-0 terms of $g(x)$ must be nonzero, it follows that it is necessary and sufficient that the coefficients of the degree-0 and $(b - r)$ terms of a degree-$(b - r)$ message polynomial be nonzero for the resulting code polynomial to correspond to an undetectable error burst of length b. Let m_0 be the coefficient of the degree-0 term in a message polynomial and p_0 the corresponding term in a polynomial of degree b. The analysis proceeds as follows.

$$\gamma_{b,b>r+1} = \frac{\left|\left\{\begin{matrix}\text{detectable burst-error}\\ \text{patterns of length } b\end{matrix}\right\}\right|}{\left|\left\{\begin{matrix}\text{burst-error}\\ \text{patterns of length } b\end{matrix}\right\}\right|}$$

$$= 1 - \frac{(n-b+1)\left|\left\{\begin{matrix}\text{polynomials of degree}\\ b-r-1 \text{ with } m_0 \neq 0\end{matrix}\right\}\right|}{(n-b+1)\left|\left\{\begin{matrix}\text{polynomials of}\\ \text{degree } b-1 \text{ with } p_0 \neq 0\end{matrix}\right\}\right|} \tag{5-18}$$

$$= 1 - \frac{(q-1)^2 q^{b-r-2}}{(q-1)^2 q^{b-2}}$$

$$= 1 - \frac{1}{q^r}$$

Theorem 5-6

A q-ary cyclic or shortened cyclic code with generator polynomial $g(x)$ of degree r can detect the fraction $1 - q^{-r}$ of all burst error patterns of length $b > (r + 1)$.

Example 5-10—Burst-Error Detecting Performance of CRC Codes

Applying the above results to the binary CRC codes CRC-7, CRC-12, CRC-CCITT, and CRC-ANSI, we obtain the following characterization of their burst-error detection performance. Note that this performance is independent of the length of the message blocks encoded.

CRC-7

- Detects all single bursts of length 7 or less.
- Detects 98.44% of all error bursts of length 8.
- Detects 99.22% of all error bursts of length > 8.

CRC-12

- Detects all single bursts of length 12 or less.
- Detects 99.95% of all error bursts of length 13.
- Detects 99.976% of all error bursts of length > 13.

CRC-CCITT, CRC-ANSI

- Detects all single bursts of length 16 or less.
- Detects 99.997% of all error bursts of length 17.
- Detects 99.9985% of all error bursts of length > 17. ■

Performance over the binary symmetric channel. An exact determination of the performance of a CRC code over the binary symmetric channel requires knowledge of the weight distribution of the code. As this is usually unavailable, the following result can be used as a lower bound on performance under noisy conditions.

Theorem 5-7

For any CRC code of length p used for error detection on the binary symmetric channel, the undetected error probability approaches 2^{-p} as the crossover probability and the dimension k of the code increase.

Proof. See [Wit] and the earlier discussion of error pattern coverage.

SUGGESTIONS FOR FURTHER READING

The following texts provide substantial discussions of the theory and implementation of cyclic and shortened cyclic codes.

E. R. BERLEKAMP, *Algebraic Coding Theory*, rev. ed., Laguna Hills: Aegean Park Press, 1984.

R. E. BLAHUT, *Theory and Practice of Error Control Codes*, Reading, MA: Addison-Wesley, 1984.

G. CLARK and J. CAIN, *Error-Correction Coding for Digital Communications*, New York: Plenum Press, 1981.

S. LIN and D. J. COSTELLO JR., *Error Control Coding: Fundamentals and Applications*, Englewood Cliffs, NJ: Prentice Hall, 1983.

There are very few texts devoted exclusively to the subject of shift-register circuits. Fortunately there is one that is very good. Those interested in pursuing the theory of shift registers are referred to the following.

S. W. GOLOMB, *Shift Register Sequences*, San Francisco: Holden-Day, 1967 (rev. edition, Laguna Hills: Aegean Park Press, 1982).

PROBLEMS

Let $\mathbf{C}_1$ be the binary cyclic code of length 15 generated by $g(x) = x^5 + x^4 + x^2 + 1$.

1. Compute the parity-check polynomial for $\mathbf{C}_1$ and show that $g(x)$ is a valid generator polynomial.

2. Determine the dimension of $\mathbf{C}_1$ and compute the number of code words in $\mathbf{C}_1$.
3. Construct parity check and generator matrices for $\mathbf{C}_1$.
4. Compute the code polynomial in $\mathbf{C}_1$ and the associated code word for the following message polynomials using the polynomial multiplication encoding technique.
   (a) x^2 (b) $x^9 + x^4 + x^2 + 1$
   (c) $x^7 + x^3 + x$ (d) $x^8 + x^7 + x^6 + x^5 + x^4$
5. Compute the code polynomial in $\mathbf{C}_1$ and the associated code word for the following message polynomials using the systematic encoding technique. Verify that the messages have been systematically encoded.
   (a) x^2 (b) $x^9 + x^4 + x^2 + 1$
   (c) $x^7 + x^3 + x$ (d) $x^8 + x^7 + x^6 + x^5 + x^4$
6. Given the code $\mathbf{C}_1$ and the parity check polynomial $h(x)$ derived in Problem 1, compute the syndrome $s(x)$ for the following received polynomials.
   (a) x^{10} (b) $x^3 + x^2$
   (c) $x^{14} + x^{10} + x^5 + x^3$ (d) $x^8 + x^6 + x + 1$
7. Design a systematic encoding circuit for $\mathbf{C}_1$.
8. Design a syndrome computation circuit for $\mathbf{C}_1$.
9. Compute the fraction of all burst errors of the following lengths detected by $\mathbf{C}_1$.
   (a) 4 (b) 5 (c) 6 (d) 7 (e) 8

Let $\mathbf{C}_2$ be the 4-ary cyclic code of length 15 generated by $g(x) = x^6 + \beta^2 x^5 + \beta x^4 + \beta x^2 + \beta$, where $\beta = \beta^2 + 1$ is primitive in GF(4).

10. Compute the parity-check polynomial for $\mathbf{C}_2$ and show that $g(x)$ is a valid generator polynomial.
11. Determine the dimension of $\mathbf{C}_2$ and compute the number of code words in $\mathbf{C}_2$.
12. Construct parity-check and generator matrices for $\mathbf{C}_2$.
13. Given the code $\mathbf{C}_2$ and the parity-check polynomial $h(x)$ derived in Problem 10, compute the syndrome $s(x)$ for the following received polynomials.
    (a) x^6 (b) $\beta x^4 + \beta^2 x^2$
14. Design a systematic encoding circuit for $\mathbf{C}_2$.
15. Design a syndrome computation circuit for $\mathbf{C}_2$.
16. Find the highest-degree generator polynomial for the binary cyclic code containing the following code words.
    (a) 1111111 (b) 1101001
    (c) 1100101 (d) 1010101
    (e) 111111111111111 (f) 001011100110000
    (g) 000111001010100 (h) 001010110110000
17. Let $\mathbf{C}_1$ and $\mathbf{C}_2$ be two cyclic codes of length n generated by $g_1(x)$ and $g_2(x)$, respectively. What is the generator polynomial for the smallest cyclic code containing the set $\mathbf{C}_1 \cup \mathbf{C}_2$?
18. List by dimension all of the binary cyclic codes of length 31.
19. List by dimension all of the binary cyclic codes of length 63.
20. List by dimension all of the binary cyclic codes of length 19.
21. List by dimension all of the 8-ary cyclic codes of length 33.

Hadamard, Quadratic Residue, and Golay Codes

Hadamard matrices are named after their discoverer, J. Hadamard, who described them in a paper published in 1893 [Had]. Hadamard showed that, when they exist for a given order n, Hadamard matrices have the maximum possible determinant given real matrix elements in the range $[-1, 1]$. Hadamard matrices have since been found useful in a variety of applications. For example, MacWilliams and Sloane [Mac] show that Hadamard matrices provide the solution to weighing design problems in which the measure of several objects is equal to the sum of the measures of the individual objects. Hadamard matrices also play a role in the definition of the Hadamard transform, which we consider later in conjunction with Reed-Muller codes. The Hadamard transform is equivalent to the Walsh, or discrete Fourier, transform. Hadamard codes are obtained through a set of trivial decompositions of Hadamard matrices. These codes include several small, nonlinear, cyclic codes that have nice properties. In this chapter we consider two constructions for Hadamard matrices and introduce three types of Hadamard codes that can be obtained from Hadamard matrices.

The quadratic residues *modulo* a prime p have applications throughout the fields of number theory, combinatorics, and, of course, the theory of block codes. In this chapter we have room for only a superficial treatment, but other, more detailed references are indicated for the interested reader. Wc see that quadratic residues can be used to construct Hadamard matrices and quadratic residue codes. Quadratic residue codes were introduced in 1958 by Prange [Pra2]. They are linear cyclic codes that generally have rates close to 1/2 and have large minimum distances. A great deal of recent work on algebraic decoding algorithms for quadratic residue codes has increased their utility in a number of applications (see, for example, [Elia] (1987), [Ree3] (1990), and [Ree4] (1992)).

A quadratic residue construction and a Hadamard matrix construction are used in the final section of this chapter to introduce what are probably the two most elegant codes that will ever be discovered: the binary and ternary Golay codes.

The Golay codes appeared at the dawn of the history of coding theory. In his 1948 paper that, among other things, gave birth to the field of information theory, Shannon gave a brief description of Hamming's perfect (7, 4) binary code [Sha1]. A number of people began searching for similar codes, with varying degrees of success. Golay, an engineer at the Signal Corps Engineering Laboratories in Fort Monmouth, NJ, published one of the first follow-up papers in June, 1949 [Gol2]. In what is almost certainly the best short paper ever written (it fits on one side of an $8\frac{1}{2} \times 11$ sheet of paper, with room to spare), Golay extended the (7, 4) Hamming code to a general class of p-ary codes of length $(p^n - 1)/(p - 1)$, where p is a prime.[1] In this same paper, Golay went on to describe a binary triple-error-correcting code and a ternary double-error-correcting code, both of which are perfect. Golay deduced the existence of the Golay codes through an examination of Pascal's triangle and the recognition of the relationship of the triangle's entries to perfect codes (see Eq. (4-5)). He then proceeded to find the Golay codes through a "limited search" of the triangle [Gol2]. All possible sets of parameters for perfect codes were thus discovered before 1950 (see Theorem 4-4).

Golay codes have seen frequent application in the United States space program, most notably with the *Voyager I* and *II* spacecraft. The extended binary Golay code served as the primary *Voyager* error control system, providing clear color pictures of Jupiter and Saturn between 1979 and 1981.[2]

Several efficient decoding algorithms for the Golay codes have been discovered. For example, in 1964 Kasami described a shift-register-based "error trapping" decoder that is quite fast and works well with Golay codes [Kas4]. The error trapping decoder is an extension of the shift-register decoding techniques discussed in Chapter 5. In 1987 Elia described an algebraic method based on a BCH decoding algorithm [Elia]. This decoding algorithm and an arithmetic decoding algorithm are presented at the end of this chapter.

6.1 HADAMARD MATRICES AND HADAMARD CODES

A Hadamard matrix of order n is an $(n \times n)$ matrix of $+1$s and -1s such that any pair of distinct rows is orthogonal (i.e., their inner product is zero). The following serves as our formal definition.

[1] It is a virtual certainty that this result was already known to Hamming [Ber2]. A more detailed discussion of the general class of codes now known as Hamming codes is included in Hamming's 1950 paper [Ham], which was probably delayed due to patent considerations.

[2] A secondary error control system based on Reed-Solomon codes was substituted for the Golay system for the *Voyager II* fly-bys of Uranus and Neptune in the mid-1980s [Lae].

Definition 6-1—Hadamard Matrices

A Hadamard matrix $\mathbf{H}$ of order n is an $(n \times n)$ matrix of $+1$s and -1s such that $\mathbf{H}\mathbf{H}^T = n\mathbf{I}$.

We shall adopt the notation $\mathbf{H}_n$ throughout this chapter to denote a Hadamard matrix of order n. Hadamard matrices have a number of interesting properties. For example, the inverse of a Hadamard matrix can be found quite easily.

$$\mathbf{H}\mathbf{H}^T = n\mathbf{I}$$
$$\Rightarrow \left(\frac{1}{n}\mathbf{H}^{-1}\right)(\mathbf{H}\mathbf{H}^T) = \left(\frac{1}{n}\mathbf{H}^{-1}\right)n\mathbf{I}$$
$$\Rightarrow \frac{1}{n}\mathbf{H}^T = \mathbf{H}^{-1}$$

It follows that

$$\frac{1}{n}\mathbf{H}^T = \mathbf{H}^{-1}$$
$$\Rightarrow \mathbf{H}^T = n\mathbf{H}^{-1}$$
$$\Rightarrow \mathbf{H}^T\mathbf{H} = n\mathbf{I}$$

The columns of a Hadamard matrix thus have the same properties as the rows.

$\mathbf{H}_n$ does not exist for all positive integers n. In fact, it is not yet known for which values of n an $\mathbf{H}_n$ does exist. The following result, however, somewhat limits the search.

Theorem 6-2—Valid Orders for Hadamard Matrices

If a Hadamard matrix of order n exists, then n is 1, 2, 4, or a multiple of 4.

Proof. See [Mac, p. 44].

We now consider two methods for constructing Hadamard matrices: the Sylvester construction and the Paley construction.

6.1.1 The Sylvester Construction

Theorem 6-3—The Sylvester Construction

If $\mathbf{H}_n$ is a Hadamard matrix, then so is $\mathbf{H}_{2n} = \begin{bmatrix} \mathbf{H}_n & \mathbf{H}_n \\ \mathbf{H}_n & -\mathbf{H}_n \end{bmatrix}$.

Proof. Let $\mathbf{H}_n$ be a Hadamard matrix of order n. The construction in the hypothesis yields the matrix

$$\mathbf{H}_{2n} = \begin{bmatrix} \mathbf{H}_n & \mathbf{H}_n \\ \mathbf{H}_n & -\mathbf{H}_n \end{bmatrix}$$

Let the row vectors of $\mathbf{H}_n$ be $\{\mathbf{h}_1, \mathbf{h}_2, \ldots, \mathbf{h}_n\}$ and proceed by computing $\mathbf{H}_{2n}\mathbf{H}_{2n}^T$. Note that since $\mathbf{H}_n$ is Hadamard, $\mathbf{h}_i \cdot \mathbf{h}_j = n$ when $i = j$ and $\mathbf{h}_i \cdot \mathbf{h}_j = 0$ when $i \neq$ j.

$$\mathbf{H}_{2n}\mathbf{H}_{2n}^T$$

$$= \begin{bmatrix} \mathbf{H}_n & \mathbf{H}_n \\ \mathbf{H}_n & -\mathbf{H}_n \end{bmatrix}\begin{bmatrix} \mathbf{H}_n & \mathbf{H}_n \\ \mathbf{H}_n & -\mathbf{H}_n \end{bmatrix}^T$$

$$= \begin{bmatrix} \mathbf{h}_1 & \mathbf{h}_1 \\ \mathbf{h}_2 & \mathbf{h}_2 \\ \vdots & \vdots \\ \mathbf{h}_n & \mathbf{h}_n \\ \mathbf{h}_1 & -\mathbf{h}_1 \\ \mathbf{h}_2 & -\mathbf{h}_2 \\ \vdots & \vdots \\ \mathbf{h}_n & -\mathbf{h}_n \end{bmatrix} \begin{bmatrix} \mathbf{h}_1^T & \mathbf{h}_2^T & \cdots & \mathbf{h}_n^T & \mathbf{h}_1^T & \mathbf{h}_2^T & \cdots & \mathbf{h}_n^T \\ \mathbf{h}_1^T & \mathbf{h}_2^T & \cdots & \mathbf{h}_n^T & -\mathbf{h}_1^T & -\mathbf{h}_2^T & \cdots & -\mathbf{h}_n^T \end{bmatrix}$$

$$= \begin{bmatrix} \mathbf{h}_1\cdot\mathbf{h}_1 + \mathbf{h}_1\cdot\mathbf{h}_1 & & \mathbf{h}_1\cdot\mathbf{h}_n + \mathbf{h}_1\cdot\mathbf{h}_n & \mathbf{h}_1\cdot\mathbf{h}_1 - \mathbf{h}_1\cdot\mathbf{h}_1 & & \mathbf{h}_1\cdot\mathbf{h}_n - \mathbf{h}_1\cdot\mathbf{h}_n \\ \mathbf{h}_2\cdot\mathbf{h}_1 + \mathbf{h}_2\cdot\mathbf{h}_1 & & \mathbf{h}_2\cdot\mathbf{h}_n + \mathbf{h}_2\cdot\mathbf{h}_n & \mathbf{h}_2\cdot\mathbf{h}_1 - \mathbf{h}_2\cdot\mathbf{h}_1 & & \mathbf{h}_2\cdot\mathbf{h}_n - \mathbf{h}_2\cdot\mathbf{h}_n \\ \vdots & \cdots & \vdots & \vdots & \cdots & \vdots \\ \mathbf{h}_n\cdot\mathbf{h}_1 + \mathbf{h}_n\cdot\mathbf{h}_1 & & \mathbf{h}_n\cdot\mathbf{h}_n + \mathbf{h}_n\cdot\mathbf{h}_n & \mathbf{h}_n\cdot\mathbf{h}_1 - \mathbf{h}_n\cdot\mathbf{h}_1 & & \mathbf{h}_n\cdot\mathbf{h}_n - \mathbf{h}_n\cdot\mathbf{h}_n \\ \mathbf{h}_1\cdot\mathbf{h}_1 - \mathbf{h}_1\cdot\mathbf{h}_1 & & \mathbf{h}_1\cdot\mathbf{h}_n - \mathbf{h}_1\cdot\mathbf{h}_n & \mathbf{h}_1\cdot\mathbf{h}_1 + \mathbf{h}_1\cdot\mathbf{h}_1 & & \mathbf{h}_1\cdot\mathbf{h}_n + \mathbf{h}_1\cdot\mathbf{h}_n \\ \mathbf{h}_2\cdot\mathbf{h}_1 - \mathbf{h}_2\cdot\mathbf{h}_1 & & \mathbf{h}_2\cdot\mathbf{h}_n - \mathbf{h}_2\cdot\mathbf{h}_n & \mathbf{h}_2\cdot\mathbf{h}_1 + \mathbf{h}_2\cdot\mathbf{h}_1 & & \mathbf{h}_2\cdot\mathbf{h}_n + \mathbf{h}_2\cdot\mathbf{h}_n \\ \vdots & \cdots & \vdots & \vdots & \cdots & \vdots \\ \mathbf{h}_n\cdot\mathbf{h}_1 - \mathbf{h}_n\cdot\mathbf{h}_1 & & \mathbf{h}_n\cdot\mathbf{h}_n - \mathbf{h}_n\cdot\mathbf{h}_n & \mathbf{h}_n\cdot\mathbf{h}_1 + \mathbf{h}_n\cdot\mathbf{h}_1 & & \mathbf{h}_n\cdot\mathbf{h}_n + \mathbf{h}_n\cdot\mathbf{h}_n \end{bmatrix}$$

$$= \begin{bmatrix} 2n & 0 & \cdots & 0 & 0 & 0 & \cdots & 0 & 0 \\ 0 & 2n & \cdots & 0 & 0 & 0 & \cdots & 0 & 0 \\ \vdots & \vdots & \ddots & \vdots & \vdots & \vdots & \ddots & \vdots & \vdots \\ 0 & 0 & \cdots & 2n & 0 & 0 & \cdots & 0 & 0 \\ 0 & 0 & \cdots & 0 & 2n & 0 & \cdots & 0 & 0 \\ 0 & 0 & \cdots & 0 & 0 & 2n & \cdots & 0 & 0 \\ \vdots & \vdots & \ddots & \vdots & \vdots & \vdots & \ddots & \vdots & \vdots \\ 0 & 0 & \cdots & 0 & 0 & 0 & \cdots & 2n & 0 \\ 0 & 0 & \cdots & 0 & 0 & 0 & \cdots & 0 & 2n \end{bmatrix} = 2n\mathbf{I}_{2n}$$

QED

Matrices of the form in Theorem 6-3 are called Sylvester-type Hadamard matrices. The first matrices that result from the Sylvester construction are as follows.

$$\mathbf{H}_1 = [1], \quad \mathbf{H}_2 = \begin{bmatrix} 1 & 1 \\ 1 & - \end{bmatrix}, \quad \mathbf{H}_4 = \begin{bmatrix} \mathbf{H}_2 & \mathbf{H}_2 \\ \mathbf{H}_2 & -\mathbf{H}_2 \end{bmatrix} = \begin{bmatrix} 1 & 1 & 1 & 1 \\ 1 & - & 1 & - \\ 1 & 1 & - & - \\ 1 & - & - & 1 \end{bmatrix}$$

$$\mathbf{H}_8 = \begin{bmatrix} \mathbf{H}_4 & \mathbf{H}_4 \\ \mathbf{H}_4 & -\mathbf{H}_4 \end{bmatrix} = \begin{bmatrix} 1 & 1 & 1 & 1 & 1 & 1 & 1 & 1 \\ 1 & - & 1 & - & 1 & - & 1 & - \\ 1 & 1 & - & - & 1 & 1 & - & - \\ 1 & - & - & 1 & 1 & - & - & 1 \\ 1 & 1 & 1 & 1 & - & - & - & - \\ 1 & - & 1 & - & - & 1 & - & 1 \\ 1 & 1 & - & - & - & - & 1 & 1 \\ 1 & - & - & 1 & - & 1 & 1 & - \end{bmatrix}$$

In these matrices and those that follow, -1s are represented solely by a minus sign. The above sequence generates $\mathbf{H}_n$ for all n of the form 2^m, m a positive integer. Clearly this does not exhaust the possibilities suggested (but not guaranteed) by Theorem 6-2. Given the existence of a Hadamard matrix of arbitrary order n, however, the Sylvester construction guarantees a sequence of matrices of order $n2^m$, m a positive integer.

6.1.2 The Paley Construction and Quadratic Residues

The Paley construction was described by R. E. A. C. Paley in 1933 [Pal]. It makes use of **quadratic residues**, which, like Hadamard matrices, have found application in a number of interesting areas.

Definition 6-2—Quadratic Residues

The nonzero squares *modulo p*, p a prime, are called the **quadratic residues *modulo p*.**

As in Chapter 2, we label the equivalence classes *modulo p* with the smallest nonnegative integer in each class. The complete set of labels for the classes modulo p is thus $\{0, 1, 2, \ldots, p-1\}$. Note that every element in an equivalence class x is of the form $(x + mp)$ for some integer m. The square of any element in the class x is in the same equivalence class as the square of every other element in x, for $(x + mp)^2 = x^2 + 2xmp + m^2p^2 \equiv x^2$ *modulo p*. The quadratic residues are thus well-defined.

Quadratic residues can be found by simply squaring the label for every nonzero class *modulo p*.

Example 6-1—Quadratic Residues

- The squares of the equivalence classes *modulo* 5 are $\{1^2, 2^2, 3^2, 4^2\} = \{1, 4, 4, 1\} = \{1, 4\}$. 1 and 4 are thus the two quadratic residues *modulo* 5.

- The squares of the equivalence classes *modulo* 7 are $\{1^2, 2^2, 3^2, 4^2, 5^2, 6^2\} = \{1, 4, 2, 2, 4, 1\} = \{1, 2, 4\}$. 1, 2 and 4 are thus the three quadratic residues *modulo* 7.
- The squares of the equivalence classes *modulo* 11 are $\{1^2, 2^2, 3^2, 4^2, 5^2, 6^2, 7^2, 8^2, 9^2, 10^2\} = \{1, 4, 9, 5, 3, 5, 9, 4, 1\} = \{1, 3, 4, 5, 9\}$. ■

The elements *modulo p* that are not quadratic residues are called, understandably, quadratic nonresidues.

The example highlights the first two of the following properties of quadratic residues.

- A complete listing of the quadratic residues *modulo p* is obtained by computing $1^2, 2^2, \ldots, [(p-1)/2]^2$. All subsequent squares repeat the first $(p-1)/2$ squares in reverse order. This follows from the fact that $(p-x)^2 = p^2 - 2px + x^2 \equiv x^2$ *modulo p*.
- There are exactly $(p-1)/2$ quadratic residues *modulo p*, p an odd prime. This follows from the previous result.
- The product of two quadratic residues or of two quadratic nonresidues is always a quadratic residue, while the product of a residue and a nonresidue is a nonresidue.

The **Legendre symbol** is a function of some importance in number theory and public key cryptography, as well as coding theory.

Definition 6-3—Legendre Symbol

The Legendre symbol $\chi(x)$, referenced to a prime p, has the following values.

$$\chi(x) = 0 \quad \text{if } x \text{ is a multiple of } p$$

$$\chi(x) = 1 \quad \text{if } x \text{ is a quadratic residue } modulo\ p$$

$$\chi(x) = -1 \quad \text{if } x \text{ is a quadratic nonresidue } modulo\ p$$

The Paley construction uses quadratic residues to construct Hadamard matrices of order n, where n is of the form $(p+1)$, p a prime, *and* n is a multiple of 4. The construction begins by building a $(p \times p)$ **Jacobsthal matrix** $\mathbf{Q}_p$, where p is a prime.[3] The elements $\{q_{ij}\}$ of $\mathbf{Q}_p$ are determined by the Legendre symbol as follows.

$$q_{ij} = \chi(j - i)$$

It is assumed that the rows and columns of $\mathbf{Q}_p$ begin their numbering at 0. The Paley Hadamard matrix of order n is then constructed in the following manner:

$$\mathbf{H}_n = \begin{bmatrix} 1 & \mathbf{1} \\ \mathbf{1}^T & \mathbf{Q}_{n-1} - \mathbf{I}_{n-1} \end{bmatrix}$$

where **1** is a row vector containing all ones.

[3] Jacobsthal matrices are sometimes called Paley matrices.

Example 6-2—A Paley Hadamard Matrix of Order 12

We begin by listing the five quadratic residues *modulo* 11.

$$\{1^2 \equiv 1, 2^2 \equiv 4, 3^2 \equiv 9, 4^2 \equiv 5, 5^2 \equiv 3\} = \{1, 3, 4, 5, 9\}$$

The first row ($i = 0$) of the Jacobsthal matrix $\mathbf{Q}_{11}$ is then

$$[0 \;\; 1 \;\; - \;\; 1 \;\; 1 \;\; 1 \;\; - \;\; - \;\; - \;\; 1 \;\; -]$$

The remaining rows of $\mathbf{Q}_{11}$ are right cyclic shifts of the first row, giving us the following matrix.

$$\mathbf{Q}_{11} = \begin{bmatrix} 0 & 1 & - & 1 & 1 & 1 & - & - & - & 1 & - \\ - & 0 & 1 & - & 1 & 1 & 1 & - & - & - & 1 \\ 1 & - & 0 & 1 & - & 1 & 1 & 1 & - & - & - \\ - & 1 & - & 0 & 1 & - & 1 & 1 & 1 & - & - \\ - & - & 1 & - & 0 & 1 & - & 1 & 1 & 1 & - \\ - & - & - & 1 & - & 0 & 1 & - & 1 & 1 & 1 \\ 1 & - & - & - & 1 & - & 0 & 1 & - & 1 & 1 \\ 1 & 1 & - & - & - & 1 & - & 0 & 1 & - & 1 \\ 1 & 1 & 1 & - & - & - & 1 & - & 0 & 1 & - \\ - & 1 & 1 & 1 & - & - & - & 1 & - & 0 & 1 \\ 1 & - & 1 & 1 & 1 & - & - & - & 1 & - & 0 \end{bmatrix}$$

The Paley Hadamard matrix $\mathbf{H}_{12}$ is then formed by inserting $\mathbf{Q}_{11}$ into the form presented earlier.

$$\mathbf{H}_{12} = \begin{bmatrix} 1 & \mathbf{1} \\ \mathbf{1}^T & \mathbf{Q}_{11} - \mathbf{I}_{11} \end{bmatrix} = \left[\begin{array}{c|ccccccccccc} 1 & 1 & 1 & 1 & 1 & 1 & 1 & 1 & 1 & 1 & 1 & 1 \\ \hline 1 & - & 1 & - & 1 & 1 & 1 & - & - & - & 1 & - \\ 1 & - & - & 1 & - & 1 & 1 & 1 & - & - & - & 1 \\ 1 & 1 & - & - & 1 & - & 1 & 1 & 1 & - & - & - \\ 1 & - & 1 & - & - & 1 & - & 1 & 1 & 1 & - & - \\ 1 & - & - & 1 & - & - & 1 & - & 1 & 1 & 1 & - \\ 1 & - & - & - & 1 & - & - & 1 & - & 1 & 1 & 1 \\ 1 & 1 & - & - & - & 1 & - & - & 1 & - & 1 & 1 \\ 1 & 1 & 1 & - & - & - & 1 & - & - & 1 & - & 1 \\ 1 & 1 & 1 & 1 & - & - & - & 1 & - & - & 1 & - \\ 1 & - & 1 & 1 & 1 & - & - & - & 1 & - & - & 1 \\ 1 & 1 & - & 1 & 1 & 1 & - & - & - & 1 & - & - \end{array}\right] \quad \blacksquare$$

A Hadamard matrix is said to be **normalized** if the first row and column contain all ones (e.g., see $\mathbf{H}_{12}$ above). Any Hadamard matrix can be reduced to this form by multiplying all rows and columns beginning with a -1 by -1. Since this has the effect of multiplying the inner product of a row or column with itself by $(-1)^2 = 1$, the resulting matrix is still Hadamard. Any two Hadamard matrices of the same order are said to be **equivalent** if the normalized form of one can be reduced to that of the other by permuting rows and columns. There is only one equivalence class of Hadamard matrices of orders 1, 2, 4, 8, and 12. There are five equivalence classes of order 16 and three of order 20 [Mac].

6.1.3 Binary Hadamard Matrices and Hadamard Codes

Binary Hadamard matrices $\mathbf{A}_n$ are obtained by converting the 1s in a Hadamard matrix $\mathbf{H}_n$ to 0s and the -1s to 1s. $\mathbf{A}_n$ can be used to construct three different types of **Hadamard codes** [Mac].

- Since the rows of $\mathbf{H}_n$ are orthogonal, any pair of distinct rows in $\mathbf{A}_n$ must agree in $n/2$ places and differ in $n/2$ places. If we delete the leftmost column of $\mathbf{A}_n$, the remaining rows form a length-$(n-1)$ Hadamard code $\mathcal{A}_n$ with n code words and minimum distance $n/2$. Because the minimum distance is $n/2$, $\mathcal{A}_n$ is also known as a **simplex code**.
- If the n code words in $\mathcal{A}_n$ are supplemented by their n complements, we get the code $\mathcal{B}_n$, which consists of $2n$ code words of length $(n-1)$ and has minimum distance $\frac{1}{2}n - 1$.
- If the n rows of $\mathbf{A}_n$ are supplemented by their n complements, we get the code $\mathcal{C}_n$, which consists of $2n$ code words of length n and has minimum distance $n/2$.

If n is of the form 2^m, then the Hadamard codes resulting from a Sylvester Hadamard matrix $\mathbf{H}_n$ are always linear. On the other hand, the codes resulting from a Paley matrix of order $n > 8$ are always nonlinear [Mac]. A nonlinear code constructed from a Paley matrix can be made into a linear code by taking all possible linear combinations of the code words (i.e., taking the linear span of the nonlinear code). The resulting code is a **quadratic residue code**, the subject of the next section.

If a Paley Hadamard matrix is used in the construction of $\mathcal{A}_n$ and $\mathcal{B}_n$, then these codes will be cyclic, but not necessarily linear.

Example 6-3—The Hadamard Codes $\mathcal{A}_{12}$, $\mathcal{B}_{12}$, and $\mathcal{C}_{12}$

We begin with the binary Hadamard matrix $\mathbf{A}_{12}$, as shown below. The empty spaces indicate the locations of zeros.

$$\mathbf{A}_{12} = \begin{bmatrix} & & & & & & & & & & & \\ & 1 & & 1 & & & & 1 & 1 & 1 & & 1 \\ & 1 & 1 & & 1 & & & & 1 & 1 & 1 & \\ & & 1 & 1 & & 1 & & & & 1 & 1 & 1 \\ & 1 & & 1 & 1 & & 1 & & & & 1 & 1 \\ & 1 & 1 & & 1 & 1 & & 1 & & & & 1 \\ & 1 & 1 & 1 & & 1 & 1 & & 1 & & & \\ & & 1 & 1 & 1 & & 1 & 1 & & 1 & & \\ & & & 1 & 1 & 1 & & 1 & 1 & & 1 & \\ & & & & 1 & 1 & 1 & & 1 & 1 & & 1 \\ & 1 & & & & 1 & 1 & 1 & & 1 & 1 & \\ & & 1 & & & & 1 & 1 & 1 & & 1 & 1 \end{bmatrix}$$

$\mathcal{A}_{12}$ is obtained by deleting the leftmost column. The rows of the resulting matrix give us the following code words.

(00000000000) (10100011101) (11010001110) (01101000111)
(10110100011) (11011010001) (11101101000) (01110110100)
(00111011010) (00011101101) (10001110110) (01000111011)

This is an excellent example of a binary nonlinear cyclic code. (The cyclicity is clear by inspection. Nonlinearity follows from the fact that there are 12 code words, and 12 is not a power of 2.) This code has length 11 and minimum distance 6.

$\mathcal{B}_{12}$ is obtained by adding to the above code words their complements. The resulting code has 24 code words of length 11 and has minimum distance 5. The reader should verify that $\mathcal{B}_{12}$ is also nonlinear and cyclic.

$\mathcal{C}_{12}$ consists of the rows of $\mathbf{A}_{12}$ and their complements. It has 24 code words of length 12 and minimum distance 6. The reader should verify that $\mathcal{C}_{12}$ is not cyclic. ■

6.2 QUADRATIC RESIDUE CODES

In Chapter 3 we examined the properties of polynomials over Galois fields in some detail. One of the principal results to emerge from that chapter dealt with conjugates. Let β be an element in the Galois field GF(q^m). The conjugates of β with respect to the subfield GF(q) are the elements

$$\beta, \beta^q, \beta^{q^2}, \beta^{q^3}, \ldots$$

These conjugates form a conjugacy class. The exponents of the elements of a conjugacy class, given a common base that is a primitive pth root of unity, form a cyclotomic coset *modulo p* with respect to GF(q). The elements of any Galois field GF(q^m) can be separated into distinct conjugacy classes. The key result in Chapter 3 was that the set of roots of a polynomial with coefficients in GF(q) must be the union of one or more conjugacy classes with respect to GF(q).

Quadratic residues *modulo p* display some interesting structural properties when viewed in the light of the above theory. The set of integers *modulo p*, p a prime, form the field GF(p) under *modulo p* addition and multiplication. Let Q be the set of quadratic residues *modulo p* and N the set of corresponding nonresidues. Since GF(p) is a Galois field, there must exist at least one primitive element $\gamma \in$ GF(p) that generates all of the elements in Q and N. It follows that γ must be a quadratic nonresidue, otherwise there exists some element $\sqrt{\gamma} \in$ GF(p) that generates $2(p-1)$ distinct elements in GF(p), contradicting the order of GF(p). We can then see that $\gamma^e \in Q$ if and only if e is even; otherwise, $\gamma^e \in N$. We conclude that all of the elements in Q correspond to the first $(p-1)/2$ consecutive powers of γ^2, and that Q is a cyclic group under *modulo p* multiplication.

Now consider a field GF(s^m) that contains a primitive pth root of unity. Such a field exists for a given s, m, and p whenever $p \mid s^m - 1$ (see Theorem 2-12). We add the further restriction that s must be a quadratic residue *modulo p*. This can be somewhat restrictive; for example, if $s = 2$, then p must be of the form $p = (8k \pm 1)$ [Mac]. Since Q is a cyclic group, multiplication of any element in Q by any other element in Q must result in an element in Q. It follows that the conjugates with respect to GF(s) of any element in Q must also be in Q. We conclude that Q is the union of one or more cyclotomic cosets *modulo p* with respect to GF(s).

Let α be primitive in $GF(s^m)$. The above results show that the following polynomials have coefficients in the subfield $GF(s)$.

$$q(x) = \prod_{i \in Q} (x - \alpha^i), \qquad n(x) = \prod_{i \in N} (x - \alpha^i)$$

Furthermore

$$x^p - 1 = (x - 1) \cdot q(x) \cdot n(x)$$

The **quadratic residue codes $\mathfrak{Q}, \overline{\mathfrak{Q}}, \mathcal{N}, \overline{\mathcal{N}}$** of length p are defined by the generator polynomials $q(x)$, $(x - 1)q(x)$, $n(x)$, and $(x - 1)n(x)$, respectively.

Example 6-4—A Binary Quadratic Residue Code of Length 7

In Example 6-1 it was shown that 2 is a QR *modulo* 7. We can thus look for binary quadratic residue codes of length 7. A *primitive* 7th root of unity can be found in GF(8). Let α be a root of $x^3 + x + 1$ and thus a primitive element in GF(8).

The cyclotomic cosets *modulo* 7 with respect to GF(2) are as follows.

$$\{0\}$$
$$\{1, 2, 4\} = Q$$
$$\{3, 5, 6\} = N$$

$\mathfrak{Q}$ thus has generator polynomial $g(x) = (x - \alpha)(x - \alpha^2)(x - \alpha^4) = x^3 + x + 1$. The $(7, 4, 3)$ cyclic Hamming code (see Example 5-8) is thus also a quadratic residue code. $\mathcal{N}$ has generator polynomial $g(x) = x^3 + x^2 + 1$. The codes $\overline{\mathfrak{Q}}$ and $\overline{\mathcal{N}}$ have generator polynomials $x^4 + x^3 + x^2 + 1$ and $x^4 + x^2 + x + 1$, respectively. Note that if α was chosen to be a root of $x^3 + x^2 + 1$, then $\mathfrak{Q}$ and $\mathcal{N}$ would exchange generator polynomials, as would $\overline{\mathfrak{Q}}$ and $\overline{\mathcal{N}}$. ∎

In general, quadratic residue codes tend to be very good block codes with rates approximating 1/2. The following result is offered without proof. The interested reader is directed to [Mac, p. 483].

Theorem 6-4—The Minimum Distance of Quadratic Residue Codes

The minimum distance d of $\mathfrak{Q}$ or $\mathcal{N}$ satisfies $d^2 \geq p$. Furthermore, if p is of the form $4k - 1$, then $d^2 - d + 1 \geq p$.

Quadratic residue codes are extended by adding a parity-check bit to each code word. The resulting codes are generally quite good. A listing of several binary and ternary examples is provided in Table 6-1. Bounds are provided where the exact value is not yet known.

There are a number of very powerful methods for decoding quadratic residue codes, including **permutation decoding** and **covering polynomials** [Mac, Chap. 16]. Recent efforts by Elia and Reed et al. [Elia], [Ree3], [Ree4] have uncovered a series of powerful algebraic decoding algorithms for several specific quadratic residue codes. In this chapter we consider an algebraic decoding algorithm for only one of the quadratic residue codes, the binary $(23, 12, 7)$ Golay code, the subject of the next section.

TABLE 6-1 Some good binary and ternary extended quadratic residue codes [Mac]

BINARY CODES

n	k	d	n	k	d	n	k	d
8	4	4*	74	37	14	138	69	14–22
18	9	6*	80	40	16*	152	76	20
24	12	8*	90	45	18*	168	84	16–24
32	16	8*	98	49	16	192	96	16–28
42	21	10*	104	52	20*	194	97	16–28
48	24	12*	114	57	12–16	200	100	16–32
72	36	12	128	64	20			

* Indicates that the code is as good as the best known for this n and k.

TERNARY CODES

n	k	d	n	k	d	n	k	d
12	6	6	48	24	15	74	37	?
14	7	6	60	30	18	84	42	?
24	12	9	62	31	?	98	49	?
38	19	?	72	36	?			

6.3 GOLAY CODES

The **binary Golay code $\mathcal{G}_{23}$** is the (23, 12, 7) quadratic residue code with $p = 23$ and $s = 2$. The quadratic residues *modulo* 23 are $Q = \{1, 2, 3, 4, 6, 8, 9, 12, 13, 16, 18\}$. Let β be a primitive 23rd root of unity (22 different β can be found in GF(2^{11}) and its extensions). The distinct powers of β form two cyclotomic cosets *modulo* 23 with respect to GF(2).

$$C_1 = \{1, 2, 3, 4, 6, 8, 9, 12, 13, 16, 18\}$$

$$C_5 = \{5, 7, 10, 11, 14, 15, 17, 19, 20, 21, 22\}$$

$x^{23} + 1$ thus factors into three binary irreducible polynomials.

$$x^{23} + 1 = (x + 1)(x^{11} + x^{10} + x^6 + x^5 + x^4 + x^2 + 1) \cdot (x^{11} + x^9 + x^7 + x^6 + x^5 + x + 1)$$

Depending on the selection of β, there are two possible generator polynomials for $\mathcal{G}_{23}$:

$$g_1(x) = x^{11} + x^{10} + x^6 + x^5 + x^4 + x^2 + 1$$

$$g_2(x) = x^{11} + x^9 + x^7 + x^6 + x^5 + x + 1$$

Using either of these generator polynomials, the resulting code can be shown to have the following properties [Mac], [Bla2]:

- Every code word in $\mathcal{G}_{23}$ whose weight is even has weight divisible by 4.
- The minimum distance of $\mathcal{G}_{23}$ is 7, so the code is triple-error-correcting.

Given that $\mathcal{G}_{23}$ has dimension 12 and minimum distance 7, it is readily observed that $\mathcal{G}_{23}$ is perfect (see Sec. 4.1). Each code word is associated with a decoding sphere containing all vectors that are Hamming distance ≤ 3 from the code word. Since the vectors have length 23, the decoding spheres have cardinality

$$\begin{aligned} V_2(23,3) &= \binom{23}{0} + \binom{23}{1} + \binom{23}{2} + \binom{23}{3} \\ &= 1 + 23 + 253 + 1771 \\ &= 2^{11} \end{aligned} \tag{6-1}$$

Since $\mathcal{G}_{23}$ has redundancy $n - k = 11$, it satisfies the Hamming bound (Theorem 4-1) with equality and is thus perfect.

Through the addition of a parity-check bit, $\mathcal{G}_{23}$ can be extended to form $\mathcal{G}_{24}$, the (24, 12, 8) code listed in Table 6-1. As noted earlier, $\mathcal{G}_{24}$ was used to provide error control on the *Voyager* spacecraft. The source encoder in the *Voyager* generated a series of 12-tuples, each corresponding to one of $2^{12} = 4096$ possible color shades for each pixel in a photograph. A $\mathcal{G}_{24}$ encoder then mapped these 12-tuples onto 24-tuples before they were transmitted back to a ground station on the Earth. The now familiar clear images of Jupiter, Saturn, and their moons were the result.

The generator polynomial definition of the Golay codes hides a great deal of their inherent structure. $\mathcal{G}_{24}$ can also be defined as shown in Figure 6-1. $\mathbf{A}_{11}$ is obtained from the normalized binary Hadamard matrix $\mathbf{A}_{12}$ by removing the top row and the leftmost column.

$$\mathbf{G}_{24} = \left[\begin{array}{c|c|c|c} \mathbf{1}^T & \mathbf{I}_{11} & \mathbf{0}^T & \mathbf{A}_{11} \\ \hline 0 & \mathbf{0} & 1 & \mathbf{1} \end{array}\right]$$

$$= \left[\begin{array}{c|ccccccccccc|c|ccccccccccc}
1 & 1&&&&&&&&&& & & 1&1&&1&1&1&&&&1& \\
1 & &1&&&&&&&&& & & &1&1&&1&1&1&&&1 \\
1 & &&1&&&&&&&& & & 1&&1&1&&1&1&1&&& \\
1 & &&&1&&&&&&& & & &1&&1&1&&1&1&1&& \\
1 & &&&&1&&&&&& & & &&1&&1&1&&1&1&1& \\
1 & &&&&&1&&&&& & & &&&1&&1&1&&1&1&1 \\
1 & &&&&&&1&&&& & & 1&&&&1&&1&1&&1&1 \\
1 & &&&&&&&1&&& & & 1&1&&&&1&&1&1&&1 \\
1 & &&&&&&&&1&& & & 1&1&1&&&&1&&1&1& \\
1 & &&&&&&&&&1& & & &1&1&1&&&&1&&1&1 \\
1 & &&&&&&&&&&1 & & 1&&1&1&1&&&&1&&1 \\
\hline
 & &&&&&&&&&& & 1 & 1&1&1&1&1&1&1&1&1&1&1
\end{array}\right]$$

Figure 6-1. A generator matrix for $\mathcal{G}_{24}$ that is based on a binary Hadamard matrix [Mac]

In this generator matrix it is clear that the binary inner product of each pair of distinct rows is zero (note that the number of coordinates in which both rows in a pair have a one is always even). This implies that every row in the generator matrix is in the dual space of $\mathcal{G}_{24}$. But since $\mathcal{G}_{24}$ has dimension 12, its dual space must have dimension $24 - 12 = 12$ (see Theorem 2-9). The generator matrix has 12 linearly independent rows that lie in the dual space, so it follows that the rows of the generator matrix form a basis set for the dual space. The generator matrix for $\mathcal{G}_{24}$ thus also serves as a parity-check matrix for $\mathcal{G}_{24}$. Any code whose code space is equal to its dual space is called **self-dual**. If the code space is simply contained in the dual space, then the code is said to be **weakly self-dual**.

Figure 6-1 is but one of the many possible generator matrices for $\mathcal{G}_{24}$. The systematic generator matrix shown in Figure 6-2 is of the form $\mathcal{G}_{24} = [\mathbf{I}_{12} | \mathbf{B}]$. This structure leads to a simple arithmetic decoding algorithm which is discussed in Sec. 6.3.2. Any generator matrix for $\mathcal{G}_{24}$ can be converted to a generator matrix for $\mathcal{G}_{23}$ through the deletion of an arbitrarily selected column.

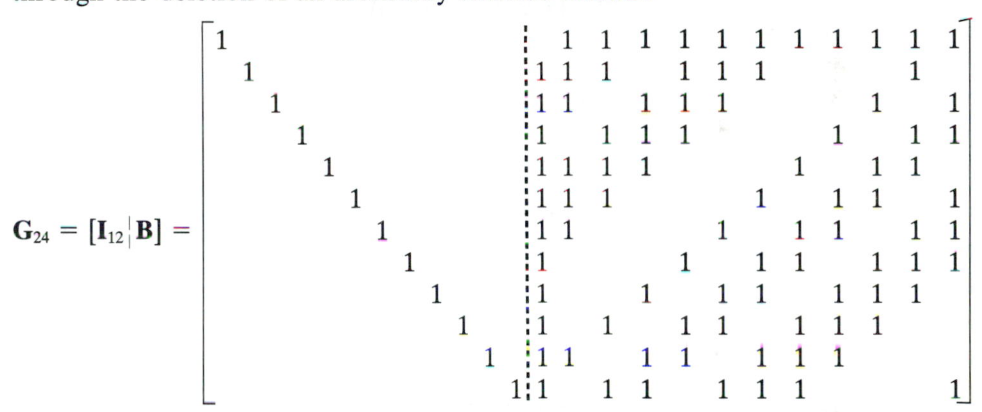

Figure 6-2. A systematic generator matrix for $\mathcal{G}_{24}$ [VanS]

The weight distributions of $\mathcal{G}_{23}$ and $\mathcal{G}_{24}$ are shown in Table 6-2.

The **ternary Golay code $\mathcal{G}_{11}$** is the quadratic residue code with $p = 11$ and $s = 3$. An analysis similar to that for $\mathcal{G}_{23}$ yields the factorization of $x^{11} - 1$ into three irreducible polynomials in GF(3)[x].

$$x^{11} - 1 = (x - 1)(x^5 + x^4 - x^3 + x^2 - 1)(x^5 - x^3 + x^2 - x - 1)$$

TABLE 6-2 Weight distributions for $\mathcal{G}_{23}$ and $\mathcal{G}_{24}$

Weight distribution for the (23, 12, 7) code $\mathcal{G}_{23}$								
i:	0	7	8	11	12	15	16	23
A_i:	1	253	506	1288	1288	506	253	1

Weight distribution for the (24, 12, 8) code $\mathcal{G}_{24}$					
i:	0	8	12	16	24
A_i:	1	759	2576	759	1

Again there are two possible generator polynomials:

$$g_1(x) = x^5 + x^4 - x^3 + x^2 - 1$$
$$g_2(x) = x^5 - x^3 + x^2 - x - 1$$

$\mathcal{G}_{11}$ has length 11, dimension 6, and minimum distance 5, and is thus perfect. A generator matrix for $\mathcal{G}_{11}$ can be constructed from the Jacobsthal matrix $\mathbf{Q}_5$. We begin by noting that the two quadratic residues *modulo* 5 are 1 and 4. The first row of $\mathbf{Q}_5$ is then

$$[0 \quad 1 \quad - \quad - \quad 1]$$

The remaining rows of $\mathbf{Q}_5$ are right cyclic shifts of the first row, giving us the following.

$$\mathbf{Q}_5 = \begin{bmatrix} 0 & 1 & - & - & 1 \\ 1 & 0 & 1 & - & - \\ - & 1 & 0 & 1 & - \\ - & - & 1 & 0 & 1 \\ 1 & - & - & 1 & 0 \end{bmatrix}$$

A $\mathcal{G}_{11}$ generator matrix is then constructed as shown in Figure 6-3.

$$\mathbf{G}_{11} = \left[\mathbf{I}_6 \;\middle|\; \begin{matrix} \mathbf{1} \\ \mathbf{Q}_5 \end{matrix} \right] = \left[\begin{array}{cccccc|ccccc} 1 & 0 & 0 & 0 & 0 & 0 & 1 & 1 & 1 & 1 & 1 \\ 0 & 1 & 0 & 0 & 0 & 0 & 0 & 1 & - & - & 1 \\ 0 & 0 & 1 & 0 & 0 & 0 & 1 & 0 & 1 & - & - \\ 0 & 0 & 0 & 1 & 0 & 0 & - & 1 & 0 & 1 & - \\ 0 & 0 & 0 & 0 & 1 & 0 & - & - & 1 & 0 & 1 \\ 0 & 0 & 0 & 0 & 0 & 1 & 1 & - & - & 1 & 0 \end{array} \right]$$

Figure 6-3. A systematic generator matrix for $\mathcal{G}_{11}$

$\mathcal{G}_{11}$ is extended through the addition of a parity check to form $\mathcal{G}_{12}$, the (12, 6, 6) ternary code listed in Table 6-1. $\mathcal{G}_{12}$ is also self-dual.

6.4 DECODING THE BINARY GOLAY CODE

Since $\mathcal{G}_{23}$ and $\mathcal{G}_{11}$ are cyclic, the shift-register error correction decoding techniques discussed in Chapter 5 can be applied. The key to the shift-register decoding of cyclic codes in general is the structure of the syndromes for received words that are cyclic shifts of a single received word (see Theorem 5-3). The decoder is simplified by minimizing the number of syndromes that the decoder must be able to recognize. Kasami used this approach to develop an efficient shift-register decoder for the Golay codes. The Kasami decoder "traps" recognizable error patterns within a short segment of the shift register. A good discussion of this "error-trapping" decoder can be found in [Kas4] and [Lin1].

Several arithmetic and algebraic decoding techniques have also been discovered. A simple arithmetic technique, similar to that developed for Hamming codes in Chapter 4, is presented below. It is followed by an algebraic technique developed by Elia [Elia]. Though it is not necessarily as fast as Kasami's error-trapping algorithm, Elia's technique is quite easy to understand.

6.4.1 An Arithmetic Decoding Algorithm

Consider the Golay generator matrix $\mathbf{G} = [\mathbf{I}_{12} | \mathbf{B}]$ in Figure 6-2. The following arithmetic decoding algorithm and its proof are adapted from [VanS, Chap. 4]. We first introduce some notation.

Let $\{\mathbf{I}_0, \mathbf{I}_1, \ldots, \mathbf{I}_{11}\}$ and $\{\mathbf{R}_0, \mathbf{R}_1, \ldots, \mathbf{R}_{11}\}$ be the columns and rows, respectively, of $\mathbf{B}$. Let $\mathbf{0}$ denote the all-zero 12-tuple. Let $\{\mathbf{x}_i\}$ and $\{\mathbf{y}_i\}$ be sets of binary 12-tuples in which only the ith coordinate contains a nonzero value. For example, $\mathbf{x}_6 = \mathbf{y}_6 = (000000100000)$.

Given a received vector $\mathbf{r}$, decoding proceeds as follows.

1. Compute $\mathbf{S} = \mathbf{G}\mathbf{r}^T$. (Note that since $\mathscr{G}_{12}$ is self-dual, the generator matrix $\mathbf{G}$ is also a parity-check matrix and $\mathbf{S}$ is the corresponding syndrome.)
2. If $w(\mathbf{S}) \leq 3$, then set the error vector $\mathbf{e} = (\mathbf{S}^T, \mathbf{0})$ and go to step 8.
3. If $w(\mathbf{S} + \mathbf{I}_i) \leq 2$ for some column vector $\mathbf{I}_i$, then set $\mathbf{e} = ((\mathbf{S} + \mathbf{I}_i)^T, \mathbf{y}_i)$ and go to step 8.
4. Compute $\mathbf{B}^T\mathbf{S}$.
5. If $w(\mathbf{B}^T\mathbf{S}) \leq 3$, then set the error vector $\mathbf{e} = (\mathbf{0}, (\mathbf{B}^T\mathbf{S})^T)$ and go to step 8.
6. If $w(\mathbf{B}^T\mathbf{S} + \mathbf{R}_i^T) \leq 2$, for some row vector $\mathbf{R}_i$, then set $\mathbf{e} = (\mathbf{x}_i, (\mathbf{B}^T\mathbf{S})^T + \mathbf{R}_i)$ and go to step 8.
7. $\mathbf{r}$ is corrupted by an uncorrectable error pattern (i.e., $w(\mathbf{e}) \geq 4$), STOP.
8. Set $\mathbf{c} = \mathbf{r} + \mathbf{e}$, STOP.

To prove that this is a valid decoding algorithm, let $\mathbf{c}$ be the transmitted code word, $\mathbf{e} = (\mathbf{x}, \mathbf{y})$ be the corrupting error pattern, where $\mathbf{x}$ and $\mathbf{y}$ are binary 12-tuples, and let $\mathbf{r} = \mathbf{c} + \mathbf{e}$.

The proof uses the systematic form of $\mathbf{G}$ and the orthonormality of $\mathbf{B}$ to deduce $\mathbf{x}$ and $\mathbf{y}$ from the syndrome $\mathbf{S}$. First consider the possible weight distributions for $\mathbf{x}$ and $\mathbf{y}$ given that $\mathbf{e}$ has weight less than or equal to 3.

$$
\begin{aligned}
w(\mathbf{x}) &\leq 3, \quad & w(\mathbf{y}) &= 0 \\
w(\mathbf{x}) &\leq 2, \quad & w(\mathbf{y}) &= 1 \\
w(\mathbf{x}) &\leq 1, \quad & w(\mathbf{y}) &= 2 \\
w(\mathbf{x}) &= 0, \quad & w(\mathbf{y}) &= 3
\end{aligned}
$$

Since $\mathbf{S} = \mathbf{G}\mathbf{r}^T = \mathbf{G}(\mathbf{c} + \mathbf{e})^T = \mathbf{G}\mathbf{e}^T$, the syndrome is only a function of the error pattern. It follows that if $\mathbf{y} = \mathbf{0}$, then

$$\mathbf{S} = [\mathbf{I}_{12} | \mathbf{B}](\mathbf{x}, \mathbf{y})^T = \mathbf{I}_{12}\mathbf{x}^T = \mathbf{x}^T$$

The error pattern is thus $\mathbf{e} = (\mathbf{x}, \mathbf{y}) = (\mathbf{S}^T, \mathbf{0})$, as seen in step 2 of the decoding algorithm.

Now consider the case in which $w(\mathbf{x}) \leq 2$ and $w(\mathbf{y}) = 1$. Suppose that the ith coordinate of $\mathbf{y}$ is the sole nonzero coordinate in y. Then

$$\mathbf{S} = [\mathbf{I}_{12} | \mathbf{B}](\mathbf{x}, \mathbf{y})^T = \mathbf{I}_{12}\mathbf{x}^T + \mathbf{B}\mathbf{y}^T = \mathbf{x}^T + \mathbf{I}_i$$

i is found by identifying the value for which $w(\mathbf{S} + \mathbf{I}_i) = w(\mathbf{x}) \leq 2$. The desired error pattern is then $\mathbf{e} = ((\mathbf{S} + \mathbf{I}_i)^T, \mathbf{y}_i)$, as noted in step 3.

When $w(\mathbf{y}) = 2$ or 3 and $w(\mathbf{x}) = 0$, then $\mathbf{S} = \mathbf{I}_{12}\mathbf{x}^T + \mathbf{B}\mathbf{y}^T = \mathbf{I}_i + \mathbf{I}_j$ $(w(\mathbf{y}) = 2)$ or $\mathbf{S} = \mathbf{I}_i + \mathbf{I}_j + \mathbf{I}_k$ $(w(\mathbf{y}) = 3)$. Since $\mathbf{B}$ is orthonormal,

$$\mathbf{B}^T\mathbf{S} = \mathbf{B}^T(\mathbf{B}\mathbf{y}^T) = \mathbf{y}^T$$

The error pattern is thus $\mathbf{e} = (\mathbf{0}, (\mathbf{B}^T\mathbf{S})^T)$, as noted in step 5.

For the last case we have $w(\mathbf{x}) = 1$ and $w(\mathbf{y}) = 2$. If the nonzero coordinate of $\mathbf{x}$ is in the ith position, then

$$\begin{aligned}\mathbf{B}^T\mathbf{S} &= \mathbf{B}^T(\mathbf{I}_{12}\mathbf{x}^T + \mathbf{B}\mathbf{y}^T) \\ &= \mathbf{B}^T\mathbf{x}^T + \mathbf{B}^T\mathbf{B}\mathbf{y}^T \\ &= \mathbf{R}_i^T + \mathbf{y}^T\end{aligned}$$

It follows that $\mathbf{y}^T = \mathbf{B}^T\mathbf{S} + \mathbf{R}_i^T$ and $\mathbf{e} = (\mathbf{x}_i, (\mathbf{B}^T\mathbf{S})^T + \mathbf{R}_i)$, as noted in step 6.

Example 6-5—Decoding the Extended Binary Golay Code (Arithmetic)

Let the received vector $\mathbf{r} = (0000\ 0010\ 0000\ 1000\ 0000\ 0000)$.

Step 1: We compute $\mathbf{S} = \mathbf{G}\mathbf{r}^T$, obtaining

$$\begin{aligned}\mathbf{S}^T &= (0000\ 0010\ 0000) + (0111\ 1111\ 1111) \\ &= (0111\ 1101\ 1111)\end{aligned}$$

Step 2: We note that $w(\mathbf{S})$ is not less than or equal to 3 and proceed to Step 3.

Step 3: We note that $\mathbf{S} + \mathbf{I}_0 = (0000\ 0010\ 0000)$ has weight less than 2. We set $\mathbf{e} = ((\mathbf{S} + \mathbf{I}_0)^T, \mathbf{y}_0) = (0000\ 0010\ 0000\ 1000\ 0000\ 0000)$ and go to Step 8.

Step 8: $\mathbf{c} = \mathbf{r} + \mathbf{e} = (0000\ 0000\ 0000\ 0000\ 0000\ 0000\ 0000)$, a result that we can anticipate, given that the code is linear and triple-error-correcting and $w(\mathbf{r}) = 2 \leq 3$. ■

Example 6-6—Decoding the Extended Binary Golay Code (Arithmetic)

Let the received vector be $\mathbf{r} = (0011\ 1101\ 0000\ 0011\ 0000\ 1010)$.

Step 1: We compute $\mathbf{S} = \mathbf{G}\mathbf{r}^T$, obtaining

$$\begin{aligned}\mathbf{S}^T &= (0011\ 1101\ 0000) + (0010\ 1001\ 1000) \\ &= (0001\ 0100\ 1000)\end{aligned}$$

Step 2: Since $w(\mathbf{S}) = 3$, we set $\mathbf{e} = (\mathbf{S}^T, \mathbf{0}) = (0001\ 0100\ 1000\ 0000\ 0000\ 0000)$ and go to Step 8.

Step 8: $\mathbf{c} = \mathbf{r} + \mathbf{e} = (0010\ 1001\ 1000\ 0011\ 0000\ 1010)$. ■

Example 6-7—Decoding the Extended Binary Golay Code (Arithmetic)

Let the received vector be $\mathbf{r}$ = (1111 0110 1111 0011 0000 1110).

Step 1: We compute $\mathbf{S} = \mathbf{G}\mathbf{r}^T$, obtaining

$$\mathbf{S}^T = (1111\ 0110\ 1111) + (1000\ 0100\ 0100)$$
$$= (0111\ 0010\ 1011).$$

Step 2: We note that $w(\mathbf{S})$ is not less than or equal to 3 and proceed to Step 3.

Step 3: We note that none of the $\mathbf{I}_i$ satisfy $w(\mathbf{S} + \mathbf{I}_i) \leq 2$, and proceed to Step 4.

Step 4: We compute $\mathbf{B}^T\mathbf{S}$.

$$\mathbf{B}^T\mathbf{S} = (1011\ 0101\ 0000)^T$$

Step 5: We note that $w(\mathbf{B}^T\mathbf{S})$ is not less than or equal to 3 and proceed to Step 6.

Step 6: Since $w(\mathbf{B}^T\mathbf{S} + \mathbf{R}_{11}^T) = 2$, we set $\mathbf{e} = (\mathbf{x}_{11}, (\mathbf{B}^T\mathbf{S})^T + \mathbf{R}_{11})$ = (0000 0000 0001 0000 0010 0001) and proceed to Step 8.

Step 8: $\mathbf{c} = \mathbf{r} + \mathbf{e}$ = (1111 0110 1110 0011 0010 1111). ∎

6.4.2 An Algebraic Decoding Algorithm

The following decoding algorithm is due to Elia [Elia]. Let $f(x)$ be the message polynomial $m_0 + m_1x + \cdots + m_{11}x^{11}$ corresponding to a 12-bit information block $(m_0, m_1, \ldots, m_{11})$. These blocks are encoded by multiplication by the generator polynomial $g(x)$ to form code words with associated code polynomials $c(x) = m(x)g(x)$. These code words are transmitted across a noisy channel, arriving in noise-corrupted form as the received polynomial $r(x) = c(x) + e(x)$.

Let β be a primitive 23rd root of unity in GF(2^{11}). If $g(x)$ is the generator polynomial associated with the cyclotomic coset $C_1 = \{1, 2, 3, 4, 6, 8, 9, 12, 13, 16, 18\}$, then β, β^3, and β^9 must all be roots of $g(x)$, and thus roots of all code polynomials $c(x)$. We can thus define three syndromes,

$$S_1 = r(\beta), \quad S_2 = r(\beta^3), \text{ and } S_3 = r(\beta^9)$$

which can be used to compute the coefficients of an error locator polynomial $L(z)$. An error locator polynomial is a polynomial whose roots indicate the positions of the errors in the received word. The error locator polynomial is generally designed so that a root β^j indicates an error in the jth coordinate of the received word. This is a powerful concept that is used in the decoding of BCH and Reed-Solomon codes.

If there are no errors in the received word, $r(\beta) = r(\beta^3) = r(\beta^9) = 0$, and thus $\mathbf{S} = (S_1, S_2, S_3) = \mathbf{0}$ and $L(z) = 0$. If a single error has occurred, $e(x)$ is a polynomial of the form x^j, and the syndromes take on the values

$$\mathbf{S} = (\beta^j, (\beta^j)^3, (\beta^j)^9)$$

The error locator polynomial is thus

$$L(z) = z + S_1 \tag{6-3}$$

Suppose that two or three errors occur in the positions z_1, z_2, and z_3, with z_3 set to zero if two errors have occurred. We can then define a generalized syndrome S_a that takes on the value $S_a = z_1^a + z_2^a + z_3^a$. The error locator polynomial has the form

$$L(z) = \prod_{i=1}^{3} (z - z_i) = z^3 + \sigma_1 z^2 + \sigma_2 z + \sigma_3 \tag{6-4}$$

The $\{\sigma_j\}$ are the **elementary symmetric functions** of the $\{z_i\}$, as listed below.

$$\sigma_1 = z_1 + z_2 + z_3$$

$$\sigma_2 = z_1 z_2 + z_1 z_3 + z_2 z_3$$

$$\sigma_3 = z_1 z_2 z_3$$

Now let $\sum_{a,b}$ denote the sum of all six terms of the form $z_i^a z_j^b$, where the indices i and j take on all pairs of values from the set $\{1,2,3\}$. For example,

$$\begin{aligned}\textstyle\sum_{1,8} &= z_1 z_1^8 + z_1 z_2^8 + z_1 z_3^8 + z_2 z_2^8 + z_2 z_3^8 + z_3 z_3^8 \\ &= z_1^9 + z_2^9 + z_3^9 + z_1 z_2^8 + z_1 z_3^8 + z_2 z_3^8\end{aligned}$$

It is then a simple matter to verify the following identities.

$$S_9 + S_1^9 = \sum_{1,8} = \sigma_2 S_7 + \sigma_3 S_3^2 \tag{6-5}$$

$$S_7 + S_1 S_3^2 = \sum_{1,6} = \sigma_2 S_5 + \sigma_3 S_1^4 \tag{6-6}$$

$$S_5 + S_1^5 = \sum_{1,4} = \sigma_2 S_3 + \sigma_3 S_1^2 \tag{6-7}$$

$$\sigma_2 + S_3 = \sigma_2 S_1 + S_1^3 \tag{6-8}$$

These four identities can be combined to show the following.

$$\begin{aligned} D &\triangleq (S_1^3 + S_3)^2 + \frac{(S_1^9 + S_9)}{(S_1^3 + S_3)} \\ &= (\sigma_2 + S_1^2)^3 \end{aligned} \tag{6-9}$$

The last term in Eq. (6-9) shows that D is guaranteed to be a cubic residue (i.e., its cube root exists and, in this case, can be found in GF(2^{11})). Combining Eqs. (6-8) and (6-9), we obtain

$$\begin{aligned} \sigma_1 &= S_1 \\ \sigma_2 &= S_1^2 + \sqrt[3]{D} \\ \sigma_3 &= S_3 + S_1 \sqrt[3]{D} \end{aligned} \tag{6-10}$$

Let x be the cube root of D. Note that since $D \in$ GF(2^{11}), the order of D is a divisor of $(2^{11} - 1) = 2047$. x can then be evaluated by raising it to the 1365th power.

$$D^{1365} = (x^3)^{1365} = x^{4095} = x^{2(2047)+1} = x \tag{6-11}$$

The values $\sigma_1, \sigma_2, \sigma_3$ computed in Eq. (6-10) are then plugged into Eq. (6-4) and the resulting error locator polynomial factored to obtain the error locations.

SUGGESTIONS FOR FURTHER READING

Hadamard, Quadratic Residue, and Golay codes are treated extensively in MacWilliams and Sloane [Mac, Chaps. 2, 16, and 20]. The majority of this material is highly mathematical in nature (e.g., symmetry groups and Steiner systems), and some of it does not have an obvious practical application for engineers. This material uncovers, however, the wealth of structure contained in these codes. The Golay codes have proved particularly interesting to the mathematician. Conway and Sloane [Con1] discuss in some detail the relationship between the binary Golay code, the Leech lattice, and some extremely large and complex finite groups.

Kasami's error-trapping decoding algorithm, on the other hand, is an extremely practical decoding algorithm. A detailed discussion can be found in Lin and Costello [Lin1], as well as Kasami's original paper [Kas4].

PROBLEMS

Hadamard matrices and Hadamard codes

1. Prove that the order n of a Hadamard matrix must be even or 1.
2. Construct a Sylvester-type Hadamard matrix of order 16.
3. Find the quadratic residues *modulo* 17.
4. Find the quadratic residues *modulo* 19.
5. Construct a Paley-type Hadamard matrix of order 20.
6. Construct three Hadamard codes $\mathcal{A}_8$, $\mathcal{B}_8$, and $\mathcal{C}_8$ through the use of a Hadamard matrix of order 8.
7. Construct three Hadamard codes $\mathcal{A}_{20}$, $\mathcal{B}_{20}$, and $\mathcal{C}_{20}$ through the use of a Hadamard matrix of order 20.
8. Show that the Paley construction fails to produce a Hadamard matrix of order 18.

Quadratic residue codes

9. Find the generator polynomial for a binary quadratic residue code of length 7.
10. Find the generator polynomial for a quadratic residue code of length 17.

11. Find the generator polynomial for a ternary quadratic residue code of length 11.

12. Find the generator polynomial for a ternary quadratic residue code of length 13.

Golay codes

13. Prove that a code is self-dual only if its length n is twice its dimension k.

14. Prove that all of the code words in $\mathcal{G}_{23}$ have weights greater than 4.

15. Prove that all of the code words in $\mathcal{G}_{11}$ have weights greater than 3.

16. Use the arithmetic decoding algorithm to decode the following received words.
- **(a)** $\mathbf{r} = (1000\ 1000\ 0000\ 1001\ 0001\ 1101)$
- **(b)** $\mathbf{r} = (1011\ 1000\ 0000\ 1110\ 0000\ 1100)$
- **(c)** $\mathbf{r} = (1000\ 1110\ 0011\ 0100\ 0110\ 1000)$
- **(d)** $\mathbf{r} = (1111\ 0111\ 1110\ 0011\ 1010\ 0111)$

17. Use the algebraic decoding algorithm to decode the following received words.
- **(a)** $\mathbf{r} = (1100\ 0000\ 0000\ 1100\ 0000\ 0000)$
- **(b)** $\mathbf{r} = (0000\ 0001\ 1111\ 0000\ 0000\ 0000)$
- **(c)** $\mathbf{r} = (1010\ 1010\ 0000\ 1000\ 1100\ 0100)$
- **(d)** $\mathbf{r} = (1101\ 1111\ 0001\ 1111\ 0100\ 1111)$

7

Reed-Muller Codes

The codes that are now called Reed-Muller (RM) codes were first described by Muller in 1954 using a "Boolean net function" language. That same year Reed [Ree1] recognized that Muller's codes could be represented as multinomials over the binary field. The resulting "Reed-Muller" (RM) codes were an important step beyond the Hamming and Golay codes of 1949 and 1950 because of their flexibility in correcting varying numbers of errors per code word. Since their discovery, RM codes have been used in a number of interesting applications. For example, a first-order RM code of length 32 provided error control on all of the United States' *Mariner*-class deep space probes flown between 1969 and 1977. RM codes have also enjoyed a great deal of attention from the more theoretically inclined researchers. RM codes have an enormous wealth of algebraic and combinatorial structure that has kept mathematicians busy for almost forty years. Part of this theoretical work has led to the discovery of other interesting codes, including the Kerdock and Preparata codes.

For the past twenty years RM codes have not received the frequent application they enjoyed between 1954 and 1968. They lost their hold on the space program when convolutional codes and sequential decoders were adopted for the *Pioneer* missions [Mas2]. It had also been recognized that RM codes do not perform as well as long BCH and Reed-Solomon codes. When an efficient decoding algorithm was discovered for the latter in 1968 [Ber1], RM codes were no longer as attractive to design engineers. It would be a great mistake, however, to believe that BCH and Reed-Solomon codes are a superior choice relative to RM codes for all applications. The short low-rate RM codes and the first-order RM codes have roughly the same minimum distances as binary BCH codes of the same length. RM codes also enjoy a tremendous benefit in that they have an extremely fast maximum likelihood decoding algorithm (the Reed decoding algorithm [Ree1]). No such algorithm has been discovered for BCH and Reed-Solomon codes. As data rates on optical channels push electronic encoders and decoders to the limits of device technology, Reed-Muller codes may once again see extensive application.

The next section describes the basic structure of RM codes. This is followed by a discussion of the Reed decoding algorithm, which, among other things, leads to some nice four-dimensional art work. In the last section we examine the first-order RM codes in depth. A high-speed soft-decision decoding algorithm related to the fast-Fourier-transform is presented.

7.1 THE CONSTRUCTION OF REED-MULLER CODES

Reed-Muller codes are most easily described through the use of Boolean functions. A Boolean function in m variables $f(x_1, x_2, \ldots, x_m)$ is defined as a mapping from the vector space V_m of binary m-tuples $\{(x_1, \ldots, x_m)\}$ into the set of binary numbers $\{0, 1\}$. Boolean functions are completely described by a truth table containing $(m + 1)$ rows. The first m rows form an $(m \times 2^m)$ matrix that contains as columns all 2^m binary m-tuples. The bottom row contains the binary value assigned to each of the m-tuples by the Boolean function. There exist a number of techniques for using this truth table to develop an algebraic expression for the function in terms of the m variables in its argument. Some readers may have become familiar with Boolean functions while studying sequential digital circuits. The problem of deriving algebraic representations for Boolean functions from their truth tables is related to the problem of simplifying digital circuits.[1]

Example 7-1—Truth Tables for a Boolean Function

Let f_1 be a Boolean function in four variables $\{v_1, v_2, v_3, v_4\}$ with the following truth table.

$v_4 =$	0	0	0	0	0	0	0	0	1	1	1	1	1	1	1	1
$v_3 =$	0	0	0	0	1	1	1	1	0	0	0	0	1	1	1	1
$v_2 =$	0	0	1	1	0	0	1	1	0	0	1	1	0	0	1	1
$v_1 =$	0	1	0	1	0	1	0	1	0	1	0	1	0	1	0	1
$f_1 =$	0	1	1	0	1	0	0	1	0	1	1	0	1	0	0	1

A quick inspection shows that this Boolean function can be represented by the expression $f_1 = v_3 + v_2 + v_1$. For a slightly more complicated example, consider the following.

[1] Muller was investigating applications of Boolean algebra to the problem of simplifying digital switching circuits when he discovered RM codes [Mul].

$v_4 =$	0	0	0	0	0	0	0	0	1	1	1	1	1	1	1	1
$v_3 =$	0	0	0	0	1	1	1	1	0	0	0	0	1	1	1	1
$v_2 =$	0	0	1	1	0	0	1	1	0	0	1	1	0	0	1	1
$v_1 =$	0	1	0	1	0	1	0	1	0	1	0	1	0	1	0	1
$f_2 =$	1	0	1	0	1	0	1	1	1	0	1	0	0	1	0	0

The reader may wish to verify that this Boolean function can be expressed as $f_2 = v_3 v_2 v_1 + v_4 v_3 + v_1 + 1$. ■

In Example 7-1 we have adopted a convention that will be maintained throughout the rest of the chapter. When viewed as the radix-2 representations for the integers $\{0, 1, 2, \ldots, 2^m - 1\}$, the columns of the truth table are seen to run in increasing order from left to right. The most significant bit of the binary expansion (the first row of the table) corresponds to the variable v with the largest subscript. This convention allows us to unambiguously associate each Boolean function f with a unique binary vector $\mathbf{f}$. $\mathbf{f}$ is simply the bottom row of the truth table for f. For the above example we have the following.

$$\mathbf{f}_1 = (0110100101101001)$$

$$\mathbf{f}_2 = (1010101110100100)$$

Since $\mathbf{f}$ is binary with length 2^m, there must be 2^{2^m} distinct Boolean functions in m variables. Under coordinate-by-coordinate binary addition of the associated vectors, the Boolean functions form the vector space V_{2^m} over GF(2).

Let the set M consist of all Boolean functions in m variables that can be represented by a single monomial term. We need not consider squares and higher-order powers of the individual variables, for v_i and v_i^2 represent the same Boolean function. M thus consists of the Boolean function 1 and the products of all combinations of one or more variables in the set $\{v_1, v_2, \ldots, v_m\}$.

$$M = \{1, \ldots, v_m, v_1 v_2, \ldots, v_{m-1} v_m, v_1 v_2 v_3, \ldots, v_{m-2} v_{m-1} v_m, \ldots, v_1 v_2 \cdots v_m\}$$

Since the Boolean functions in M are linearly independent, it follows that the vectors with which they are associated are also linearly independent. There is thus a unique Boolean function f for every vector $\mathbf{f}$ of the form

$$\mathbf{f} = a_0 \mathbf{1} + \cdots + a_m \mathbf{v}_m + a_{12} \mathbf{v}_1 \mathbf{v}_2 + \cdots + a_{12\ldots m} \mathbf{v}_1 \mathbf{v}_2 \cdots \mathbf{v}_m \qquad (7\text{-}1)$$

Since there are a total of 2^{2^m} such vectors, the Boolean functions in M form a basis for the vector space of Boolean functions in m variables. We now have sufficient machinery at our disposal to define Reed-Muller codes.

Definition 7-1—Reed-Muller Codes

The binary Reed-Muller code $\mathfrak{R}(r, m)$ of order r and length 2^m consists of the vectors $\mathbf{f}$ associated with all Boolean functions f that are polynomials of degree less than or equal to r in m variables.

Example 7-2—$\mathfrak{R}(1,3)$: The First-Order RM Code of Length 8

The monomials in three variables of degree 1 or less are $\{1, v_1, v_2, v_3\}$. Each of these monomials is associated with a vector as shown below.

$$\begin{array}{r@{\;=\;}l}
\mathbf{1} & (1\ \ 1\ \ 1\ \ 1\ \ 1\ \ 1\ \ 1\ \ 1) \\ \hline
\mathbf{v}_3 & (0\ \ 0\ \ 0\ \ 0\ \ 1\ \ 1\ \ 1\ \ 1) \\
\mathbf{v}_2 & (0\ \ 0\ \ 1\ \ 1\ \ 0\ \ 0\ \ 1\ \ 1) \\
\mathbf{v}_1 & (0\ \ 1\ \ 0\ \ 1\ \ 0\ \ 1\ \ 0\ \ 1)
\end{array}$$

The code words in $\mathfrak{R}(1,3)$ consist of the 16 distinct linear combinations of these vectors. Since the four vectors form a basis set for $\mathfrak{R}(1,3)$, we can employ them as the rows of a generator matrix.

$$\mathbf{G} = \begin{bmatrix} 1 & 1 & 1 & 1 & 1 & 1 & 1 & 1 \\ 0 & 0 & 0 & 0 & 1 & 1 & 1 & 1 \\ 0 & 0 & 1 & 1 & 0 & 0 & 1 & 1 \\ 0 & 1 & 0 & 1 & 0 & 1 & 0 & 1 \end{bmatrix}$$

$\mathfrak{R}(1,3)$ has length 8, dimension 4, and minimum distance 4. It is thus single-error-correcting and double-error-detecting.

This generator matrix may look vaguely familiar. It is also the parity-check matrix for the (8, 4) extended Hamming code (see Ex. 4-8). First-order Reed-Muller codes are the duals of extended Hamming codes. ■

Example 7-3—$\mathfrak{R}(2,4)$: The Second-Order RM Code of Length 16

The monomials in four variables of degree 2 or less are as follows.

$$\{1, v_1, v_2, v_3, v_4, v_1v_2, v_1v_3, v_1v_4, v_2v_3, v_2v_4, v_3v_4\}$$

The binary vectors associated with these functions are shown in Figure 7-1.

$$\begin{array}{r@{\;=\;}l}
\mathbf{1} & (1\ \ 1\ \ 1\ \ 1\ \ 1\ \ 1\ \ 1\ \ 1\ \ 1\ \ 1\ \ 1\ \ 1\ \ 1\ \ 1\ \ 1\ \ 1) \\ \hline
\mathbf{v}_4 & (0\ \ 0\ \ 0\ \ 0\ \ 0\ \ 0\ \ 0\ \ 0\ \ 1\ \ 1\ \ 1\ \ 1\ \ 1\ \ 1\ \ 1\ \ 1) \\
\mathbf{v}_3 & (0\ \ 0\ \ 0\ \ 0\ \ 1\ \ 1\ \ 1\ \ 1\ \ 0\ \ 0\ \ 0\ \ 0\ \ 1\ \ 1\ \ 1\ \ 1) \\
\mathbf{v}_2 & (0\ \ 0\ \ 1\ \ 1\ \ 0\ \ 0\ \ 1\ \ 1\ \ 0\ \ 0\ \ 1\ \ 1\ \ 0\ \ 0\ \ 1\ \ 1) \\
\mathbf{v}_1 & (0\ \ 1\ \ 0\ \ 1\ \ 0\ \ 1\ \ 0\ \ 1\ \ 0\ \ 1\ \ 0\ \ 1\ \ 0\ \ 1\ \ 0\ \ 1) \\ \hline
\mathbf{v}_3\mathbf{v}_4 & (0\ \ 0\ \ 0\ \ 0\ \ 0\ \ 0\ \ 0\ \ 0\ \ 0\ \ 0\ \ 0\ \ 0\ \ 1\ \ 1\ \ 1\ \ 1) \\
\mathbf{v}_2\mathbf{v}_4 & (0\ \ 0\ \ 0\ \ 0\ \ 0\ \ 0\ \ 0\ \ 0\ \ 0\ \ 0\ \ 1\ \ 1\ \ 0\ \ 0\ \ 1\ \ 1) \\
\mathbf{v}_1\mathbf{v}_4 & (0\ \ 0\ \ 0\ \ 0\ \ 0\ \ 0\ \ 0\ \ 0\ \ 0\ \ 1\ \ 0\ \ 1\ \ 0\ \ 1\ \ 0\ \ 1) \\
\mathbf{v}_2\mathbf{v}_3 & (0\ \ 0\ \ 0\ \ 0\ \ 0\ \ 0\ \ 1\ \ 1\ \ 0\ \ 0\ \ 0\ \ 0\ \ 0\ \ 0\ \ 1\ \ 1) \\
\mathbf{v}_1\mathbf{v}_3 & (0\ \ 0\ \ 0\ \ 0\ \ 0\ \ 1\ \ 0\ \ 1\ \ 0\ \ 0\ \ 0\ \ 0\ \ 0\ \ 1\ \ 0\ \ 1) \\
\mathbf{v}_1\mathbf{v}_2 & (0\ \ 0\ \ 0\ \ 1\ \ 0\ \ 0\ \ 0\ \ 1\ \ 0\ \ 0\ \ 0\ \ 1\ \ 0\ \ 0\ \ 0\ \ 1)
\end{array}$$

Figure 7-1. Basis Vectors for $\mathfrak{R}(2,4)$

$\mathfrak{R}(2,4)$ is a (16, 11) code. As in the previous example, this code has minimum distance 4. The $\mathfrak{R}(2,4)$ code, however, has a higher rate than the $\mathfrak{R}(1,3)$ code. ■

In general the space of Boolean functions of degree r or less has as a basis all monomial functions of degree r or less. It is a simple matter to count the number of elements k in this basis set for a given r and m, and thus determine the dimension k of $\mathfrak{R}(r,m)$.

$$k = 1 + \binom{m}{1} + \binom{m}{2} + \cdots + \binom{m}{r} \tag{7-2}$$

The following theorem shows that $\mathfrak{R}(r+1,m+1)$ can be constructed from $\mathfrak{R}(r,m)$ and $\mathfrak{R}(r+1,m)$. It is useful in determining the minimum distance of RM codes. Note that given $\mathbf{x} = (x_0,\ldots,x_{n-1})$ and $\mathbf{y} = (y_0,\ldots,y_{n-1})$, $(\mathbf{x}\,|\,\mathbf{y})$ is defined to be the concatenation of $\mathbf{x}$ and $\mathbf{y}$ (i.e., $(\mathbf{x}\,|\,\mathbf{y}) = (x_0,\ldots,x_{n-1},y_0,\ldots,y_{n-1})$).

Theorem 7-1

$$\mathfrak{R}(r+1,m+1) = \{(\mathbf{f}\,|\,\mathbf{f}+\mathbf{g}), \text{ for all } \mathbf{f} \in \mathfrak{R}(r+1,m) \text{ and } \mathbf{g} \in \mathfrak{R}(r,m)\}$$

Proof. All code words $\mathbf{c} \in \mathfrak{R}(r+1,m+1)$ are associated with Boolean functions $c(v_1,\ldots,v_{m+1})$ of degree $\leq r+1$. Such functions can be rewritten as $c(v_1,\ldots,v_{m+1}) = f(v_1,\ldots,v_m) + v_{m+1}\cdot g(v_1,\ldots,v_m)$. f has degree $\leq r+1$ and g has degree $\leq r$, so the corresponding vectors $\mathbf{f}$ and $\mathbf{g}$ can be found in $\mathfrak{R}(r+1,m)$ and $\mathfrak{R}(r,m)$, respectively.

Let $f' = f(v_1,\ldots,v_m) + 0\cdot v_{m+1}$ and $g' = v_{m+1}\cdot g(v_1,\ldots,v_m)$. The associated vectors have the form $\mathbf{f}' = (\mathbf{f}\,|\,\mathbf{f})$ and $\mathbf{g}' = (\mathbf{0}\,|\,\mathbf{g})$, and both are code words in $\mathfrak{R}(r+1,m+1)$. It follows that $\mathbf{c} = (\mathbf{f}\,|\,\mathbf{f}+\mathbf{g}) \in \mathfrak{R}(r+1,m+1)$. **QED**

Theorem 7-2

The minimum distance of $\mathfrak{R}(r,m)$ is 2^{m-r}.

Proof. We proceed by induction on m.

For the case $m=1$, $\mathfrak{R}(0,1)$ is the length-2 repetition code with $d_{\min} = 2 = 2^{(1-0)}$. $\mathfrak{R}(1,1)$ consists of all 2-tuples and thus has $d_{\min} = 1 = 2^{(1-1)}$.

Assume that, up to some m and for $0 \leq r \leq m$, the minimum distance of $\mathfrak{R}(r,m)$ is $2^{(m-r)}$. We now show that the minimum distance of $\mathfrak{R}(r,m+1)$ must then be $2^{(m-r+1)}$. Let $w(\mathbf{c})$ be the weight of $\mathbf{c}$ and let $d(\mathbf{c}_1,\mathbf{c}_2)$ be the Hamming distance between $\mathbf{c}_1$ and $\mathbf{c}_2$.

Let $\mathbf{f}$ and $\mathbf{f}'$ be in $\mathfrak{R}(r,m)$ and $\mathbf{g}$ and $\mathbf{g}'$ in $\mathfrak{R}(r-1,m)$. Applying Theorem 7-1, $\mathbf{c}_1 = (\mathbf{f}\,|\,\mathbf{f}+\mathbf{g})$ and $\mathbf{c}_2 = (\mathbf{f}'\,|\,\mathbf{f}'+\mathbf{g}')$ must be code words in $\mathfrak{R}(r,m+1)$.

If $\mathbf{g} = \mathbf{g}'$, then $d(\mathbf{c}_1,\mathbf{c}_2) = 2d(\mathbf{f},\mathbf{f}') \geq 2^{m-r+1}$ (twice the minimum distance of $\mathfrak{R}(r,m)$). If $\mathbf{g} \neq \mathbf{g}'$, then $d(\mathbf{c}_1,\mathbf{c}_2) = w(\mathbf{f}-\mathbf{f}') + w[(\mathbf{g}-\mathbf{g}') + (\mathbf{f}-\mathbf{f}')]$. Note that $w(\mathbf{x}+\mathbf{y}) \geq w(\mathbf{x}) - w(\mathbf{y})$, since the nonzero elements in $\mathbf{x}$ and $\mathbf{y}$ may not completely overlap. We thus have $d(\mathbf{c}_1,\mathbf{c}_2) \geq w(\mathbf{f}-\mathbf{f}') +$

$w(\mathbf{g} - \mathbf{g}') - w(\mathbf{f} - \mathbf{f}') = w(\mathbf{g} - \mathbf{g}')$. Since $(\mathbf{g} - \mathbf{g}')$ must be a code word in $\mathfrak{R}(r-1,m)$, $d(\mathbf{c}_1,\mathbf{c}_2) \geq 2^{m-r+1}$. The result follows. **QED**

Table 7-1 contains the length n, dimension k, and minimum distance $d_{\min}$ of several RM codes.

TABLE 7-1. Several $\mathfrak{R}(r,m)$ codes expressed as $(n,k,d_{\min})$

	$m =$ 2	3	4	5	6	7
$r = 0$	(4, 1, 4)	(8, 1, 8)	(16, 1, 16)	(32, 1, 32)	(64, 1, 64)	(128, 1, 128)
1	(4, 3, 2)	(8, 4, 4)	(16, 5, 8)	(32, 6, 16)	(64, 7, 32)	(128, 8, 64)
2	(4, 4, 1)	(8, 7, 2)	(16, 11, 4)	(32, 16, 8)	(64, 22, 16)	(128, 29, 32)
3		(8, 8, 1)	(16, 15, 2)	(32, 26, 4)	(64, 42, 8)	(128, 64, 16)
4			(16, 16, 1)	(32, 31, 2)	(64, 57, 4)	(128, 99, 8)
5				(32, 32, 1)	(64, 63, 2)	(128, 120, 4)
6					(64, 64, 1)	(128, 127, 2)
7						(128, 128, 1)

The table illustrates a few interesting points. The codes $\mathfrak{R}(0,m)$ are repetition codes of length 2^m. On the other end of the spectrum, the codes $\mathfrak{R}(m,m)$ correspond to the vector spaces formed by all binary 2^m-tuples for all positive integers m. The codes $\mathfrak{R}(m-1,m)$ are simple parity-check codes, containing all binary 2^m-tuples of even weight.

The duals of Reed-Muller codes are also Reed-Muller codes, as shown in the following theorem.

Theorem 7-3—Dual Codes of Reed-Muller Codes

For $0 \leq r \leq m-1$, $\mathfrak{R}(m-r-1,m)$ is the dual code to $\mathfrak{R}(r,m)$.

Proof [Mac]. Consider a pair of code words $\mathbf{a} \in \mathfrak{R}(m-r-1,m)$ and $\mathbf{b} \in \mathfrak{R}(r,m)$. $\mathbf{a}$ is associated with a polynomial $a(x_1,x_2,\ldots,x_m)$ of degree $\leq (m-r-1)$, while $\mathbf{b}$ is associated with a polynomial $b(x_1,x_2,\ldots,x_m)$ of degree $\leq r$. The polynomial product ab has degree $\leq (m-1)$ and is thus associated with a code word $\mathbf{ab}$ in the parity-check code $\mathfrak{R}(m-1,m)$. $\mathbf{ab}$ has even weight, so the dot product $\mathbf{a} \cdot \mathbf{b} \equiv 0$ *modulo* 2. $\mathfrak{R}(m-r-1,m)$ is thus contained in the dual space of $\mathfrak{R}(r,m)$. However, since

$$\dim(\mathfrak{R}(r,m)) + \dim(\mathfrak{R}(m-r-1,m)) = 2^m$$

$\mathfrak{R}(m-r-1,m)$ must be the dual code of $\mathfrak{R}(r,m)$ by the dimension theorem (Theorem 2-9). **QED**

The first order Reed-Muller codes have a very simple weight distribution.

$$A_0 = A_{2^m} = 1, \qquad A_{2^{m-1}} = 2^{m+1} - 2$$

The weight distribution for the second-order codes is also known [see Mac, p. 443]. We can use these distributions in conjunction with the MacWilliams identity

(Theorem 4-11) to obtain the weight distributions for $\mathfrak{R}(m-2,m)$ and $\mathfrak{R}(m-3,m)$. Unfortunately the weight distribution (m a variable) is not known for any code of order r greater than 2.

7.2 THE REED DECODING ALGORITHM

In his 1954 paper, Reed described a multiple-error-correcting decoding algorithm for RM codes based on sets of parity-check equations [Ree1]. This decoding algorithm was the first nontrivial example of majority logic, or "threshold" decoding. In general, majority logic techniques are fast, but suboptimal. This is particularly true in the case of convolutional codes, where the difference in performance between majority logic decoding and Viterbi (maximum likelihood) decoding is substantial. For RM codes, however, majority logic decoding provides maximum likelihood, hard-decision decoding in an efficient manner.[2]

This section begins by demonstrating the application of the Reed decoding algorithm to a particular code, the $\mathfrak{R}(2,4)$ code in Example 7-3. A description of the general algorithm then follows.

The set of basis vectors for $\mathfrak{R}(2,4)$ (see Figure 7-1) can be divided into three groups. The first group consists of the vector **1**, corresponding to the single monomial Boolean function of degree zero. The second group contains the vectors associated with the monomials of degree one, and the third group the vectors associated with monomials of degree two. The vectors in these three groups form the rows of a generator matrix for $\mathfrak{R}(2,4)$ as follows.

$$\mathbf{G} = \begin{bmatrix} \mathbf{1} \\ \hline \mathbf{v}_4 \\ \vdots \\ \mathbf{v}_1 \\ \hline \mathbf{v}_3\mathbf{v}_4 \\ \vdots \\ \mathbf{v}_1\mathbf{v}_2 \end{bmatrix} = \begin{bmatrix} \mathbf{G}_0 \\ \hline \mathbf{G}_1 \\ \hline \mathbf{G}_2 \end{bmatrix} \tag{7-3}$$

G can be used to implement a nonsystematic encoder for $\mathfrak{R}(2,4)$ through simple matrix multiplication, as discussed in Chapter 4. Let the 11 bits in the message block **m** be written $\mathbf{m} = (m_0, m_4, m_3, m_2, m_1, m_{34}, \ldots, m_{12})$. The subscripts associate each message bit with a row in **G**. **m** is encoded as follows.

$$\begin{aligned} \mathbf{c} &= (c_0, c_1, \ldots, c_{15}) \\ &= m_0\mathbf{1} + m_4\mathbf{v}_4 + \cdots + m_1\mathbf{v}_1 + m_{34}\mathbf{v}_3\mathbf{v}_4 + \cdots + m_{12}\mathbf{v}_1\mathbf{v}_2 \\ &= [\mathbf{m}_0 \mid \mathbf{m}_1 \mid \mathbf{m}_2] \begin{bmatrix} \mathbf{G}_0 \\ \hline \mathbf{G}_1 \\ \hline \mathbf{G}_2 \end{bmatrix} \end{aligned} \tag{7-4}$$

[2] For a thorough treatment of majority logic decoding of both block and convolutional codes, see [Mas1].

In this expression the message bits are grouped according to the order of the vectors they weight. The block $\mathbf{m}_0$ consists of the single message bit associated with the zeroth-order term $\mathbf{1}$, $\mathbf{m}_1$ the message bits associated with the first-order terms $\mathbf{v}_4$, $\mathbf{v}_3$, $\mathbf{v}_2$, and $\mathbf{v}_1$, and $\mathbf{m}_2$ the bits associated with the second-order terms $\mathbf{v}_3\mathbf{v}_4, \ldots, \mathbf{v}_1\mathbf{v}_2$. Given a received word $\mathbf{r}$, the Reed algorithm begins decoding by computing estimates for the highest-order block of message bits, in this case $\mathbf{m}_2$. The estimate for the product $\mathbf{m}_2\mathbf{G}_2$ is then subtracted from $\mathbf{r}$, leaving a noise-corrupted, lower-order code word. This process continues, with successively lower-order products $\mathbf{m}_i\mathbf{G}_i$ being estimated and removed from $\mathbf{r}$ until only the $\mathbf{m}_0\mathbf{G}_0$ term remains. It is like peeling an onion layer by layer, but without the side effects.

The key to the Reed algorithm lies in the process by which the message bits are estimated. Continuing with the $\mathfrak{R}(2,4)$ example, Figure 7-1 and Eq. (7-4) show that the first four bits of $\mathbf{c}$ are computed as follows.

$$\begin{aligned} c_0 &= m_0 \\ c_1 &= m_0 + m_1 \\ c_2 &= m_0 + m_2 \\ c_3 &= m_0 + m_1 + m_2 + m_{12} \end{aligned} \tag{7-5}$$

If these four code word bits are summed, we get the following.

$$\begin{aligned} c_0 + c_1 + c_2 + c_3 &= (m_0) + (m_0 + m_1) + (m_0 + m_2) + (m_0 + m_1 + m_2 + m_{12}) \\ &= (m_0 + m_0 + m_0 + m_0) + (m_1 + m_1) \\ &\quad + (m_2 + m_2) + (m_{12}) \\ &= m_{12} \end{aligned} \tag{7-6}$$

We get the same result if we sum the next four bits of the code word.

$$\begin{aligned} c_4 + c_5 + c_6 + c_7 &= (m_0 + m_3) + (m_0 + m_1 + m_3 + m_{13}) + (m_0 + m_2 + m_3 + m_{23}) \\ &\quad + (m_0 + m_1 + m_2 + m_3 + m_{23} + m_{13} + m_{12}) \\ &= (m_0 + m_0 + m_0 + m_0) + (m_1 + m_1) + (m_2 + m_2) \\ &\quad + (m_3 + m_3 + m_3 + m_3) + (m_{13} + m_{13}) \\ &\quad + (m_{23} + m_{23}) + (m_{12}) \\ &= m_{12} \end{aligned} \tag{7-7}$$

There are a total of four such expressions for computing the value of m_{12} by summing four bits from the code word. These four expressions are as follows.

$$\left\{\begin{aligned} m_{12} &= c_0 + c_1 + c_2 + c_3 \\ m_{12} &= c_4 + c_5 + c_6 + c_7 \\ m_{12} &= c_8 + c_9 + c_{10} + c_{11} \\ m_{12} &= c_{12} + c_{13} + c_{14} + c_{15} \end{aligned}\right\} \tag{7-8}$$

These expressions have the value m_{12} in common, but no other value appears in more than one of the sums. They are said to form a set of check sums that are **orthogonal** on the message bit m_{12}. A similar set of four expressions can be obtained for the other message bits comprising the block $\mathbf{m}_2$.

$$\left\{\begin{matrix} m_{13} = c_0 + c_1 + c_4 + c_5 \\ m_{13} = c_2 + c_3 + c_6 + c_7 \\ m_{13} = c_8 + c_9 + c_{12} + c_{13} \\ m_{13} = c_{10} + c_{11} + c_{14} + c_{15} \end{matrix}\right\} \cdots \left\{\begin{matrix} m_{34} = c_0 + c_4 + c_8 + c_{12} \\ m_{34} = c_1 + c_5 + c_9 + c_{13} \\ m_{34} = c_2 + c_6 + c_{10} + c_{14} \\ m_{34} = c_3 + c_7 + c_{11} + c_{15} \end{matrix}\right\} \tag{7-9}$$

A systematic method for obtaining these expressions will be described momentarily.

These groups of equations can be used to estimate the value of the message bits in a noise-corrupted received code word. Let the received word at the output of the receiver demodulator (see Figure 1-1) be denoted by

$$\mathbf{r} = (r_0, r_1, \ldots, r_{15}) = (c_0, c_1, \ldots, c_{15}) + (e_0, e_1, \ldots, e_{15}) \tag{7-10}$$

The 15-bit vector $\mathbf{e}$ denotes the error pattern that has corrupted the transmitted $\mathfrak{R}(2,4)$ code word $\mathbf{c}$ during transmission.

Each of the sums in Eq. (7-8) is used to provide an estimate $\hat{m}_{12}^{(i)}$ of the information bit m_{12}.

$$\begin{aligned} \hat{m}_{12}^{(1)} &= r_0 + r_1 + r_2 + r_3 \\ \hat{m}_{12}^{(2)} &= r_4 + r_5 + r_6 + r_7 \\ \hat{m}_{12}^{(3)} &= r_8 + r_9 + r_{10} + r_{11} \\ \hat{m}_{12}^{(4)} &= r_{12} + r_{13} + r_{14} + r_{15} \end{aligned} \tag{7-11}$$

The four estimates are then used to obtain a final message-bit estimate $\hat{m}_{12}$ by majority vote.

$$\hat{m}_{12} = \text{maj}\,\{\hat{m}_{12}^{(1)}, \hat{m}_{12}^{(2)}, \hat{m}_{12}^{(3)}, \hat{m}_{12}^{(4)}\} \tag{7-12}$$

If there are no errors in the received word (i.e., $\mathbf{e}$ has weight zero), then all four of the expressions in Eq. (7-11) will sum to the correct value m_{12}. Equation (7-12) will then provide the correct value for the message bit. Since the expressions use all of the code bits and are orthogonal on m_{12}, a single error causes exactly one of the expressions in Eq. (7-12) to be incorrect. Since the other three are still correct, the majority logic operation in Eq. (7-12) corrects the error. If there are two errors, then at most two of the expressions in Eq. (7-11) are incorrect, and at least two are correct. In the worst case (two expressions incorrect, two correct), Eq. (7-12) will not correct the errors, but it will indicate that the received code word has been corrupted by an uncorrectable error pattern. Finally, if there are three or more errors in the received word, then it is possible that Eq. (7-12) will indicate a majority value that is incorrect. We are thus guaranteed the ability to simultaneously correct all single-error patterns and detect all double-error patterns affecting the second-order message bits. Looking ahead, we will see that it is possible to create sets of eight or more check sums that are orthogonal on each of the lower-order message bits. The decoding of the highest-order block of message bits is always the limiting factor. The Reed algorithm thus provides simultaneous single-error correction and double-error detection when used with $\mathfrak{R}(2,4)$.

Once all of the second-order message bits have been estimated, the second-order terms are removed from the received word.

$$\hat{\mathbf{m}}_2 = (\hat{m}_{34}, \hat{m}_{24}, \hat{m}_{14}, \hat{m}_{23}, \hat{m}_{13}, \hat{m}_{12})$$
$$\mathbf{r}' = \mathbf{r} - \hat{\mathbf{m}}_2 \mathbf{G}_2 \tag{7-13}$$

We then proceed to the estimation of the first-order bits. It is possible to generate sets of eight orthogonal check sums for each of the first-order message bits. For example, we get the following for m_1.

$$\begin{aligned} m_1 &= c_0 + c_1, & m_1 &= c_8 + c_9 \\ m_1 &= c_2 + c_3, & m_1 &= c_{10} + c_{11} \\ m_1 &= c_4 + c_5, & m_1 &= c_{12} + c_{13} \\ m_1 &= c_6 + c_7, & m_1 &= c_{14} + c_{15} \end{aligned} \tag{7-14}$$

As in the previous case, an estimate for m_1 is obtained through a voting procedure.

$$\hat{m}_1 = \text{maj}\,\{\hat{m}_1^{(1)}, \hat{m}_1^{(2)}, \hat{m}_1^{(3)}, \hat{m}_1^{(4)}, \hat{m}_1^{(5)}, \hat{m}_1^{(6)}, \hat{m}_1^{(7)}, \hat{m}_1^{(8)}\} \tag{7-15}$$

Once all of the first-order message bits have been estimated, their effect is also removed from the received word.

$$\hat{\mathbf{m}}_1 = (\hat{m}_4, \hat{m}_3, \hat{m}_2, \hat{m}_1)$$
$$\mathbf{r}'' = \mathbf{r}' - \hat{\mathbf{m}}_1 \mathbf{G}_1 \tag{7-16}$$
$$= m_0 \mathbf{1} + \mathbf{e}$$

Finally we are left with the estimation of the last of the message bits, m_0. Here we have 16 check sums, all of the form $m_0 = c_i$, where i runs from 0 to 15.

$$\hat{m}_0 = \text{maj}\,\{r''_0, r''_1, \ldots, r''_{15}\} \tag{7-17}$$

Example 7-4—Decoding $\mathfrak{R}(2,4)$ Using the Reed Algorithm

Let the received vector be $\mathbf{r} = (0111010000010100)$. The first step is to obtain the estimates for the second-order message bit m_{12} using Eqs. (7-11) and (7-12).

$$\hat{m}_{12}^{(1)} = r_0 + r_1 + r_2 + r_3 = 1$$
$$\hat{m}_{12}^{(2)} = r_4 + r_5 + r_6 + r_7 = 1$$
$$\hat{m}_{12}^{(3)} = r_8 + r_9 + r_{10} + r_{11} = 1$$
$$\hat{m}_{12}^{(4)} = r_{12} + r_{13} + r_{14} + r_{15} = 1$$
$$\hat{m}_{12} = \text{maj}\,\{1, 1, 1, 1\} = 1$$

We next proceed to the estimation of m_{13}.

$$\hat{m}_{13}^{(1)} = r_0 + r_1 + r_4 + r_5 = 0$$
$$\hat{m}_{13}^{(2)} = r_2 + r_3 + r_6 + r_7 = 0$$
$$\hat{m}_{13}^{(3)} = r_8 + r_9 + r_{12} + r_{13} = 1$$
$$\hat{m}_{13}^{(4)} = r_{10} + r_{11} + r_{14} + r_{15} = 1$$

Decoding terminates with the estimation of the message bit m_{13}, for there is no majority value. We now know that the received word contains an error pattern of weight greater than or equal to 2.

Now let the received vector be $\mathbf{r} = (0101101100011011)$.

The estimate of m_{12} is $\hat{m}_{12} = \text{maj}\{0, 1, 1, 1\} = 1$. The majority logic operation has indicated the presence of a single error. The estimate of m_{13} is $\hat{m}_{13} = \text{maj}\{1, 1, 1, 1\} = 1$, while all the rest of the second-order estimates are zero. We thus have $\hat{\mathbf{m}}_2 = (000011)$.

We now subtract the second-order terms from the received word (Eq. (7-13)).

$$\mathbf{r}' = \mathbf{r} - \hat{\mathbf{m}}_2 \mathbf{G}_2$$
$$\begin{array}{rcccccccccccccccc} & 0 & 1 & 0 & 1 & 1 & 0 & 1 & 1 & 0 & 0 & 0 & 1 & 1 & 0 & 1 & 1 \\ = - & 0 & 0 & 0 & 1 & 0 & 1 & 0 & 0 & 0 & 0 & 0 & 1 & 0 & 1 & 0 & 0 \\ \hline & 0 & 1 & 0 & 0 & 1 & 1 & 1 & 1 & 0 & 0 & 0 & 0 & 1 & 1 & 1 & 1 \end{array}$$

The first-order estimates are all zero except for the estimate of m_3, which is 1, giving us $\hat{\mathbf{m}}_1 = (0100)$. We now finish decoding by applying Eqs. (7-16) and (7-17).

$$\mathbf{r}'' = \mathbf{r}' - \hat{\mathbf{m}}_1 \mathbf{G}_1$$
$$\begin{array}{rcccccccccccccccc} & 0 & 1 & 0 & 0 & 1 & 1 & 1 & 1 & 0 & 0 & 0 & 0 & 1 & 1 & 1 & 1 \\ = - & 0 & 0 & 0 & 0 & 1 & 1 & 1 & 1 & 0 & 0 & 0 & 0 & 1 & 1 & 1 & 1 \\ \hline & 0 & 1 & 0 & 0 & 0 & 0 & 0 & 0 & 0 & 0 & 0 & 0 & 0 & 0 & 0 & 0 \end{array}$$

The estimate for m_0 is clearly zero. A single bit error in coordinate c_1 has been corrected. The resulting information word is

$$\mathbf{m} = (m_0, m_4, m_3, m_2, m_1, m_{34}, m_{24}, m_{14}, m_{23}, m_{13}, m_{12})$$
$$= (00100000011)$$ ■

There is a general method for obtaining the sets of orthogonal check sums used in computing the message bit estimates. Let $\mathbf{c} = (c_0, c_1, \ldots, c_{n-1})$ be a code word in the RM code $\mathfrak{R}(r, m)$. Let the coordinate c_i be associated with the m-tuple P_i that is the one's-complement of the binary equivalent of the index i. For the codes $\mathfrak{R}(r, 4)$ we have Table 7-2. Note that the rows of the resulting table form the complements for the vectors $\mathbf{v}_4$, $\mathbf{v}_3$, $\mathbf{v}_2$, and $\mathbf{v}_1$ in Example 7-3.

The points $\{P_i\}$ form the m-dimensional Euclidean geometry EG$(m, 2)$. EG$(m, 2)$ is commonly represented as an m-dimensional square. For example, the points in EG$(4, 2)$ form the four-dimensional cube, or tesseract, shown in Figure 7-2. Each vertex on the cube is connected by an edge to all other vertices that differ from the vertex in exactly one coordinate.

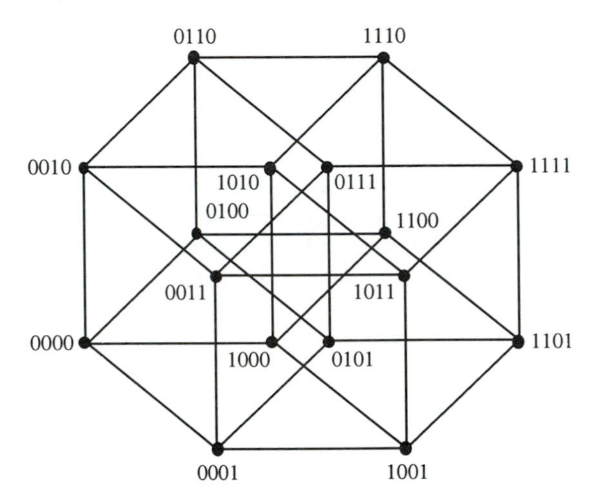

Figure 7-2. A Four-Dimensional Cube (Tesseract)

Each code word in $\mathfrak{R}(r, m)$ is an **incidence vector** that defines a subspace in EG$(m, 2)$. The coordinates in the incidence vector that contain ones correspond to EG$(m, 2)$ points that lie in the subspace. For example, the code word $\mathbf{c} =$ (1010101110100100) in $\mathfrak{R}(3, 4)$ (see Example 7-1) is an incidence vector for the subspace containing the points $\{P_0, P_2, P_4, P_6, P_7, P_8, P_{10}, P_{13}\}$.

Let $I = \{1, 2, \ldots, m\}$ be the set of indices for the first-order basis vectors $\{\mathbf{v}_1, \mathbf{v}_2, \ldots, \mathbf{v}_m\}$. All of the other basis vectors for $\mathfrak{R}(r, m)$ have as indices some combination of the indices in I (e.g., $\{1, 3\}$ for $\mathbf{v}_1 \mathbf{v}_3$ and $\{1, 2, 4\}$ for $\mathbf{v}_1 \mathbf{v}_2 \mathbf{v}_4$). Suppose that we want to find the set of orthogonal check sums that provide an estimate for the kth order message bit $m_{i_1 i_2 \cdots i_k}$ corresponding to the basis vector $\mathbf{v}_{i_1} \mathbf{v}_{i_2} \cdots \mathbf{v}_{i_k}$. The construction proceeds as follows.

1. Let S be the subspace of points associated with the incidence vector $\mathbf{v}_{i_1} \mathbf{v}_{i_2} \cdots \mathbf{v}_{i_k}$.
2. Let T be the subspace of points associated with the incidence vector $\mathbf{v}_{j_1} \mathbf{v}_{j_2} \cdots \mathbf{v}_{j_{m-k}}$, where $\{j_1, j_2, \ldots, j_{m-k}\}$ is the set difference $\{1, 2, \ldots, m\} - \{i_1, i_2, \ldots, i_k\}$. T is said to be the "complementary subspace to S."
3. The first check sum consists of the sum of the code-word coordinates specified by the points in T.

4. Let the **translation of T with respect to P** be the set of points whose coordinates are obtained by adding (without carry) the coordinates of P to the coordinates of each point of T. The rest of the check sums are the sums of the code word coordinates indicated by the translations of T with respect to the nonzero points in S.

Example 7-5—Reed-Algorithm Check Sums for $\mathfrak{R}(2,4)$

First consider the message bit associated with the second-order basis vector $\mathbf{v}_3\mathbf{v}_4 = (0000000000001111)$. $\mathbf{v}_3\mathbf{v}_4$ is the incidence vector for the subspace containing the points $S = \{P_{12}, P_{13}, P_{14}, P_{15}\}$. In the upper left tesseract in Figure 7-3(a), the points of S are joined by a dotted line.

The complementary space T is defined by the incidence vector $\mathbf{v}_1\mathbf{v}_2 = (0001000100010001)$ $(\{1,2\} = \{1,2,3,4\} - \{3,4\})$. The points $T = \{P_3, P_7, P_{11}, P_{15}\}$ are the vertices of the darker of the shaded regions in the figure. The translations of T with respect to the nonzero points in S are denoted by the lightly shaded regions. These translations are obtained by adding the coordinate representations for the points in S to each of the coordinate representations of the points in T. Using Table 7-2, we see the following.

$$S = \{(0011), (0010), (0001), (0000)\}$$

$$T = \{(1100), (1000), (0100), (0000)\} = \{P_3, P_7, P_{11}, P_{15}\}$$

$$\text{Translation of } T \text{ with respect to } (0011) = \{(1111), (1011), (0111), (0011)\} = \{P_0, P_4, P_8, P_{12}\}$$

$$\text{Translation of } T \text{ with respect to } (0010) = \{(1110), (1010), (0110), (0010)\} = \{P_1, P_5, P_9, P_{13}\}$$

$$\text{Translation of } T \text{ with respect to } (0001) = \{(1101), (1001), (0101), (0001)\} = \{P_2, P_6, P_{10}, P_{14}\}$$

T and its translations indicate which code-word coordinates are to be summed in the check sums that estimate m_{34}, as shown in Eq. (7-9).

TABLE 7-2 Representation of the code word coordinates as points in a Euclidean space

	$c_0 \updownarrow P_0$	$c_1 \updownarrow P_1$	$c_2 \updownarrow P_2$	$c_3 \updownarrow P_3$	$c_4 \updownarrow P_4$	$c_5 \updownarrow P_5$	$c_6 \updownarrow P_6$	$c_7 \updownarrow P_7$	$c_8 \updownarrow P_8$	$c_9 \updownarrow P_9$	$c_{10} \updownarrow P_{10}$	$c_{11} \updownarrow P_{11}$	$c_{12} \updownarrow P_{12}$	$c_{13} \updownarrow P_{13}$	$c_{14} \updownarrow P_{14}$	$c_{15} \updownarrow P_{15}$
$\overline{\mathbf{v}}_4$	1	1	1	1	1	1	1	1	0	0	0	0	0	0	0	0
$\overline{\mathbf{v}}_3$	1	1	1	1	0	0	0	0	1	1	1	1	0	0	0	0
$\overline{\mathbf{v}}_2$	1	1	0	0	1	1	0	0	1	1	0	0	1	1	0	0
$\overline{\mathbf{v}}_1$	1	0	1	0	1	0	1	0	1	0	1	0	1	0	1	0

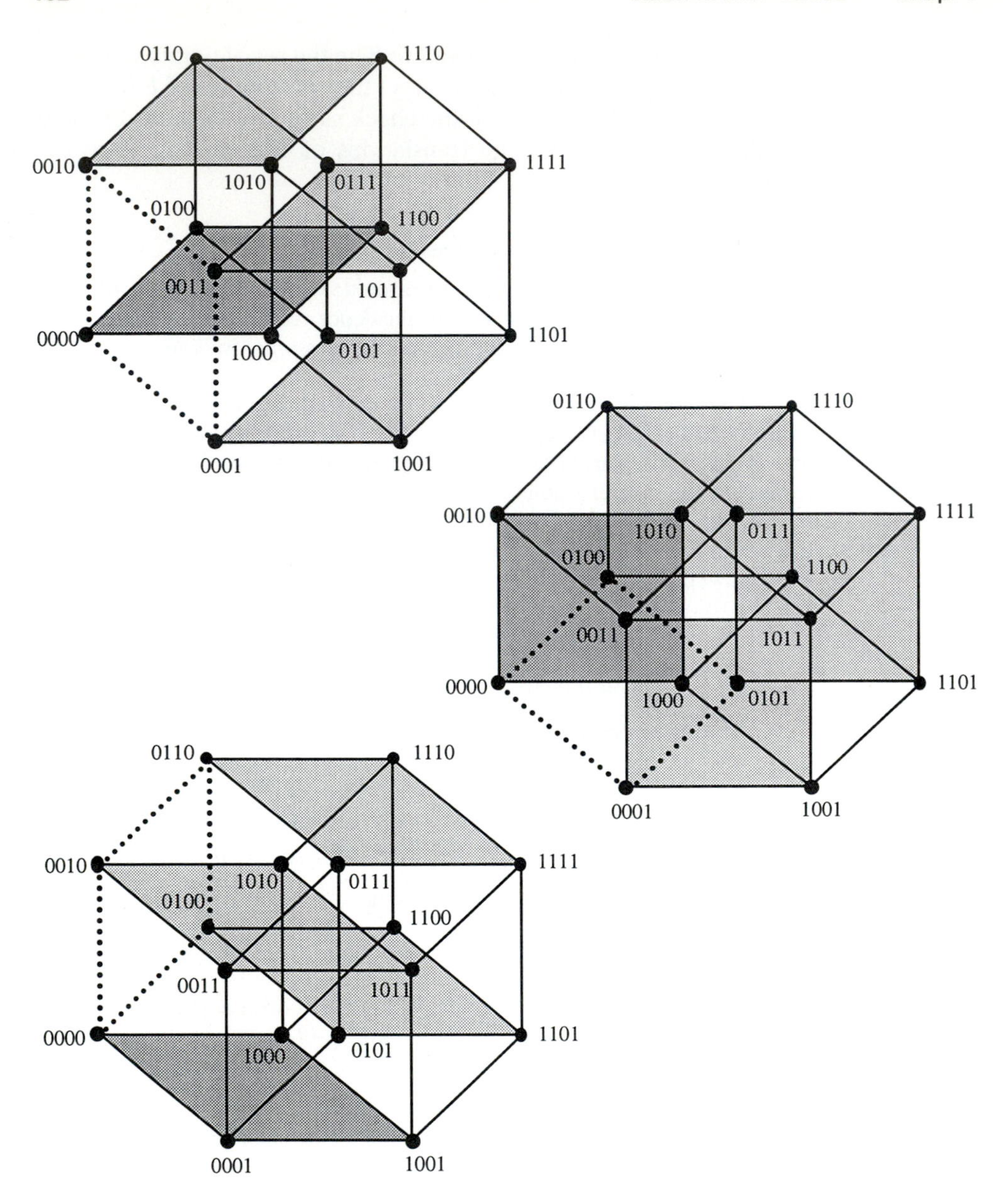

Figure 7-3(a). Parity-Check Equations for $\Re(2,4)$ Basis Vectors $\mathbf{v}_3\mathbf{v}_4$, $\mathbf{v}_2\mathbf{v}_4$, and $\mathbf{v}_1\mathbf{v}_4$

Figures 7-3(a) and 7-3(b) show the various spaces corresponding to the check sums for the other second-order message bits in $\Re(2,4)$.

The first-order terms are obtained in exactly the same manner. Consider the check sums used to estimate m_4. The corresponding basis vector is $\mathbf{v}_4 = (0000000011111111)$. $\mathbf{v}_4$ is the incidence vector for a space of eight points, as shown by the dotted lines in

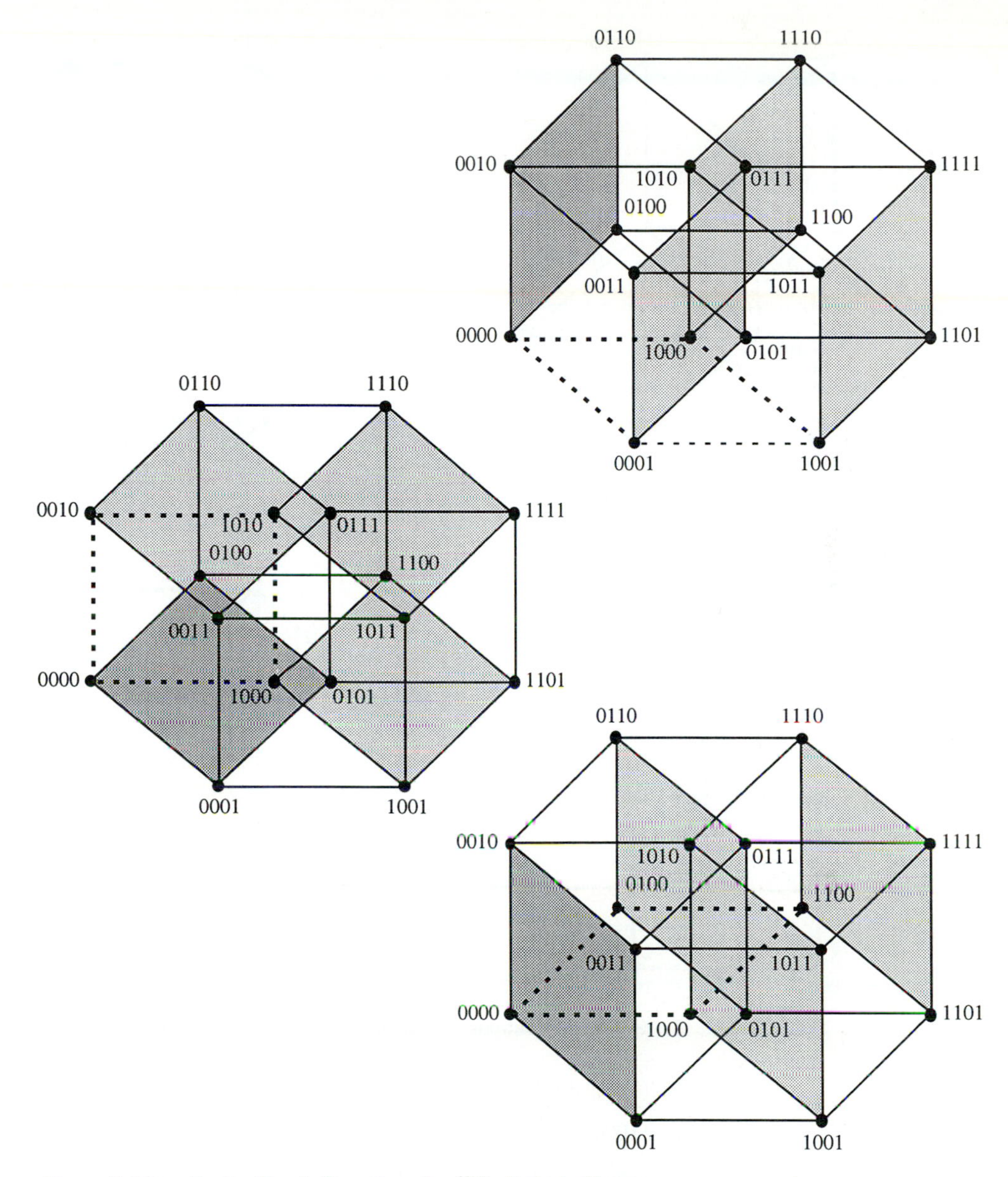

Figure 7-3(b). Parity-Check Equations for $\Re(2,4)$ Basis Vectors $\mathbf{v}_2\mathbf{v}_3$, $\mathbf{v}_1\mathbf{v}_3$, and $\mathbf{v}_1\mathbf{v}_2$

the upper-left tesseract in Fig. 7-3(c). The complementary space *T* is defined by the basis vector $\mathbf{v}_1\mathbf{v}_2\mathbf{v}_3 = (0000000100000001)$ (obtained by multiplying $\mathbf{v}_1$, $\mathbf{v}_2$, and $\mathbf{v}_3$ together coordinate by coordinate). *T* contains only two points and is indicated by the dark line segment connecting (0000) and (1000). The other line segments indicate the translations of *T* with respect to the points in *S*.

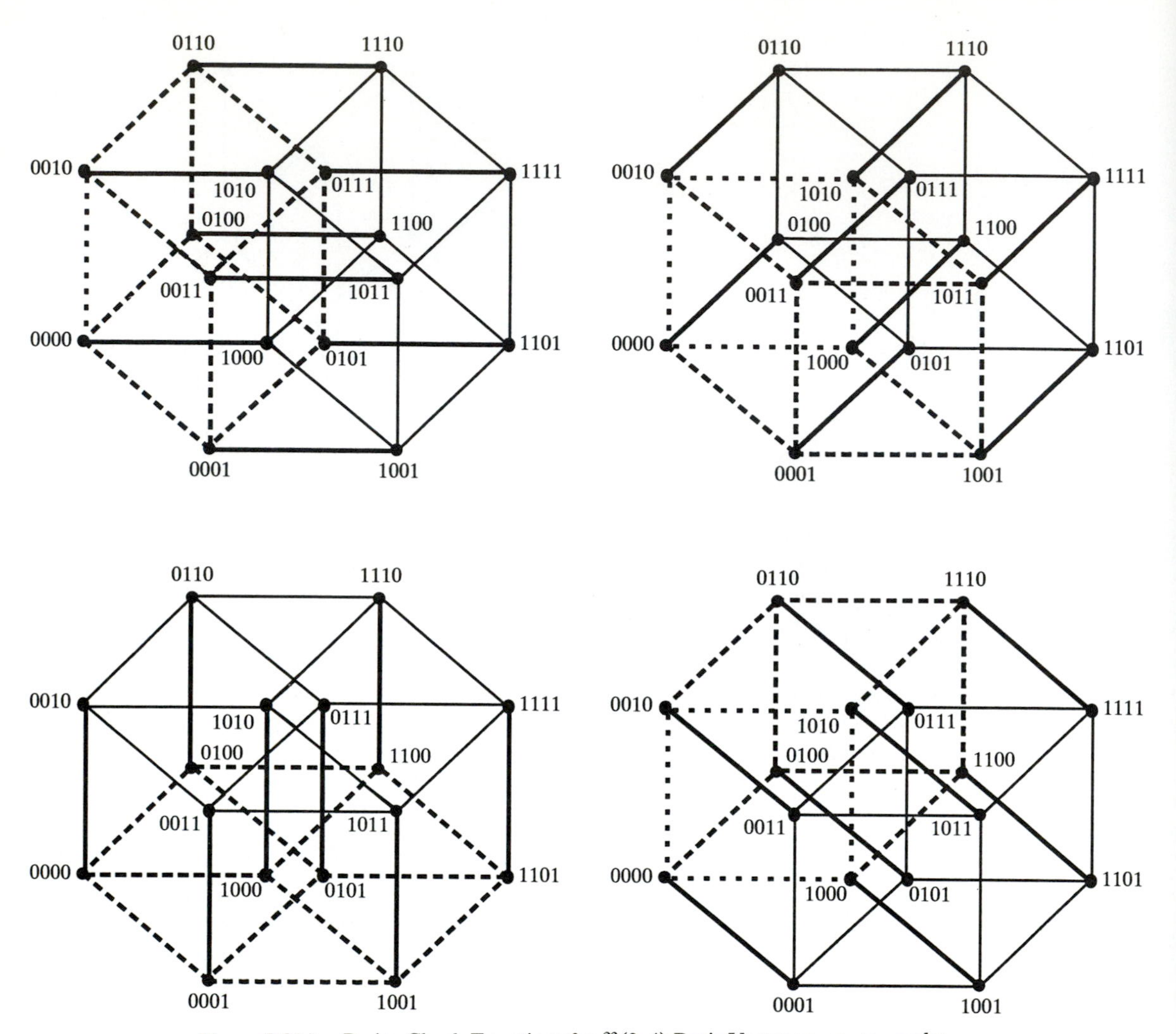

Figure 7-3(c). Parity-Check Equations for $\Re(2,4)$ Basis Vectors $\mathbf{v}_4$, $\mathbf{v}_3$, $\mathbf{v}_2$, and $\mathbf{v}_1$

$$S = \{(0111), (0110), (0101), (0100), (0011), (0010), (0001), (0000)\}$$

$$T = \{(1000), (0000)\} = \{P_7, P_{15}\}$$

$$\text{Translation of } T \text{ with respect to } (0111) = \{(1111), (0111)\}$$
$$= \{P_0, P_8\}$$

$$\text{Translation of } T \text{ with respect to } (0110) = \{(1110), (0110)\}$$
$$= \{P_1, P_9\}$$

$$\vdots$$

$$\text{Translation of } T \text{ with respect to } (0001) = \{(1001), (0001)\}$$
$$= \{P_6, P_{14}\}$$

This provides the check sums $m_4 = c_0 + c_8 = c_1 + c_9 = \cdots = c_7 + c_{15}$.

Figure 7-3(c) contains tesseracts showing the various check sums for all of the first-order message bits. ■

7.3 FIRST-ORDER REED-MULLER CODES

The first order Reed-Muller codes $\mathfrak{R}(1,m)$ have length 2^m, dimension $m+1$, and minimum distance $d_{\min} = 2^{m-1}$. The best-known application of the first-order codes is probably the use of $\mathfrak{R}(1,5)$ in the design of the *Mariner* deep space probes flown between 1969 and 1977. $\mathfrak{R}(1,5)$ provided error control on a 16.2 kbps downlink that conveyed photographs of Mars and its moons back to ground stations on earth. Each photograph was composed of a (600×600) grid of pixels, with each pixel assigned 1 of 64 possible shades of gray. Error control was provided by mapping each shade of gray to 1 of the 64 $\mathfrak{R}(1,5)$ code words. Each photograph was thus composed of 360,000 code words, with each code word capable of correcting up to 7 bit errors. This design provided 2.2 dB of coding gain, saving an estimated $2,200,000 that would have otherwise been spent in boosting the output power of the *Mariner* transmitter [Mas2]. This cost savings was achieved at the expense of a substantial increase in bandwidth, for $\mathfrak{R}(1,5)$ has an extremely low code rate $(6/32 \approx 0.19)$. Bandwidth was one of the few things that were available in great quantity in the early days of space exploration.

In this section we examine the techniques used to implement the *Mariner* '69 encoder and decoder, and highlight the special properties of first-order Reed-Muller codes that made the resulting design so efficient.

7.3.1 Encoding $\mathfrak{R}(1,m)$

The generator matrix for $\mathfrak{R}(1,m)$ has the form $\mathbf{G} = [\mathbf{1}\, \mathbf{v}_m\, \mathbf{v}_{m-1} \cdots \mathbf{v}_1]^T$. Figure 7-4 shows the corresponding encoding operation for $\mathfrak{R}(1,4)$.

$$\mathbf{c} = (c_0 \quad c_1 \quad \cdots \quad c_{14} \quad c_{15}) = \mathbf{mG} = (m_0 \quad m_4 \quad m_3 \quad m_2 \quad m_1)\begin{bmatrix} \mathbf{1} \\ \mathbf{v}_4 \\ \mathbf{v}_3 \\ \mathbf{v}_2 \\ \mathbf{v}_1 \end{bmatrix}$$

$$= (m_0 \quad m_4 \quad m_3 \quad m_2 \quad m_1)\begin{bmatrix} 1&1&1&1&1&1&1&1&1&1&1&1&1&1&1&1 \\ 0&0&0&0&0&0&0&0&1&1&1&1&1&1&1&1 \\ 0&0&0&0&1&1&1&1&0&0&0&0&1&1&1&1 \\ 0&0&1&1&0&0&1&1&0&0&1&1&0&0&1&1 \\ 0&1&0&1&0&1&0&1&0&1&0&1&0&1&0&1 \end{bmatrix}$$

Figure 7-4. Encoding Operation for $\mathfrak{R}(1,4)$

Since they have been obtained from a truth table, the columns of $\mathbf{G}$ have a very regular form: they consist of the binary 5-tuples beginning with $(10000)^T$ and running in increasing order to $(1\,1\,1\,1\,1)^T$. These 5-tuples can be generated by a simple digital counter, giving us the extremely simple encoding circuit shown in Figure 7-5.

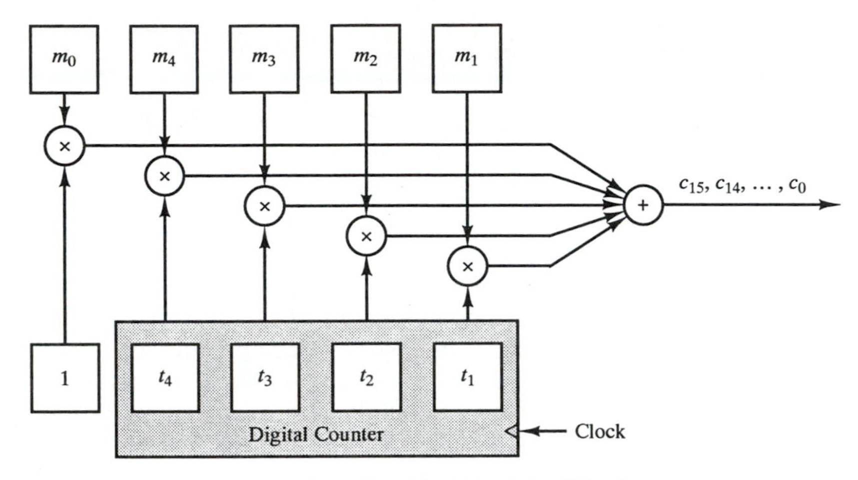

Figure 7-5. Encoding Circuit for $\Re(1,4)$

This encoder design has the dual benefit of being extremely fast and using very little hardware.

7.3.2 Decoding $\Re(1, m)$: The Green Machine

We begin by considering the relationship between a maximum likelihood decoder and a correlation decoder, and then show how a correlation decoder for $\Re(1,m)$ is obtained through the fast Hadamard transform.

Consider the following transformation $\mathcal{F}$ of a binary word $\mathbf{r}$ into a word $\mathbf{F}$ containing $+1$s and -1s.

$$\mathcal{F}(\mathbf{r}) = \mathcal{F}((r_0, r_1, \ldots, r_{2^m-1})) = \mathbf{F} = ((-1)^{r_0}, (-1)^{r_1}, \ldots, (-1)^{r_{2^m-1}}) \quad (7\text{-}18)$$

Given a set of code words $\mathbf{C} = \{\mathbf{c}_0, \ldots, \mathbf{c}_{n-1}\}$, the code word $\mathbf{c} \in \mathbf{C}$ that *minimizes* $d(\mathbf{r}, \mathbf{c})$ is also the code word that *maximizes* the correlation between $\mathcal{F}(\mathbf{r})$ and $\mathcal{F}(\mathbf{c})$. The correlation function $\text{cor}\,(\mathcal{F}(\mathbf{r}), \mathcal{F}(\mathbf{c}))$ reduces to an inner product for vector inputs.

$$\text{cor}\,(\mathbf{F}, \mathbf{G}) = \text{cor}\,((F_0, F_1, \ldots, F_{2^m-1}), (G_0, G_1, \ldots, G_{2^m-1})) = \sum_{i=0}^{2^m-1} F_i G_i \quad (7\text{-}19)$$

Note that the operations in Eq. (7-19) are real addition and multiplication. The *Mariner* '69 decoder computed the correlation between a received vector and each of the code words in $\Re(1,5)$ through the use of the Hadamard transform. The Hadamard transform is a Fourier transform that has been generalized to operate over finite abelian groups. The general technique was developed by Welch in 1966

[Wel], though the special case used in the *Mariner* '69 decoder was developed by Green at roughly the same time (and at roughly the same place: Jet Propulsion Laboratory in Pasadena, California). The decoder for the *Mariner* '69 mission will thus always be known as the "Green machine."

The Hadamard transform of $\mathbf{F}$ is computed through the matrix multiplication

$$\hat{\mathbf{F}} = (\hat{F}_0, \hat{F}_1, \ldots, \hat{F}_{2^m-1}) = \mathbf{F}\mathbf{H}_{2^m} \tag{7-20}$$

where $\mathbf{H}_{2^m}$ is the Hadamard matrix of order 2^m obtained through the repeated application of Sylvester's construction technique (see Section 6.1.1) to the Hadamard matrix

$$\mathbf{H}_2 = \begin{bmatrix} 1 & 1 \\ 1 & - \end{bmatrix} \tag{7-21}$$

For example, in decoding $\mathfrak{R}(1,4)$ we would use the Sylvester Hadamard matrix

$$\mathbf{H}_{16} = \begin{bmatrix}
1 & 1 & 1 & 1 & 1 & 1 & 1 & 1 & 1 & 1 & 1 & 1 & 1 & 1 & 1 & 1 \\
1 & - & 1 & - & 1 & - & 1 & - & 1 & - & 1 & - & 1 & - & 1 & - \\
1 & 1 & - & - & 1 & 1 & - & - & 1 & 1 & - & - & 1 & 1 & - & - \\
1 & - & - & 1 & 1 & - & - & 1 & 1 & - & - & 1 & 1 & - & - & 1 \\
1 & 1 & 1 & 1 & - & - & - & - & 1 & 1 & 1 & 1 & - & - & - & - \\
1 & - & 1 & - & - & 1 & - & 1 & 1 & - & 1 & - & - & 1 & - & 1 \\
1 & 1 & - & - & - & - & 1 & 1 & 1 & 1 & - & - & - & - & 1 & 1 \\
1 & - & - & 1 & - & 1 & 1 & - & 1 & - & - & 1 & - & 1 & 1 & - \\
1 & 1 & 1 & 1 & 1 & 1 & 1 & 1 & - & - & - & - & - & - & - & - \\
1 & - & 1 & - & 1 & - & 1 & - & - & 1 & - & 1 & - & 1 & - & 1 \\
1 & 1 & - & - & 1 & 1 & - & - & - & - & 1 & 1 & - & - & 1 & 1 \\
1 & - & - & 1 & 1 & - & - & 1 & - & 1 & 1 & - & - & 1 & 1 & - \\
1 & 1 & 1 & 1 & - & - & - & - & - & - & - & - & 1 & 1 & 1 & 1 \\
1 & - & 1 & - & - & 1 & - & 1 & - & 1 & - & 1 & 1 & - & 1 & - \\
1 & 1 & - & - & - & - & 1 & 1 & - & - & 1 & 1 & 1 & 1 & - & - \\
1 & - & - & 1 & - & 1 & 1 & - & - & 1 & 1 & - & 1 & - & - & 1
\end{bmatrix} \tag{7-22}$$

Recall that the generator matrix for $\mathfrak{R}(1,4)$ in Figure 7-4 has as rows the vectors $\mathbf{1}$, $\mathbf{v}_4$, $\mathbf{v}_3$, $\mathbf{v}_2$, and $\mathbf{v}_1$. A close examination of the above matrix reveals that the first, second, fourth, and eighth columns (counting begins with zero on the left-hand side) are $\mathcal{F}(\mathbf{v}_1)$, $\mathcal{F}(\mathbf{v}_2)$, $\mathcal{F}(\mathbf{v}_3)$, and $\mathcal{F}(\mathbf{v}_4)$ respectively. In general the $\mathbf{a}$th column of $\mathbf{H}_{16}$, where $\mathbf{a}$ has the binary representation $\mathbf{a} = (a_4 a_3 a_2 a_1)$, is the vector $\mathcal{F}(\mathbf{c}_a) = \mathcal{F}(a_1\mathbf{v}_1 + a_2\mathbf{v}_2 + a_3\mathbf{v}_3 + a_4\mathbf{v}_4)$. For example, column 12 (12 has the binary, or radix-2 representation 1100) consists of the vector

$$\begin{aligned}\mathcal{F}(0\cdot\mathbf{v}_1 + 0\cdot\mathbf{v}_2 + 1\cdot\mathbf{v}_3 + 1\cdot\mathbf{v}_4) &= \mathcal{F}(0000111111110000) \\ &= (1111--------1111)\end{aligned}$$

The Hadamard transform of the vector $\mathbf{F} = \mathcal{F}(\mathbf{r})$ thus computes the correlation of $\mathcal{F}(\mathbf{r})$ with all words $\mathcal{F}(\mathbf{c}_a)$, where $\mathbf{c}_a$ is a linear combination of the vectors $\{\mathbf{v}_1, \mathbf{v}_2, \mathbf{v}_3, \mathbf{v}_4\}$. It follows that each coordinate of the transform has the value

$$\hat{F}_a = 2^m - 2d(\mathbf{r}, \mathbf{c}_a) \tag{7-23}$$

Since $-\hat{F}_a$ is the correlation of $\mathcal{F}(\mathbf{r})$ with $\mathcal{F}(\mathbf{1} + \mathbf{c}_a) = \mathcal{F}(\mathbf{1} + a_1\mathbf{v}_1 + a_2\mathbf{v}_2 + a_3\mathbf{v}_3 + a_4\mathbf{v}_4)$, we need only compute the correlation of $\mathcal{F}(\mathbf{r})$ with half of the code words in $\mathfrak{R}(1, m)$ to determine the maximum likelihood code word. The Hadamard transform decoding algorithm is thus summarized as follows.

Hadamard Transform Decoding Algorithm for $\mathfrak{R}(1, m)$

1. Given a received vector $\mathbf{r}$, compute $\mathbf{F} = \mathcal{F}(\mathbf{r})$.
2. Compute the Hadamard transform $\hat{\mathbf{F}} = \mathbf{F}\mathbf{H}_{2^m}$.
3. Find the coordinate $\mathbf{a} = (a_m \cdots a_3 a_2 a_1)$ where $\hat{\mathbf{F}}$ has the greatest magnitude.
4. If $\hat{F}_\mathbf{a}$ is positive, the ML code word is $\mathbf{c} = (a_1\mathbf{v}_1 + a_2\mathbf{v}_2 + \cdots + a_m\mathbf{v}_m)$. If $\hat{F}_\mathbf{a}$ is negative, the ML code word is $\mathbf{c} = (\mathbf{1} + a_1\mathbf{v}_1 + a_2\mathbf{v}_2 + \cdots + a_m\mathbf{v}_m)$. ■

Example 7-6—Decoding $\mathfrak{R}(1,3)$

The generator matrix for $\mathfrak{R}(1,3)$ is as follows.

$$\mathbf{G} = \begin{bmatrix} \mathbf{1} \\ \mathbf{v}_3 \\ \mathbf{v}_2 \\ \mathbf{v}_1 \end{bmatrix} = \begin{bmatrix} 1&1&1&1&1&1&1&1 \\ 0&0&0&0&1&1&1&1 \\ 0&0&1&1&0&0&1&1 \\ 0&1&0&1&0&1&0&1 \end{bmatrix}$$

The Hadamard matrix $\mathbf{H}_8$ is

$$\mathbf{H}_8 = \begin{bmatrix} 1&1&1&1&1&1&1&1 \\ 1&-&1&-&1&-&1&- \\ 1&1&-&-&1&1&-&- \\ 1&-&-&1&1&-&-&1 \\ 1&1&1&1&-&-&-&- \\ 1&-&1&-&-&1&-&1 \\ 1&1&-&-&-&-&1&1 \\ 1&-&-&1&-&1&1&- \end{bmatrix}$$

The receiver demodulator sends the vector $\mathbf{r} = (10000011)$ to the decoder.

1. $\mathbf{F} = (-11111--)$
2. $\hat{\mathbf{F}} = \mathbf{F}\mathbf{H}_8 = (2, -2, 2, -2, 2, -2, -6, -2)$.
3. The sixth coordinate $(6 \leftrightarrow (110))$ has the greatest magnitude $(|-6|)$.
4. The ML code word is $(\mathbf{1} + (1)\mathbf{v}_3 + (1)\mathbf{v}_2 + (0)\mathbf{v}_1) = (11000011)$. ■

The complexity of the implementation of the above algorithm is greatly reduced through the application of the fast Hadamard transform theorem. We must first introduce the Kronecker product.

Definition 7-2—Kronecker Product

Let **A** be an $(m \times m)$ matrix and **B** an $(n \times n)$ matrix with elements $\{a_{ij}\}$ and $\{b_{ij}\}$, respectively. The **Kronecker product** $(\mathbf{A} \otimes \mathbf{B})$ is the $(mn \times mn)$ matrix obtained by replacing every entry a_{ij} in **A** by the matrix $a_{ij}\,\mathbf{B}$.

Example 7-6

$$\begin{bmatrix} a & b \\ c & d \end{bmatrix} \otimes \begin{bmatrix} 1 & 1 \\ 1 & -1 \end{bmatrix} = \begin{bmatrix} a & a & b & b \\ a & -a & b & -b \\ c & c & d & d \\ c & -c & d & -d \end{bmatrix}$$

■

The Kronecker product has a number of interesting properties, including associativity and distributivity. The Kronecker product is not, however, commutative in all cases. We make use of the following property.

Lemma 7-1

Let A, B, C, and D be $(n \times n)$ matrices. The following is true.

$$(\mathbf{A} \otimes \mathbf{B})(\mathbf{C} \otimes \mathbf{D}) = (\mathbf{AC}) \otimes (\mathbf{BD})$$

Proof

$$(\mathbf{A} \otimes \mathbf{B})(\mathbf{C} \otimes \mathbf{D}) = \begin{bmatrix} a_{11}\mathbf{B} & a_{12}\mathbf{B} & \cdots & a_{1n}\mathbf{B} \\ a_{21}\mathbf{B} & a_{22}\mathbf{B} & \cdots & a_{2n}\mathbf{B} \\ \vdots & \vdots & \ddots & \vdots \\ a_{n1}\mathbf{B} & a_{n2}\mathbf{B} & \cdots & a_{nn}\mathbf{B} \end{bmatrix} \begin{bmatrix} c_{11}\mathbf{D} & c_{12}\mathbf{D} & \cdots & c_{1n}\mathbf{D} \\ c_{21}\mathbf{D} & c_{22}\mathbf{D} & \cdots & c_{2n}\mathbf{D} \\ \vdots & \vdots & \ddots & \vdots \\ c_{n1}\mathbf{D} & c_{n2}\mathbf{D} & \cdots & c_{nn}\mathbf{D} \end{bmatrix}$$

$$= \begin{bmatrix} a_{11}c_{11}\mathbf{BD} + \cdots + a_{1n}c_{n1}\mathbf{BD} & a_{11}c_{12}\mathbf{BD} + \cdots + a_{1n}c_{n2}\mathbf{BD} \\ a_{21}c_{11}\mathbf{BD} + \cdots + a_{2n}c_{n1}\mathbf{BD} & a_{21}c_{12}\mathbf{BD} + \cdots + a_{2n}c_{n2}\mathbf{BD} \\ \vdots & \vdots \\ a_{n1}c_{11}\mathbf{BD} + \cdots + a_{nn}c_{n1}\mathbf{BD} & a_{n1}c_{12}\mathbf{BD} + \cdots + a_{nn}c_{n2}\mathbf{BD} \end{bmatrix.$$

$$\begin{matrix} \cdots & a_{11}c_{1n}\mathbf{BD} + \cdots + a_{1n}c_{nn}\mathbf{BD} \\ \cdots & a_{21}c_{1n}\mathbf{BD} + \cdots + a_{2n}c_{nn}\mathbf{BD} \\ \ddots & \vdots \\ \cdots & a_{n1}c_{1n}\mathbf{BD} + \cdots + a_{nn}c_{nn}\mathbf{BD} \end{matrix}\Bigg]$$

$$= \begin{bmatrix} a_{11}c_{11} + \cdots + a_{1n}c_{n1} & a_{11}c_{12} + \cdots + a_{1n}c_{n2} \\ a_{21}c_{11} + \cdots + a_{2n}c_{n1} & a_{21}c_{12} + \cdots + a_{2n}c_{n2} \\ \vdots & \vdots \\ a_{n1}c_{11} + \cdots + a_{nn}c_{n1} & a_{n1}c_{12} + \cdots + a_{nn}c_{n2} \end{bmatrix.$$

$$\begin{matrix} \cdots & a_{11}c_{1n} + \cdots + a_{1n}c_{nn} \\ \cdots & a_{21}c_{1n} + \cdots + a_{2n}c_{nn} \\ \ddots & \vdots \\ \cdots & a_{n1}c_{1n} + \cdots + a_{nn}c_{nn} \end{matrix}\Bigg] \otimes (\mathbf{BD})$$

$$= (\mathbf{AC}) \otimes (\mathbf{BD})$$

QED

The Sylvester construction for Hadamard matrices can be expressed in terms of a series of Kronecker products.

$$\mathbf{H}_{2^m} = \begin{bmatrix} 1 & 1 \\ 1 & -1 \end{bmatrix} \otimes \mathbf{H}_{2^{m-1}}$$
$$= \underbrace{\begin{bmatrix} 1 & 1 \\ 1 & -1 \end{bmatrix} \otimes \begin{bmatrix} 1 & 1 \\ 1 & -1 \end{bmatrix} \otimes \cdots \otimes \begin{bmatrix} 1 & 1 \\ 1 & -1 \end{bmatrix}}_{m \text{ terms}} \tag{7-24}$$

There is another representation for $\mathbf{H}_{2^m}$ which is more useful; it forms the basis for the Green machine.

Theorem 7-4—The Fast Hadamard Transform Theorem

$\mathbf{H}_{2^m} = \mathbf{M}_{2^m}^{(1)} \mathbf{M}_{2^m}^{(2)} \cdots \mathbf{M}_{2^m}^{(m)}$, where $\mathbf{M}_{2^m}^{(i)} = \mathbf{I}_{2^{m-i}} \otimes \mathbf{H}_2 \otimes \mathbf{I}_{2^{i-1}}, 1 \leq i \leq m$, $\mathbf{I}_n$ is an $n \times n$ identity matrix, and

$$\mathbf{H}_2 = \begin{bmatrix} 1 & 1 \\ 1 & -1 \end{bmatrix}$$

Proof [Mac]. We use induction on the index m. The case $m = 1$ is clearly true, for

$$\mathbf{H}_{2^1} = \mathbf{M}_2^{(1)} = (\mathbf{I}_{2^0} \otimes \mathbf{H}_2 \otimes \mathbf{I}_{2^0}) = \left([1] \otimes \begin{bmatrix} 1 & 1 \\ 1 & - \end{bmatrix} \otimes [1] \right) = \begin{bmatrix} 1 & 1 \\ 1 & - \end{bmatrix}$$

We now assume that the hypothesis is true for m, and then show that it must therefore hold for $m + 1$.

For $1 \leq i \leq m$ we obtain the following from the hypothesis.

$$\begin{aligned} \mathbf{M}_{2^{m+1}}^{(i)} &= \mathbf{I}_{2^{m+1-i}} \otimes \mathbf{H}_2 \otimes \mathbf{I}_{2^{i-1}} \\ &= (\mathbf{I}_2 \otimes \mathbf{I}_{2^{m-i}}) \otimes \mathbf{H}_2 \otimes \mathbf{I}_{2^{i-1}} \\ &= \mathbf{I}_2 \otimes (\mathbf{I}_{2^{m-i}} \otimes \mathbf{H}_2 \otimes \mathbf{I}_{2^{i-1}}) \\ &= \mathbf{I}_2 \otimes \mathbf{M}_{2^m}^{(i)} \end{aligned}$$

and

$$\mathbf{M}_{2^{m+1}}^{(m+1)} = \mathbf{H}_2 \otimes \mathbf{I}_{2^m}$$

It follows that

$$\begin{aligned} \mathbf{H}_{2^{m+1}} &= \mathbf{M}_{2^{m+1}}^{(1)} \mathbf{M}_{2^{m+1}}^{(2)} \cdots \mathbf{M}_{2^{m+1}}^{(m+1)} \\ &= (\mathbf{I}_2 \otimes \mathbf{M}_{2^m}^{(1)})(\mathbf{I}_2 \otimes \mathbf{M}_{2^m}^{(2)}) \cdots (\mathbf{I}_2 \otimes \mathbf{M}_{2^m}^{(m)})(\mathbf{H}_2 \otimes \mathbf{I}_{2^m}) \\ &= \mathbf{H}_2 \otimes (\mathbf{M}_{2^m}^{(1)} \mathbf{M}_{2^m}^{(2)} \cdots \mathbf{M}_{2^m}^{(m)} \mathbf{I}_{2^m}) \qquad \text{(Lemma 7-1)} \\ &= \mathbf{H}_2 \otimes \mathbf{H}_{2^m} \qquad \text{(by the induction hypothesis)} \end{aligned}$$

$\mathbf{H}_2 \otimes \mathbf{H}_{2^m}$ is simply the Sylvester construction for $\mathbf{H}_{2^{m+1}}$ (see Theorem 6-3).
QED

Example 7-7—The Fast Hadamard Transform Theorem

$$
\begin{aligned}
\mathbf{H}_8 &= \mathbf{M}_8^{(1)}\mathbf{M}_8^{(2)}\mathbf{M}_8^{(3)} \\
&= (\mathbf{I}_{2^2}\otimes\mathbf{H}_2\otimes\mathbf{I}_{2^0})(\mathbf{I}_{2^1}\otimes\mathbf{H}_2\otimes\mathbf{I}_{2^1})(\mathbf{I}_{2^0}\otimes\mathbf{H}_2\otimes\mathbf{I}_{2^2}) \\
&= \left(\begin{bmatrix} 1 & & & \\ & 1 & & \\ & & 1 & \\ & & & 1 \end{bmatrix}\otimes\begin{bmatrix} 1 & 1 \\ 1 & - \end{bmatrix}\otimes[1]\right)\left(\begin{bmatrix} 1 & \\ & 1 \end{bmatrix}\otimes\begin{bmatrix} 1 & 1 \\ 1 & - \end{bmatrix}\otimes\begin{bmatrix} 1 & \\ & 1 \end{bmatrix}\right) \\
&\quad\cdot\left([1]\otimes\begin{bmatrix} 1 & 1 \\ 1 & - \end{bmatrix}\otimes\begin{bmatrix} 1 & & & \\ & 1 & & \\ & & 1 & \\ & & & 1 \end{bmatrix}\right) \\
&= \begin{bmatrix}
1 & 1 & & & & & & \\
1 & - & & & & & & \\
 & & 1 & 1 & & & & \\
 & & 1 & - & & & & \\
 & & & & 1 & 1 & & \\
 & & & & 1 & - & & \\
 & & & & & & 1 & 1 \\
 & & & & & & 1 & -
\end{bmatrix}\cdot\begin{bmatrix}
1 & & 1 & & & & & \\
 & 1 & & 1 & & & & \\
1 & & - & & & & & \\
 & 1 & & - & & & & \\
 & & & & 1 & & 1 & \\
 & & & & & 1 & & 1 \\
 & & & & 1 & & - & \\
 & & & & & 1 & & -
\end{bmatrix} \\
&\quad\cdot\begin{bmatrix}
1 & & & & 1 & & & \\
 & 1 & & & & 1 & & \\
 & & 1 & & & & 1 & \\
 & & & 1 & & & & 1 \\
1 & & & & - & & & \\
 & 1 & & & & - & & \\
 & & 1 & & & & - & \\
 & & & 1 & & & & -
\end{bmatrix}
\end{aligned}
$$

■

The Green machine decoder performs the Hadamard transform operation $\mathbf{F}\cdot\mathbf{H}_{2^m}$ through the series of matrix multiplications $(((\mathbf{F}\cdot\mathbf{M}_{2^m}^{(1)})\cdot\mathbf{M}_{2^m}^{(2)})\cdots)\cdot\mathbf{M}_{2^m}^{(m)}$. The decoder is built in m stages, the jth of which is used to perform matrix multiplication by $M_{2^m}^{(j)}$.

Let $\mathbf{F} = (F_0, F_1, \ldots, F_n)$ be the input to the decoder. The matrix product computations in the $\Re(1, m)$ decoding operation are noted below along with the associated stage of the Green machine.

$$
\begin{aligned}
\text{1st stage} &\leftrightarrow \mathbf{F}\cdot\mathbf{M}_{2^m}^{(1)} = \mathbf{P}^{(1)} \\
\text{2nd stage} &\leftrightarrow \mathbf{P}^{(1)}\cdot\mathbf{M}_{2^m}^{(2)} = \mathbf{P}^{(2)} \\
&\vdots \\
m\text{th stage} &\leftrightarrow \mathbf{P}^{(m-1)}\cdot\mathbf{M}_{2^m}^{(m)} = \mathbf{P}^{(m)} = \hat{\mathbf{F}}
\end{aligned}
$$

As indicated by the matrices in the previous example, each coordinate in the intermediate product vectors $\{\mathbf{P}^{(i)}\}$ is simply the sum or difference of a pair of coordinates computed in the previous stage. Figure 7-6 shows the typical layout for a fast Hadamard transform circuit. This particular circuit has been designed for the simple $\Re(1,2)$ code. It is easily shown that the decoding circuit for a lower-order code can be used in the decoder for a higher-order code.

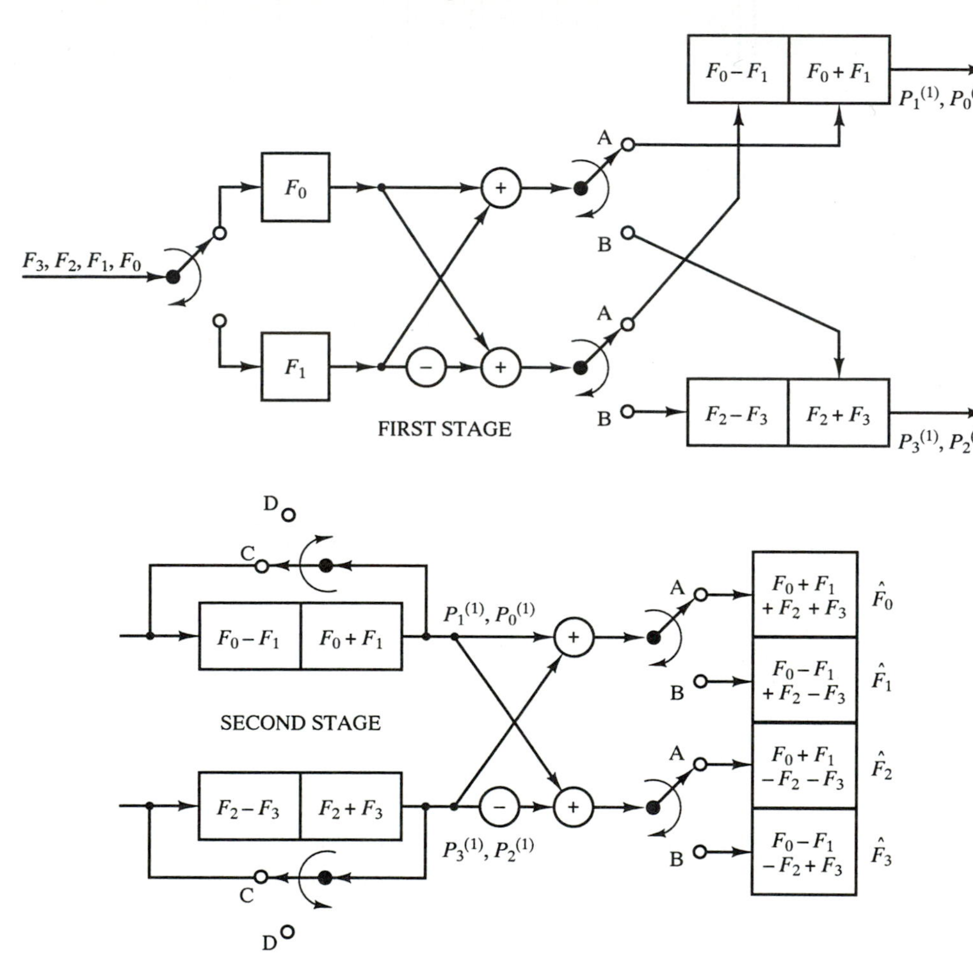

Figure 7-6. A Two-Stage Fast Hadamard Transform Circuit for $\Re(1,2)$

In general, the m stages of the $\Re(1,m)$ decoder are closely related to the first m stages of the $\Re(1,m+1)$ decoder by the following lemma.

Lemma 7-3

$$\boldsymbol{M}_{2^{m+1}}^{(i)} = \begin{bmatrix} \boldsymbol{M}_{2^m}^{(i)} & \\ & \boldsymbol{M}_{2^m}^{(i)} \end{bmatrix} \qquad \textbf{for } i = 1, \ldots, m$$

Proof. This result follows immediately from the first step of the proof of the fast Hadamard transform theorem (Theorem 7-4).

Let $\mathbf{F}' = (F_0, F_1, \ldots, F_{2n-1})$ be the input to the $\mathfrak{R}(1, m+1)$ decoder. By separating $\mathbf{F}'$ into two n-bit segments $\mathbf{F}_1' = (F_0, F_1, \ldots, F_{n-1})$ and $\mathbf{F}_2' = (F_n, F_{n+1}, \ldots, F_{2n-1})$, the first m stages of decoding can be performed by an $\mathfrak{R}(1, m)$ decoder through the exploitation of the lemma.

$$\text{1st stage} \leftrightarrow F' \cdot M_{2^{m+1}}^{(1)} = F' \cdot \begin{pmatrix} M_{2^m}^{(1)} & \\ & M_{2^m}^{(1)} \end{pmatrix} = (F_1' \cdot M_{2^m}^{(1)} \,|\, F_2' \cdot M_{2^m}^{(1)}) = (P_1^{(1)} \,|\, P_2^{(1)})$$

$$\text{2nd stage} \leftrightarrow (P_1^{(1)} \,|\, P_2^{(1)}) \cdot M_{2^{m+1}}^{(2)} = (P_1^{(1)} \cdot M_{2^m}^{(2)} \,|\, P_2^{(1)} \cdot M_{2^m}^{(2)}) = (P_1^{(2)} \,|\, P_2^{(2)})$$

$$\vdots$$

$$m\text{th stage} \leftrightarrow (P_1^{(m-1)} \,|\, P_2^{(m-1)}) \cdot M_{2^{m+1}}^{(m)} = (P_1^{(m-1)} \cdot M_{2^m}^{(m)} \,|\, P_2^{(m-1)} \cdot M_{2^m}^{(m)}) = (P_1^{(m)} \,|\, P_2^{(m)})$$

$$(m+1)\text{st stage} \leftrightarrow (P_1^{(m)} \,|\, P_2^{(m)}) \cdot M_{2^{m+1}}^{(m+1)} = (P_1^{(m+1)} \,|\, P_2^{(m+1)}) = \hat{F}$$

The Green machine is thus a modular decoder. Stages can be added or subtracted to fit codes of varying length.

PROBLEMS

The general theory of Reed-Muller codes

1. Construct a generator and parity-check matrix for $\mathfrak{R}(2, 4)$.

2. Show that all code words in $\mathfrak{R}(r, m)$ are also code words in $\mathfrak{R}(r+1, m)$. $\mathfrak{R}(r, m)$ is thus a *subcode* of $\mathfrak{R}(r+1, m)$.

3. Show that $\mathfrak{R}(0, m)$ is a repetition code.

4. Show that $\mathfrak{R}(m-1, m)$ is composed of all 2^m-tuples of even weight. It is thus a simple parity-check code.

5. The punctured Reed-Muller code $\mathfrak{R}^*(r, m)$ is obtained by deleting the first coordinate (that corresponding to $v_1 = v_2 = \cdots = v_m = 0$) from every code word in $\mathfrak{R}(r, m)$. Determine the length, dimension, and minimum distance of $\mathfrak{R}^*(r, m)$.

6. Show that for all code words $\mathbf{c} = (c_0, c_1, \ldots, c_{2^m-1}) \in \mathfrak{R}(r, m)$, the vector $\mathbf{c}' = (\mathbf{c} \,|\, \mathbf{c}) = (c_0, c_1, \ldots, c_{2^m-1}, c_0, c_1, \ldots, c_{2^m-1})$ is a code word in $\mathfrak{R}(r, m+1)$.

7. Prove that all code words in $\mathfrak{R}(1, m)$ have weight 0, 2^{m-1}, or 2^m.

The Reed algorithm

8. Find the Reed algorithm checksums for $\mathfrak{R}(1, 4)$.

9. Find the Reed algorithm checksums for $\mathfrak{R}(2, 5)$.

10. Use the Reed algorithm to decode the following noise-corrupted code words from $\mathfrak{R}(2, 4)$.
(a) (0011, 0111, 0001, 0111) **(b)** (1100, 1100, 1011, 1100)
(c) (0011, 0011, 0001, 1000) **(d)** (0101, 0110, 1110, 0110)

The Kronecker product and the fast Hadamard transform

11. Prove that the Kronecker product is associative.

12. Prove that the Kronecker product is not commutative.

13. Prove that, for a Hadamard matrix $\mathbf{H}_n$, the following are also Hadamard matrices.

$$\left(\begin{bmatrix} 1 & 1 \\ 1 & - \end{bmatrix} \otimes \mathbf{H}_n\right) \quad \left(\mathbf{H}_n \otimes \begin{bmatrix} 1 & 1 \\ 1 & - \end{bmatrix}\right)$$

14. Compute $\mathbf{M}_4^{(1)}$ and $\mathbf{M}_4^{(2)}$. Show that their product is $\mathbf{H}_4$.

15. Prove that the Hadamard transform can be used to perform maximum likelihood soft-decision decoding.

Encoding and decoding $\mathfrak{R}(1, m)$

16. Design a Mariner-type encoder for $\mathfrak{R}(1,5)$.

17. Design a Green-machine decoder for $\mathfrak{R}(1,5)$.

18. Decode the following noise-corrupted $\mathfrak{R}(1,4)$ code words.

(a) $(1111, 1010, 1010, 1101)$
(b) $(1100, 0011, 1111, 1100)$
(c) $(1001, 0000, 1001, 1001)$
(d) $(1011, 1010, 0011, 1111)$

19. A demodulator is designed to output values in the set $\{-4, -3, -2, -1, 0, 1, 2, 3\}$. The first four values correspond to received zeros of decreasing strength (-4 is a strong zero, -1 a weak zero). The second four values correspond to received ones of increasing strength. Design a corresponding $\mathfrak{R}(1,3)$ Green machine and decode the following received words.

(a) $(3, -1, 0, 1, 2, -2, -3, 1)$
(b) $(-3, 0, 2, -3, -2, 3, 0, 0)$
(c) $(-4, -3, -1, 3, -2, 0, 3, 2)$
(d) $(-2, -3, 2, 0, -3, -2, 3, 3)$

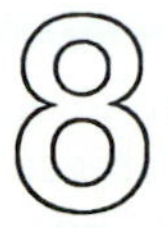

BCH and Reed-Solomon Codes

BCH and Reed-Solomon codes form the core of the most powerful known algebraic codes and have seen widespread application in the past thirty years. The fundamental work on BCH codes was conducted by two independent research teams that published their results at roughly the same time. Binary BCH codes were discussed as "a generalization of Hamming's work" by A. Hocquenghem in a 1959 paper entitled "Codes correcteur d'erreurs" [Hoc]. This was followed in March and September of 1960 by Bose and Ray-Chaudhuri's publications on "Error Correcting Binary Group Codes" [Bos1], [Bos2]. Given their simultaneous discovery of these codes, all three gentlemen have given their name to what are now called BCH codes. Shortly after these initial publications, Peterson proved that BCH codes were cyclic and presented a moderately efficient decoding algorithm [Pet3]. Gorenstein and Zierler then extended BCH codes to arbitrary fields of size p^m [Gor1].

Reed-Solomon codes were first described in a June 1960 paper in the *SIAM Journal on Applied Mathematics* by Reed and Solomon. This paper, "Polynomial Codes over Certain Finite Fields," discussed an extremely powerful set of nonbinary codes that are now known as Reed-Solomon codes. Through the work of Gorenstein and Zierler it was later discovered that Reed-Solomon codes and BCH codes are closely related, and that Reed-Solomon codes can be described as nonbinary BCH codes.

The application of Fourier transforms over finite fields to BCH and Reed-Solomon codes was first discussed by Gore in 1973 [Gor2]. The transform approach was then pursued by a number of other authors. The most extensive discussion of this area to date is provided by Blahut in [Bla2] (1984).

In this chapter we investigate the fundamental properties of BCH and Reed-Solomon codes. Two approaches are considered: The first uses the notion of consecutive powers of a primitive root as zeros in the generator polynomial, the second focuses on consecutive zero coordinates in the Galois field Fourier transform of the code words. These two approaches lead to the same codes, for a time-domain signal and its Fourier transform are two different descriptions of a single object. The two techniques provide different insights, however, into the problems of encoding and decoding.

Reed-Solomon codes are defined first as a special case of BCH codes. It is shown, however, that Reed-Solomon codes display properties that are not found in any of the other BCH codes. In particular Reed-Solomon codes are *maximum distance separable* (*MDS*). The implications of this are explored in some detail both here and in later chapters. The MDS nature of Reed-Solomon codes comes into play later in this chapter when we consider modifications to Reed-Solomon codes (puncturing and shortening) that greatly extend their utility in a variety of applications.

The chapter concludes with a discussion of the Galois field Fourier transform and its application to the analysis of cyclic codes in general and BCH and Reed-Solomon codes in particular. Reed-Solomon encoders based on the Galois field Fourier transform are examined.

8.1 THE GENERATOR POLYNOMIAL APPROACH TO BCH CODES

When constructing an arbitrary cyclic code, there is no guarantee as to the resulting minimum distance. Given an arbitrary generator polynomial $g(x)$, we must conduct a computer search of all corresponding nonzero code words $c(x)$ to determine the minimum-weight code word and thus the minimum distance of the code. BCH codes, on the other hand, take advantage of a useful result that ensures a minimum "design distance" given a particular constraint on the generator polynomial. This result is known as the BCH bound.

Theorem 8-1—The BCH Bound

Let C be a q-ary (n, k) cyclic code with generator polynomial $g(x)$. Let m be the multiplicative order of q *modulo* n ($GF(q^m)$ is thus the smallest extension field of $GF(q)$ that contains a primitive nth root of unity). Let α be a primitive nth root of unity.

Select $g(x)$ to be a minimal-degree polynomial in $GF(q)[x]$ such that $g(\alpha^b) = g(\alpha^{b+1}) = g(\alpha^{b+2}) = \cdots = g(\alpha^{b+\delta-2}) = 0$ for some integers $b \geq 0$ and $\delta \geq 1$. $g(x)$ thus has $(\delta - 1)$ consecutive powers of α as zeros.

$\Rightarrow$ The code C defined by $g(x)$ has minimum distance $d_{\min} \geq \delta$.

Definition 8-1—Design Distance

The parameter δ in Theorem 8-1 is the **design distance** of the BCH code defined by the generator polynomial $g(x)$.

Proof of Theorem 8-1. The following proof is a bit long, but well worth pursuing. However, the reader who is endowed with a great deal of faith but considerably less fondness for linear algebra may skip the proof without suffering too much damage to his or her understanding of the practical issues surrounding BCH and Reed-Solomon codes.

It is first shown that the constraint placed on the generator polynomial in the premise ensures that all of the square submatrices of a BCH parity-check matrix are Vandermonde. It is then shown that Vandermonde matrices are nonsingular, thus placing a lower bound on the minimum distance of the code.

A vector **c** is a code word in **C** if and only if the associated code polynomial $c(x)$ is a multiple of the generator polynomial $g(x)$. It follows from the premise that **c** is a code word if and only if $c(\alpha^b) = c(\alpha^{b+1}) = c(\alpha^{b+2}) = \cdots = c(\alpha^{b+\delta-2}) = 0$.

Parity-check matrices can be created that check received words **r** to see if they have the required zeros.

$$\mathbf{s}^T = \mathbf{H}\mathbf{r}^T$$

$$= \begin{bmatrix} 1 & \alpha^b & \alpha^{2b} & \cdots & \alpha^{(n-1)b} \\ 1 & \alpha^{b+1} & \alpha^{2(b+1)} & \cdots & \alpha^{(n-1)(b+1)} \\ \vdots & \vdots & \vdots & \ddots & \vdots \\ 1 & \alpha^{b+\delta-3} & \alpha^{2(b+\delta-3)} & \cdots & \alpha^{(n-1)(b+\delta-3)} \\ 1 & \alpha^{b+\delta-2} & \alpha^{2(b+\delta-2)} & \cdots & \alpha^{(n-1)(b+\delta-2)} \end{bmatrix} \begin{bmatrix} r_0 \\ r_1 \\ \vdots \\ r_{n-2} \\ r_{n-1} \end{bmatrix}$$

$$= \begin{bmatrix} r_0 + r_1\alpha^b + r_2\alpha^{2b} + \cdots + r_{n-2}\alpha^{(n-2)b} + r_{n-1}\alpha^{(n-1)b} \\ r_0 + r_1\alpha^{b+1} + r_2\alpha^{2(b+1)} + \cdots + r_{n-2}\alpha^{(n-2)(b+1)} + r_{n-1}\alpha^{(n-1)(b+1)} \\ \vdots \\ r_0 + r_1\alpha^{b+\delta-3} + r_2\alpha^{2(b+\delta-3)} + \cdots + r_{n-2}\alpha^{(n-2)(b+\delta-3)} + r_{n-1}\alpha^{(n-1)(b+\delta-3)} \\ r_0 + r_1\alpha^{b+\delta-2} + r_2\alpha^{2(b+\delta-2)} + \cdots + r_{n-2}\alpha^{(n-2)(b+\delta-2)} + r_{n-1}\alpha^{(n-1)(b+\delta-2)} \end{bmatrix} \quad (8\text{-}1)$$

H is a valid parity-check matrix because any vector **r** whose corresponding polynomial $r(x)$ has $\{\alpha^b, \alpha^{b+1}, \alpha^{b+2}, \ldots, \alpha^{b+\delta-2}\}$ as zeros is a code polynomial and vice versa.

Theorem 4-9 is now used to show that the code **C** has a minimum distance $\geq \delta$. Suppose there exists a nonzero code word **c** with weight $w < \delta$. Let $\mathbf{c}' = (c_{a_1}, c_{a_2}, c_{a_w})$ be the nonzero coordinates of **c**. It follows that

$$\begin{bmatrix} \alpha^{a_1 b} & \alpha^{a_2 b} & \alpha^{a_3 b} & \cdots & \alpha^{a_w b} \\ \alpha^{a_1(b+1)} & \alpha^{a_2(b+1)} & \alpha^{a_3(b+1)} & \cdots & \alpha^{a_w(b+1)} \\ \vdots & \vdots & \vdots & \ddots & \vdots \\ \alpha^{a_1(b+\delta-3)} & \alpha^{a_2(b+\delta-3)} & \alpha^{a_3(b+\delta-3)} & \cdots & \alpha^{a_w(b+\delta-3)} \\ \alpha^{a_1(b+\delta-2)} & \alpha^{a_2(b+\delta-2)} & \alpha^{a_3(b+\delta-2)} & \cdots & \alpha^{a_w(b+\delta-2)} \end{bmatrix} \begin{bmatrix} c_{a_1} \\ c_{a_2} \\ \vdots \\ c_{a_{w-1}} \\ c_{a_w} \end{bmatrix} = \mathbf{0}_{(\delta-1)\times 1} \quad (8\text{-}2)$$

A square $(w \times w)$ matrix $\mathbf{H}'$ can be formed such that $\mathbf{H}'\mathbf{c}'^T = \mathbf{0}$.

$$\mathbf{H}'\mathbf{c}'^T = \begin{bmatrix} \alpha^{a_1 b} & \alpha^{a_2 b} & \alpha^{a_3 b} & \cdots & \alpha^{a_w b} \\ \alpha^{a_1(b+1)} & \alpha^{a_2(b+1)} & \alpha^{a_3(b+1)} & \cdots & \alpha^{a_w(b+1)} \\ \vdots & \vdots & \vdots & \ddots & \vdots \\ \alpha^{a_1(b+w-2)} & \alpha^{a_2(b+w-2)} & \alpha^{a_3(b+w-2)} & \cdots & \alpha^{a_w(b+w-2)} \\ \alpha^{a_1(b+w-1)} & \alpha^{a_2(b+w-1)} & \alpha^{a_3(b+w-1)} & \cdots & \alpha^{a_w(b+w-1)} \end{bmatrix} \begin{bmatrix} c_{a_1} \\ c_{a_2} \\ \vdots \\ c_{a_{w-1}} \\ c_{a_w} \end{bmatrix} = \mathbf{0}_{w\times 1} \quad (8\text{-}3)$$

Since $\mathbf{c}'$ is a nonzero vector, the columns of the $(w \times w)$ matrix $\mathbf{H}'$ must be linearly dependent, and $\mathbf{H}'$ is thus a singular matrix. We now reduce the expression $\det(\mathbf{H}') = 0$ to an identical (and contradictory) result concerning Vandermonde matrices. Using row and column reduction operations, it can be shown that

$$
\begin{aligned}
\det(\mathbf{H}') &= \alpha^{(a_1+a_2+\cdots+a_w)b} \det \begin{bmatrix} 1 & 1 & 1 & \cdots & 1 \\ \alpha^{a_1} & \alpha^{a_2} & \alpha^{a_3} & \cdots & \alpha^{a_w} \\ \vdots & \vdots & \vdots & \ddots & \vdots \\ \alpha^{a_1(w-2)} & \alpha^{a_2(w-2)} & \alpha^{a_3(w-2)} & \cdots & \alpha^{a_w(w-2)} \\ \alpha^{a_1(w-1)} & \alpha^{a_2(w-1)} & \alpha^{a_3(w-1)} & \cdots & \alpha^{a_w(w-1)} \end{bmatrix} \\
&= \alpha^{(a_1+a_2+\cdots+a_w)b} \det \mathbf{A} = 0
\end{aligned}
\tag{8-4}
$$

Since $\alpha^x \neq 0$ for any finite integer x, the determinant of $\mathbf{A}$ must be zero. $\mathbf{A}$ is a $(w \times w)$ Vandermonde matrix. Vandermonde matrices have the general form

$$
\mathbf{V} = \begin{bmatrix} 1 & 1 & 1 & \cdots & 1 \\ x_0 & x_1 & x_2 & \cdots & x_{n-1} \\ x_0^2 & x_1^2 & x_2^2 & \cdots & x_{n-1}^2 \\ \vdots & \vdots & \vdots & \vdots & \vdots \\ x_0^{n-1} & x_1^{n-1} & x_2^{n-1} & \cdots & x_{n-1}^{n-1} \end{bmatrix},
$$

where $\{x_0, x_1, x_2, \ldots, x_{n-1}\}$ are distinct values. The determinant of this general Vandermonde matrix can be reduced to a simple expression through column reduction and cofactor expansion. Note that

$$
\begin{aligned}
\det(\mathbf{V}) &= \det\left(\begin{bmatrix} 1 & 1 & 1 & \cdots & 1 \\ x_0 & x_1 & x_2 & \cdots & x_{n-1} \\ x_0^2 & x_1^2 & x_2^2 & \cdots & x_{n-1}^2 \\ \vdots & \vdots & \vdots & \ddots & \vdots \\ x_0^{n-1} & x_1^{n-1} & x_2^{n-1} & \cdots & x_{n-1}^{n-1} \end{bmatrix}\right) \\
&= \det\left(\begin{bmatrix} 1 & 1-1 & 1-1 & \cdots & 1-1 \\ x_0 & x_1 - x_0 & x_2 - x_0 & \cdots & x_{n-1} - x_0 \\ x_0^2 & x_1^2 - x_0^2 & x_2^2 - x_0^2 & \cdots & x_{n-1}^2 - x_0^2 \\ \vdots & \vdots & \vdots & \ddots & \vdots \\ x_0^{n-1} & x_1^{n-1} - x_0^{n-1} & x_2^{n-1} - x_0^{n-1} & \cdots & x_{n-1}^{n-1} - x_0^{n-1} \end{bmatrix}\right) \\
&= \det\left(\begin{bmatrix} 1 & 0 & \cdots & 0 \\ x_0 & x_1 - x_0 & \cdots & x_{n-1} - x_0 \\ x_0^2 & (x_1 + x_0)(x_1 - x_0) & \cdots & (x_{n-1} + x_0)(x_{n-1} - x_0) \\ \vdots & \vdots & \ddots & \vdots \\ x_0^{n-1} & (x_1^{n-2} + x_1^{n-3}x_0 + \cdots + x_0^{n-2})(x_1 - x_0) & \cdots & (x_{n-1}^{n-2} + x_{n-1}^{n-3}x_0 + \cdots + x_0^{n-2})(x_{n-1} - x_0) \end{bmatrix}\right)
\end{aligned}
\tag{8-5}
$$

Using cofactor expansion, the last expression reduces to

$$= (1)\det\left(\begin{bmatrix} x_1 - x_0 & \cdots & x_{n-1} - x_0 \\ (x_1 + x_0)(x_1 - x_0) & \cdots & (x_{n-1} + x_0)(x_{n-1} - x_0) \\ \vdots & \ddots & \vdots \\ (x_1^{n-2} + x_1^{n-3}x_0 + \cdots + x_0^{n-2})(x_1 - x_0) & \cdots & (x_{n-1}^{n-2} + x_{n-1}^{n-3}x_0 + \cdots + x_0^{n-2})(x_{n-1} - x_0) \end{bmatrix}\right) \tag{8-6}$$

$$= \det(\mathbf{V}')$$

Since the first column of $\mathbf{V}'$ is divisible by $(x_1 - x_0)$, the determinant must also be divisible by $(x_1 - x_0)$. In general, since the jth column is divisible by $(x_j - x_0)$, then the determinant must be divisible by $(x_j - x_0)$. We have thus shown that

$$\prod_{k=1}^{n} (x_k - x_0) \,|\, \det(\mathbf{V}) \tag{8-7}$$

Now return to the original form of $\mathbf{V}$. Note that all of the elements in the first row are 1. We did not have to start the computation of the determinant by subtracting the first column from all of the other columns. We could just as easily have selected the $(j + 1)$st column and subtracted it from all of the other columns. Repeating the above analysis, we would then come to the conclusion that

$$\prod_{\substack{k=0,\\ k\neq j}}^{n} (x_k - x_j) \,|\, \det(\mathbf{V}) \tag{8-8}$$

We can thus conclude that the determinant of $\mathbf{V}$ must be divisible by all expressions of the form $(x_k - x_j)$ for all $j < k, j \neq k$, i.e.,

$$\prod_{0\leq j<k\leq n-1} (x_k - x_j) \,|\, \det(\mathbf{V}) \tag{8-9}$$

Now consider what happens when one tries to calculate the determinant of $\mathbf{V}$ using the definition of a determinant.

$$\det\left(\begin{bmatrix} a_{1,1} & a_{1,2} & \cdots & a_{1,n} \\ a_{2,1} & a_{2,2} & \cdots & a_{2,n} \\ \vdots & \vdots & \ddots & \vdots \\ a_{n,1} & a_{n,2} & \cdots & a_{n,n} \end{bmatrix}\right) = \sum_{\substack{(s_1,s_2,\ldots,s_n)=S \\ S\in\{\sigma(1,2,3,\ldots,n)\}}} \operatorname{sgn}(\sigma)(a_{1,s_1}\cdot a_{2,s_2}\cdot a_{3,s_3}\cdot \ldots \cdot a_{n,s_n}) \tag{8-10}$$

where $\{\sigma(1,2,3,\ldots,n)\}$ is the set of permutations of $\{1,2,3,\ldots,n\}$. Every product term in the above summation has exactly one element from each row and each column. Given the Vandermonde matrix $\mathbf{V}$, each product term in the summation that determines the determinant of $\mathbf{V}$ will be of degree $(0 + 1 + 2 + 3 + \cdots + n - 1) = n(n - 1)/2$, for

$$\det(\mathbf{V}) = \det\left(\begin{bmatrix} 1 & 1 & 1 & \cdots & 1 \\ x_0 & x_1 & x_2 & \cdots & x_{n-1} \\ x_0^2 & x_1^2 & x_2^2 & \cdots & x_{n-1}^2 \\ \vdots & \vdots & \vdots & \ddots & \vdots \\ x_0^{n-1} & x_1^{n-1} & x_2^{n-1} & \cdots & x_{n-1}^{n-1} \end{bmatrix}\right) \tag{8-11}$$

$$= \sum_{\substack{(s_0, s_1, \ldots, s_{n-1}) = S, \\ S \in \{\sigma(0,1,2,\ldots,n-1)\}}} \operatorname{sgn}(\sigma)(x_0^{s_0} \cdot x_1^{s_1} \cdot x_2^{s_2} \cdot \cdots \cdot x_{n-1}^{s_{n-1}})$$

Since each of the terms in the summation is different, the determinant of $\mathbf{V}$ is a monic multinomial of degree $n(n - 1)/2$. We also know that

$$\prod_{0 \le j < k \le n-1} (x_k - x_j) \mid \det(\mathbf{V}) \tag{8-12}$$

where $\prod_{0 \le j < k \le n-1} (x_k - x_j)$ is also a monic multinomial of degree $n(n - 1)/2$. Since both are monic of degree $n(n - 1)/2$ and one divides the other, they must be equal:

$$\det(\mathbf{V}) = \prod_{0 \le j < k \le n-1} (x_k - x_j) \tag{8-13}$$

Since the $\{x_j\}$ are all distinct, all of the terms in the product expression above must be nonzero, therefore the product is nonzero and $\mathbf{V}$ is nonsingular. But we showed earlier that if there exists a nonzero code word $\mathbf{c}$ of weight $w < \delta$, then there exists a $(w \times w)$ Vandermonde matrix that is singular. Since all Vandermonde matrices are nonsingular, no such code word can exist. **QED**

Using the BCH bound, the following design procedure for BCH codes is established.

The BCH code design procedure

To construct a t-error correcting q-ary BCH code of length n:

1. Find a primitive nth root of unity α in a field $GF(q^m)$, where m is minimal.
2. Select $(\delta - 1) = 2t$ consecutive powers of α, starting with α^b for some nonnegative integer b.
3. Let $g(x)$ be the least common multiple of the minimal polynomials for the selected powers of α with respect to $GF(q)$. (Each of the minimal polynomials should appear only once in the product.)

Step 1 follows from our design procedure for general cyclic codes. Steps 2 and

3 ensure, through the BCH bound, that the minimum distance of the resulting code equals or exceeds δ and that the generator polynomial has the minimal possible degree. Since $g(x)$ is a product of minimal polynomials with respect to GF(q), $g(x)$ must be in GF(q)[x] and the corresponding code is thus q-ary with $d_{\min} \geq \delta$.

Definition 8-2—Narrow-Sense and Primitive BCH Codes

If $b = 1$, then the BCH code is **narrow-sense**. If $n = q^m - 1$ for some positive integer m, then the BCH code is **primitive**, for the nth root of unity α is a primitive element in GF(q^m).

Appendix E contains a list of the generator polynomials for binary, narrow-sense, primitive BCH codes of lengths 7 through 255.

The BCH design procedure creates an interesting relationship between certain sets of BCH codes. Fix b and note the following: a BCH code of length n and design distance δ_1 contains as linear subspaces all length-n BCH codes with design distance $\delta_2 \geq \delta_1$. This can be proven as follows.

$$d_{\min} = \delta_1 \quad \Rightarrow g_1(x) = (x - \alpha^b) \cdot (x - \alpha^{b+1}) \cdot \cdots \cdot (x - \alpha^{b+\delta_1-2})$$

$$d_{\min} = \delta_2 > \delta_1 \Rightarrow g_2(x) = (x - \alpha^b) \cdot (x - \alpha^{b+1}) \cdot \cdots \cdot (x - \alpha^{b+\delta_1-2}) \cdot \cdots \cdot (x - \alpha^{b+\delta_2-2})$$

$$\Rightarrow g_2(x) = m(x) g_1(x)$$

$\Rightarrow$ all multiples of $g_2(x)$ are multiples of $g_1(x)$

$\Rightarrow$ all code words in $\mathbf{C}_2$ are code words in $\mathbf{C}_1$

Example 8-1—Binary BCH Codes of Length 31

Let α be a root of the primitive polynomial $x^5 + x^2 + 1$. It is thus a primitive element in the field GF(32). Since 31 is of the form $2^m - 1$, our BCH codes in this example are primitive. We begin by determining the cyclotomic cosets *modulo* 31 with respect to GF(2) and the associated minimal polynomials. All of the computations necessary to obtain the following results can be performed using the representation for GF(32) in Appendix B, or one can simply look up the polynomials in Appendix D.

CYCLOTOMIC COSETS		MINIMAL POLYNOMIALS
$C_0 = \{0\}$	$\leftrightarrow$	$M_{(0)}(x) = x + 1$
$C_1 = \{1, 2, 4, 8, 16\}$	$\leftrightarrow$	$M_{(1)}(x) = x^5 + x^2 + 1$
$C_3 = \{3, 6, 12, 24, 17\}$	$\leftrightarrow$	$M_{(3)}(x) = x^5 + x^4 + x^3 + x^2 + 1$
$C_5 = \{5, 10, 20, 9, 18\}$	$\leftrightarrow$	$M_{(5)}(x) = x^5 + x^4 + x^2 + x + 1$
$C_7 = \{7, 14, 28, 25, 19\}$	$\leftrightarrow$	$M_{(7)}(x) = x^5 + x^3 + x^2 + x + 1$
$C_{11} = \{11, 22, 13, 26, 21\}$	$\leftrightarrow$	$M_{(11)}(x) = x^5 + x^4 + x^3 + x + 1$
$C_{15} = \{15, 30, 29, 27, 23\}$	$\leftrightarrow$	$M_{(15)}(x) = x^5 + x^3 + 1$

If $\mathbf{C}$ is to be a binary cyclic code, then it must have a generator polynomial $g(x)$ that factors into one or more of the above minimal polynomials. If $\mathbf{C}$ is to be t-error correcting BCH code, then $g(x)$ must have as zeros $2t$ consecutive powers of α.

• ***One-error-correcting Narrow-Sense Primitive BCH Code***

Since the code is to be narrow-sense and single-error-correcting, $b = 1$ and $\delta = 3$. The generator polynomial must thus have α and α^2 as zeros. $M_{(1)}(x)$ is the minimal polynomial of both α and α^2. The generator polynomial is thus

$$g(x) = \text{LCM}(M_{(1)}(x)M_{(2)}(x)) = M_{(1)}(x) = x^5 + x^2 + 1$$

Since the degree of the generator polynomial $g(x)$ is 5, the dimension of the resulting code is $31 - 5 = 26$. $g(x)$ thus defines a (31, 26) binary single-error-correcting BCH code.

A parity-check matrix for this code is obtained by applying the general form in Eq. (8-1).

$$\mathbf{H} = \begin{bmatrix} 1 & \alpha & \alpha^2 & \cdots & \alpha^{29} & \alpha^{30} \\ 1 & \alpha^2 & \alpha^4 & \cdots & \alpha^{27} & \alpha^{29} \end{bmatrix}$$

This parity-check matrix is actually guilty of redundancy, for any binary polynomial having α as a zero must also have α^2 and the other conjugates of α as zeros by Theorem 3-3.

• ***Two-error-correcting Narrow-Sense Primitive BCH Code***

Again $b = 1$, but δ has been increased to 5. $g(x)$ must thus have as roots α, α^2, α^3, and α^4.

$$\begin{aligned} g(x) &= \text{LCM}(M_{(1)}(x), M_{(2)}(x), M_{(3)}(x), M_{(4)}(x)) = M_{(1)}(x)M_{(3)}(x) \\ &= (x^5 + x^2 + 1)(x^5 + x^4 + x^3 + x^2 + 1) \\ &= x^{10} + x^9 + x^8 + x^6 + x^5 + x^3 + 1 \end{aligned}$$

Since the degree of $g(x)$ is 10, it defines a (31, 21) binary double-error-correcting code. The corresponding parity-check matrix is shown below.

$$\mathbf{H} = \begin{bmatrix} 1 & \alpha & \alpha^2 & \cdots & \alpha^{29} & \alpha^{30} \\ 1 & \alpha^2 & \alpha^4 & \cdots & \alpha^{27} & \alpha^{29} \\ 1 & \alpha^3 & \alpha^6 & \cdots & \alpha^{25} & \alpha^{28} \\ 1 & \alpha^4 & \alpha^8 & \cdots & \alpha^{23} & \alpha^{27} \end{bmatrix}$$

■

Example 8-2—Nonbinary BCH Codes of Length 21

Here we consider 4-ary BCH codes of length 21. We begin by identifying the smallest field in which we can find γ, a primitive 21st root of unity. The smallest m such that $21 \mid 4^m - 1$ is 3, and γ can thus be found in $GF(4^3) = GF(64)$, but not in a smaller field. Appendix B provides the information necessary for the computations. Let α be a root of the primitive polynomial $x^6 + x + 1$. Set $\gamma = \alpha^3$, which is thus a 21st root of unity. Let $\{0, 1, \gamma^7 = \beta, \gamma^{14} = \beta^2\}$ represent the elements in GF(4). The cyclotomic cosets *modulo* 21 with respect to GF(4) and the associated minimal polynomials are as follows.

CYCLOTOMIC COSETS		MINIMAL POLYNOMIALS
$C_0 = \{0\}$	$\leftrightarrow$	$M_{(0)}(x) = x + 1$
$C_1 = \{1, 4, 16\}$	$\leftrightarrow$	$M_{(1)}(x) = x^3 + \beta^2 x + 1$
$C_2 = \{2, 8, 11\}$	$\leftrightarrow$	$M_{(2)}(x) = x^3 + \beta x + 1$

$$\begin{array}{lcl}
C_3 = \{3, 12, 6\} & \leftrightarrow & M_{(3)}(x) = x^3 + x^2 + 1 \\
C_5 = \{5, 20, 17\} & \leftrightarrow & M_{(5)}(x) = x^3 + \beta^2 x^2 + 1 \\
C_7 = \{7\} & \leftrightarrow & M_{(7)}(x) = x + \beta \\
C_9 = \{9, 15, 18\} & \leftrightarrow & M_{(9)}(x) = x^3 + x + 1 \\
C_{10} = \{10, 19, 13\} & \leftrightarrow & M_{(10)}(x) = x^3 + \beta x^2 + 1 \\
C_{14} = \{14\} & \leftrightarrow & M_{(14)}(x) = x + \beta^2
\end{array}$$

- ***One-error-correcting Narrow-Sense 4-ary BCH Code***

γ and γ^2 are the required roots.

$$\begin{aligned}
g(x) &= \text{LCM}(M_{(1)}(x), M_{(2)}(x)) = M_{(1)}(x)M_{(2)}(x) \\
&= (x^3 + \beta^2 x + 1)(x^3 + \beta x + 1) \\
&= x^6 + x^4 + x^2 + x + 1
\end{aligned}$$

The degree of $g(x)$ is 6, giving us a (21, 15) 4-ary single-error-correcting BCH code. The corresponding parity check matrix is shown below.

$$\mathbf{H} = \begin{bmatrix} 1 & \gamma & \gamma^2 & \cdots & \gamma^{19} & \gamma^{20} \\ 1 & \gamma^2 & \gamma^4 & \cdots & \gamma^{17} & \gamma^{19} \end{bmatrix} = \begin{bmatrix} 1 & \alpha^3 & \alpha^6 & \cdots & \alpha^{57} & \alpha^{60} \\ 1 & \alpha^6 & \alpha^{12} & \cdots & \alpha^{51} & \alpha^{57} \end{bmatrix}$$

- ***Two-error-correcting 4-ary BCH Code***

γ, γ^2, γ^3, and γ^4 are the required roots.

$$\begin{aligned}
g(x) &= \text{LCM}(M_{(1)}(x), M_{(2)}(x), M_{(3)}(x), M_{(4)}(x)) = M_{(1)}(x)M_{(2)}(x)M_{(3)}(x) \\
&= (x^3 + \beta^2 x + 1)(x^3 + \beta x + 1)(x^3 + x^2 + 1) \\
&= x^9 + x^8 + x^7 + x^5 + x^4 + x + 1
\end{aligned}$$

The degree of $g(x)$ is 9, giving us a (21, 12) 4-ary double-error-correcting BCH code. The corresponding parity-check matrix is shown below.

$$\mathbf{H} = \begin{bmatrix} 1 & \gamma & \gamma^2 & \cdots & \gamma^{19} & \gamma^{20} \\ 1 & \gamma^2 & \gamma^4 & \cdots & \gamma^{17} & \gamma^{19} \\ 1 & \gamma^3 & \gamma^6 & \cdots & \gamma^{15} & \gamma^{18} \\ 1 & \gamma^4 & \gamma^8 & \cdots & \gamma^{13} & \gamma^{17} \end{bmatrix} = \begin{bmatrix} 1 & \alpha^3 & \alpha^6 & \cdots & \alpha^{57} & \alpha^{60} \\ 1 & \alpha^6 & \alpha^{12} & \cdots & \alpha^{51} & \alpha^{57} \\ 1 & \alpha^9 & \alpha^{18} & \cdots & \alpha^{45} & \alpha^{54} \\ 1 & \alpha^{12} & \alpha^{24} & \cdots & \alpha^{39} & \alpha^{51} \end{bmatrix}$$ ■

When constructing error control codes, as in the previous examples, we generally want to achieve a given error correction capability (usually expressed in terms of minimum distance when dealing with algebraic codes) while simultaneously minimizing the redundancy that will be added to the data (i.e., maximizing the code rate). In BCH codes this translates into ensuring the required number of consecutive powers of a primitive nth root of unity as zeros while minimizing the number of "extraneous" zeros. The extraneous zeros are, of course, the conjugates of the desired zeros. The conjugates must be included if the generator polynomial is to have coefficients in the desired subfield (recall Theorem 3-4). Through careful selection of the parameter b it is often possible to find a code that has the same minimum distance as the narrow-sense code ($b = 1$) but has greater dimension and thus higher rate. Consider the single-error-correcting code designed in Example 8-2. If $b = 1$, we obtain a $(21, \underline{15})$ single-error-correcting code. If we set $b = 7$, however, we obtain a $(21, \underline{17})$ single-error-correcting code $(g(x) = M_{(7)}(x)M_{(8)}(x) = M_{(7)}(x)M_{(2)}(x))$. The code rate is thus improved from 0.71 to 0.81. It should be noted, however, that the BCH bound is an inequality; the $(21, 15)$ code may thus have a higher minimum distance than the $(21, 17)$ code. Unfortunately the most common BCH decoding techniques cannot take advantage of error correction capabilities beyond that predicted by the BCH bound (Theorem 8-1). The higher rate code is thus preferred.

After constructing a number of codes using various symbol alphabets GF(q) and code lengths n, one begins to notice a general trend which may be summarized as follows.

1. For a fixed code-symbol alphabet GF(q), the cardinality of cyclotomic cosets *modulo n* is generally smaller (providing more resolution in code design and fewer extraneous zeros) for $n = q^m - 1$ (primitive codes).
2. The cardinality of cyclotomic cosets *modulo n* grows smaller as the code-symbol alphabet size increases for fixed n.

The asymptotic result of the above trends is obtained by considering codes of length $(q^m - 1)$ in conjunction with code-symbol alphabet GF(q^m). Such codes are called Reed-Solomon codes and are the subject of Sections 8.3, 8.4, and 8.5.

8.2 WEIGHT DISTRIBUTIONS FOR SOME BINARY BCH CODES

The weight distributions for most BCH codes are not known. There are, however, a few useful results to be considered. In particular, the weight distributions for all double- and triple-error-correcting binary primitive BCH codes have been found [Kas3], [Lin1].

In Tables 8-1(a) through 8-2(b) the weight distributions for the duals of the desired codes are given. To obtain the desired weight distribution, the MacWilliams identity is applied (Theorem 4-11). This is demonstrated in Example 8-2.

TABLE 8-1a. Weight Distributions of the Duals of Double-Error-Correcting Primitive Narrow-Sense Binary BCH Codes ($n = 2^m - 1, m \geq 3, m$ odd) [Lin1][1]

Code-word weight i	Number of code words of weight i
0	1
$2^{m-1} - 2^{[(m+1)/2]-1}$	$[2^{m-2} + 2^{[(m-1)/2]-1}](2^m - 1)$
2^{m-1}	$(2^{2m-1} + 2^{m-1} - 1)$
$2^{m-1} + 2^{[(m+1)/2]-1}$	$[2^{m-2} - 2^{[(m-1)/2]-1}](2^m - 1)$

[1] Tables 8-1(a), 8-1(b), 8-2(a), and 8-2(b) are reprinted, with permission, from S. Lin and D. J. Costello, Jr., *Error Control Coding: Fundamentals and Applications* (Englewood Cliffs, NJ: Prentice Hall), pp. 177–178.

TABLE 8-1(b). Weight Distributions of the Duals of Double-Error-Correcting Primitive Narrow-Sense Binary BCH Codes ($n = 2^m - 1, m \geq 4, m$ even) [Lin1]

Code-word weight i	Number of code words of weight i
0	1
$2^{m-1} - 2^{[(m+2)/2]-1}$	$\frac{1}{3} \cdot 2^{[(m-2)/2]-1}[2^{(m-2)/2} + 1](2^m - 1)$
$2^{m-1} - 2^{(m/2)-1}$	$\frac{1}{3} \cdot 2^{[(m+2)/2]-1}(2^{m/2} + 1)(2^m - 1)$
2^{m-1}	$(2^{m-2} + 1)(2^m - 1)$
$2^{m-1} + 2^{(m/2)-1}$	$\frac{1}{3} \cdot 2^{[(m+2)/2]-1}(2^{m/2} - 1)(2^m - 1)$
$2^{m-1} + 2^{[(m+2)/2]-1}$	$\frac{1}{3} \cdot 2^{[(m-2)/2]-1}[2^{(m-2)/2} - 1](2^m - 1)$

TABLE 8-2(a). Weight Distributions of the Duals of Triple-Error-Correcting Primitive Narrow-Sense Binary BCH Codes ($n = 2^m - 1, m \geq 5, m$ odd) [Lin1]

Code-word weight i	Number of code words of weight i
0	1
$2^{m-1} - 2^{(m+1)/2}$	$\frac{1}{3} \cdot 2^{(m-5)/2}[2^{(m-3)/2} + 1](2^{m-1} - 1)(2^m - 1)$
$2^{m-1} - 2^{(m-1)/2}$	$\frac{1}{3} \cdot 2^{(m-3)/2}[2^{(m-1)/2} + 1](5 \cdot 2^{m-1} + 4)(2^m - 1)$
2^{m-1}	$(9 \cdot 2^{2m-4} + 3 \cdot 2^{m-3} + 1)(2^m - 1)$
$2^{m-1} + 2^{(m-1)/2}$	$\frac{1}{3} \cdot 2^{(m-3)/2}[2^{(m-1)/2} - 1](5 \cdot 2^{m-1} + 4)(2^m - 1)$
$2^{m-1} + 2^{(m+1)/2}$	$\frac{1}{3} \cdot 2^{(m-5)/2}[2^{(m-3)/2} - 1](2^{m-1} - 1)(2^m - 1)$

TABLE 8-2(b). Weight Distributions of the Duals of Triple-Error-Correcting Primitive Narrow-Sense Binary BCH Codes ($n = 2^m - 1, m \geq 6, m$ even) [Lin1]

Code-word weight i	Number of code words of weight i
0	1
$2^{m-1} - 2^{[(m+4)/2]-1}$	$\frac{1}{960}[2^{m-1} + 2^{[(m+4)/2]-1}](2^m - 4)(2^m - 1)$
$2^{m-1} - 2^{[(m+2)/2]-1}$	$\frac{7 \cdot 2^m}{48}[2^{m-1} + 2^{[(m+2)/2]-1}](2^m - 1)$
$2^{m-1} - 2^{(m/2)-1}$	$\frac{2}{15}(2^{m-1} + 2^{(m/2)-1})(3 \cdot 2^m + 8)(2^m - 1)$
2^{m-1}	$\frac{1}{64}(29 \cdot 2^{2m} - 4 \cdot 2^m + 64)(2^m - 1)$
$2^{m-1} + 2^{(m/2)-1}$	$\frac{2}{15}(2^{m-1} - 2^{(m/2)-1})(3 \cdot 2^m + 8)(2^m - 1)$
$2^{m-1} + 2^{[(m+2)/2]-1}$	$\frac{7 \cdot 2^m}{48}[2^{m-1} - 2^{[(m+2)/2]-1}](2^m - 1)$
$2^{m-1} + 2^{[(m+4)/2]-1}$	$\frac{1}{960}[2^{m-1} - 2^{[(m+4)/2]-1}](2^m - 4)(2^m - 1)$

Example 8-3—Weight Distribution of the (31, 21) Narrow-Sense Primitive Double-error-correcting BCH Code

In Example 8-1 we obtained the following generator polynomial for a (31,21) binary double-error-correcting code.

$$g(x) = x^{10} + x^9 + x^8 + x^6 + x^5 + x^3 + 1$$

Since $31 = 2^5 - 1$ and $m = 5$ is odd, we use Table 8-1(a) to obtain the weight distribution for the (31, 10) code that is dual to the (31,21) code. The weight distribution for the dual code is then rewritten as the weight enumerator

$$A(x) = 1 + 310x^{12} + 527x^{16} + 186x^{20}$$

The MacWilliams Identity (Theorem 4-11) is used to determine the weight enumerator $B(x)$ for the dual of the (31, 10) code.

$$\begin{aligned}
B(x) &= 2^{-k}(1 + x)^n A\left[\frac{(1 - x)}{(1 + x)}\right] \\
&= 2^{-10}(1 + x)^{31}\left[1 + 310\frac{(1 - x)^{12}}{(1 + x)^{12}} + 527\frac{(1 - x)^{16}}{(1 + x)^{16}} + 186\frac{(1 - x)^{20}}{(1 + x)^{20}}\right] \\
&= 2^{-10}[(1 + x)^{31} + 310(1 - x)^{12}(1 + x)^{19} + 527(1 - x)^{16}(1 + x)^{15} \\
&\quad +186(1 - x)^{20}(1 + x)^{11}] \\
&= 1 + 186x^5 + 806x^6 + 2635x^7 + 7905x^8 + 18910x^9 + 41602x^{10} + 85560x^{11} \\
&\quad + 142600x^{12} + 195300x^{13} + 251100x^{14} + 301971x^{15} + 301971x^{16} + 251100x^{17} \\
&\quad + 195300x^{18} + 142600x^{19} + 85560x^{20} + 41602x^{21} + 18910x^{22} + 7905x^{23} \\
&\quad + 2635x^{24} + 806x^{25} + 186x^{26} + x^{31}
\end{aligned}$$

The polynomial $B(x)$ has the form $B_0 + B_1 x + B_2 x^2 + \cdots + B_{31} x^{31}$. The coefficients $\{B_i\}$ comprise the weight distribution for the (31,21) code, with B_i being the number of code words of weight i. It should now be clear why the tables are compiled for the dual codes and not the codes themselves. ■

8.3 BASIC PROPERTIES OF REED-SOLOMON CODES

There are a number of ways to define Reed-Solomon codes. Reed and Solomon's initial definition focused on the evaluation of polynomials over the elements in a finite field [Ree2]. This approach has been generalized to an algebro-geometric definition involving rational curves [Wic1]. Others have preferred to examine Reed-Solomon codes in light of the Galois field Fourier transform [Bla2]. Finally, Reed-Solomon codes can be viewed as a natural extension of BCH codes [Lin1], [Mac]. In this book we focus on the last two approaches, for they lead to the efficient decoding algorithms discussed in Chapter 9. We begin with the BCH extension approach, returning to the Galois field Fourier transform in Section 8.5.

Definition 8-3—Reed-Solomon Codes

A **Reed-Solomon code** is a q^m-ary BCH code of length $q^m - 1$.

Consider the construction of a t-error correcting Reed-Solomon code of length $(q^m - 1)$. The first step is to note that the required primitive $(q^m - 1)$st root of unity α can be found in GF(q^m) (Theorem 2-12). Since the code symbols are to be from GF(q^m), the next step is to construct the cyclotomic cosets *modulo* $(q^m - 1)$ with respect to GF(q^m). This is a trivial task, for $(s \cdot q^m) \equiv s$ *modulo* $(q^m - 1)$. The cyclotomic cosets are singleton sets of the form $\{s\}$ and the associated minimal polynomials are of the form $(x - \alpha^s)$.

The BCH bound indicates that $2t$ consecutive powers of α are required as zeros of the generator polynomial $g(x)$ for a t-error-correcting Reed-Solomon code. The generator polynomial is the product of the associated minimal polynomials:

$$g(x) = (x - \alpha^b)(x - \alpha^{b+1})(x - \alpha^{b+2}) \cdots (x - \alpha^{b+2t-1})$$

Example 8-4—Two-error-correcting 8-ary Reed-Solomon Code, Length 7

Let α be a root of the primitive binary polynomial $x^3 + x + 1$ and thus a primitive 7th of unity. The Galois field GF(8) can be represented as consecutive powers of α.

$$\begin{array}{ll} \alpha = \alpha & \alpha^5 = \alpha^2 + \alpha + 1 \\ \alpha^2 = \alpha^2 & \alpha^6 = \alpha^2 + 1 \\ \alpha^3 = \alpha + 1 & \alpha^7 = 1 \\ \alpha^4 = \alpha^2 + \alpha & 0 = 0 \end{array}$$

If the resulting code is to be double-error-correcting, it must have $2t = 4$ consecutive powers of α as zeros. A narrow-sense generator polynomial is constructed as follows.

$$g(x) = (x - \alpha)(x - \alpha^2)(x - \alpha^3)(x - \alpha^4) = x^4 + \alpha^3 x^3 + x^2 + \alpha x + \alpha^3$$

Since the generator polynomial has degree 4, the (7, 3) Reed-Solomon code it defines has dimension 3 over GF(8) and thus $8^3 = 512$ code words. The required roots lead to the following parity-check matrix.

$$\mathbf{H} = \begin{bmatrix} 1 & \alpha & \alpha^2 & \alpha^3 & \alpha^4 & \alpha^5 & \alpha^6 \\ 1 & \alpha^2 & \alpha^4 & \alpha^6 & \alpha & \alpha^3 & \alpha^5 \\ 1 & \alpha^3 & \alpha^6 & \alpha^2 & \alpha^5 & \alpha & \alpha^4 \\ 1 & \alpha^4 & \alpha & \alpha^5 & \alpha^2 & \alpha^6 & \alpha^3 \end{bmatrix}$$ ■

Example 8-5—Three-error-correcting 64-ary Reed-Solomon Code, Length 63

Let α be a root of the primitive polynomial $x^6 + x + 1$. The corresponding representation of GF(64) as powers of α can be found in Appendix B. To be triple-error-correcting, $g(x)$ must have six consecutive powers of α as zeros.

$$\begin{aligned} g(x) &= (x - \alpha)(x - \alpha^2)(x - \alpha^3)(x - \alpha^4)(x - \alpha^5)(x - \alpha^6) \\ &= x^6 + \alpha^{59}x^5 + \alpha^{48}x^4 + \alpha^{43}x^3 + \alpha^{55}x^2 + \alpha^{10}x + \alpha^{21} \end{aligned}$$

$g(x)$ defines a (63, 57) triple-error-correcting Reed-Solomon code. This code has $64^{57} = 8.96 \times 10^{102}$ code words. ■

Example 8-6—Two-error-correcting 7-ary Reed-Solomon Code, Length 6

The GF(7) operations are simple *modulo* 7 addition and multiplication. The labels $\{0, 1, 2, 3, 4, 5, 6\}$ are used to represent the field elements. It was shown in Example 2-20 that 5 is a primitive element in GF(7). Given the requirement for 4 consecutive powers of 5 as zeros, the following generator polynomial is constructed.

$$\begin{aligned} g(x) &= (x - 5)(x - 5^2)(x - 5^3)(x - 5^4) \\ &= (x - 5)(x - 4)(x - 6)(x - 2) \\ &= x^4 + 4x^3 + 6x^2 + 5x + 2 \end{aligned}$$

$g(x)$ defines a (6, 2) double-error-correcting Reed-Solomon code. This code has $7^2 = 49$ code words. ■

Reed-Solomon codes have a number of interesting properties that are not shared by the other BCH codes. Recall that the minimum distance for BCH codes is in general lower-bounded by the design distance, but in many cases the actual minimum distance exceeds the design distance. One of the most significant properties of Reed-Solomon codes is that an (n, k) Reed-Solomon code always has minimum distance exactly equal to $(n - k + 1)$. We begin by proving this result using the Singleton bound (Theorem 4-10) and then proceed to explore its consequences.

Theorem 8-2—The Minimum Distance of Reed-Solomon Codes

An (n, k) Reed-Solomon code has minimum distance $(n - k + 1)$.

Proof. The Singleton bound places an upper bound of $(n - k + 1)$ on the minimum distance of all (n, k) codes. In the case of Reed-Solomon codes, however, we can construct a lower bound on minimum distance as well. Let **C** be an (n, k) Reed-Solomon code. The degree of the generator polynomial is $(n - k)$, so it must contain $(n - k) = (\delta - 1)$ consecutive powers of a primitive nth root of unity. The BCH bound implies that $d_{\min} \geq \delta = (n - k + 1)$. The Singleton bound implies that $d_{\min} \leq (n - k + 1)$. The only way both can be true is if $d_{\min} = (n - k + 1)$. **QED**

An (n, k) code that satisfies the Singleton bound with equality is called **maximum-distance separable (MDS)**. Though Reed-Solomon codes are the most commonly used MDS codes, there are a number of other interesting cases. In the next section we consider several MDS codes that are obtained through modifications of Reed-Solomon codes.

MDS codes have several properties that are of particular interest.

Theorem 8-3—Duals of MDS Codes

If C is MDS, then so is its dual $\mathbf{C}^{\perp}$.

Proof. Every (n, k) MDS code **C** can be associated with an $(n - k) \times n$ parity-check matrix **H**. Since **C** is MDS, all combinations of $(n - k)$ columns of **H** are linearly independent (Theorem 4-9).

H is the generator matrix for the $(n, n-k)$ dual code $\mathbf{C}^{\perp}$. Let nonzero $\mathbf{c} \in \mathbf{C}^{\perp}$ have weight $\leq k$. **c** thus has zeros in some set of coordinates $\{w_1, w_2, \ldots, w_{n-k}\}$. Since **c** is by definition a linear combination of the rows of **H**, the $(n-k) \times (n-k)$ square submatrix formed by the w_1st, w_2nd, ..., w_{n-k}th columns in **H** must be singular. The column rank of **H** is thus less than $(n-k)$, contradicting the premise that **C** is MDS. The nonzero code words of $\mathbf{C}^{\perp}$ must thus have weight $> k$. **QED**

Theorem 8-4—Systematic Representations

Any combination of *k* code-word coordinates in an MDS code may be used as message coordinates in a systematic representation.

Proof. Every (n, k) MDS code **C** can be associated with a $k \times n$ generator matrix **G**. By Theorem 8-3, **G** must be the parity-check matrix for an $(n, n-k)$ MDS code $\mathbf{C}^{\perp}$. Since $\mathbf{C}^{\perp}$ has minimum distance $(k+1)$, any combination of k columns of **G** must be linearly independent. Since the row rank equals the column rank, any $k \times k$ submatrix of **G** must thus be nonsingular. This implies that by using row reduction on **G**, we can reduce an arbitrary $k \times k$ submatrix to the identity matrix $\mathbf{I}_k$. The result follows. **QED**

Theorem 8-5—Weight Distribution of MDS Codes

The number of weight-*j* code words in a *q*-ary (n, k) MDS code is

$$A_j = \binom{n}{j}(q-1) \sum_{i=0}^{j-d_{\min}} (-1)^i \binom{j-1}{i} q^{j-i-d_{\min}}$$

Proof (Partial). Using Theorem 8-4, we can select an arbitrary set of k coordinates as the message positions for a weight-1 message block. Since the minimum distance is $(n-k+1)$, the $(n-k)$ redundant positions must all be nonzero. There is thus a weight-$(n-k+1)$ code word for every combination of $(n-k+1)$ coordinates, and given those coordinates, every weight-1 message block over GF(q).

$$A_{n-k+1} = \binom{n}{n-k+1}(q-1) = \binom{n}{k-1}(q-1)$$

The rest of the weight distribution can be obtained by increasing the weight of the message blocks and subtracting lower index enumerators to prevent multiple-counting. The general result follows through induction. A virtually complete proof using a variant of the MacWilliams identities can be found in [Mac]. [Ber1] also has a nice proof. **QED (Partial)**

Example 8-7—Weight Distributions for Reed-Solomon Codes

- (7, 3) Reed-Solomon code (Example 8-4)

j	A_j
0	1
1–4	0
5	147
6	147
7	217

- (31, 15) Reed-Solomon code[2]

j	A_j	j	A_j	j	A_j
0	1	21	7.64648×10^{14}	27	4.81054×10^{20}
1–16	0	22	1.07649×10^{16}	28	2.13038×10^{21}
17	8.22066×10^{9}	23	1.30596×10^{17}	29	6.83191×10^{21}
18	9.59077×10^{10}	24	1.34948×10^{18}	30	1.41193×10^{22}
19	2.62922×10^{12}	25	1.17135×10^{19}	31	1.41193×10^{22}
20	4.67616×10^{13}	26	8.37965×10^{19}		

■

8.4 MODIFIED REED-SOLOMON CODES

In Section 4.6 six techniques were presented for modifying linear block codes. In this section we investigate the application of these techniques to Reed-Solomon codes.

Punctured Reed-Solomon codes. In Theorem 8-4 it was shown that any combination of k coordinates in an (n, k) Reed-Solomon code can be treated as message positions in a systematic representation. An (n, k) Reed-Solomon code is thus punctured by deleting any one of its coordinates. The resulting $(n - 1, k)$ code is, in general, no longer cyclic, but it is MDS. We prove a slightly stronger result.

Theorem 8-6

Punctured MDS codes are MDS.

Proof. An (n, k) MDS code $\mathbf{C}$ is punctured by deleting a redundant coordinate, creating an $(n - 1, k)$ code $\mathbf{C}'$. Since $\mathbf{C}$ is MDS, it has minimum distance $(n - k + 1)$. When a coordinate is deleted from each code word in $\mathbf{C}$, the minimum distance between code words is reduced by at most one. $\mathbf{C}'$ thus has minimum distance $d_{\min} \geq (n - k)$. This must be an equality (Singleton bound), and thus $d_{\min} = (n - k)$ and $\mathbf{C}'$ is MDS. **QED**

[2] This code has been adopted for the US Joint Tactical Information Distribution System (JTIDS).

Punctured Reed-Solomon codes will prove extremely useful in Chapter 15, when we consider Reed-Solomon code combining techniques for channels with feedback.

Shortened Reed-Solomon codes. Since any k coordinates of a Reed-Solomon code can be treated as message positions, an RS code is shortened by taking a cross section of the code at an arbitrary coordinate. This is done by collecting all of the code words for which a given coordinate is zero and deleting that coordinate. The resulting collection of vectors is a shortened Reed-Solomon code. This $(n-1, k-1)$ code is generally not cyclic, but it is MDS.

Theorem 8-7

Shortened MDS codes are MDS.

Proof. Let $\mathbf{C}$ be an (n, k) MDS code. Let $\mathbf{C}'$ be the collection of code words in $\mathbf{C}$ for which the ith coordinate is zero. Since $\mathbf{C}$ has minimum distance $(n-k+1)$, the subset of vectors $\mathbf{C}'$ must have minimum distance equal to or greater than $(n-k+1)$. If we delete the ith coordinate from the code words in $\mathbf{C}'$, the minimum distance between any pair of the resulting vectors does not change, for they all had the same value at the deleted position. The $(n-1, k-1)$ shortened code thus has minimum distance $d_{\min} \geq (n-k+1)$. This must be an equality by the Singleton bound, and the shortened code is thus MDS. **QED**

Extended Reed-Solomon codes. Any code can be extended multiple times through the addition of parity checks; the key, however, lies in making sure that the resulting code has a higher minimum distance than the original code. There are some interesting results in the area of extended Reed-Solomon codes that are extremely easy to state but less so to prove. We begin with the easiest result.

Theorem 8-8—Singly-Extended Reed-Solomon Codes

Any narrow-sense $(q-1, k)$ q-ary Reed-Solomon code $\mathbf{C}$ can be extended to form a noncyclic (q, k) q-ary MDS code by adding a parity check. Each code word $(c_0, c_1, \ldots, c_{q-2})$ thus becomes $(c_0, c_1, \ldots, c_{q-1})$, where

$$c_{q-1} = -\sum_{j=0}^{q-2} c_j$$

Proof. Let $\mathbf{C}$ be a $(q-1, k)$ narrow-sense t-error-correcting RS code with generator polynomial $g(x)$. $g(x)$ has the form

$$g(x) = (x-\alpha)(x-\alpha^2)\cdots(x-\alpha^{2t})$$

where α is primitive in GF(q). Let $\mathbf{c} = (c_0, c_1, \ldots, c_{q-2})$ be a code word of weight $d_{\min}$ in $\mathbf{C}$. If the corresponding code polynomial $c(x)$ is not zero at $x = 1$, then the parity check c_{q-1} is nonzero and the corresponding word in the extended code has weight $(d_{\min} + 1)$. If $c(1) = 0$, then $c(x)$ must be of the form $a(x)g(x)$, where $(x - 1)$ is a factor of $a(x)$. $c(x)$ is thus also a code polynomial in the code defined by

$$g'(x) = (x - 1)(x - \alpha)\cdots(x - \alpha^{2t})$$

which by the BCH bound must have minimum distance $(d_{\min} + 1)$. Since the length and minimum distance have both increased by one over the original RS code, the extended code is still MDS. **QED**

The impact of the single extension of a $(q - 1, k)$ narrow-sense Reed-Solomon code to a (q, k) MDS code can be seen in terms of the respective parity-check matrices as follows.

$$\mathbf{H}_{(q-1,k)} = \begin{bmatrix} 1 & \alpha & \alpha^2 & \cdots & \alpha^{(q-2)} \\ 1 & \alpha^2 & \alpha^4 & \cdots & \alpha^{(q-2)2} \\ \vdots & \vdots & \vdots & \ddots & \vdots \\ 1 & \alpha^{\delta-2} & \alpha^{2(\delta-2)} & \cdots & \alpha^{(q-2)(\delta-2)} \\ 1 & \alpha^{\delta-1} & \alpha^{2(\delta-1)} & \cdots & \alpha^{(q-2)(\delta-1)} \end{bmatrix}$$

Single Extension

$$\Downarrow$$

$$\mathbf{H}_{(q,k)} = \begin{bmatrix} 1 & 1 & 1 & \cdots & 1 & 1 \\ 1 & \alpha & \alpha^2 & \cdots & \alpha^{(q-2)} & 0 \\ 1 & \alpha^2 & \alpha^4 & \cdots & \alpha^{(q-2)2} & 0 \\ \vdots & \vdots & \vdots & \vdots & \ddots & \vdots \\ 1 & \alpha^{\delta-2} & \alpha^{2(\delta-2)} & \cdots & \alpha^{(q-2)(\delta-2)} & 0 \\ 1 & \alpha^{\delta-1} & \alpha^{2(\delta-1)} & \cdots & \alpha^{(q-2)(\delta-1)} & 0 \end{bmatrix} \tag{8-14}$$

Theorem 8-9—Doubly-Extended Reed-Solomon Codes

Any narrow-sense singly-extended (q, k) q-ary Reed-Solomon code C can be extended to form a noncyclic $(q + 1, k)$ q-ary MDS code by adding a $(q + 1)$st element c_q to each code word $(c_0, c_1, \ldots, c_{q-1})$, where

$$c_q = -\sum_{j=0}^{q-2} c_j\,\alpha^{j\delta}$$

The resulting parity-check matrix is as follows.

$$\mathbf{H}_{(q+1,k)} = \begin{bmatrix} 1 & 1 & 1 & \cdots & 1 & 1 & 0 \\ 1 & \alpha & \alpha^2 & \cdots & \alpha^{(q-2)} & 0 & 0 \\ 1 & \alpha^2 & \alpha^4 & \cdots & \alpha^{(q-2)2} & 0 & 0 \\ \vdots & \vdots & \vdots & \ddots & \vdots & \vdots & \vdots \\ 1 & \alpha^{\delta-2} & \alpha^{2(\delta-2)} & \cdots & \alpha^{(q-2)(\delta-2)} & 0 & 0 \\ 1 & \alpha^{\delta-1} & \alpha^{2(\delta-1)} & \cdots & \alpha^{(q-2)(\delta-1)} & 0 & 0 \\ 1 & \alpha^{\delta} & \alpha^{2\delta} & \cdots & \alpha^{(q-2)\delta} & 0 & 1 \end{bmatrix} \tag{8-15}$$

Proof. To show that the code defined by $\mathbf{H}_{(q+1,k)}$ has minimum distance $\delta + 2$, we must show that any $(\delta + 1)$ of the columns are linearly independent. We begin by noting that any $(\delta + 1) \times (\delta + 1)$ submatrix constructed from the first q columns is Vandermonde, and thus nonsingular (see the proof to Theorem 8-1). Furthermore, any $(\delta + 1) \times (\delta + 1)$ submatrix using one or both of the last two columns has Vandermonde matrices in the cofactor expansion, and thus has a nonzero determinant. Since any $(\delta + 1) \times (\delta + 1)$ submatrix of $\mathbf{H}_{(q+1,k)}$ is nonsingular, $\mathbf{H}_{(q+1,k)}$ defines a code with $d_{\min} \geq \delta + 2$. Singleton's bound then provides the conclusion that $d_{\min} = \delta + 2$. **QED**

The singly- and doubly-extended RS codes are in general not cyclic. MacWilliams and Sloane have demonstrated a method for generating cyclic q-ary MDS codes of length $q + 1$ [Mac].

There is a very limited class of triply-extended Reed-Solomon codes that are MDS. These codes have either dimension or redundancy 3 and are the only known examples of q-ary MDS codes of length $(q + 2)$. These codes are constructed using matrices of the following form as either a parity-check or generator matrix [Wic1].

$$\begin{bmatrix} 1 & 1 & \cdots & 1 & 1 & 0 & 0 \\ 1 & \alpha & \cdots & \alpha^{q-2} & 0 & 1 & 0 \\ 1 & \alpha^2 & \cdots & \alpha^{2(q-2)} & 0 & 0 & 1 \end{bmatrix}$$

The columns of this matrix can be viewed as points in the projective plane over GF(q) (points in a $(k - 1)$-dimensional projective space are represented using k-tuples and are identified under multiplication by a scalar). This particular collection of points forms an **oval** (i.e., any three points are linearly independent).

Lurking behind these results one can find one of the more interesting open problems in projective geometry over Galois fields. The problem can be roughly stated as follows: What is the maximum number of points in a projective space of dimension $(k - 1)$ over GF(q) such that any k points are always linearly independent? Given the results discussed in this chapter, n is clearly greater than or equal to $q + 1$ in general, and $q + 2$ for the specific case above. Given a relation between MDS codes and rational curves discussed in [Wic1], the author believes that the answer to the open question is $q + 1$, the number of points on the projective line over GF(q), with the known exception being the only exception. We are drawing too near the subject of projective geometries over Galois fields, however, so we had best change the subject.

8.5 THE GALOIS FIELD FOURIER TRANSFORM APPROACH TO BCH AND REED-SOLOMON CODES

The application of the Galois field Fourier transform to the study of cyclic codes provides some interesting insights into the structure of these codes. In the particular case of Reed-Solomon codes, the transform leads to efficient encoder and decoder designs. The encoder is discussed here, while the decoder is treated in the next chapter.

Definition 8-4—Galois Field Fourier Transform (GFFT)

Let $\mathbf{v} = (v_0, v_1, \ldots, v_{n-1})$ be a vector over GF(q) whose length n divides $q^m - 1$ for some positive integer m. Let α be an element of order n in GF(q^m). The Galois field Fourier transform of $\mathbf{v}$ is the vector $\mathcal{F}(\mathbf{v}) = \mathbf{V} = (V_0, V_1, \ldots, V_{n-1})$, where the $\{V_j\}$ are computed as follows.

$$V_j = \sum_{i=0}^{n-1} \alpha^{ij} v_i, \qquad j = 0, 1, \ldots, n-1$$

The inverse transform is provided in the following theorem.

Theorem 8-10—The GFFT Transform Pair

Let v be a vector over GF(q), a field with characteristic p. v and its transform are related by the following expressions.

$$\begin{aligned} V_j &= \sum_{i=0}^{n-1} \alpha^{ij} v_i, \qquad j = 0, 1, \ldots, n-1 \\ v_i &= \frac{1}{n} \sum_{j=0}^{n-1} \alpha^{-ij} V_j, \qquad i = 0, 1, \ldots, n-1 \end{aligned} \tag{8-16}$$

Proof [Bla2]. Since α has order n, α and its integer powers α^r are zeros of the expression

$$x^n - 1 = (x-1)(x^{n-1} + x^{n-2} + \cdots + x + 1)$$

It follows that if $r \not\equiv 0 \bmod n$, then α^r must be a zero of $(x^{n-1} + x^{n-2} + \cdots + x + 1)$, and thus satisfies the equality

$$\sum_{j=0}^{n-1} \alpha^{rj} = 0, \qquad r \not\equiv 0 \bmod n$$

When $r \equiv 0 \bmod n$, we get the result

$$\sum_{j=0}^{n-1} \alpha^{rj} = \sum_{j=0}^{n-1} 1 \equiv n \bmod p$$

Combining these two results, we get

$$\sum_{j=0}^{n-1} \alpha^{-ij} V_j = \sum_{j=0}^{n-1} \alpha^{-ij} \left(\sum_{k=0}^{n-1} \alpha^{kj} v_k \right) = \sum_{k=0}^{n-1} v_k \sum_{j=0}^{n-1} \alpha^{(k-i)j} \equiv v_i n \bmod p \qquad \textbf{QED}$$

The GFFT is a generalization of the familiar discrete Fourier transform (DFT) to finite fields. All of the general results for the DFT (e.g., time and frequency translation, the convolution theorem) also hold for the GFFT. The convolution theorem is of particular interest in the study of cyclic codes.

Theorem 8-11—The GFFT Convolution Theorem

Consider the following GFFT pairs.

$$\mathbf{a} = (a_0, a_1, \ldots, a_{n-1}) \leftrightarrow \mathcal{F}(\mathbf{a}) = \mathbf{A} = (A_0, A_1, \ldots, A_{n-1})$$

$$\mathbf{b} = (b_0, b_1, \ldots, b_{n-1}) \leftrightarrow \mathcal{F}(\mathbf{b}) = \mathbf{B} = (B_0, B_1, \ldots, B_{n-1})$$

$$\mathbf{c} = (c_0, c_1, \ldots, c_{n-1}) \leftrightarrow \mathcal{F}(\mathbf{c}) = \mathbf{C} = (C_0, C_1, \ldots, C_{n-1})$$

$C_j = A_j B_j \quad (j = 0, 1, \ldots, n-1)$ **if and only if**

$$c_i = \sum_{k=0}^{n-1} a_k b_{i-k} \quad (i = 0, 1, \ldots, n-1)$$

Proof. Apply the inverse GFFT to $\mathbf{C} = (A_0 B_0, A_1 B_1, \ldots, A_{n-1} B_{n-1})$.

$$\begin{aligned} c_i &= \frac{1}{n} \sum_{j=0}^{n-1} \alpha^{-ij} C_j = \frac{1}{n} \sum_{j=0}^{n-1} \alpha^{-ij} A_j B_j \\ &= \frac{1}{n} \sum_{j=0}^{n-1} \alpha^{-ij} \left(\sum_{k=0}^{n-1} \alpha^{kj} a_k \right) B_j \\ &= \frac{1}{n} \sum_{k=0}^{n-1} a_k \left(\sum_{j=0}^{n-1} \alpha^{-(i-k)j} B_j \right) \\ &= \sum_{k=0}^{n-1} a_k b_{i-k} \end{aligned}$$

The reverse direction of this "if and only if" result follows from Theorem 8-10.

QED

Corollary 8-12

Consider the following GFFT pairs.

$$\mathbf{a} = (a_0, a_1, \ldots, a_{n-1}) \leftrightarrow \mathcal{F}(\mathbf{a}) = \mathbf{A} = (A_0, A_1, \ldots, A_{n-1})$$

$$\mathbf{b} = (b_0, b_1, \ldots, b_{n-1}) \leftrightarrow \mathcal{F}(\mathbf{b}) = \mathbf{B} = (B_0, B_1, \ldots, B_{n-1})$$

$$\mathbf{c} = (c_0, c_1, \ldots, c_{n-1}) \leftrightarrow \mathcal{F}(\mathbf{c}) = \mathbf{C} = (C_0, C_1, \ldots, C_{n-1})$$

$c_j = a_j b_j \quad (j = 0, 1, \ldots, n-1)$ **if and only if** $C_i = \frac{1}{n} \sum_{k=0}^{n-1} A_k B_{i-k}$

Proof. Apply the same argument used in the proof of Theorem 8-11.

In studying communication systems, one learns that the continuous and discrete Fourier transforms provide a convenient mechanism for translating time-domain characteristics into frequency-domain, or spectral characteristics. Since code words are generally transmitted as a sequence of symbols indexed in time, we can make use of a similar interpretation for the GFFT. We thus adopt the terms "time domain" and "frequency domain," or "spectrum," and use them to describe the coding theoretic relationship that makes the GFFT so useful.

Definition 8-5

The **spectrum** of the polynomial $v(x) = v_0 + v_1 x + \cdots + v_{n-1}x^{n-1}$ is the GFFT of the vector $\mathbf{v} = (v_0, v_1, \ldots, v_{n-1})$.

Theorem 8-13

1. **α^j is a zero of the polynomial $v(x)$ if and only if the jth frequency component of the spectrum of $v(x)$ equals zero.**
2. **α^{-i} is a zero of the polynomial $V(x)$ if and only if the ith time component v_i of the inverse transform v of V equals zero.**

Proof. For the first part we note that

$$v(\alpha^j) = v_0 + v_1\alpha^j + v_2\alpha^{2j} + \cdots + v_{n-1}\alpha^{(n-1)j} = \sum_{i=0}^{n-1} v_i\alpha^{ij} = V_j$$

The second part follows in a similar manner.

$$V(\alpha^{-i}) = V_0 + V_1\alpha^{-i} + V_2\alpha^{-2i} + \cdots + V_{n-1}\alpha^{-(n-1)i} = \sum_{j=0}^{n-1} v_j\alpha^{-ij} = nv_i$$

QED

In Theorem 3-4 it was shown that a polynomial with coefficients in GF(q) has as roots the union of one or more conjugacy classes with respect to GF(q). The GFFT provides an analogous result in terms of the spectrum of the polynomial.

Theorem 8-14

Let V be a vector of length n over GF(q^m), where n is a divisor of $q^m - 1$ and GF(q^m) has characteristic p. The inverse transform v of V contains elements exclusively from the subfield GF(q) if and only if the following conditions are satisfied.

$$V_j^q \bmod p \equiv V_{qj \bmod n}, \qquad j = 0, \ldots, n-1$$

Proof [Bla2]. Recall that for fields with characteristic p, $(a+b)^{p^r} = a^{p^r} + b^{p^r}$. It was also shown in Theorem 2-16 that an element β in GF(q^m) lies in the subfield GF(q) if and only if $\beta^q = \beta$. It follows that

$$V_j^q = \left(\sum_{i=0}^{n-1} \alpha^{ij} v_i\right)^q = \sum_{i=0}^{n-1} \alpha^{qij} v_i^q = \sum_{i=0}^{n-1} \alpha^{iqj} v_i = V_{qj \bmod n}$$

To prove the reverse case, assume that $V_j^q = V_{qj \bmod n}$ for all j. We obtain by Definition 8-4

$$\sum_{i=0}^{n-1} \alpha^{iqj} v_i^q = \sum_{i=0}^{n-1} \alpha^{iqj} v_i, \qquad j = 0, \ldots, n-1$$

Let $k = qj \bmod n$. Since $n = q^m - 1$, q and n must be relatively prime. Thus as j ranges from 0 to $n - 1$, k takes on all values in the same range.

$$\sum_{i=0}^{n-1} \alpha^{ik} v_i^q = \sum_{i=0}^{n-1} \alpha^{ik} v_i, \qquad k = 0, \ldots, n - 1$$

Since v_i and v_i^q have the same transform, they must be equal by Theorem 8-10.

QED

Theorems 8-13 and 8-14 provide a series of frequency-domain analogs for the time-domain results in Chapters 3 and 5.

Example 8-8—Minimal Polynomials and the GFFT

In Example 3-5 the eight elements in GF(8) were arranged in conjugacy classes and their minimal polynomials computed as follows.

CONJUGACY CLASS	ASSOCIATED MINIMAL POLYNOMIAL
$\{0\}$	$M_*(x) = (x - 0) = x$
$\{\alpha^0\}$	$M_0(x) = (x - 1) = x + 1$
$\{\alpha, \alpha^2, \alpha^4\}$	$M_1(x) = (x - \alpha)(x - \alpha^2)(x - \alpha^4) = x^3 + x + 1$
$\{\alpha^3, \alpha^6, \alpha^5\}$	$M_3(x) = (x - \alpha^3)(x - \alpha^6)(x - \alpha^5) = x^3 + x^2 + 1$

When we compute the GFFT of the coefficients of the minimal polynomials, we obtain the following. We begin by writing out the coefficients of each polynomial as a vector. The spectrum of each vector is then computed with the following results.

$M_*(x)$: $\mathcal{F}(0100000) = \{\alpha^j\} = (1, \alpha, \alpha^2, \alpha^3, \alpha^4, \alpha^5, \alpha^6)$

$M_0(x)$: $\mathcal{F}(1100000) = \{1 + \alpha^j\} = (0, \alpha^3, \alpha^6, \alpha, \alpha^5, \alpha^4, \alpha^2)$

$M_1(x)$: $\mathcal{F}(1101000) = (1 + \alpha^j + \alpha^{3j}) = (1, 0, 0, \alpha^4, 0, \alpha^2, \alpha)$

$M_3(x)$: $\mathcal{F}(1011000) = (1 + \alpha^{2j} + \alpha^{3j}) = (1, \alpha^4, \alpha, 0, \alpha^2, 0, 0)$

Note that the positions of the zero coordinates in the spectra correspond with the roots of the minimal polynomials. ■

In Chapters 5 and 6 and in earlier sections of this chapter we constructed several cyclic codes by placing restrictions on the roots of the generator polynomial. All of these results can now be reexpressed as restrictions on the zeros of the spectrum of the generator polynomial. The most important result in this chapter, the BCH bound (Theorem 8-1), is restated and reproved as follows.

Theorem 8-15

Let n divide $q^m - 1$ for some positive integer m. A q-ary n-tuple with weight $\leq (\delta - 1)$ that also has $\delta - 1$ consecutive zeros in its spectrum must be the all-zero vector.

Proof. Let $\mathbf{c}$ be a vector with exactly ν nonzero coordinates, these coordinates being in positions $i_1, i_2, \ldots, i_\nu$. We now define a "locator polynomial" $\Lambda(x)$ in the frequency domain whose zeros correspond to the nonzero coordinates of **c** in the time domain.

$$\Lambda(x) = \prod_{j=1}^{\nu}(1 - x\alpha^{-i_j}) = \Lambda_0 + \Lambda_1 x + \cdots + \Lambda_\nu x^\nu$$

The inverse transform of $\Lambda(x)$ is thus a time-domain vector $\boldsymbol{\lambda}$ that has zero coordinates in the exact positions where **c** has nonzero coordinates and vice versa. The time-domain products $\{c_i \lambda_i\}$ are thus zero for all indices i. The transform of these products must also be zero. Corollary 8-12 then implies that the convolution of the transform of **c** and the transform of $\boldsymbol{\lambda}$, $\mathbf{C} * \boldsymbol{\Lambda}$, is zero as well.

Now assume that **c** has weight less than or equal to $\delta - 1$, while its transform **C** has $\delta - 1$ consecutive zeros. The coordinates $\{\Lambda_k\}$ are thus equal to zero for all $k > \delta - 1$. We also know that $\Lambda_0 = 1$ by its definition above (x cannot be a factor). The frequency-domain convolution of **C** and $\boldsymbol{\Lambda}$ can then be expressed as follows.

$$\frac{1}{n}\sum_{k=0}^{n-1} \Lambda_k C_{i-k} = 0$$

$$\Rightarrow C_i + \sum_{k=1}^{\delta-1} \Lambda_k C_{i-k} = 0$$

$$\Rightarrow C_i = -\sum_{k=1}^{\delta-1} \Lambda_k C_{i-k}$$

By substituting the $\delta - 1$ consecutive zero coordinates of **C** into the last expression, the recursion shows that **C** is all-zero, and thus its inverse transform **c** is as well. **QED**

To construct a code with minimum distance greater than or equal to δ, we require that the spectra of all of our code words have $\delta - 1$ consecutive zeros. This is an exact analog of the BCH bound as stated in Theorem 8-1. We now note that the coefficients of the product of two polynomials can be expressed as the convolution of the coefficients of the two polynomials:

$$m(x) = m_0 + m_1 x + m_2 x^2 + \cdots + m_k x^k$$

$$g(x) = g_0 + g_1 x + g_2 x^2 + \cdots + g_r x^r$$

$$c(x) = m(x) \cdot g(x) = c_0 + c_1 x + c_2 x^2 + \cdots + c_{k+r} x^{k+r}$$

where $c_j = \sum_{i=0}^{k+r-1} m_i g_{j-i}$. It follows from the convolution theorem (Theorem 8-11) that the spectrum of the product of two polynomials is simply the coordinatewise product of the individual spectra. We can now develop a frequency-domain interpretation of the encoding process for a cyclic code.

Example 8-9—Frequency-Domain Encoding of Cyclic Codes

In Example 8-8 we computed the transforms of the minimal polynomials of the elements in GF(8). We now choose as a generator polynomial $g(x) = M_1(x) = x^3 + x + 1$ (this is the (7, 4) cyclic Hamming code discussed in Example 5-8). $g(x)$ has spectrum

$\mathbf{G} = (1,0,0,\alpha^4,0,\alpha^2,\alpha)$. The following table contains several message polynomials, their transforms, the code words obtained through multiplication by $g(x)$, and the transforms of the code words. Note that each code-word spectrum is the coordinate-by-coordinate product of the associated message spectrum and the spectrum of $g(x)$.

$m(x)$	$\mathbf{M}$	$c(x)$	$\mathbf{C}$
1	$(1,1,1,1,1,1,1)$	x^3+x+1	$(1,0,0,\alpha^4,0,\alpha^2,\alpha)$
x	$(1,\alpha,\alpha^2,\alpha^3,\alpha^4,\alpha^5,\alpha^6)$	x^4+x^2+x	$(1,0,0,1,0,1,1)$
$x+1$	$(0,\alpha^3,\alpha^6,\alpha,\alpha^5,\alpha^4,\alpha^2)$	$x^4+x^3+x^2+1$	$(0,0,0,\alpha^5,0,\alpha^6,\alpha^3)$
x^2+x+1	$(1,\alpha^5,\alpha^3,\alpha^5,\alpha^6,\alpha^6,\alpha^3)$	x^5+x^4+1	$(1,0,0,\alpha^2,0,\alpha,\alpha^4)$ ■

Given the version of the BCH bound in Theorem 8-15, the design rule for Reed-Solomon codes can be reexpressed as follows.

Reed-Solomon Design Rule (Frequency Domain)

To construct a t-error-correcting Reed-Solomon code of length $q^m - 1$, select as code words all vectors whose transforms have $2t$ consecutive zeros.

This design rule leads to an interesting nonsystematic encoder design for Reed-Solomon codes. Consider Figure 8-1. We begin in the frequency domain by constraining $2t$ consecutive transform coordinates to be zero. Since the minimum distance of an (n,k) Reed-Solomon code is $2t+1 = n-k+1$ (Theorem 8-2), this constraint leaves k coordinates in the transform unspecified. These k coordinates are filled with the k information symbols provided by the source. The resulting transform is then inverted to provide the desired code word.

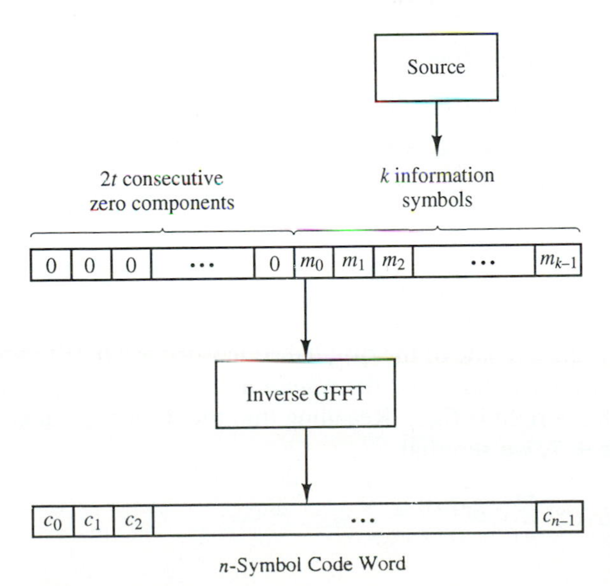

Figure 8-1. A GFFT Encoder for a t-Error-Correcting Reed-Solomon Code

The frequency-domain encoder in Figure 8-1 can be extended to an encoder for doubly-extended Reed-Solomon codes [Bla2]. In Figure 8-2 we see that the information symbols on each side of the $2t$ consecutive zeros are appended to each end of the inverse transform, creating a code word of length $(n + 2)$. The minimum distance of the code is increased by two, thus increasing the error correcting capability by one.

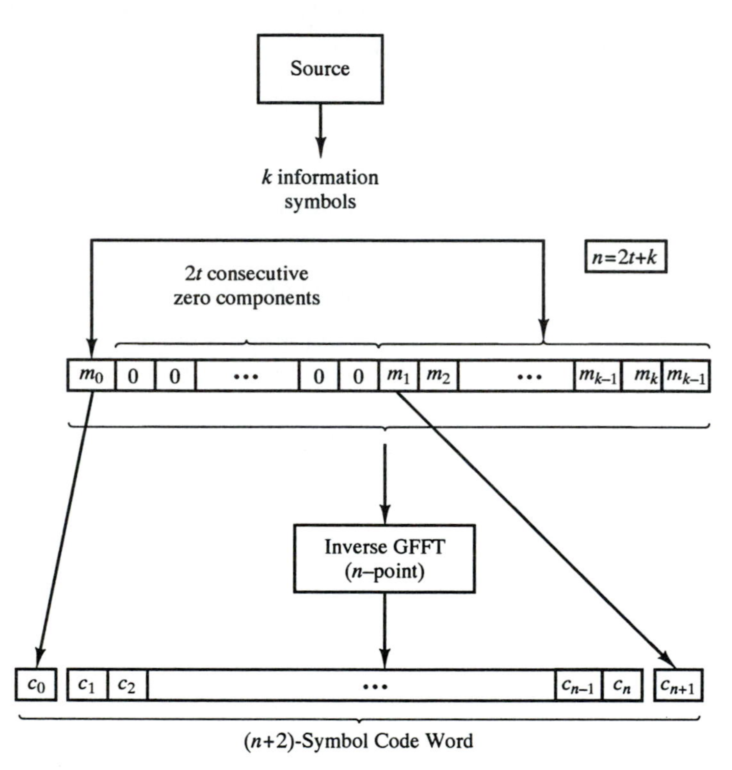

Figure 8-2. A GFFT Encoder for a $(t + 1)$-Error-Correcting, Doubly-Extended Reed-Solomon Code

The impact of the first extension is clear if we note that the first transform coordinate is

$$C_0 = \sum_{j=1}^{n} c_j \alpha^{0j} = \sum_{j=1}^{n} c_j = c_0$$

The symbol added on the left side of the original code word is a parity symbol. Theorem 8-8 implies that we have increased the minimum distance of the code by one. The symbol added on the right is C_{2t+1}. Recalling that the design distance δ for the unextended code is $2t + 1$, we see that

$$C_{2t+1} = \sum_{j=1}^{n} c_j \alpha^{j(2t+1)} = \sum_{j=1}^{n} c_j \alpha^{j\delta} = c_{n+1}$$

Theorem 8-9 then shows that the minimum distance is increased by one once again.

SUGGESTIONS FOR FURTHER READING

The following book by Berlekamp contains the most extensive treatment of BCH and Reed-Solomon codes of which this author is aware. It includes material of interest to the mathematician as well as the practicing engineer.

E. R. BERLEKAMP, *Algebraic Coding Theory*, rev. ed., Laguna Hills: Aegean Park Press, 1984.

MacWilliams and Sloane take a more theoretical approach than Berlekamp. They provide an excellent treatment of the theory of BCH and Reed-Solomon codes, as well as a complete chapter on the general class of MDS codes.

F. J. MACWILLIAMS and N. J. A. SLOANE, *The Theory of Error Correcting Codes*, Amsterdam: North Holland, 1977.

Blahut treats Reed-Solomon codes and the other alternant codes from a frequency-domain perspective. His treatment emphasizes the digital signal-processing aspects of coding theory, and thus tends to emphasize encoder and decoder design.

R. E. BLAHUT, *Theory and Practice of Error Control Codes*, Reading, MA: Addison-Wesley, 1984.

And finally, the most thorough discussion of the theory and application of Reed-Solomon codes can be found in an edited collection of papers that was recently published as a tribute to Irving Reed and Gus Solomon.

S. B. WICKER and V. K. BHARGAVA (eds.), *Reed-Solomon Codes and Their Applications*, Piscataway, N.J.: IEEE Press, 1994.

PROBLEMS

BCH code design

1. Consider a binary narrow-sense BCH code of length 15 and design distance 3.
   (a) Compute a generator polynomial for this code.
   (b) Determine the rate of this code.
   (c) Construct generator and parity-check matrices for this code.

2. Compute a generator polynomial for a binary narrow-sense BCH code of length 15 and design distance 4. What is the rate of this code?

3. Compute a generator polynomial for a binary BCH code of length 15 and design distance 4. Maximize the rate of the code through appropriate selection of the consecutive roots, then compare the results to the solution to Problem 8.2.

4. Compute a generator polynomial for a double-error-correcting binary BCH code of length 21.

5. Compute a generator polynomial for a single-error-correcting 4-ary BCH code of length 63.
6. Are there any good binary BCH codes of length 19? Provide support for your answer.
7. Are there any good 2^9-ary BCH codes of length 19? Provide support for your answer.
8. Determine the weight distribution for a double-error-correcting narrow-sense binary BCH code of length 15.

Reed-Solomon code design

9. Consider a narrow-sense Reed-Solomon code of length 15 and design distance 3.
   (a) Compute a generator polynomial for this code.
   (b) Determine the rate of this code.
   (c) Construct generator and parity-check matrices for this code.
10. Compute a generator polynomial for a narrow-sense double-error-correcting Reed-Solomon code of length 31.
11. Compute the weight distribution for an $(8, 3)$ 8-ary MDS code.
12. Consider a $(15, 10)$ narrow-sense Reed-Solomon code.
    (a) Construct a parity-check matrix for this code.
    (b) Construct a parity-check matrix for the $(8, 3)$ shortened Reed-Solomon code obtained by deleting 7 message coordinates from the encoding process of the $(15, 10)$ code.
    (c) Compute the weight distribution for the shortened code in (b).
    (d) Describe how one can obtain the code in (b) by puncturing a Reed-Solomon code.

9

The Decoding of BCH and Reed-Solomon Codes

In 1960 Peterson provided the first explicit description of a decoding algorithm for binary BCH codes [Pet2]. His "direct-solution" algorithm is quite useful for correcting small numbers of errors but becomes computationally intractable as the number of errors increases. Reed and Solomon discussed a decoding algorithm in their original paper on Reed-Solomon codes (1960) [Ree2], but that algorithm was also inefficient for large codes and large numbers of errors corrected. Peterson's algorithm was improved and extended to nonbinary codes by Gorenstein and Zierler (1961) [Gor1], Chien (1964) [Chi], and Forney (1965) [For7], but it was not until 1967 that Berlekamp introduced the first truly efficient decoding algorithm for both binary and nonbinary BCH codes [Ber6]. In 1969 Massey showed that the BCH decoding problem is equivalent to the problem of synthesizing the shortest linear feedback shift register capable of generating a given sequence [Mas4]. Massey then demonstrated a fast shift-register-based decoding algorithm for BCH and Reed-Solomon codes that is equivalent to Berlekamp's algorithm.

In 1975 Sugiyama et al. showed that Euclid's algorithm can be used to decode BCH and Reed-Solomon codes [Sug]. Reed et al. then showed in 1978 that a related technique based on continued fractions and Fermat-theoretic transforms resulted in a fast decoding algorithm for Reed-Solomon codes [Ree5].

In concluding this brief historical survey, we note that in 1973 Gore described a BCH decoding algorithm from the frequency-domain perspective [Gor2]. Blahut provided a more general discussion of spectral decoding techniques in 1979 [Bla4].

In the first section of this chapter we examine Peterson's direct-solution decoding algorithm for binary BCH codes. Peterson expressed the decoding problem in terms of a series of equations whose coefficients are provided by the syndromes of the received word. Berlekamp's algorithm is then introduced as a fast solution to the same problem. We proceed in the next section to a discussion of the decoding of nonbinary codes, beginning with a generalization of Peterson's technique. Berlekamp's algorithm is then presented in its general form using Massey's shift-register-based interpretation. We conclude the second section with a discussion of the decoding algorithm based on Euclid's algorithm. The next section considers frequency-domain decoding algorithms, and the chapter concludes with a discussion of erasure decoding.

9.1 DECODING ALGORITHMS FOR BINARY BCH CODES

In Chapter 8 the BCH bound (Theorem 8-1) was used as the basis for a design procedure for binary BCH codes. A t-error-correcting BCH code is constructed by first identifying the smallest field GF(2^m) that contains α, a primitive nth root of unity. A binary generator polynomial $g(x)$ is then selected so that it has as zeros some $2t$ consecutive powers of α.

$$g(\alpha^b) = g(\alpha^{b+1}) = \cdots = g(\alpha^{b+2t-1}) = 0$$

A binary vector $\mathbf{c} = (c_0, c_1, \ldots, c_{n-1})$ is a code word if and only if its associated polynomial $c(x) = c_0 + c_1 x + \cdots + c_{n-1}x^{n-1}$ has as zeros these same $2t$ consecutive powers of α. Now consider a received polynomial $r(x)$. $r(x)$ can be expressed as the sum of the transmitted code polynomial $c(x)$ and an error polynomial $e(x) = e_0 + e_1 x + \cdots + e_{n-1}x^{n-1}$. A series of syndromes is obtained by evaluating the received polynomial at the $2t$ zeros. To minimize the complexity of our notation, it is assumed henceforth that all codes under discussion are narrow-sense ($b = 1$).

$$S_j = r(\alpha^j) = c(\alpha^j) + e(\alpha^j) = e(\alpha^j) = \sum_{k=0}^{n-1} e_k(\alpha^j)^k, \qquad j = 1,2,\ldots,2t \tag{9-1}$$

The computations in Eq. (9-1) are performed in GF(2^m), the field containing the primitive nth root of unity. Now assume that the received word $\mathbf{r}$ has ν errors in positions $i_1, i_2, \ldots, i_\nu$. Since the code is binary, the errors in these positions have value $e_{i_j} = 1$. The syndrome sequence can be reexpressed in terms of these error locations.

$$S_j = \sum_{l=1}^{\nu} e_{i_l}(\alpha^j)^{i_l} = \sum_{l=1}^{\nu} (\alpha^{i_l})^j = \sum_{l=1}^{\nu} X_l^j, \qquad j = 1,\ldots,2t \tag{9-2}$$

The $\{X_l\}$ are *error locators*, for their values indicate the positions of the errors in the received word. Expanding Eq. (9-2), we obtain a sequence of $2t$ algebraic *syndrome equations* in the ν unknown error locations.

$$\begin{aligned} S_1 &= X_1 + X_2 + \cdots + X_\nu \\ S_2 &= X_1^2 + X_2^2 + \cdots + X_\nu^2 \\ S_3 &= X_1^3 + X_2^3 + \cdots + X_\nu^3 \\ &\vdots \\ S_{2t} &= X_1^{2t} + X_2^{2t} + \cdots + X_\nu^{2t} \end{aligned} \tag{9-3}$$

Equations of this form are called *power-sum symmetric functions*. Since they form a system of nonlinear algebraic equations in multiple variables, they are somewhat difficult to solve in a direct manner. Peterson showed, however, that the BCH syndrome equations can be translated into a series of linear equations that are much easier to work with [Pet2]. Let $\Lambda(x)$ be the *error locator polynomial* that has as its roots the inverses of the ν error locators $\{X_l\}$.

$$\Lambda(x) = \prod_{l=1}^{\nu}(1 - X_l x) = \Lambda_\nu x^\nu + \Lambda_{\nu-1}x^{\nu-1} + \cdots + \Lambda_1 x + \Lambda_0 \tag{9-4}$$

Equation (9-4) can be used to express the coefficients of $\Lambda(x)$ directly in terms of the $\{X_l\}$.

$$\begin{aligned}
\Lambda_0 &= 1\\
\Lambda_1 &= \sum_{i=1}^{\nu} X_i = X_1 + X_2 + \cdots + X_{\nu-1} + X_\nu\\
\Lambda_2 &= \sum_{i<j} X_i X_j = X_1X_2 + X_1X_3 + \cdots + X_{\nu-2}X_\nu + X_{\nu-1}X_\nu\\
\Lambda_3 &= \sum_{i<j<k} X_iX_jX_k = X_1X_2X_3 + X_1X_2X_4 + \cdots + X_{\nu-3}X_{\nu-1}X_\nu + X_{\nu-2}X_{\nu-1}X_\nu\\
&\ \ \vdots\\
\Lambda_\nu &= \prod_{i=1}^{\nu} X_i = X_1X_2\cdots X_{\nu-1}X_\nu
\end{aligned} \tag{9-5}$$

The expressions in Eq. (9-5) are the *elementary symmetric functions* of the error locators. Power-sum symmetric functions and elementary symmetric functions are related by *Newton's identities* [Rio], which are generally expressed as follows for polynomials over arbitrary fields.

$$\begin{aligned}
S_1 + \Lambda_1 &= 0\\
S_2 + \Lambda_1 S_1 + 2\Lambda_2 &= 0\\
S_3 + \Lambda_1 S_2 + \Lambda_2 S_1 + 3\Lambda_3 &= 0\\
&\ \ \vdots\\
S_\nu + \Lambda_1 S_{\nu-1} + \Lambda_2 S_{\nu-2} + \cdots + \Lambda_{\nu-1}S_1 + \nu\Lambda_\nu &= 0\\
S_{\nu+1} + \Lambda_1 S_\nu + \Lambda_2 S_{\nu-1} + \cdots + \Lambda_\nu S_1 &= 0\\
&\ \ \vdots\\
S_{2t} + \Lambda_1 S_{2t-1} + \Lambda_2 S_{2t-2} + \cdots + \Lambda_\nu S_{2t-\nu} &= 0
\end{aligned} \tag{9-6}$$

Newton's identities are linear in the ν unknown coefficients of the error locator polynomial. Since we are working in a field of characteristic 2, Eq. (9-6) can be greatly simplified. First note that $j\Lambda_j = \Lambda_j$ if j is odd, and $j\Lambda_j = 0$ if j is even. The syndromes for the binary case have some additional useful structure. Recall the following lemma from Chapter 3.

Lemma 3-2

Let $\alpha_1, \alpha_2, \ldots, \alpha_t$ be elements in the field GF(p^m). $(\alpha_1 + \alpha_2 + \cdots + \alpha_t)^{p^r} = \alpha_1^{p^r} + \alpha_2^{p^r} + \cdots + \alpha_t^{p^r}$ for $r = 1, 2, 3, \ldots$.

The proof is in Chapter 3. Applying the lemma to the syndromes as defined in Eq. (9-2), we obtain

$$S_{2j} = \sum_{l=1}^{\nu} X_l^{2j} = \left(\sum_{l=1}^{\nu} X_l^j\right)^2 = S_j^2$$

The syndrome sequence for binary codes is thus highly constrained, with even-indexed syndromes being the squares of earlier-indexed syndromes. Given this constraint, we need not make use of all of Newton's identities in order to obtain the coefficients of the error locator polynomial.

Assume that $\nu = t$ errors have occurred, where t is the error correcting capability of the code. Newton's identities can be reduced to a system of t equations in t unknowns, as shown below.

$$\begin{aligned} S_1 + \Lambda_1 &= 0 \\ S_3 + \Lambda_1 S_2 + \Lambda_2 S_1 + \Lambda_3 &= 0 \\ S_5 + \Lambda_1 S_4 + \Lambda_2 S_3 + \Lambda_3 S_2 + \Lambda_4 S_1 + \Lambda_5 &= 0 \\ &\vdots \\ S_{2t-1} + \Lambda_1 S_{2t-2} + \Lambda_2 S_{2t-3} + \cdots + \Lambda_t S_{t-1} &= 0 \end{aligned} \tag{9-7}$$

In the next two subsections we investigate two techniques for solving the equations in Eq. (9-7) for the coefficients of the locator polynomial $\Lambda(x)$. The first is due to Peterson [Pet2], the second to Berlekamp [Ber1].

9.1.1 Peterson's Direct-Solution Decoding Algorithm

Equation (9-7) can be expressed in matrix form as follows.

$$\mathbf{A\Lambda} = \begin{bmatrix} 1 & 0 & 0 & 0 & \cdots & 0 & 0 \\ S_2 & S_1 & 1 & 0 & \cdots & 0 & 0 \\ S_4 & S_3 & S_2 & S_1 & \cdots & 0 & 0 \\ S_6 & S_5 & S_4 & S_3 & \cdots & 0 & 0 \\ \vdots & \vdots & \vdots & \vdots & \ddots & \vdots & \vdots \\ S_{2t-4} & S_{2t-5} & S_{2t-6} & S_{2t-7} & \cdots & S_{t-2} & S_{t-3} \\ S_{2t-2} & S_{2t-3} & S_{2t-4} & S_{2t-5} & \cdots & S_t & S_{t-1} \end{bmatrix} \begin{bmatrix} \Lambda_1 \\ \Lambda_2 \\ \Lambda_3 \\ \Lambda_4 \\ \vdots \\ \Lambda_{t-1} \\ \Lambda_t \end{bmatrix} = \begin{bmatrix} -S_1 \\ -S_3 \\ -S_5 \\ -S_7 \\ \vdots \\ -S_{2t-3} \\ -S_{2t-1} \end{bmatrix} \tag{9-8}$$

The system in Eq. (9-8) has a unique solution if and only if $\mathbf{A}$ is nonsingular, and thus has a nonzero determinant. Given that $\mathbf{A}$ is nonsingular, we can solve for $\mathbf{\Lambda}$ using the standard techniques of linear algebra. Peterson showed that, owing to the constraints on the syndrome sequence when the code is binary, $\mathbf{A}$ has a nonzero determinant if there are t or $t - 1$ errors in the received word [Pet2]. The general result is as follows.

Theorem 9-1

The dimension of the null space of the linear transformation defined by A in Eq. (9-8) is

$$\textit{Nullity}\ \mathbf{A} = \left\| \left\lfloor \frac{t - \deg[\Lambda(x)]}{2} \right\rfloor \right\|$$

Proof. See [Ber1, Sec. 7.6] for a sketch of the proof.

If fewer than $t - 1$ errors have occurred, we eliminate the two bottom rows and the two rightmost columns of **A** and check to see if the remaining matrix is nonsingular. Continue until the remaining matrix is nonsingular.

It should be noted that though there may be a solution to Eq. (9-8), it may not lead to the correct error locator polynomial. There are two other possibilities.

- If the received word is within Hamming distance t of an incorrect code word, $\Lambda(x)$ will "correct" the received word onto the incorrect code word. This condition is undetectable, and is called a *decoder error*.
- If the received word is not within distance t of any code word, the error locator polynomial may have repeated roots, or roots that do not lie in the smallest field containing the primitive nth root of unity used to construct the code. This condition is detectable and is used to trigger the declaration of a *decoder failure*.

There are some cases in which a valid error locator polynomial can be constructed that corrects more than t errors, but such cases are regrettably rare. A received word **r** is decoded as follows.

Peterson's direct-solution decoding algorithm for a binary *t*-error-correcting BCH code

1. Compute the syndromes for **r**: $\{S_j\} = \{r(\alpha^j)\}, j = 1, 2, 3, \ldots, 2t$.
2. Construct the syndrome matrix **A** in Eq. (9-8).
3. Compute the determinant of the syndrome matrix. If the determinant is nonzero, go to step 5.
4. Construct a new syndrome matrix by deleting the two rightmost columns and the two bottom rows from the old syndrome matrix. Go to step 3.
5. Solve for $\mathbf{\Lambda}$ and construct $\Lambda(x)$.
6. Find the roots of $\Lambda(x)$. If the roots are not distinct or $\Lambda(x)$ does not have roots in the desired field, go to step 9.
7. Complement the bit positions in **r** indicated by $\Lambda(x)$. If the number of errors corrected is less than t, verify that the resulting corrected word satisfies all $2t$ syndrome equations. If it does not, got to step 9.

8. Output the corrected word and STOP.
9. Declare a decoding failure and STOP.

For simple cases we can solve Eq. (9-8) and express the coefficients of $\Lambda(x)$ as general functions of the syndromes [Mic].

Single-Error Correction

$$\Lambda_1 = S_1 \tag{9-9}$$

Double-Error Correction

$$\begin{aligned} \Lambda_1 &= S_1 \\ \Lambda_2 &= \frac{S_3 + S_1^3}{S_1} \end{aligned} \tag{9-10}$$

Triple-Error Correction

$$\begin{aligned} \Lambda_1 &= S_1 \\ \Lambda_2 &= \frac{S_1^2 S_3 + S_5}{S_1^3 + S_3} \\ \Lambda_3 &= (S_1^3 + S_3) + S_1 \Lambda_2 \end{aligned} \tag{9-11}$$

Quadruple-Error Correction

$$\begin{aligned} \Lambda_1 &= S_1 \\ \Lambda_2 &= \frac{S_1(S_7 + S_1^7) + S_3(S_1^5 + S_5)}{S_3(S_1^3 + S_3) + S_1(S_1^5 + S_5)} \\ \Lambda_3 &= (S_1^3 + S_3) + S_1 \Lambda_2 \\ \Lambda_4 &= \frac{(S_5 + S_1^2 S_3) + (S_1^3 + S_3)\Lambda_2}{S_1} \end{aligned} \tag{9-12}$$

Quintuple-Error Correction

$$\begin{aligned} \Lambda_1 &= S_1 \\ \Lambda_2 &= \frac{(S_1^3+S_3)[(S_1^9+S_9)+S_1^4(S_5+S_1^2 S_3)+S_3^2(S_1^3+S_3)]+[(S_1^5+S_5)(S_7+S_1^7)+S_1(S_3^2+S_1 S_5)]}{(S_1^3+S_3)[(S_7+S_1^7)+S_1 S_3(S_1^3+S_3)]+(S_5+S_1^2 S_3)(S_1^5+S_5)} \\ \Lambda_3 &= (S_1^3 + S_3) + S_1 \Lambda_2 \\ \Lambda_4 &= \frac{(S_1^9 + S_9) + S_3^2(S_1^3 + S_3) + S_1^4(S_5 + S_1^2 S_3) + [(S_7 + S_1^7) + S_1 S_3(S_1^3 + S_3)]\Lambda_2}{(S_1^5 + S_5)} \\ \Lambda_5 &= (S_5 + S_1^2 S_3) + S_1 \Lambda_4 + (S_1^3 + S_3)\Lambda_2 \end{aligned} \tag{9-13}$$

Once the error locator polynomial is known, the roots can be located through the use of the *Chien search* [Chi]. The Chien search is a systematic means of evaluating the error locator polynomial at all elements in a field GF(2^m). A Chien search circuit is shown in Figure 9-1. Each coefficient Λ_i of the error locator polynomial is repeatedly multiplied by α^i, where α is primitive in GF(2^m). Each set of products is then summed to obtain $A_i = \Lambda(\alpha^i) - 1$. If α^i is a root of $\Lambda(x)$, then $A_i = \Lambda(\alpha^i) - 1 = 1$ and an error is indicated at the coordinate associated with $\alpha^{-i} = \alpha^{n-i}$. The correction bit B_i is then set to 1 and added to the received bit r_{n-i}. If $A_i \neq 1$, then B_i is set to zero and the corresponding received bit is left unchanged. Note that verification is performed after error correction to ensure that the resulting word is a code word. Code word verification can be performed by using a circuit similar to that used in the Chien search, as shown in Figure 9-2. The bits $c_0, c_1, \ldots, c_{n-1}$ of the corrected word take the place of the error locator polynomial coefficients. The syndromes $\{S_i'\}$ for the corrected word are computed one at a time. If they are all zero, the corrected word is accepted as a valid code word; otherwise, a decoder failure is indicated.

Example 9-1 is an example of direct decoding using a length-31 double-error-correcting BCH code.

Example 9-1—Double-Error Correction Using Peterson's Algorithm

In Example 8-1 we examined several BCH codes of length 31. Consider the narrow-sense double-error-correcting code of length 31 with generator polynomial $= 1 + x^3 + x^5 + x^6 + x^8 + x^9 + x^{10}$. The roots of this polynomial include four consecutive powers

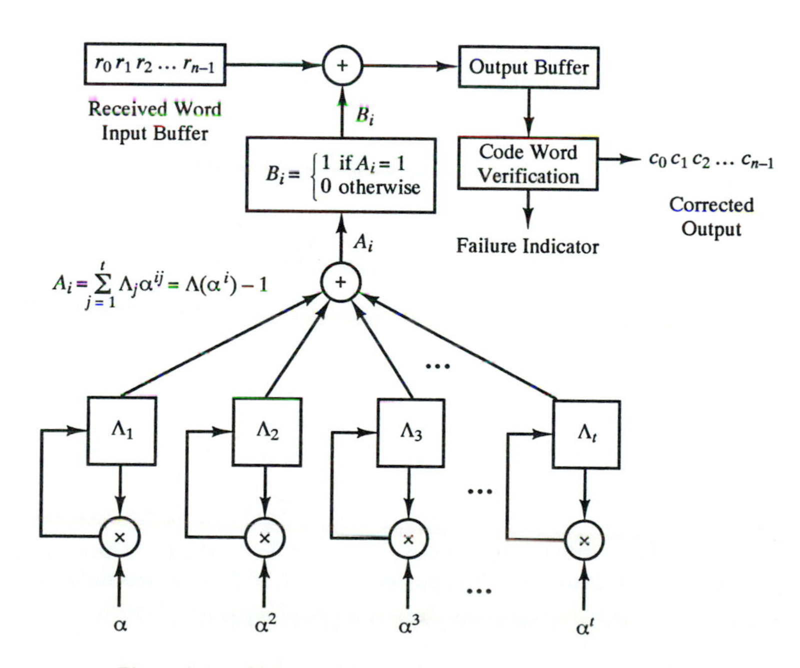

Figure 9-1. Chien Search and Error Correction Hardware

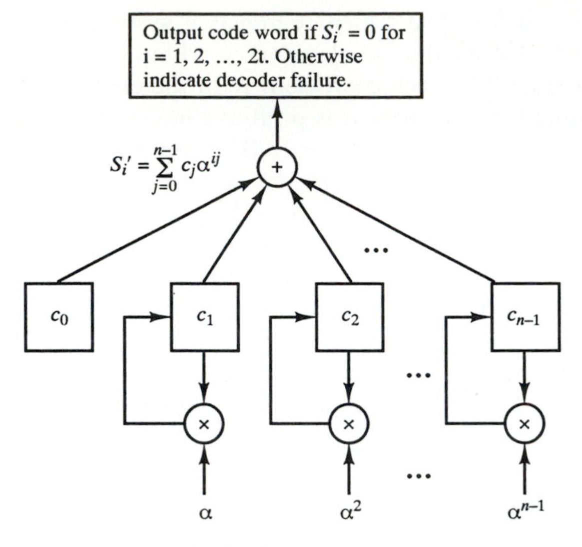

Figure 9-2. Code Word Verification Circuit

of α: $\{\alpha, \alpha^2, \alpha^3, \alpha^4\}$, where α is primitive in GF(32). The following binary vector is received.

$$\mathbf{r} = (0010000110011000000000000000000)$$

$$\updownarrow$$

$$r(x) = x^2 + x^7 + x^8 + x^{11} + x^{12}$$

The syndrome computations are performed using the representation for GF(32) in Appendix B, giving the following syndromes. Note that $S_4 = (S_2)^2 = (S_1)^4$.

$$S_1 = r(\alpha) = \alpha^7$$

$$S_2 = r(\alpha^2) = \alpha^{14}$$

$$S_3 = r(\alpha^3) = \alpha^8$$

$$S_4 = r(\alpha^4) = \alpha^{28}$$

Using Eq. (9-10), the following error locator polynomial is obtained.

$$\Lambda_1 = \alpha^7$$

$$\Lambda_2 = \frac{\alpha^8 + (\alpha^7)^3}{\alpha^7} = \alpha^{15}$$

$$\Lambda(x) = 1 + \alpha^7 x + \alpha^{15} x^2$$

$$= (1 + \alpha^5 x)(1 + \alpha^{10} x)$$

The error locators are $X_1 = \alpha^5$ and $X_2 = \alpha^{10}$, indicating errors at the fifth and tenth coordinates of **r**. The corrected word, with the corrected positions underlined, is

$$\mathbf{c} = (00100\underline{1}0110\underline{1}11000000000000000000)$$

$$\updownarrow$$

$$c(x) = x^2 + x^5 + x^7 + x^8 + x^{10} + x^{11} + x^{12} = x^2 g(x)$$ ■

Example 9-2—Single-Error Correction Using Peterson's Algorithm and a Triple-Error-Correcting Code

In this example we demonstrate what happens when the number of errors in a received word is less than the error correcting capability of the code. Let the code **C** be the triple-error-correcting narrow-sense binary BCH code with length 31. The generator polynomial

$$g(x) = 1 + x + x^2 + x^3 + x^5 + x^7 + x^8 + x^9 + x^{10} + x^{11} + x^{15}$$

has six consecutive roots $\{\alpha, \alpha^2, \alpha^3, \alpha^4, \alpha^5, \alpha^6\}$, where α is primitive in GF(32). Let the received polynomial be

$$r(x) = x^{10}$$

Since the received word has weight 1, the corresponding code word is clearly the all-zero code word. We now verify this using Peterson's technique. The syndromes are readily found to be

$$S_1 = \alpha^{10}, \quad S_2 = \alpha^{20}, \quad S_3 = \alpha^{30}, \quad S_4 = \alpha^9, \quad S_5 = \alpha^{19}, \quad S_6 = \alpha^{29}$$

When the matrix **A** in Eq. (9-8) is constructed, we find that **A** is singular (the third row is equal to the second row weighted by α^{20}).

$$\mathbf{A} = \begin{bmatrix} 1 & 0 & 0 \\ \alpha^{20} & \alpha^{10} & 1 \\ \alpha^9 & \alpha^{30} & \alpha^{20} \end{bmatrix}$$

Removing the second and third columns and rows, we obtain the matrix [1]. Equation (9-8) is thus reduced in this example to $\Lambda_1 = \alpha^{10}$, giving us the error locator $X_1 = \alpha^{10}$. The tenth received bit is thus in error, and the decoded code word is the all-zero word.

Had we attempted to use the triple-error-correction formulae in Eq. (9-11), we would have found that Λ_2 and Λ_3 were undefined (0/0). This result indicates that the matrix **A** is singular. ∎

Wolf has pointed out that Peterson's decoding technique is closely related to a curve-fitting algorithm described by Prony in 1795 [Wol]. That makes Peterson's algorithm a bit of a senior citizen in the world of digital error control, but it is still a mere youth in comparison to Euclid's algorithm, which is used later in the chapter to decode nonbinary BCH and Reed-Solomon codes.

9.1.2 Berlekamp's Algorithm for Binary BCH Codes

Berlekamp's algorithm is much more difficult to understand than Peterson's approach, but results in a substantially more efficient implementation. The complexity of Peterson's technique increases with the square of the number of errors corrected, providing an efficient implementation for binary BCH decoders that correct up to 6 or 7 errors. The complexity of Berlekamp's algorithm increases linearly, allowing for construction of efficient decoders that correct dozens of errors. In this section Berlekamp's decoding algorithm for binary BCH codes is presented without proof. The reader who is interested in the full treatment is referred to Berlekamp [Ber1].

We begin by defining an infinite-degree syndrome polynomial

$$S(x) = S_1 x + S_2 x^2 + \cdots + S_{2t} x^{2t} + S_{2t+1} x^{2t+1} + \cdots \tag{9-14}$$

Clearly we do not know all of the coefficients of $S(x)$, but fortunately the first $2t$ coefficients are entirely sufficient. $S(x)$ is made into an infinite-degree polynomial so that it can be treated as a **generating function**. Define a third polynomial as follows.

$$\begin{aligned}\Omega(x) &\triangleq [1 + S(x)]\Lambda(x)\\ &= (1 + S_1 x + S_2 x^2 + \cdots)(1 + \Lambda_1 x + \Lambda_2 x^2 + \cdots)\\ &= 1 + (S_1 + \Lambda_1)x + (S_2 + \Lambda_1 S_1 + \Lambda_2)x^2 + (S_3 + \Lambda_1 S_2 + \Lambda_2 S_1 + \Lambda_3)x^3 + \cdots\\ &= 1 + \Omega_1 x + \Omega_2 x^2 + \cdots\end{aligned} \tag{9-15}$$

$\Omega(x)$ is called the **error magnitude polynomial** and is useful in nonbinary decoding. For now we will simply note that if the syndrome and error locator polynomials are to satisfy Eq. (9-7), then the odd-indexed coefficients of $\Omega(x)$ must be zero. Given that we know only the first $2t$ coefficients of $S(x)$, the decoding problem then becomes one of finding a polynomial $\Lambda(x)$ of degree less than or equal to t that satisfies

$$[1 + S(x)]\Lambda(x) \equiv (1 + \Omega_2 x^2 + \Omega_4 x^4 + \cdots + \Omega_{2t} x^{2t}) \; mod\, x^{2t+1} \tag{9-16}$$

Berlekamp's algorithm proceeds iteratively by breaking Eq. (9-16) down into a series of smaller problems of the form

$$[1 + S(x)]\Lambda^{(2k)}(x) \equiv (1 + \Omega_2 x^2 + \Omega_4 x^4 + \cdots + \Omega_{2k} x^{2k}) \; mod\, x^{2k+1} \tag{9-17}$$

where k runs from 1 to t. A solution $\Lambda^{(0)}(x) = 1$ is first assumed and tested to see if it works for the case $k = 1$. If it does work, we proceed to $k = 2$; otherwise, a correction factor is computed and added to $\Lambda^{(0)}$, creating a new solution $\Lambda^{(2)}(x)$. The genius of the algorithm lies in the computation of the correction factor. It is designed so that the new solution will work not only for the current case, but for all previous values of k as well. Once the algorithm concludes, the polynomial $\Lambda^{(2t)}(x)$ is a solution for all t of the expressions in Eq. (9-7).

Berlekamp's algorithm for decoding binary BCH codes [Ber1]

1. Set the initial conditions: $k = 0, \Lambda^{(0)}(x) = 1, T^{(0)} = 1$.
2. Let $\Delta^{(2k)}$ be the coefficient of x^{2k+1} in the product $\Lambda^{(2k)}(x)[1 + S(x)]$.
3. Compute

$$\Lambda^{(2k+2)}(x) = \Lambda^{(2k)}(x) + \Delta^{(2k)}[x \cdot T^{(2k)}(x)]$$

4. Compute

$$T^{(2k+2)}(x) = \begin{cases} x^2 T^{(2k)}(x) & \text{if } \Delta^{(2k)} = 0 \text{ or if } \deg[\Lambda^{(2k)}(x)] > k \\ \dfrac{x\Lambda^{(2k)}(x)}{\Delta^{(2k)}} & \text{if } \Delta^{(2k)} \neq 0 \text{ and } \deg[\Lambda^{(2k)}(x)] \leq k \end{cases}$$

5. Set $k = k + 1$. If $k < t$, then go to 2.
6. Determine the roots of $\Lambda(x) = \Lambda^{(2t)}(x)$. If the roots are distinct and lie in the right field, then correct the corresponding locations in the received word and STOP.
7. Declare a decoding failure and STOP.

Example 9-3—Double-Error Correction Using Berlekamp's Algorithm

In Example 9-1 we considered a narrow-sense double-error-correcting code of length 31 with generator polynomial $= 1 + x^3 + x^5 + x^6 + x^8 + x^9 + x^{10}$. Let's repeat the example, but this time use Berlekamp's algorithm. The binary vector and associated polynomial

$$\mathbf{r} = (0010000110011000000000000000000)$$

$$\updownarrow$$

$$r(x) = x^2 + x^7 + x^8 + x^{11} + x^{12}$$

provide the following syndrome polynomial.

$$S(x) = \alpha^7 x + \alpha^{14} x^2 + \alpha^8 x^3 + \alpha^{28} x^4$$

Applying Berlekamp's algorithm, we obtain the following sequence of solutions to Eq. (9-17).

k	$\Lambda^{(2k)}(x)$	$T^{(2k)}(x)$	$\Delta^{(2k)}$
0	1	1	α^7
1	$1 + \alpha^7 x$	$\alpha^{24} x$	α^{22}
2	$1 + \alpha^7 x + \alpha^{15} x^2$	—	—

$\Lambda^{(4)}(x) = 1 + \alpha^7 x + \alpha^{15} x^2$ is, of course, the same error locator polynomial obtained in Example 9-1. ■

Example 9-4—Triple-Error Correction Using Berlekamp's Algorithm

As in Example 9-2, we let the code **C** be the triple-error-correcting narrow-sense binary BCH code with length 31. The generator polynomial

$$g(x) = 1 + x + x^2 + x^3 + x^5 + x^7 + x^8 + x^9 + x^{10} + x^{11} + x^{15}$$

has six consecutive roots $\{\alpha, \alpha^2, \alpha^3, \alpha^4, \alpha^5, \alpha^6\}$, where α is primitive in GF(32). Let the received polynomial be

$$r(x) = 1 + x^9 + x^{11} + x^{14}$$

With a bit of effort the syndrome polynomial is seen to be

$$S(x) = x + x^2 + \alpha^{29} x^3 + x^4 + \alpha^{23} x^5 + \alpha^{27} x^6$$

Berlekamp's algorithm then proceeds as follows.

k	$\Lambda^{(2k)}(x)$	$T^{(2k)}(x)$	$\Delta^{(2k)}$
0	1	1	1
1	$1 + x$	x	α^3
2	$1 + x + \alpha^3 x^2$	$\alpha^{28} x + \alpha^{28} x^2$	α^{20}
3	$1 + x + \alpha^{16} x^2 + \alpha^{17} x^3$	—	—

The error locator polynomial is then

$$\Lambda(x) = 1 + x + \alpha^{16}x^2 + \alpha^{17}x^3$$
$$= (1 + \alpha^{13}x)(1 + \alpha^{16}x)(1 + \alpha^{19}x)$$

indicating errors at the positions corresponding to α^{13}, α^{16}, and α^{19}. The corrected received word is then

$$r(x) = 1 + x^9 + x^{11} + x^{13} + x^{14} + x^{16} + x^{19} = (x^4 + x + 1)g(x)$$ ■

9.2 DECODING ALGORITHMS FOR NONBINARY BCH AND REED-SOLOMON CODES

We now consider the decoding of nonbinary BCH and Reed-Solomon codes. In such cases we have to determine not only the locations of the errors, but also their magnitude. We begin by generalizing Peterson's and Berlekamp's algorithms for the nonbinary case. In the latter case Massey's shift-register synthesis interpretation is used. This section concludes with a demonstration of the decoding of nonbinary BCH and Reed-Solomon codes using Euclid's algorithm.

9.2.1 Peterson-Gorenstein-Zierler Decoding Algorithm

Gorenstein and Zierler generalized Peterson's decoding algorithm for use with nonbinary codes [Gor1]. Note that the syndromes are now a function of the magnitude of the errors as well as their locations. Assuming that some ν errors have corrupted the received word, the syndromes are as follows.

$$S_j = e(\alpha^j) = \sum_{k=0}^{n-1} e_k(\alpha^j)^k = \sum_{l=1}^{\nu} e_{i_l}X_l^j \tag{9-18}$$

Equation (9-18) defines a series of $2t$ algebraic equations in 2ν unknowns.

$$\begin{aligned} S_1 &= e_{i_1}X_1 + e_{i_2}X_2 + \cdots + e_{i_\nu}X_\nu \\ S_2 &= e_{i_1}X_1^2 + e_{i_2}X_2^2 + \cdots + e_{i_\nu}X_\nu^2 \\ S_3 &= e_{i_1}X_1^3 + e_{i_2}X_2^3 + \cdots + e_{i_\nu}X_\nu^3 \\ &\vdots \\ S_{2t} &= e_{i_1}X_1^{2t} + e_{i_2}X_2^{2t} + \cdots + e_{i_\nu}X_\nu^{2t} \end{aligned} \tag{9-19}$$

Unlike the binary case, the syndrome equations are not power-sum symmetric functions. It is still possible, however, to reduce the system in Eq. (9-19) to a set of linear functions in the unknown quantities. Once again we make use of an error locator polynomial $\Lambda(x)$ whose zeros are the inverses of the error locators $\{X_i\}$.

$$\Lambda(x) = \prod_{l=1}^{\nu}(1 - X_l x) = \Lambda_\nu x^\nu + \Lambda_{\nu-1}x^{\nu-1} + \cdots + \Lambda_1 x + \Lambda_0 \tag{9-20}$$

It follows immediately that for some error locator X_l,

$$\Lambda(X_l^{-1}) = \Lambda_\nu X_l^{-\nu} + \Lambda_{\nu-1} X_l^{-\nu+1} + \cdots + \Lambda_1 X_l^{-1} + \Lambda_0 = 0 \tag{9-21}$$

Since the expression sums to zero, we can multiply through by a constant.

$$\begin{aligned} e_{i_l} X_l^j (\Lambda_\nu X_l^{-\nu} + \Lambda_{\nu-1} X_l^{-\nu+1} + \cdots + \Lambda_1 X_l^{-1} + \Lambda_0) \\ = e_{i_l}(\Lambda_\nu X_l^{-\nu+j} + \Lambda_{\nu-1} X_l^{-\nu+j+1} + \cdots + \Lambda_1 X_l^{j-1} + \Lambda_0 X_l^j) = 0 \end{aligned} \tag{9-22}$$

Sum Eq. (9-22) over all indices l, obtaining an expression from which Newton's identities can be constructed.

$$\begin{aligned} &\sum_{l=1}^{\nu} e_{i_l}(\Lambda_\nu X_l^{j-\nu} + \Lambda_{\nu-1} X_l^{j-\nu+1} + \cdots + \Lambda_1 X_l^{j-1} + \Lambda_0 X_l^j) \\ &= \Lambda_\nu \sum_{l=1}^{\nu} e_{i_l} X_l^{j-\nu} + \Lambda_{\nu-1} \sum_{l=1}^{\nu} e_{i_l} X_l^{j-\nu+1} + \cdots + \Lambda_1 \sum_{l=1}^{\nu} e_{i_l} X_l^{j-1} + \Lambda_0 \sum_{l=1}^{\nu} e_{i_l} X_l^j \\ &= \Lambda_\nu S_{j-\nu} + \Lambda_{\nu-1} S_{j-\nu+1} + \cdots + \Lambda_1 S_{j-1} + \Lambda_0 S_j = 0 \end{aligned} \tag{9-23}$$

From Eq. (9-20) it is clear that Λ_0 is always one. Equation (9-23) can thus be reexpressed as

$$\Lambda_\nu S_{j-\nu} + \Lambda_{\nu-1} S_{j-\nu+1} + \cdots + \Lambda_1 S_{j-1} = -S_j \tag{9-24}$$

If we assume that $\nu = t$, we obtain the following matrix equation.

$$\mathbf{A}'\mathbf{\Lambda} = \begin{bmatrix} S_1 & S_2 & S_3 & S_4 & \cdots & S_{t-1} & S_t \\ S_2 & S_3 & S_4 & S_5 & \cdots & S_t & S_{t+1} \\ S_3 & S_4 & S_5 & S_6 & \cdots & S_{t+1} & S_{t+2} \\ S_4 & S_5 & S_6 & S_7 & \cdots & S_{t+2} & S_{t+3} \\ \vdots & \vdots & \vdots & \vdots & \ddots & \vdots & \vdots \\ S_{t-1} & S_t & S_{t+1} & S_{t+2} & \cdots & S_{2t-3} & S_{2t-2} \\ S_t & S_{t+1} & S_{t+2} & S_{t+3} & \cdots & S_{2t-2} & S_{2t-1} \end{bmatrix} \begin{bmatrix} \Lambda_t \\ \Lambda_{t-1} \\ \Lambda_{t-2} \\ \Lambda_{t-3} \\ \vdots \\ \Lambda_2 \\ \Lambda_1 \end{bmatrix} = \begin{bmatrix} -S_{t+1} \\ -S_{t+2} \\ -S_{t+3} \\ -S_{t+4} \\ \vdots \\ -S_{2t-1} \\ -S_{2t} \end{bmatrix} \tag{9-25}$$

It can be shown that $\mathbf{A}'$ is nonsingular if the received word contains exactly t errors. It can also be shown that $\mathbf{A}'$ is singular if fewer than t errors have occurred (see, for example, [Bla2, Ch. 7]). If $\mathbf{A}'$ is singular, then the rightmost column and the bottom row are removed and the determinant of the resulting matrix computed. This process is repeated until the resulting matrix is nonsingular. The coefficients of the error locator polynomial are then found through the use of the standard linear algebraic techniques (with computations performed in GF(2^m)).

Once the ν error locations are known, Eq. (9-19) becomes a system of 2ν equations in ν unknowns (the error magnitudes). This system can be reduced to form the matrix relation below. Since the $\{X_i\}$ are nonzero and distinct, the matrix $\mathbf{B}$ is Vandermonde and thus nonsingular (see the proof of Theorem 8-1).

$$\mathbf{B}e = \begin{bmatrix} X_1 & X_2 & \cdots & X_\nu \\ X_1^2 & X_2^2 & \cdots & X_\nu^2 \\ \vdots & \vdots & \ddots & \vdots \\ X_1^\nu & X_2^\nu & \cdots & X_\nu^\nu \end{bmatrix} \begin{bmatrix} e_{i_1} \\ e_{i_2} \\ \vdots \\ e_{i_\nu} \end{bmatrix} = \begin{bmatrix} S_1 \\ S_2 \\ \vdots \\ S_\nu \end{bmatrix} \tag{9-26}$$

Decoding is completed by solving for the $\{e_{i_j}\}$.

Peterson-Gorenstein-Zierler decoding algorithm

1. Compute the syndromes for **r**: $\{S_l\} = \{r(\alpha^l)\}, l = 1,2,3,\ldots,2t$.
2. Construct the syndrome matrix $\mathbf{A}'$ in Eq. (9-25).
3. Compute the determinant of the syndrome matrix. If the determinant is nonzero, go to 5.
4. Construct a new syndrome matrix by deleting the rightmost column and the bottom row from the old syndrome matrix. Shorten $\mathbf{\Lambda}$ by one coordinate position by deleting Λ_t for the largest remaining t. Go to 3.
5. Solve for $\mathbf{\Lambda}$ and construct $\Lambda(x)$.
6. Find the roots of $\Lambda(x)$. If the roots are not distinct or $\Lambda(x)$ does not have roots in the desired field, go to step 10.
7. Construct the matrix **B** in Eq. (9-26) and solve for the error magnitudes.
8. Subtract the error magnitudes from the values at the appropriate coordinates of the received word.
9. Output the corrected word and STOP.
10. Declare a decoding failure and STOP.

Example 9-5—Double-Error Correction Using the Peterson-Gorenstein-Zierler Decoding Algorithm and a (7, 3) Reed-Solomon Code

In Example 8-3 a double-error-correcting narrow-sense Reed-Solomon code of length 7 was constructed over GF(8) using the GF(8) representation in Appendix B and the generator polynomial

$$g(x) = (x - \alpha)(x - \alpha^2)(x - \alpha^3)(x - \alpha^4) = x^4 + \alpha^3 x^3 + x^2 + \alpha x + \alpha^3$$

Let the received polynomial be $r(x) = \alpha^2 x^6 + \alpha^2 x^4 + x^3 + \alpha^5 x^2$, giving the syndrome sequence $S_1 = \alpha^6, S_2 = \alpha^3, S_3 = \alpha^4, S_4 = \alpha^3$. Equation (9-25) reduces to the following system of two equations in two unknowns.

$$\mathbf{A}'\mathbf{\Lambda} = \begin{bmatrix} \alpha^6 & \alpha^3 \\ \alpha^3 & \alpha^4 \end{bmatrix} \begin{bmatrix} \Lambda_2 \\ \Lambda_1 \end{bmatrix} = \begin{bmatrix} \alpha^4 \\ \alpha^3 \end{bmatrix}$$

We verify by inspection that the matrix $\mathbf{A}'$ is nonsingular. Simple substitution then yields the following solution for the coefficients of the error locator polynomial.

$$\Lambda_1 = \alpha^2, \quad \Lambda_2 = \alpha$$

$$\Rightarrow \Lambda(x) = \alpha x^2 + \alpha^2 x + 1$$

$$= (1 + \alpha^3 x)(1 + \alpha^5 x)$$

The factorization of the error locator polynomial indicates that there are errors in the third and fifth positions of the received word.

Equation (9-26) provides the following system of equations for determining the error magnitudes.

$$\mathbf{B}e = \begin{bmatrix} \alpha^3 & \alpha^5 \\ \alpha^6 & \alpha^3 \end{bmatrix} \begin{bmatrix} e_3 \\ e_5 \end{bmatrix} = \begin{bmatrix} \alpha^6 \\ \alpha^3 \end{bmatrix}$$

Solving for the error magnitudes, we obtain the following error polynomial.

$$e(x) = e_{i_1} x^{\log_\alpha X_1} + e_{i_2} x^{\log_\alpha X_2} = \alpha x^3 + \alpha^5 x^5$$

The error polynomial is added to the received polynomial to obtain the closest code word. Remember that addition and subtraction are the same operation in fields of the form $GF(2^m)$.

$$\begin{aligned} c(x) &= r(x) + e(x) \\ &= (\alpha^2 x^6 + \alpha^2 x^4 + x^3 + \alpha^5 x^2) + (\alpha x^3 + \alpha^5 x^5) \\ &= \alpha^2 x^6 + \alpha^5 x^5 + \alpha^2 x^4 + \alpha^3 x^3 + \alpha^5 x^2 \\ &= \alpha^2 x^2 g(x) \end{aligned}$$

■

Example 9-6—Single-Error Correction Using the Peterson-Gorenstein-Zierler Decoding Algorithm and a (7, 3) Reed-Solomon Code

Here we consider an example in which the full error correcting capability of the code is not used. The double-error-correcting RS code from the previous example ($g(x) = x^4 + \alpha^3 x^3 + x^2 + \alpha x + \alpha^3$) is used to correct the received word

$$r(x) = x^4 + x^2 + \alpha x + \alpha^3$$

The syndromes are $S_1 = \alpha^6$, $S_2 = \alpha^2$, $S_3 = \alpha^5$, $S_4 = \alpha$, and the matrix $\mathbf{A}'$ in Eq. (9-25) is seen to have the following form.

$$\mathbf{A}' = \begin{bmatrix} \alpha^6 & \alpha^2 \\ \alpha^2 & \alpha^5 \end{bmatrix}$$

Since the second row is α^3 times the first row, the matrix is singular. When we remove the right column and bottom row from $\mathbf{A}'$, the system of equations in Eq. (9-25) reduces to the following.

$$[\alpha^6][\Lambda_1] = [\alpha^2]$$

Λ_1 is thus α^3, as is the error locator X_1. Equation (9-26) then reduces to

$$[\alpha^3][e_3] = [\alpha^6]$$

giving the error magnitude α^3. The error polynomial is thus $e(x) = \alpha^3 x^3$, and the code word corresponding to the received word is the generator polynomial itself. ■

9.2.2 The Berlekamp-Massey Algorithm

The Peterson-Gorenstein-Zierler decoding algorithm is a bit more complex than the Peterson algorithm (to obtain the error magnitudes, one must solve an additional system of equations). It is thus more limited in the number of errors it can correct before its implementation becomes too complex. Once again Berlekamp's algorithm offers a much more efficient alternative for the correction of large numbers of errors. The binary version in Section 9.1.2 is readily modified for use in the nonbinary case without a substantial increase in complexity. In this section Berlekamp's algorithm for decoding nonbinary BCH and Reed-Solomon codes is explained using Massey's shift-register-based interpretation [Mas4].

Equation (9-24) (repeated below as Eq. (9-27)) showed that the syndrome S_j can be expressed in recursive form as a function of the coefficients of the error locator polynomial $\Lambda(x)$ and the earlier syndromes $S_{j-1}, \ldots, S_{j-\nu}$.

$$\Lambda_\nu S_{j-\nu} + \Lambda_{\nu-1} S_{j-\nu+1} + \cdots + \Lambda_1 S_{j-1} = -S_j \tag{9-27}$$

Figure 9-3 shows that expressions of this form can be given a physical interpretation through the use of a linear feedback shift register (LFSR) [Gol1]. The double-lined elements denote storage of and operations on nonbinary field elements (see Fig. 5-1).

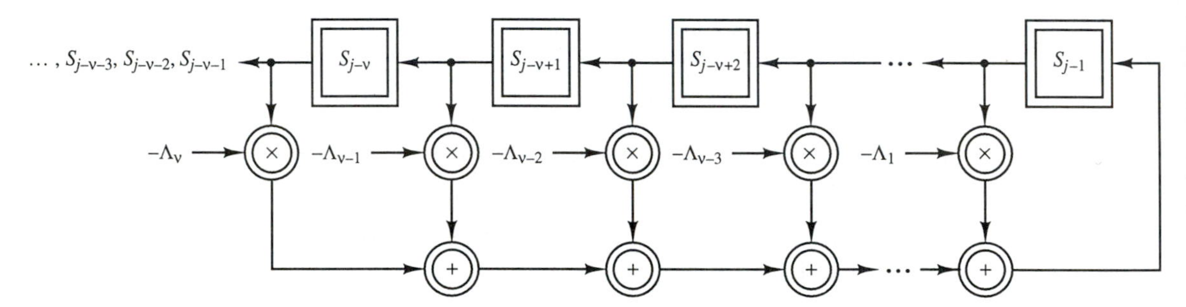

Figure 9-3. LFSR Interpretation of Eq. (9-27)

The problem of decoding BCH and Reed-Solomon codes can thus be re-expressed as follows: find an LFSR of minimal length such that the first $2t$ elements in the LFSR output sequence are the syndromes $S_1, S_2, \ldots, S_{2t}$. The taps of this shift register provide the desired error locator polynomial $\Lambda(x)$.

Let $\Lambda^{(k)}(x) = \Lambda_k x^k + \Lambda_{k-1} x^{k-1} + \cdots + \Lambda_1 x + 1$ be the *connection polynomial* of length k whose coefficients specify the taps of a length-k LFSR. The Berlekamp-Massey algorithm starts by finding $\Lambda^{(1)}$ such that the first element output by the corresponding LFSR is the first syndrome S_1. The second output of this LFSR is then compared to the second syndrome. If the two do not have the same value, then the *discrepancy* between the two is used to construct a modified connection polynomial. If there is no discrepancy, then the same connection polynomial is used to generate a third sequence element, which is compared to the third syndrome. The process continues until a connection polynomial is obtained that specifies an LFSR capable of generating all $2t$ elements of the syndrome sequence.

Massey showed that, given an error pattern of weight $\nu \leq t$, the connection polynomial resulting from the Berlekamp-Massey algorithm uniquely specifies the correct error locator polynomial. For detailed proofs of this assertion and the overall validity of the algorithm, the diligent reader is referred to [Mas4] and [Ber1].

The algorithm has five basic parameters: the connection polynomial $\Lambda^{(k)}(x)$, the correction polynomial $T(x)$, the discrepancy $\Delta^{(k)}$, the length L of the shift register, and the indexing variable k. The algorithm proceeds as follows.

The Berlekamp-Massey shift register synthesis decoding algorithm

1. Compute the syndrome sequence $S_1, \ldots, S_{2t}$ for the received word.
2. Initialize the algorithm variables as follows: $k = 0$, $\Lambda^{(0)}(x) = 1$, $L = 0$, and $T(x) = x$.
3. Set $k = k + 1$. Compute the discrepancy $\Delta^{(k)}$ by subtracting the kth output of the LFSR defined by $\Lambda^{(k-1)}(x)$ from the kth syndrome.
$$\Delta^{(k)} = S_k - \sum_{i=1}^{L} \Lambda_i^{(k-1)} S_{k-i}$$
4. If $\Delta^{(k)} = 0$, then go to step 8.
5. Modify the connection polynomial: $\Lambda^{(k)}(x) = \Lambda^{(k-1)}(x) - \Delta^{(k)} T(x)$.
6. If $2L \geq k$, then go to step 8.
7. Set $L = k - L$ and $T(x) = \Lambda^{(k-1)}(x)/\Delta^{(k)}$.
8. Set $T(x) = x \cdot T(x)$
9. If $k < 2t$, then go to step 3.
10. Determine the roots of $\Lambda(x) = \Lambda^{(2t)}(x)$. If the roots are distinct and lie in the right field, then determine the error magnitudes, correct the corresponding locations in the received word, and STOP.
11. Declare a decoding failure and STOP.

Example 9-7—Double-Error Correction Using the Berlekamp-Massey Algorithm and a (7, 3) Reed-Solomon Code: Part 1: Finding the Error Locator Polynomial

We repeat the decoding problem in Example 9-5, but this time we use the Berlekamp-Massey algorithm. Let the received polynomial be $r(x) = \alpha^2 x^6 + \alpha^2 x^4 + x^3 + \alpha^5 x^2$, giving the syndrome sequence $S_1 = \alpha^6, S_2 = \alpha^3, S_3 = \alpha^4, S_4 = \alpha^3$. The algorithm generates the following set of connection polynomials, discrepancies, and correction polynomials.

k	S_k	$\Lambda^{(k)}(x)$	$\Delta^{(k)}$	L	$T(x)$*
0	—	1	—	0	x
1	α^6	$1 + \alpha^6 x$	$S_1 - 0 = \alpha^6$	1	αx
2	α^3	$1 + \alpha^4 x$	$S_2 - \alpha^5 = \alpha^2$	1	αx^2
3	α^4	$1 + \alpha^4 x + \alpha^6 x^2$	$S_3 - 1 = \alpha^5$	2	$\alpha^2 x + \alpha^6 x^2$
4	α^3	$1 + \alpha^2 x + \alpha x^2$	$S_4 - \alpha^4 = \alpha^6$	—	—

* At the conclusion of step 8.

As before, we obtain the error locator polynomial $\Lambda(x) = 1 + \alpha^2 x + \alpha x^2$. The LFSRs corresponding to the connection polynomials $\Lambda^{(1)}(x)$ through $\Lambda^{(4)}(x)$ are drawn below, along with the initial conditions and the generated output sequence. Consider the LFSR with connection polynomial $\Lambda^{(3)}(x)$. This LFSR correctly generates the first three syndromes, but its fourth output, α^4, is not equal to $S_4 = \alpha^3$. The discrepancy $\Delta^{(4)} = \alpha^6$ is the difference between the two values, as shown in the chart. This discrepancy is used to determine the connection polynomial $\Lambda^{(4)}(x)$, which, as shown in the accompanying drawing, correctly generates all four syndromes.

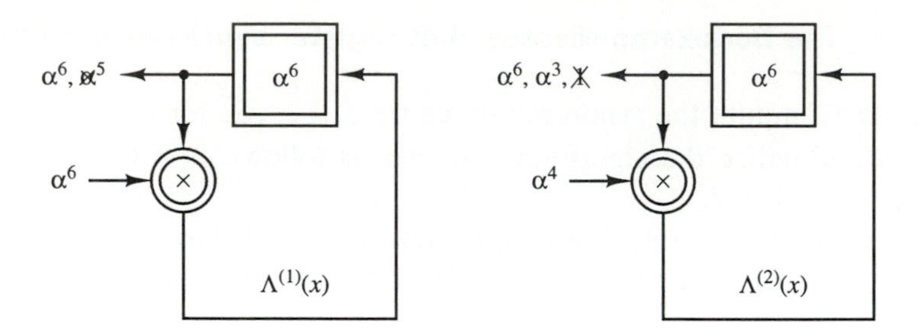

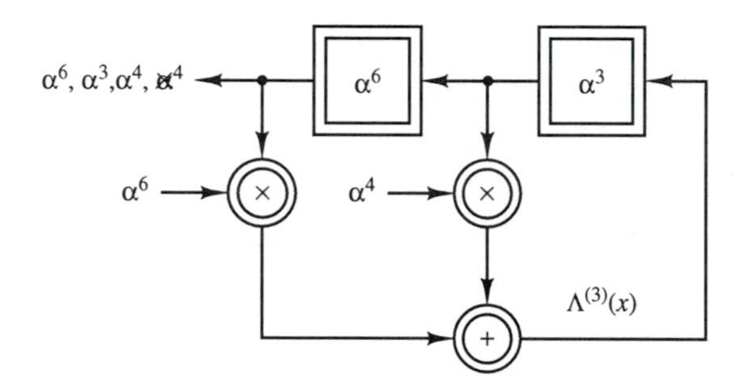

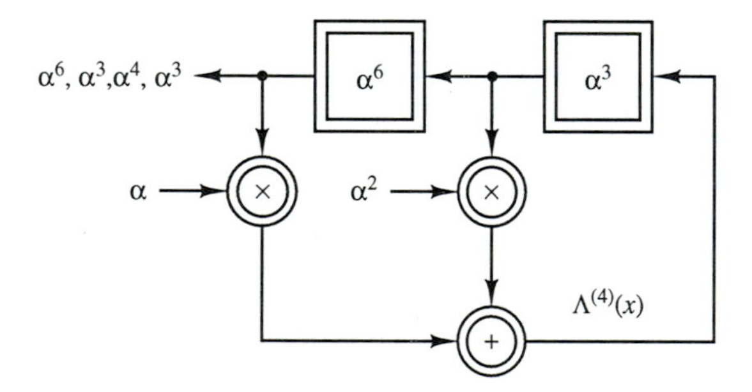

The Berlekamp-Massey algorithm allows us to find the error locator polynomial, but there remains the problem of finding the error magnitudes. To do so we reexpress the decoding problem once again, this time making full use of the syndrome polynomial as a generating function.

Let the error vector $\mathbf{e} = (e_0, e_1, e_2, \ldots, e_{n-1})$ corrupting the received word $\mathbf{r}$ have nonzero values at some ν coordinates $\{i_1, i_2, \ldots, i_\nu\}$. When the syndromes are computed as in Eq. (9-2) (we again assume a narrow-sense code), we obtain the following.

$$S_j = r(\alpha^j) = \sum_{i=0}^{n-1} e_i(\alpha^j)^i = \sum_{l=1}^{\nu} e_{i_l} X_l^j \tag{9-28}$$

These syndromes can be expressed in the form of an infinite-degree syndrome polynomial, then reduced to a simple rational expression.

$$\begin{aligned}1 + S(x) &= 1 + \sum_{j=1}^{\infty} S_j x^j \\ &= 1 + \sum_{j=1}^{\infty}\left(\sum_{l=1}^{\nu} e_{i_l} X_l^j\right)x^j \\ &= 1 + \sum_{l=1}^{\nu} e_{i_l} \sum_{j=1}^{\infty} (X_l x)^j \\ &= 1 + \sum_{l=1}^{\nu} e_{i_l}\left(\frac{X_l x}{1 - X_l x}\right)\end{aligned} \tag{9-29}$$

The error locator polynomial is defined as in Eq. (9-4) for the case of ν errors.

$$\Lambda(x) = \prod_{l=1}^{\nu}(1 - X_l x) \tag{9-30}$$

We now compute $\Omega(x) = \Lambda(x)[1 + S(x)]$ to obtain the following expression.

$$\begin{aligned}\Omega(x) &= \Lambda(x)[1 + S(x)] \\ &= \left[\prod_{j=1}^{\nu}(1 - X_j x)\right]\left[1 + \sum_{l=1}^{\nu} e_{i_l}\left(\frac{X_l x}{1 - X_l x}\right)\right] \\ &= \Lambda(x) + \sum_{l=1}^{\nu}\left[e_{i_l} X_l x \prod_{j\neq l}(1 - X_j x)\right]\end{aligned} \tag{9-31}$$

We know the coefficients of $S(x)$ only up through S_{2t}. Equation (9-31) is thus reduced to the following *key equation* that relates the known syndrome values to the error locator and error magnitude polynomials.

Key equation for BCH/RS decoding[1]

$$\Lambda(x)[1 + S(x)] \equiv \Omega(x) \bmod x^{2t+1} \tag{9-32}$$

In extracting the error magnitudes from $\Omega(x)$, we need to make use of an operation that looks like a derivative. However, the usual definition of a derivative cannot be applied to a finite field because we cannot apply limits. We shall therefore define a *formal* derivative that looks and acts like a derivative but does not have the corresponding interpretation.

Definition 9-1—Formal Derivatives

Let $f(x) = f_0 + f_1 x + f_2 x^2 + \cdots + f_n x^n + \cdots$ be a polynomial with coefficients in GF(q). The **formal derivative** $f'(x)$ is defined as follows.

$$f'(x) = f_1 + 2f_2 x + \cdots + nf_n x^{n-1} + \cdots$$

[1] The alternate form $\Lambda(x)S(x) \equiv \Omega(x) \bmod x^{2t+1}$ is also used.

Given this definition, the other properties that we normally associate with derivatives can be derived. For example, the reader is invited to prove the following.

Three properties of the formal derivative

- If $f^2(x)$ divides $g(x)$, then $f(x)$ divides $g'(x)$
- $[f(x)g(x)]' = f'(x)g(x) + f(x)g'(x)$
- If $f(x) \in GF(2^m)[x]$, then $f'(x)$ has no odd-powered terms.

We now express the error magnitudes in terms of $\Omega(x)$ and $\Lambda'(x)$ as follows.

Theorem 9-2—The Forney Algorithm [For7]

The error magnitudes are computed using the expression

$$e_{i_k} = \frac{-X_k\,\Omega(X_k^{-1})}{\Lambda'(X_k^{-1})}$$

Proof. When we take the formal derivative of the error locator polynomial and apply the second of the above properties of formal derivatives, we obtain

$$\Lambda'(x) = \left[\prod_{l=1}^{\nu}(1 - X_l x)\right]' = -\sum_{l=1}^{\nu}\left[X_l \prod_{j\neq l}(1 - X_j x)\right]$$

Substituting X_k^{-1} for x, we note that all of the products in the last expression above have a zero term except for the single case $l = k$.

$$\Lambda'(X_k^{-1}) = -\sum_{l=1}^{\nu}\left[X_l \prod_{j\neq l}(1 - X_j X_k^{-1})\right] = -X_k \prod_{j\neq k}(1 - X_j X_k^{-1})$$

Substituting X_k^{-1} for x in Eq. (9-31), we obtain

$$\Omega(X_k^{-1}) = \Lambda(X_k^{-1}) + \sum_{l=1}^{\nu}\left[e_{i_l} X_l X_k^{-1} \prod_{j\neq l}(1 - X_j X_k^{-1})\right] = e_{i_k} \prod_{j\neq k}(1 - X_j X_k^{-1})$$

and the result follows. **QED**

Example 9-7—Double-Error Correction Using the Berlekamp-Massey Algorithm and a (7, 3) Reed-Solomon Code: Part 2: Finding the Error Magnitudes

The syndrome sequence provides the syndrome polynomial $S(x) = \alpha^6 x + \alpha^3 x^2 + \alpha^4 x^3 + \alpha^3 x^4$. In the first part of this example we used the Berlekamp-Massey algorithm to obtain the error locator polynomial $\Lambda(x) = 1 + \alpha^2 x + \alpha x^2$. We can now compute the error magnitude polynomial.

$$\begin{aligned}\Omega(x) &\equiv \Lambda(x)[1 + S(x)] \bmod x^{2t+1} \\ &\equiv (1 + \alpha^2 x + \alpha x^2)(1 + \alpha^6 x + \alpha^3 x^2 + \alpha^4 x^3 + \alpha^3 x^4) \bmod x^5 \\ &\equiv (1 + x + \alpha^3 x^2) \bmod x^5\end{aligned}$$

The error locators are $X_1 = \alpha^3$ and $X_2 = \alpha^5$. Using the Forney algorithm, the error magnitudes are found to be

$$e_{i_k} = \frac{-X_k\,\Omega(X_k^{-1})}{\Lambda'(X_k^{-1})} = \frac{-X_k[1 + X_k^{-1} + \alpha^3 X_k^{-2}]}{\alpha^2} = \alpha^5 X_k + \alpha^5 + \alpha X_k^{-1}$$

$$e_3 = \alpha^5\alpha^3 + \alpha^5 + \alpha\alpha^4 = \alpha$$

$$e_5 = \alpha^5\alpha^5 + \alpha^5 + \alpha\alpha^2 = \alpha^5$$

The error polynomial is thus $e(x) = \alpha x^3 + \alpha^5 x^5$, as found previously in Example 9-5.

■

Example 9-8—Triple-Error Correction Using the Berlekamp-Massey Algorithm and a (31, 25) Reed-Solomon Code

Using the representation for GF(32) in Appendix B, the following generator polynomial for the narrow-sense (31, 25) Reed-Solomon code is obtained.

$$g(x) = \prod_{i=1}^{6}(x - \alpha^i) = x^6 + \alpha^{10}x^5 + \alpha^9 x^4 + \alpha^{24}x^3 + \alpha^{16}x^2 + \alpha^{24}x + \alpha^{21}$$

The received polynomial is

$$r(x) = \alpha^{23}x^6 + x^5 + \alpha^6 x^4 + \alpha^6 x^2 + \alpha^3 x$$

giving the syndrome sequence $S_1 = 0, S_2 = \alpha^3, S_3 = \alpha^{29}, S_4 = \alpha^{27}, S_5 = \alpha^8, S_6 = \alpha^{23}$. The Berlekamp-Massey algorithm generates the following series of connection polynomials.

k	S_k	$\Lambda^{(k)}(x)$	$\Delta^{(k)}$	L	$T(x)$*
0	—	1	—	0	x
1	0	1	0	0	x^2
2	α^3	$1 + \alpha^3 x^2$	α^3	2	$\alpha^{28}x$
3	α^{29}	$1 + \alpha^{26}x + \alpha^3 x^2$	α^{29}	2	$\alpha^{28}x^2$
4	α^{27}	$1 + \alpha^{26}x + \alpha^{19}x^2$	α^{15}	2	$\alpha^{28}x^3$
5	α^8	$1 + \alpha^{26}x + \alpha^{19}x^2 + \alpha^{24}x^3$	α^{27}	3	$\alpha^4 x + \alpha^{30}x^2 + \alpha^{23}x^3$
6	α^{23}	$1 + \alpha^{25}x + \alpha^{30}x^2 + \alpha^{15}x^3$	α^8		

* At the conclusion of step 8.

The error locator polynomial factors as follows, indicating the error locations $X_1 = \alpha^3$, $X_2 = \alpha^5$, and $X_3 = \alpha^7$.

$$\Lambda(x) = \Lambda^{(6)}(x) = 1 + \alpha^{25}x + \alpha^{30}x^2 + \alpha^{15}x^3 = (1 - \alpha^3 x)(1 - \alpha^5 x)(1 - \alpha^7 x)$$

The error magnitude polynomial is then computed.

$$\begin{aligned}\Omega(x) &\equiv \Lambda(x)[1 + S(x)] \bmod x^7 \\ &\equiv (1 + \alpha^{25}x + \alpha^{30}x^2 + \alpha^{15}x^3)(1 + \alpha^3 x^2 + \alpha^{29}x^3 + \alpha^{27}x^4 + \alpha^8 x^5 + \alpha^{23}x^6) \bmod x^7 \\ &\equiv (1 + \alpha^{25}x + \alpha^9 x^2) \bmod x^5\end{aligned}$$

Forney's algorithm provides the error magnitudes as follows.

$$e_{i_k} = \frac{-X_k\,\Omega(X_k^{-1})}{\Lambda'(X_k^{-1})} = \frac{-X_k[1 + \alpha^{25}X_k^{-1} + \alpha^9 X_k^{-2}]}{\alpha^{25} + \alpha^{15}X_k^{-2}} = \frac{[X_k + \alpha^{25} + \alpha^9 X_k^{-1}]}{\alpha^{25} + \alpha^{15}X_k^{-2}}$$

$$e_3 = \frac{\alpha^3 + \alpha^{25} + \alpha^9\alpha^{28}}{\alpha^{25} + \alpha^{15}\alpha^{25}} = \alpha^{29}$$

$$e_5 = \frac{\alpha^5 + \alpha^{25} + \alpha^9\alpha^{26}}{\alpha^{25} + \alpha^{15}\alpha^{21}} = \alpha^{7}$$

$$e_7 = \frac{\alpha^7 + \alpha^{25} + \alpha^9\alpha^{24}}{\alpha^{25} + \alpha^{15}\alpha^{17}} = \alpha^{13}$$

The error polynomial and the corresponding code word are thus

$$e(x) = \alpha^{13}x^7 + \alpha^7x^5 + \alpha^{29}x^3$$

$$c(x) = r(x) + e(x) = \alpha^{13}x^7 + \alpha^{23}x^6 + \alpha^{22}x^5 + \alpha^6x^4 + \alpha^{29}x^3 + \alpha^6x^2 + \alpha^3x = \alpha^{13}xg(x)$$

■

9.2.3 Euclid's Algorithm

Euclid's algorithm is a fast method for finding the greatest common divisor (GCD) of a collection of elements in a Euclidean domain. Theorem 3-1 showed that the GCD of a collection of elements can be expressed as the linear combination of those elements. The extended form of Euclid's algorithm was then introduced in Chapter 3 as a means of finding the coefficients for this linear combination.

The algorithm operates on two elements (a, b) at a time. Given the initial conditions $r_{-1} = a, r_0 = b, s_{-1} = 1, s_0 = 0, t_{-1} = 0, t_0 = 1$, it proceeds according to the following set of recursion relations.

$$r_i = r_{i-2} - q_i r_{i-1}, \qquad \text{where } r_i < r_{i-1}$$

$$s_i = s_{i-2} - q_i s_{i-1}$$

$$t_i = t_{i-2} - q_i t_{i-1}$$

The algorithm terminates when the remainder $r_n = 0$. The remainder r_{n-1} is then the GCD of a and b. The recursion relations insure that, at any given point in the algorithm, we have the relation

$$s_i a + t_i b = r_i$$

which brings us to our purpose for introducing it in the first place. The key equation for decoding BCH and Reed-Solomon codes (Eq. (9-32)) can be reexpressed as follows.

$$\begin{aligned} \Lambda(x)[1 + S(x)] &\equiv \Omega(x)\ mod\, x^{2t+1} \\ \Rightarrow \Theta(x)x^{2t+1} + \Lambda(x)[1 + S(x)] &= \Omega(x) \end{aligned} \tag{9-33}$$

If we use the extended form of Euclid's algorithm to determine the GCD of x^{2t+1} and $[1 + S(x)]$, we generate sets of solutions $(\Lambda^{(k)}(x), \Omega^{(k)}(x))$ that satisfy

$$s_k x^{2t+1} + t_k[1 + S(x)] = \Theta^{(k)}(x)x^{2t+1} + \Lambda^{(k)}(x)[1 + S(x)]$$
$$= \Omega^{(k)}(x)$$

We are not interested in determining $\Theta(x)$, but the solution pair $(\Lambda^{(k)}(x), \Omega^{(k)}(x))$ is of the greatest interest, as we know from the earlier sections. The particular solution that corresponds to the error locator and magnitude polynomials is obtained when $\Omega^{(k)}(x)$ has degree less than or equal to that of $\Lambda^{(k)}(x)$.

Euclid's algorithm nonbinary decoding

1. Compute the syndrome polynomial $S(x)$.
2. Set the following initial conditions: $r_{-1}(x) = x^{2t+1}, r_0(x) = 1 + S(x), t_{-1}(x) = 0, t_0(x) = 1$.
3. Using the extended algorithm, compute the successive remainders $r_i(x)$ and the corresponding $t_i(x)$ until the following stopping condition is reached: $\deg[r_i(x)] \leq t$.
4. Find the roots of $t_i(x) = \Lambda(x)$, thus determining the error locations.
5. Determine the magnitude of the errors.

Example 9-9—Double-Error Correction Using Euclid's Algorithm and a (7, 3) Reed-Solomon Code

We return once again to the problem in Examples 9-5 and 9-7, but this time we apply Euclid's algorithm to perform the decoding operation. Recall that the Reed-Solomon code in use is double-error-correcting, so we have $x^{2t+1} = x^5$ as the initial condition $\Omega^{(-1)}(x)$ and $[1 + S(x)] = 1 + \alpha^6 x + \alpha^3 x^2 + \alpha^4 x^3 + \alpha^3 x^4$ as $\Omega^{(0)}(x)$. The algorithm proceeds as follows.

k	r_i $\Omega^{(k)}(x)$	q_i $q_k(x)$	t_i $\Lambda^{(k)}(x)$
-1	x^5	—	0
0	$1 + \alpha^6 x + \alpha^3 x^2 + \alpha^4 x^3 + \alpha^3 x^4$	—	1
1	$\alpha^5 + x^2 + \alpha^6 x^3$	$\alpha^5 + \alpha^4 x$	$\alpha^5 + \alpha^4 x$
2	$1 + x + \alpha^3 x^2$	$\alpha^4 x$	$1 + \alpha^2 x + \alpha x^2$

$$(1 + \alpha^2 x + \alpha x^2)(1 + \alpha^6 x + \alpha^3 x^2 + \alpha^4 x^3 + \alpha^3 x^4) \equiv (1 + x + \alpha^3 x^2) \bmod x^5$$

$$\Rightarrow \Omega(x) = 1 + x + \alpha^3 x^2 \quad \text{and} \quad \Lambda(x) = 1 + \alpha^2 x + \alpha x^2$$ ■

Note that, in the above example, we are forced to work with the entire syndrome polynomial in the first computation. This was not the case with the Berlekamp-Massey algorithm in the first part of Example 9-7. The Berlekamp-Massey algorithm is in general more efficient than Euclid's algorithm, but the difference between the two is not as dramatic as that between the Euclidian approach and the direct-solution technique of Peterson et al. The principal benefit of the

Euclidean approach to decoding BCH and Reed-Solomon codes is the ease with which it is understood and applied.

9.3 FREQUENCY-DOMAIN DECODING

When we examine the BCH/Reed-Solomon decoding problem from the frequency-domain perspective, we end up with the same key equation, but the interpretation of the result and its subsequent treatment are quite different.

Theorem 8-13 showed that the zeros of a time-domain polynomial correspond to zero coordinates in the frequency spectrum of the polynomial. The requirement that a generator polynomial have as zeros some $2t$ consecutive powers of a primitive element α thus corresponds to a requirement that the spectrum of all code words have $2t$ consecutive zero coordinates.

Consider a received word $\mathbf{r} = (r_0, \ldots, r_{n-1})$. This received word can be expressed as the sum of the transmitted code word $\mathbf{c} = (c_0, \ldots, c_{n-1})$ and an error vector $\mathbf{e} = (e_0, \ldots, e_{n-1})$.

$$\mathbf{r} = \mathbf{c} + \mathbf{e}$$

Since the Galois field Fourier transform (GFFT) is linear, we have the following frequency-domain relationship.

$$\mathbf{R} = \mathbf{C} + \mathbf{E}$$

If we find the transform of the error vector, we can thus invert it, subtract the result from the received word, and recover the transmitted code word. The transform of the error vector is computed as follows.

$$E_j = \sum_{i=0}^{n-1} \alpha^{ij} e_i, \qquad j = 0, 1, \ldots, n-1$$

The first $2t$ coordinates of the transform of the error vector are the syndromes in Eq. (9-1). We need only determine the other $n - 2t$ coordinates of the transform to be able to recover the error vector itself.

Given the error locators $\{X_i\}$, let the error locator polynomial $\Lambda(x)$ be defined such that the error locators are the inverses of the zeros of $\Lambda(x)$

$$\Lambda(x) = \prod_{i=1}^{\nu} (1 - X_i x)$$

If we treat the coefficients of $\Lambda(x)$ as a spectrum, the inverse transform yields a vector $\boldsymbol{\lambda}$ that has zeros at the coordinates corresponding to the zeros of $\Lambda(x)$. $\boldsymbol{\lambda}$ has a zero value wherever $\mathbf{e}$ is nonzero, and thus $\lambda_i e_i = 0$ for all i. Since these product expressions are zero in the time domain, the corresponding convolutions must be zero in the frequency domain.

$$\sum_{k=0}^{n-1} \Lambda_k E_{j-k} = 0, \qquad j = 0, 1, \ldots, n-1 \tag{9-34}$$

If we assume that ν errors have occurred, then $\Lambda_i = 0$ for all $i > \nu$. Equation (9-34) thus takes the familiar form

$$\sum_{k=0}^{\nu} \Lambda_k E_{j-k} = 0, \qquad j = 0, 1, \ldots, n-1 \tag{9-35}$$

Using the known coordinates for **E**, we can obtain a polynomial $E(x) = E_1 x + E_2 x^2 + \cdots + E_{2t} x^{2t}$. Equation (9-35) can then be expressed as the key equation

$$\Lambda(x)[1 + E(x)] \equiv \Omega(x) \; mod \, x^{2t+1}$$

which we can solve using any of the techniques discussed in the first two sections of this chapter. Once $\Lambda(x)$ is known, we can use its coordinates to generate the remaining coordinate values for **E** through *recursive extension*. Since $\sum_{k=0}^{\nu} \Lambda_k E_{j-k} = 0$, we have $E_j = -\sum_{k=1}^{\nu} \Lambda_k E_{j-k}$. The desired values are then found as follows.

$$\begin{aligned} E_{2t+1} &= -\sum_{k=1}^{\nu} \Lambda_k E_{2t+1-k} \\ E_{2t+2} &= -\sum_{k=1}^{\nu} \Lambda_k E_{2t+2-k} \\ &\;\vdots \\ E_{n-1} &= -\sum_{k=1}^{\nu} \Lambda_k E_{n-1-k} \end{aligned} \tag{9-36}$$

We now need only compute the inverse transform of **E** to obtain the error vector corrupting the received word.

9.4 ERASURE DECODING

In a digital communication system, information is formatted into sequences of symbols that each take on one value from a finite range of possibilities. For example, in a binary system the choices are limited to zero and one. These symbols are modulated onto a carrier and the resulting signal transmitted to one or more receivers. At some point in each receiver, a detection circuit examines the received, processed signal and decides which of the possible transmitted symbols is most likely to have been sent.

In a **hard-decision receiver**, the detection circuit limits its range of choices to the same group of choices available at the transmitter. For example, if the transmitter makes binary decisions, then the receiver will do the same. Unfortunately hard decisions destroy information that can improve the overall performance of our communication system. In some cases, the received signal may not offer a clear choice as to which of the possible symbols has been transmitted. Rather than force a decision that is likely to be incorrect, the **soft-decision receiver** uses an expanded selection of choices to communicate the quality of the received symbol to the error control decoder. The simplest form of soft-decision receiver uses **erasures** to indicate the reception of a signal whose corresponding symbol value is in doubt. Assuming a q-ary transmission, the receiver has $(q + 1)$ detection choices, as shown in Figure 9-4. In order to maintain clarity, the figure assumes that the 0 symbol is sent. The $\{p_i\}$ are the corresponding probabilities of receiving the correct symbol, one of the incorrect symbols, or an erasure. These values are a function of the modulation format, transmitter power, and the channel conditions.

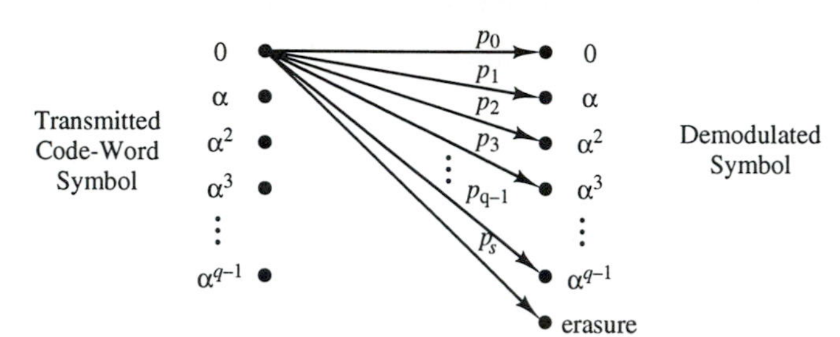

Figure 9-4. The Memoryless q-ary Erasure Channel

BCH and Reed-Solomon codes allow for a very efficient means of erasure decoding. Erasure decoding does not provide significant additional gain for additive white Gaussian noise channels, but it does provide substantial improvement over fading and bursty channels. Reed-Solomon erasure decoding plays a significant role in the error control system used in compact audio disc players (see Chapter 16). Detailed performance analyses for erasure decoding systems are provided in Chapter 10. In this section we examine the erasure decoding techniques themselves.

In Chapter 4 we characterized the error correction capabilities of block codes in terms of their minimum distance $d_{\min}$. Suppose that we have a received word with a single erased coordinate. Over the unerased coordinates, all pairs of distinct code words are separated by a Hamming distance of at least $(d_{\min} - 1)$. In general, given f erased coordinates, the code words will have an effective minimum distance of $(d_{\min} - f)$ over the unerased coordinates. It follows that we can correct

$$t_e = \frac{\lfloor d_{\min} - f - 1 \rfloor}{2}$$

errors in the unerased coordinates of the received word. In other words, we can correct e errors and f erasures so long as

$$(2e + f) < d_{\min} \tag{9-37}$$

It is interesting to note that we can correct twice as many erasures as we can errors. This can be explained intuitively by noting that we have more information about the erasures to begin with. We know exactly where the erasures are, but we have no idea as to the location of the errors.

9.4.1 Binary Erasure Decoding

Binary erasure decoding can be performed by making only the slightest modification to the standard decoder. The procedure is as follows.

The binary erasure decoding algorithm

1. Given a received word **r**, place zeros in all erased coordinates and decode normally. Label the resulting code word $\mathbf{c}_0$.
2. Now place ones in all erased coordinates and decode normally. Label the resulting code word $\mathbf{c}_1$.
3. Compare $\mathbf{c}_0$ and $\mathbf{c}_1$ to **r**, selecting as the final decoded output the code word that is closest in Hamming distance to **r**.

It is a simple matter to show that this algorithm works. We first assume that the number of errors and erasures caused by the channel satisfies the constraint $(2e + f) < d_{\min}$. If we assign a zero to the f erased coordinates, we generate some e_0 errors, making the total number of errors at the input to the decoder equal to $(e + e_0)$. When we assign ones to the erased coordinates, we end up with $(e + e_1) = (e + f - e_0)$ errors. Either e_0 or $(f - e_0)$ must be less than or equal to $f/2$. It follows that, in at least one of the decoding operations, the total number of errors e_t will satisfy $2e_t \leq 2(e + f/2) < d_{\min}$. At least one of the decoding operations yields the correct code word.

Though the hardware modifications necessary for binary erasure decoding are minimal, it should be noted that the speed of the "errors and erasures" decoder is half that of the "errors-only" decoder (or it has twice the hardware), for we must now perform two standard decoding operations for each received word containing erasures.

9.4.2 Nonbinary Erasure Decoding

Suppose that we have a received word with ν errors and f erasures. The errors occur in coordinates $i_1, i_2, \ldots, i_\nu$, while the erasures occur in coordinates $j_1, j_2, \ldots, j_f$. In the analysis that follows the coordinates for the errors and erasures are designated using the error locators $X_1 = \alpha^{i_1}, X_2 = \alpha^{i_2}, \ldots, X_\nu = \alpha^{i_\nu}$ and the *erasure locators* $Y_1 = \alpha^{j_1}, Y_2 = \alpha^{j_2}, \ldots, Y_f = \alpha^{j_f}$. Remember that the primary difference between the error locators and the erasure locators is that we have the values for the latter at the beginning of the decoding operation. The first of the two tasks for our decoding operation is to determine the values of the error locators. The second task is to find the values $\{e_{i_k}\}$ associated with the error locators and the values $\{f_{j_k}\}$ associated with the erasure locators.

An *erasure locator polynomial* is computed using the erasure locators.

$$\Gamma(x) = \prod_{l=1}^{f} (1 - Y_l x) \tag{9-38}$$

In order to compute the syndrome for the received word, we must first insert values at every coordinate where an erasure has been indicated. Naturally the computations that follow are much simpler if we select the value zero for this substitution.

Since the syndrome is only a function of the error/erasure polynomial, the syndrome computations have the following form. Assume that the code is narrow-sense, having as zeros $\alpha, \alpha^2, \ldots,$ and α^{2t}.

$$S_l = r(\alpha^l) = \sum_{k=1}^{\nu} e_{i_k} X_k^l + \sum_{k=1}^{f} f_{j_k} Y_k^l \tag{9-39}$$

Construct a syndrome polynomial

$$S(x) = \sum_{l=1}^{2t} S_l x^l \tag{9-40}$$

with which the key equation for errors and erasure decoding is obtained.

$$\Lambda(x)\Gamma(x)[1 + S(x)] \equiv \Omega(x) \; mod \, x^{2t+1} \tag{9-41}$$

Equation (9-41) can be slightly simplified by combining all information that is known at the beginning of the decoding operation into a single modified syndrome polynomial $\Xi(x)$ [For1].

$$1 + \Xi(x) \equiv \Gamma(x)[1 + S(x)] \; mod \, x^{2t+1} \tag{9-42}$$

The key equation now takes the form

$$\Lambda(x)[1 + \Xi(x)] \equiv \Omega(x) \; mod \, x^{2t+1} \tag{9-43}$$

which can be solved using Berlekamp's algorithm, the Berlekamp-Massey algorithm, or the extended version of Euclid's algorithm.

Berlekamp-Massey algorithm erasure decoding

1. Compute the erasure polynomial $\Gamma(x)$ using the erasure information provided by the receiver.
2. Replace the erased coordinates with zeros and compute the syndrome polynomial $S(x)$.
3. Compute the modified syndrome polynomial $\Xi(x) \equiv (\Gamma(x)[1 + S(x)] - 1) \; mod \, x^{2t+1}$.
4. Apply the Berlekamp-Massey algorithm to find the connection polynomial $\Lambda(x)$ for the LFSR that generates the modified syndrome coefficients $\Xi_1, \Xi_2, \ldots, \Xi_{2t}$.
5. Find the roots of $\Lambda(x)$, thus determining the error locations.
6. Determine the magnitude of the errors and erasures.

Euclid's algorithm erasure decoding

1. Compute the erasure polynomial $\Gamma(x)$ using the erasure information provided by the receiver.
2. Replace the erased coordinates with zeros and compute the syndrome polynomial $S(x)$.

3. Compute the modified syndrome polynomial $\Xi(x) = (\Gamma(x)[1 + S(x)] - 1) \bmod x^{2t+1}$.
4. Set the following initial conditions: $r_{-1}(x) = x^{2t+1}$, $r_0(x) = 1 + \Xi(x)$, $t_{-1}(x) = 0$, $t_0(x) = 1$.
5. Using the extended algorithm, compute the successive remainders $r_i(x)$ and the corresponding $t_i(x)$ until the following stopping condition is reached. t is the maximum error correcting capability of the code, while f is the number of erasures in the received word.

$$\deg[r_i(x)] \leq \begin{cases} t + \dfrac{f}{2}, & f \text{ even} \\ t + \dfrac{f-1}{2}, & f \text{ odd} \end{cases}$$

6. Find the roots of $t_i(x) = \Lambda(x)$, thus determining the error locations.
7. Determine the magnitude of the errors and erasures.

Once the error locator polynomial is known, combine it with the erasure locator polynomial to obtain a single error/erasure locator polynomial $\Psi(x)$.

$$\Psi(x) = \Lambda(x)\Gamma(x) \tag{9-44}$$

A modified version of the Forney algorithm (Theorem 9-2) can then be used to compute the error and erasure values.

$$e_{i_k} = \frac{-X_k\,\Omega(X_k^{-1})}{\Psi'(X_k^{-1})}, \qquad f_{i_k} = \frac{-Y_k\,\Omega(Y_k^{-1})}{\Psi'(Y_k^{-1})} \tag{9-45}$$

An error/erasure polynomial is then constructed and subtracted from the received polynomial to obtain the desired code polynomial. Note that before the correction is performed, the received polynomial should have the same values at the erased positions as used in the computation of the syndromes.

Example 9-10—Error/Erasure Correction Using the Berlekamp-Massey Algorithm and a (7, 3) Reed-Solomon Code

In Example 9-5 we saw that the double-error-correcting narrow-sense Reed-Solomon code of length 7 over GF(8) has the generator polynomial

$$g(x) = (x - \alpha)(x - \alpha^2)(x - \alpha^3)(x - \alpha^4) = x^4 + \alpha^3x^3 + x^2 + \alpha x + \alpha^3$$

The received vector is

$$\mathbf{r} = (0, 0, \alpha^3, f, 1, 0, 1) \leftrightarrow r(x) = \alpha^3x^2 + fx^3 + x^4 + x^6$$

where the "f" value indicates an erasure. This erasure occurs at $Y_1 = \alpha^3$, giving the erasure polynomial $\Gamma(x) = 1 + \alpha^3x$. Place a zero in the erasure location and compute the syndromes.

$$S_l = \alpha^3(\alpha^l)^2 + (\alpha^l)^4 + (\alpha^l)^6$$
$$\Rightarrow S(x) = \alpha^2x + \alpha^2x^2 + \alpha^6x^3 + x^4$$

Now compute the modified syndrome polynomial.

$$\begin{aligned}1 + \Xi(x) &\equiv \Gamma(x)[1 + S(x)]\ mod\, x^{2t+1}\\ &\equiv (1 + \alpha^3 x)(1 + \alpha^2 x + \alpha^2 x^2 + \alpha^6 x^3 + x^4)\ mod\, x^5\\ &\equiv 1 + \alpha^5 x + \alpha^3 x^2 + \alpha x^3 + \alpha^6 x^4\ mod\, x^5\end{aligned}$$

$\Xi(x)$ is thus $\alpha^5 x + \alpha^3 x^2 + \alpha x^3 + \alpha^6 x^4$. Now use the Berlekamp-Massey algorithm to find the error locator polynomial that satisfies Eq. (9-43) for the above $\Xi(x)$.

k	Ξ_k	$\Lambda^{(k)}(x)$	$\Delta^{(k)}$	L	$T(x)$
0	—	1	—	0	x
1	α^5	$1 + \alpha^5 x$	α^5	1	$\alpha^2 x$
2	α^3	$1 + \alpha^5 x$	0	1	$\alpha^2 x^2$
3	α	$1 + \alpha^5 x$	0	1	$\alpha^2 x^3$
4	α^6	$1 + \alpha^5 x$	0	—	—

We obtain $\Lambda(x) = 1 + \alpha^5 x$, indicating a single error at $X_1 = \alpha^5$. We next obtain the error magnitude polynomial and the error/erasure locator polynomial.

$$\Omega(x) \equiv (1 + \alpha^5 x)(1 + \alpha^5 x + \alpha^3 x^2 + \alpha x^3 + \alpha^6 x^4) \equiv 1\ mod\, x^5$$

$$\Psi(x) = \Lambda(x)\Gamma(x) = (1 + \alpha^5 x)(1 + \alpha^3 x) = 1 + \alpha^2 x + \alpha x^2$$

The error and erasure magnitudes follow readily.

$$e_{i_k} = \frac{-X_k\,\Omega(X_k^{-1})}{\Psi'(X_k^{-1})} = \frac{-X_k}{\alpha^2} \Rightarrow e_5 = \alpha^3, \qquad f_{i_k} = \frac{-Y_k\,\Omega(Y_k^{-1})}{\Psi'(Y_k^{-1})} = \frac{-Y_k}{\alpha^2} \Rightarrow f_3 = \alpha$$

The corrected code word is thus

$$\begin{aligned}c(x) &= r(x) + e(x) + f(x)\\ &= (\alpha^3 x^2 + x^4 + x^6) + \alpha^3 x^5 + \alpha x^3\\ &= x^2 g(x)\end{aligned}$$

■

Example 9-11—Error/Erasure Correction Using Euclid's Algorithm and a (31, 25) Reed-Solomon Code

In Example (9-8) we saw that the generator polynomial for the narrow-sense (31, 25) Reed-Solomon code is

$$g(x) = \prod_{i=1}^{6}(x - \alpha^i) = \alpha^{21} + \alpha^{24} x + \alpha^{16} x^2 + \alpha^{24} x^3 + \alpha^9 x^4 + \alpha^{10} x^5 + x^6$$

The received vector is

$$\mathbf{r} = (0, \alpha^{21}, \alpha^{24}, 0, \alpha^{24}, f, f, 1, 0, 0, 0, 0, 0, 0, 0, 0, 0, 0, 0, 0, 0, 0, 0, 0, 0, 0, 0, 0, 0, 0, 0)$$

The erasures occur at $Y_1 = \alpha^5$ and $Y_2 = \alpha^6$, giving the following erasure polynomial.

$$\Gamma(x) = \prod_{j=1}^{2}(1 - Y_j x) = (1 - \alpha^5 x)(1 - \alpha^6 x) = 1 + \alpha^{23} x + \alpha^{11} x^2$$

We now substitute zeros into those coordinates where **r** has erasures and compute the syndromes.

$$S_l = \alpha^{21}(\alpha^l) + \alpha^{24}(\alpha^l)^2 + \alpha^{24}(\alpha^l)^4 + (\alpha^l)^7$$
$$\Rightarrow S(x) = 0 + \alpha^{19}x^2 + \alpha^{10}x^3 + \alpha^{26}x^4 + \alpha^{17}x^5 + \alpha^9 x^6$$

The modified syndrome polynomial is then found as follows.

$$1 + \Xi(x) \equiv \Gamma(x)[1 + S(x)] \; mod\, x^{2t+1}$$
$$\equiv (1 + \alpha^{23}x + \alpha^{11}x^2)(1 + \alpha^{19}x^2 + \alpha^{10}x^3 + \alpha^{26}x^4 + \alpha^{17}x^5 + \alpha^9 x^6) \; mod\, x^7$$
$$\Rightarrow \Xi(x) = \alpha^{23}x + x^2 + \alpha^{28}x^3 + x^4 + \alpha^3 x^5 + \alpha^6 x^6$$

Proceed with Euclid's algorithm. Since there are $f = 2$ erasures and the error correcting capability of the code is $t = 3$, we stop when the degree of the remainder polynomial is less than or equal to 4.

k	r_i $\Omega^{(k)}(x)$	q_i $q_k(x)$	t_i $\Lambda^{(k)}(x)$
-1	x^7	—	0
0	$1 + \alpha^{23}x + x^2 + \alpha^{28}x^3 + x^4 + \alpha^3 x^5 + \alpha^6 x^6$	—	1
1	$\alpha^{22} + \alpha^2 x + \alpha^{19}x^2 + \alpha^{15}x^3$	$\alpha^{22} + \alpha^{25}x$	$\alpha^{22} + \alpha^{25}x$

The error magnitude polynomial is $\Omega(x) = \alpha^{22} + \alpha^2 x + \alpha^{19}x^2 + \alpha^{15}x^3$. The error locator polynomial $\Lambda(x) = \alpha^{22} + \alpha^{25}x = \alpha^{22}(1 + \alpha^3 x)$ indicates an error at the location corresponding to $X_1 = \alpha^3$. The error/erasure locator polynomial is

$$\Psi(x) = \Lambda(x)\Gamma(x) = (\alpha^{22} + \alpha^{25}x)(1 + \alpha^{23}x + \alpha^{11}x^2) = \alpha^{22} + \alpha^2 x + \alpha^{26}x^2 + \alpha^5 x^3$$

The error and erasure magnitudes are then obtained.

$$e_{i_k} = \frac{-X_k\,\Omega(X_k^{-1})}{\Psi'(X_k^{-1})} = \frac{\alpha^{22}X_k + \alpha^2 + \alpha^{19}X_k^{-1} + \alpha^{15}X_k^{-2}}{\alpha^2 + \alpha^5 X_k^{-2}} \Rightarrow e_3 = \alpha^{16}$$
$$f_{i_k} = \frac{-Y_k\,\Omega(Y_k^{-1})}{\Psi'(Y_k^{-1})} = \frac{\alpha^{22}Y_k + \alpha^2 + \alpha^{19}Y_k^{-1} + \alpha^{15}Y_k^{-2}}{\alpha^2 + \alpha^5 Y_k^{-2}} \Rightarrow f_5 = \alpha^9, \quad f_6 = \alpha^{10}$$

The decoded code polynomial is thus $c(x) = r(x) + \alpha^{16}x^3 + \alpha^9 x^5 + \alpha^{10}x^6 = xg(x)$. ■

9.4.3 Decoding Shortened/Punctured Reed-Solomon Codes with an Erasure Decoder

In Chapter 8 we discussed modified Reed-Solomon codes in some detail, and alluded to their utility in a number of applications. When parity or information coordinates are deleted from the code words in a Reed-Solomon code, the resulting code is (in general) no longer cyclic and does not have many of the properties we have used in this chapter. If we treat the deleted coordinates as erased positions, however, we can use the standard errors and erasure decoder for an RS code to decode shortened and punctured versions of this same code.

The reasoning behind this assertion is simple. Let **C** be a punctured Reed-Solomon code with length n, dimension k, minimum distance $d_{\min} = n - k + 1$, and symbols from GF(q). The corresponding "mother code" from which the punctured code was obtained is a Reed-Solomon code of length $(q - 1)$, dimension k, and minimum distance $d_{\min} = q - k$. The difference in minimum distance between the punctured code and its corresponding mother code is $(q - 1 - n)$, the number of punctured symbols. If we append $(q - 1 - n)$ erasures to the shortened code word and treat it as if it were from the mother code, the additional minimum distance is exactly sufficient to decode the appended erasures (see Eq. (9-37)).

This logic applies to shortened codes as well. A shortened code **C** is obtained by deleting *information* coordinates from a code **C**$'$. However, this same code **C** can be obtained by deleting *parity* coordinates from a lower-rate code **C**$''$.

9.4.4 Systematic Encoding Using an Erasure Decoder

In Theorem 8-4 we showed that any k coordinates of a Reed-Solomon code can be used as the message coordinates in a systematic representation. Let **C** be an (n, k) Reed-Solomon code and consider a corresponding errors and erasures decoder. If we give the decoder a word with k arbitrary symbols (information) and $(n - k)$ erasures, the decoder will output a word containing the k arbitrary symbols in their input positions. The erasures will have been "corrected," providing us with a valid code word.

This technique has the obvious benefit of rendering the encoder and decoder designs identical. A transceiver thus needs only one device for half-duplex coded transmission and reception.

PROBLEMS

In the following problems we adopt the convention in which the leftmost bit of the received vectors corresponds to the 1s position in the received polynomial (e.g., $\mathbf{r} = (1101000) \leftrightarrow r(x) = 1 + x + x^3$). Solutions can be verified by checking to see if the decoded word has the proper zeros. Appendix B contains a series of add-one tables that may prove useful in completing these problems.

Binary BCH decoding

1. Let the transmission code be the single-error correcting, narrow-sense, binary BCH code of length 15. The generator polynomial is $g(x) = 1 + x + x^4$. Use Peterson's technique to decode the following received vectors.
- **(a)** $\mathbf{r} = (001000000000000)$
- **(b)** $\mathbf{r} = (000001000000000)$
- **(c)** $\mathbf{r} = (111110000000000)$
- **(d)** $\mathbf{r} = (111011000000000)$
- **(e)** $\mathbf{r} = (001011111000000)$
- **(f)** $\mathbf{r} = (000001110000000)$
- **(g)** $\mathbf{r} = (000000100110100)$
- **(h)** $\mathbf{r} = (000001101101010)$

2. Let the transmission code be the double-error correcting, narrow-sense, binary BCH code of length 15. The generator polynomial is $g(x) = 1 + x^4 + x^6 + x^7 + x^8$. Use Peterson's technique to decode the following received vectors.

(a) $\mathbf{r}$ = (010000000000000)
(b) $\mathbf{r}$ = (100100000000000)
(c) $\mathbf{r}$ = (001110110000000)
(d) $\mathbf{r}$ = (001110010000000)
(e) $\mathbf{r}$ = (001111011101101)
(f) $\mathbf{r}$ = (101100011001100)
(g) $\mathbf{r}$ = (011011111001010)
(h) $\mathbf{r}$ = (010011110110100)

3. Let the transmission code be the same code used in Problem 2. Use Berlekamp's algorithm to decode the following received vectors.

(a) $\mathbf{r}$ = (001000000000000)
(b) $\mathbf{r}$ = (100000010000000)
(c) $\mathbf{r}$ = (000110001100000)
(d) $\mathbf{r}$ = (101110110010000)
(e) $\mathbf{r}$ = (101010100001000)
(f) $\mathbf{r}$ = (111011001000000)
(g) $\mathbf{r}$ = (101010000100000)
(h) $\mathbf{r}$ = (101100000010100)

4. Let the transmission code be the triple-error-correcting, narrow-sense, binary BCH code of length 31. The generator polynomial is $g(x) = 1 + x + x^2 + x^3 + x^5 + x^7 + x^8 + x^9 + x^{10} + x^{11} + x^{15}$. Use Berlekamp's algorithm to decode the following received vectors.

(a) $\mathbf{r}$ = (1010000000000000000000000000000)
(b) $\mathbf{r}$ = (0101010000000000000000000000000)
(c) $\mathbf{r}$ = (1100000101101100011000000000000)
(d) $\mathbf{r}$ = (1111011000000011011000000000000)
(e) $\mathbf{r}$ = (1011001001110100000000000000000)
(f) $\mathbf{r}$ = (1001111100001000000000000000000)
(g) $\mathbf{r}$ = (1110110111100000101010000000000)
(h) $\mathbf{r}$ = (0001100000001111101100000000000)

Nonbinary decoding

5. Let the transmission code be the double-error-correcting, narrow-sense, Reed-Solomon code of length 7 used in Example 9-5. The generator polynomial is $g(x) = \alpha^3 + \alpha x + x^2 + \alpha^3 x^3 + x^4$. Use the Peterson-Gorenstein-Zierler decoding algorithm to decode the following received vectors.

(a) $\mathbf{r} = (000\alpha^2 000)$
(b) $\mathbf{r} = (0100\alpha^5 00)$
(c) $\mathbf{r} = (0\alpha^3 01\alpha^4 10)$
(d) $\mathbf{r} = (\alpha^5 0\alpha^2 \alpha^5 \alpha^2 00)$
(e) $\mathbf{r} = (\alpha^5 1\alpha^4 \alpha^4 010)$
(f) $\mathbf{r} = (010\alpha^4 \alpha^4 01)$
(g) $\mathbf{r} = (1\alpha^5 \alpha^4 11\alpha^4 1)$
(h) $\mathbf{r} = (01\alpha^4 1\alpha^2 \alpha^5 \alpha^3)$

6. Let the transmission code be the double-error-correcting, narrow-sense, Reed-Solomon code of length 7 used in Example 9-5. The generator polynomial is $g(x) = \alpha^3 + \alpha x + x^2 + \alpha^3 x^3 + x^4$. Use the Berlekamp-Massey decoding algorithm to decode the following received vectors.
(a) $\mathbf{r} = (00010\alpha0)$
(b) $\mathbf{r} = (1001000)$
(c) $\mathbf{r} = (\alpha^6 01\alpha^3 100)$
(d) $\mathbf{r} = (0\alpha^5 11\alpha^5 \alpha^2 0)$
(e) $\mathbf{r} = (0\alpha^5 00\alpha^4 \alpha^5 0)$
(f) $\mathbf{r} = (00\alpha^4 \alpha^4 \alpha^5 10)$
(g) $\mathbf{r} = (1\alpha^5 \alpha^4 01\alpha^4 0)$
(h) $\mathbf{r} = (\alpha^3 \alpha^4 1\alpha^5 0\alpha^4 0)$

7. For each of the decoding exercises in Problem 6, draw the linear feedback shift register defined by the connection polynomial generated by the algorithm. Verify that this shift register generates the desired syndromes.

8. Describe the modifications to the Berlekamp-Massey algorithm that are necessary for the accommodation of a non-narrow-sense Reed-Solomon code.

Frequency-domain decoding

9. Let the transmission code be the double-error-correcting, narrow-sense, Reed-Solomon code of length 7 used in Example 9-5. The generator polynomial is $g(x) = \alpha^3 + \alpha x + x^2 + \alpha^3 x^3 + x^4$. Use the Berlekamp-Massey algorithm in a frequency-domain decoder to decode the following received vectors.
(a) $\mathbf{r} = (00\alpha^3 \alpha^2 000)$
(b) $\mathbf{r} = (010\alpha\alpha^4 \alpha0)$
(c) $\mathbf{r} = (1\alpha^2 \alpha^2 1\alpha^4 00)$
(d) $\mathbf{r} = (00100\alpha^6 1)$
(e) $\mathbf{r} = (\alpha^4 \alpha^4 \alpha^5 1000)$
(f) $\mathbf{r} = (1\alpha^5 \alpha^4 01\alpha^4 0)$
(g) $\mathbf{r} = (\alpha^5 \alpha^6 \alpha^2 10\alpha^6 0)$

10. Compare the complexity of the time domain decoding techniques to that of the frequency domain technique.

Erasure decoding

11. Let the transmission code be the same double-error-correcting binary code used in Problem 2. Use Berlekamp's algorithm to decode the following received vectors. The character "f" indicates the presence of an erasure
(a) $\mathbf{r} = (00f000000000000)$
(b) $\mathbf{r} = (0f000f00f000000)$
(c) $\mathbf{r} = (0001100fff00100)$
(d) $\mathbf{r} = (f0f110110010100)$
(e) $\mathbf{r} = (10ff10000000000)$
(f) $\mathbf{r} = (11111100f0f0000)$
(g) $\mathbf{r} = (f01f10f101f0000)$
(h) $\mathbf{r} = (10ff00001000100)$

12. Let the transmission code be the double-error-correcting, narrow-sense, Reed-Solomon code of length 7 used in Example 9-5. The generator polynomial is $g(x) = \alpha^3 + \alpha x +$

$x^2 + \alpha^3 x^3 + x^4$. Use the Berlekamp-Massey algorithm to decode the following received vectors.

(a) $\mathbf{r} = (00ff000)$
(b) $\mathbf{r} = (f00f00\alpha)$
(c) $\mathbf{r} = (0f\alpha 1f10)$
(d) $\mathbf{r} = (fff\alpha^5\alpha^2 00)$
(e) $\mathbf{r} = (f0\alpha^4 f\alpha^5 1\alpha^2)$
(f) $\mathbf{r} = (0\alpha^5 0\alpha^4\alpha^4 f1)$
(g) $\mathbf{r} = (f\alpha^5 001f1)$
(h) $\mathbf{r} = (01f1\alpha^5\alpha^5 0)$

Encoding with an erasure decoder

13. Encode the following messages using an error/erasure decoder for the narrow-sense Reed-Solomon code of length 7 with generator polynomial $g(x) = \alpha^3 + \alpha x + x^2 + \alpha^3 x^3 + x^4$.

(a) $\mathbf{m} = (011)$
(b) $\mathbf{m} = (\alpha^4\alpha 0)$
(c) $\mathbf{m} = (\alpha^5\alpha^2 1)$
(d) $\mathbf{m} = (\alpha^4 00)$
(e) $\mathbf{m} = (\alpha^2\alpha^4\alpha^4)$
(f) $\mathbf{m} = (\alpha\alpha\alpha)$

10

Block Code Performance Analysis

In this chapter we characterize the performance of various types of block error control systems over memoryless channels and fading channels. We begin by establishing a general framework for the analyses, and then pursue some specific cases.

Figure 10-1 shows a block diagram for a generic error control system within a digital communication system. The data source generates blocks of k **message symbols** taking values from $GF(q^m)$. It is assumed that any data compaction or encryption has taken place before the messages arrive at the error control encoder (see Figure 1-3). The message blocks are then encoded, generating code words of n **code-word symbols**, each symbol taking values from $GF(q^m)$. The code-word symbols are sent to the modulator for transmission over the channel. In many cases, the cardinality of the code-word symbol alphabet (q^m) does not match that of the modulator signal constellation (assumed here to be q^b). Before modulation can take place, therefore, the q^m-ary code-word symbols must be translated into q^b-ary **channel symbols**. These channel symbols are then mapped onto signals in the q^b-ary modulator signal constellation and transmitted over the channel.

The receiver/demodulator takes the incoming signal and recovers the channel symbols from the noise-corrupted, modulated carrier. These channel symbols are then passed along to a translator for conversion back into q^m-ary code-word symbols. It should be noted that the translators in the transmitter and receiver may have no actual physical interpretation. They merely form a convenient boundary for the system performance analysis. The received channel symbols are characterized at the translator input by the **probability of channel-symbol error** p_{ce} and the **probability of channel-symbol erasure** p_{cs}. The received code-word symbols at the output of the translator are in turn characterized by the **probability of code-word symbol error** p_e

and the **probability of code-word symbol erasure** p_s. The received code word symbols are sent in n-symbol blocks to the error control decoder. The output of the decoder is characterized in a number of ways. In the discussion that follows we consider the **probability of decoder error** (or word error rate) $P(E)$, the **probability of decoder failure** $P(F)$, the **probability of bit error** (or bit error rate) $P_b(E)$, the **probability of detected word error** $P_d(E)$, and the **probability of undetected word error** $P_u(E)$.

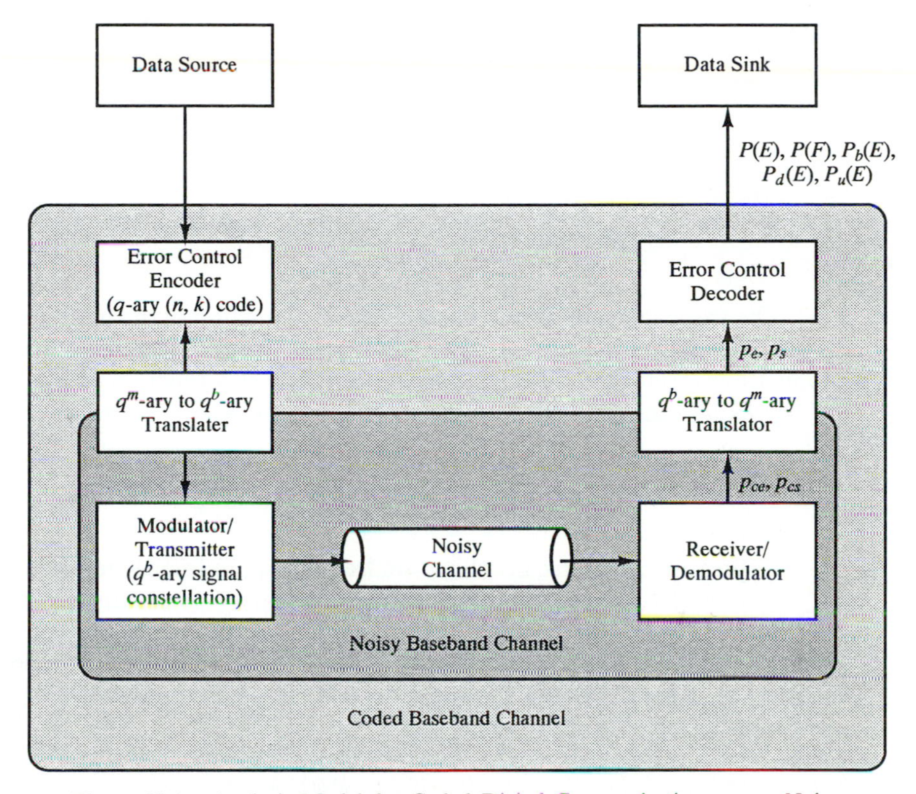

Figure 10-1. Analysis Model for Coded Digital Communication over a Noisy Channel

10.1 PERFORMANCE ANALYSIS FOR BINARY BLOCK CODES OVER THE BINARY SYMMETRIC CHANNEL

In the following we assume that the code-word symbol and channel-symbol alphabets are both binary (i.e., $q^m = q^b = 2$). The communication channel is also assumed to be symmetric and memoryless, and is thus a **binary symmetric channel** (BSC).

Definition 10-1—Symmetric Channels

A binary channel is **symmetric** if the probability that a transmitted symbol will be received incorrectly is independent of the value of the symbol.

Definition 10-2—Memoryless Channels

A channel is **memoryless** if the noise process affecting a given symbol during its transmission is independent of that affecting preceding or succeeding symbols.

The BSC is shown in Figure 10-2. It is completely characterized by the **transition**, or **crossover probability** p.

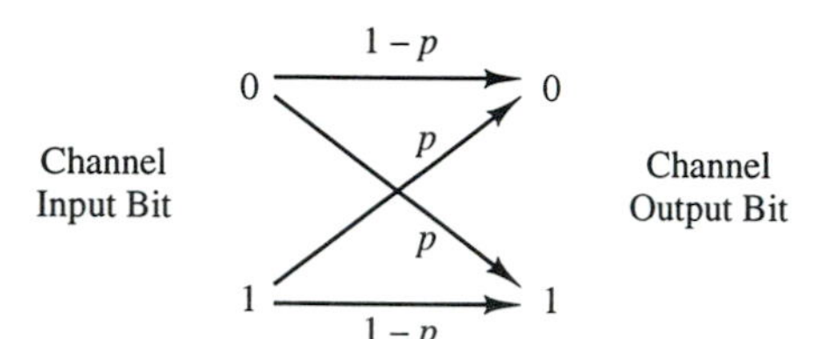

Figure 10-2. The Binary Symmetric Channel

The value taken on by the parameter p is a function of the modulation format, the channel noise level, and the transmitter power level. For example, for the BPSK modulation format the bit error rate is expressed as a function of the received bit energy E_b and the one-sided noise spectral density N_0.

$$p = Q\left(\sqrt{\frac{2E_b}{N_0}}\right), \qquad \text{where } Q(x) = \frac{1}{\sqrt{2\pi}}\int_x^{\infty} e^{-u^2/2}\,du \tag{10-1}$$

Expressions for p like the above are frequently modified when comparisons are to be drawn between multiple error control systems. It is usually assumed that the transmitter power level and the *information*-symbol transmission rate are to remain constant. The redundancy introduced by the various error control codes thus increases the channel-symbol transmission rate, reducing the received E_b/N_0. For a code with rate R, Eq. (10-1) becomes

$$p = Q\left(\sqrt{\frac{2RE_b}{N_0}}\right) \tag{10-2}$$

10.1.1 Error Detection Performance

Let **C** be a binary (n, k) linear block code with minimum distance $d_{\min}$. We assume that the all zero code word has been transmitted. This assumption does not affect the generality of our analysis because the code is linear, and thus the relative distances between the transmitted code word and all of the other code words are independent of the transmitted code word (see Property Three for linear block codes in Section 4.2). We can obtain simple bounds on the probability of undetected word error $P_u(E)$ using the minimum distance of the code. Recall that we are guaranteed to be able to detect all error patterns of weight $(d_{\min} - 1)$ or less. The probability of undetected word error is thus bounded above by the probability of occurrence of error patterns of weight $d_{\min}$ or greater.

$$P_u(E) \leq \sum_{j=d_{\min}}^{n} \binom{n}{j} p^j(1-p)^{n-j} = 1 - \left[\sum_{j=0}^{d_{\min}-1} \binom{n}{j} p^j(1-p)^{n-j}\right] \tag{10-3}$$

In Eq. (10-3) the binomial coefficient $\binom{n}{j}$ is the number of error patterns of weight j, while $p^j(1-p)^{n-j}$ is the probability of occurrence of a particular weight j error pattern. The probability of detected word error $P_d(E)$ is bounded above by the probability that one or more bit errors occur during the transmission of the n-bit code word.

$$P_d(E) \leq \sum_{j=1}^{n} \binom{n}{j} p^j(1-p)^{n-j} = 1 - (1-p)^n \tag{10-4}$$

The bound on $P_d(E)$ is usually good, but the bound on $P_u(E)$ is not very tight. Both, however, have the significant advantage of not requiring any knowledge of the code's weight distribution. If the weight distribution of the code is known, exact expressions for the probability of undetected and detected word error can be obtained. Property Three of linear block codes (Section 4.2) states that the undetectable error patterns are those error patterns that are identical to code words. The weight distribution for **C** thus lists the number of undetectable error patterns for each possible weight. The probability of occurrence for these patterns on the BSC is completely determined by their length and weight. Let A_j be the number of code words of weight j.

$$P_u(E) = \sum_{j=d_{\min}}^{n} A_j p^j(1-p)^{n-j} \tag{10-5}$$

The probability of detected word error is simply the probability that one or more bit errors occurs in the received word, subtracting the probability that the resulting pattern is undetectable.

$$\begin{aligned} P_d(E) &= \sum_{j=1}^{n} \binom{n}{j} p^j(1-p)^{n-j} - P_u(E) \\ &= 1 - (1-p)^n - P_u(E) \end{aligned} \tag{10-6}$$

Let P_{u_b} be the undetected bit error rate and P_{d_b} the detected bit error rate. We define the undetected bit error rate as the probability that a received information bit is in error and is contained within a code word corrupted by an undetectable error pattern. The detected bit error rate is the probability that a received information bit is incorrect and is contained within a code word corrupted by a detectable error pattern. The undetected and detected word error rates can be translated into lower bounds on the undetected and detected information-bit error rates by assuming that detected and undetected code word errors correspond to single information-bit errors in the corresponding message blocks. Upper bounds are obtained by assuming the worst case: undetected word errors indicate that all of the decoded message bits are incorrect. The bounds below become approximations if the upper bounds on $P_u(E)$ and $P_d(E)$ are used instead of the exact values.

$$P_u(E) \geq P_{u_b}(E) \geq \frac{1}{k} P_u(E)$$
$$P_d(E) \geq P_{d_b}(E) \geq \frac{1}{k} P_d(E) \tag{10-7}$$

The undetected bit error rate in decoded message blocks can be computed exactly if a complete weight enumerator is available for the code in use. Let B_i be the total weight of the message blocks associated with all code words of weight i.

$$P_{u_b}(E) = \sum_{j=d_{\min}}^{n} \frac{B_j}{k} p^j (1-p)^{n-j} \tag{10-8}$$

Example 10-1—Error Detection Performance of a (31, 21) Binary BCH Code

In this example we consider the error detection performance of the (31, 21) binary BCH code defined by the generator polynomial

$$g(x) = x^{10} + x^9 + x^8 + x^6 + x^5 + x^3 + 1$$

The design distance for this code is $\delta = 5$, which in this case happens to be the actual minimum distance. The code words are transmitted over an AWGN channel with a coherent BPSK modulation format. The corresponding BSC model has crossover probability

$$p = Q\left(\sqrt{\frac{2(21/31)E_b}{N_0}}\right)$$

Equations (10-3) and (10-4) provide the following upper bounds on the probability of undetected and detected word error.

$$P_u(E) \leq 1 - \sum_{j=0}^{4} \binom{31}{j} p^j (1-p)^{31-j}$$
$$P_d(E) \leq 1 - (1-p)^{31}$$

The weight distribution for this code can be obtained using Table 8-1(a) (see Example 8-3). Exact expressions for undetected and detected word error probability can thus be obtained as well.

$$P_u(E) = \sum_{j=5}^{31} A_j p^j (1-p)^{31-j}$$
$$P_d(E) = 1 - (1-p)^{31} - P_u(E)$$

Figure 10-3 contains the upper bounds and exact values for the probabilities of undetected and detected word error. An undetected word error curve for the uncoded case has been included. This curve corresponds to the probability of there being one or more bit errors within a block of 21 uncoded bits:

$$P_{u_{\text{uncoded}}}(E) = 1 - \left[1 - Q\left(\sqrt{\frac{2E_b}{N_0}}\right)\right]^{21}$$

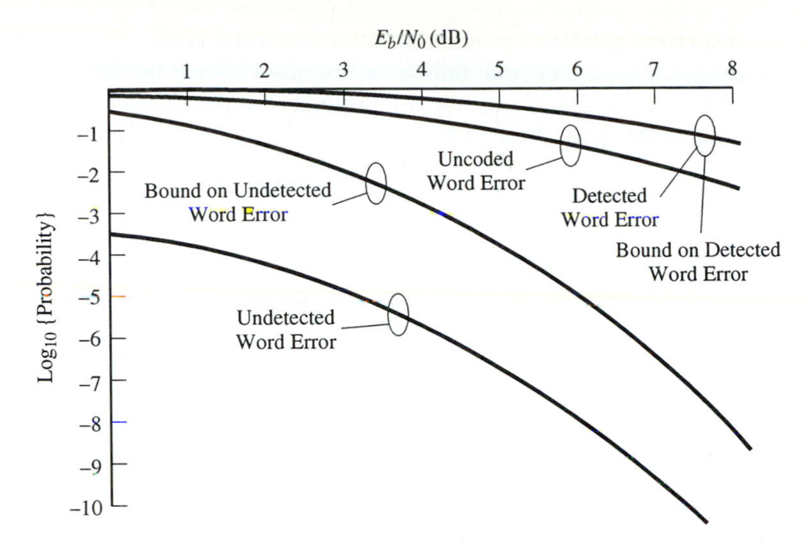

Figure 10-3. Error Detection Performance for (31, 21) BCH Code

Figure 10-3 shows that the added redundancy (10 bits for every 21 data bits) causes the detected word error probability for the coded case to be noticeably higher than the uncoded word error probability. The benefit gained, however, is made clear by the substantially lower probability of undetected word error for the coded case. Note that the bound on detected error and the exact curve are indistinguishable. This is expected, for the two curves differ by an amount reflected on the exact probability of undetected word error curve, which is several orders of magnitude less than the first two curves at all points. The bound on the probability of undetected word error is not as tight. ■

10.1.2 Error Correction Performance

In the following analysis it is assumed that a bounded-distance decoder is used in conjunction with an (n, k) code with minimum distance $d_{\min}$. If a received word is within Hamming distance $\lfloor (d_{\min} - 1)/2 \rfloor$ of a code word, the decoder selects that code word as the one most likely to have been sent. If the selected code word is not the one that was sent, we say that a **decoder error** has occurred. If there is no code word within Hamming distance $\lfloor (d_{\min} - 1)/2 \rfloor$ of a received word, then a **decoder failure** is declared. In many cases it is possible to design **complete decoders** that select the nearest code word to a received word, regardless of the Hamming distance separating them. The analysis of complete decoders is usually highly specific to the actual code used and is thus not pursued here, except to note that for a perfect code, the bounded-distance decoder is also a complete decoder.

The probability of decoder error is bounded above by the probability of occurrence of error patterns of weight greater than $\lfloor (d_{\min} - 1)/2 \rfloor$.

$$P(E) \le \sum_{j=\lfloor (d_{\min}-1)/2 \rfloor + 1}^{n} \binom{n}{j} p^j (1-p)^{n-j} = 1 - \left[\sum_{j=0}^{\lfloor (d_{\min}-1)/2 \rfloor} \binom{n}{j} p^j (1-p)^{n-j} \right] \qquad (10\text{-}9)$$

Equation (10-9) is an equality only when the code being analyzed is perfect. The probability of decoder failure is bounded above by the probability that the received word lies outside the decoding sphere for the all-zero code word, and is thus the same as the bound on $P(E)$.

$$P(F) \leq 1 - \sum_{j=0}^{\lfloor (d_{\min}-1)/2 \rfloor} \binom{n}{j} p^j (1-p)^{n-j} \tag{10-10}$$

If the weight distribution $\{A_j\}$ for $\mathbf{C}$ is known, exact expressions can be obtained for the probabilities of decoder error and failure. Decoder errors occur whenever the received word is within Hamming distance $\lfloor (d_{\min} - 1)/2 \rfloor$ of an incorrect code word. Let P_k^j be the probability that a received word is exactly Hamming distance k from a weight-j binary code word. Through a simple counting exercise, P_k^j can be shown to be

$$P_k^j = \sum_{r=0}^{k} \binom{j}{k-r} \binom{n-j}{r} p^{j-k+2r} (1-p)^{n-j+k-2r} \tag{10-11}$$

This result is proved for the more general nonbinary case in Section 10.2.2. The probability of falling within the decoding sphere of a nonzero code word (and thus the probability of decoder error for a bounded-distance decoder) is

$$P(E) = \sum_{j=d_{\min}}^{n} A_j \sum_{k=0}^{\lfloor (d_{\min}-1)/2 \rfloor} P_k^j \tag{10-12}$$

The probability of decoder failure for a bounded-distance decoder is the probability that the received word does not fall into any of the decoding spheres (correct or incorrect) for the bounded-distance decoder.

$$P(F) = 1 - \sum_{j=0}^{\lfloor (d_{\min}-1)/2 \rfloor} \binom{n}{j} p^j (1-p)^{n-j} - P(E) \tag{10-13}$$

Exact bit error rate expressions for data at the decoder output require some knowledge of the relationship between the weight of the message blocks and the weight of the corresponding code words. If we know the weight of the information block associated with each code word, we can compute B_j, the total weight of the message blocks associated with the code words of weight j. The bit error rate can then be computed through a simple modification of Eq. (10-12).

$$BER = P_b(E) = \frac{1}{k} \sum_{j=d_{\min}}^{n} B_j \sum_{k=0}^{\lfloor (d_{\min}-1)/2 \rfloor} P_k^j \tag{10-14}$$

In many cases the derivation of the information weight distribution $\{B_i\}$ is impractical. We must then rely on upper and lower bounds on the BER obtained directly from the decoder error rate $P(E)$. A lower bound is obtained by assuming that decoder errors cause a single bit error in the decoded data. The upper bound is obtained by assuming that decoder errors make all of the associated data bits incorrect.

$$\frac{1}{k} P(E) \leq P_b(E) \leq P(E) \tag{10-15}$$

Example 10-2—Error Correction Performance of a (31, 21) Binary BCH Code

We use the same code and channel as in Example 10-1, but this time we perform error correction instead of error detection. Since the design distance of the (31, 21) BCH code is 5, we can use the techniques in Chapter 9 to construct a two-error-correcting bounded-distance decoder. Figure 10-4 compares the decoder word error and failure rates to the uncoded word error rate. The decoding gain at a word error rate of 1×10^{-6} is 3 dB; the (31, 21) code thus allows a 50% reduction in transmitter power while maintaining the same word error performance. Note that the bound on decoder error and failure is quite good. ∎

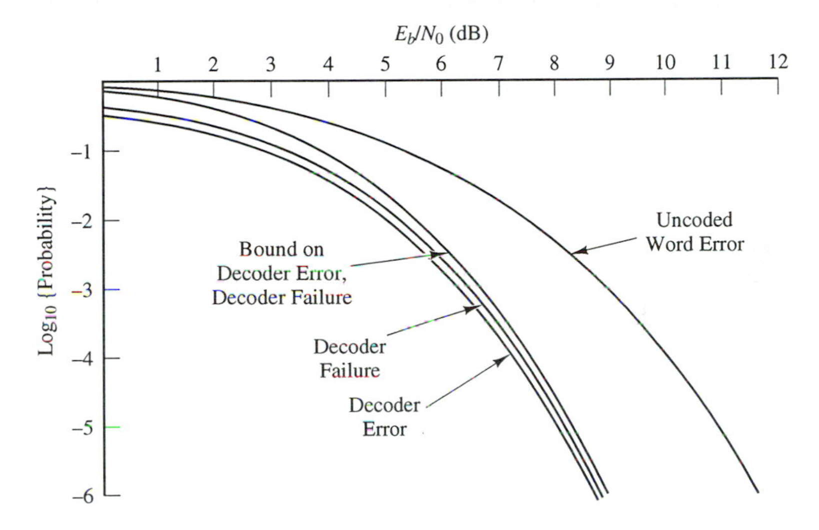

Figure 10-4. Decoder Error/Failure Performance for a (31, 21) Binary BCH Code

10.2 PERFORMANCE ANALYSIS FOR NONBINARY BLOCK CODES OVER MEMORYLESS CHANNELS

The analysis of nonbinary codes is a bit more complicated than that of binary codes. We must first consider the translation process, indicated in Figure 10-1, between the q^m-ary code-word alphabet and the q^b-ary channel-symbol alphabet. If $m \leq b$, then each channel symbol represents b/m code-word symbols. If $m \geq b$, then each code-word symbol is transmitted as m/b consecutive channel symbols.[1]

In those cases in which $b > m$, blocks of code-word symbols are mapped onto channel symbols such that the most probable channel-symbol errors cause a minimal number of code-word symbol errors. This is the basic idea behind Gray codes. In the following example we discuss a 4-ary/16-ary translation between code-word and channel symbols. In this and most other cases the design process is heuristic.

[1] Even if the code-word symbols are interleaved to combat channel fading, it is best to maintain the contiguity of the channel symbols composing each code-word symbol during transmission. The resultant burst-error-correcting capability is discussed later in this chapter.

Example 10-3—Transmission of Nonbinary Code-Word Symbols ($b > m$)

Consider a 4-ary code that is to be used in conjunction with 16-PSK over an AWGN channel. The modulation signal set is represented by 16 points equally spaced on a circle of radius $\sqrt{E}$ (Figure 10-5). When one of the signals is transmitted, the incorrect signals most likely to be received are those immediately adjacent to the transmitted signal on the circle. We thus assign 4-ary code-word symbols to the signals such that adjacent signals differ by only one symbol in their code-word symbol assignment. Most of the channel-symbol errors will thus cause only one code-word symbol error. Let GF(4) be represented by the symbols $\{0, 1, \alpha, \beta\}$. The following assignments of pairs of symbols from GF(4) to the 16-PSK constellation meet the desired requirements. ■

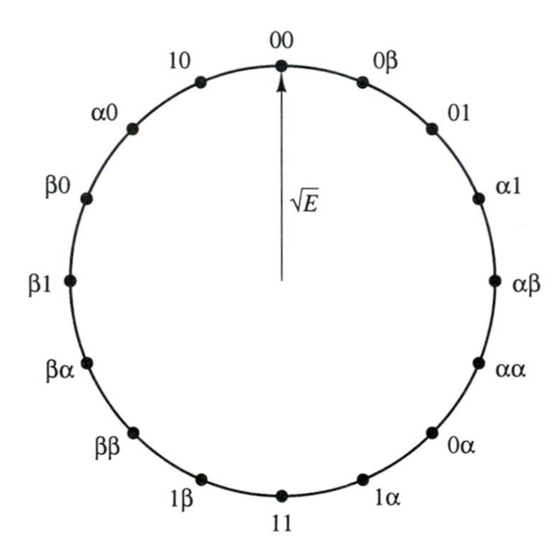

Figure 10-5. 4-ary Code-Word Symbol Assignments for 16-PSK

Suppose now that $m \geq b$. In this case the code-word symbols are represented by one or more channel symbols. A single channel-symbol error is sufficient to make a code-word symbol incorrect, and thus the mapping of blocks of channel symbols to code word symbols has no impact on performance.[2] The implementation of the communication system is facilitated, however, by mappings that preserve the vector-space structure used to implement finite field mathematics in the encoder and decoder.

Example 10-4—Transmission of Nonbinary Code-Word Symbols ($m \geq b$)

Suppose that we wish to transmit the code-word symbols for a length-7 Reed-Solomon code using a binary modulation format. Length-7 Reed-Solomon codes are defined over the field GF(8), whose elements are represented as powers of a primitive element α. The mapping of GF(8) onto 3-tuples of channel symbols follows the corresponding vector-space representation used to perform addition and multiplication in the encoder and decoder (see Example 2-25).

[2] It is assumed that the code is random-error-correcting.

CODE-WORD SYMBOLS		CHANNEL SYMBOLS
0	$\leftrightarrow$	0, 0, 0
1	$\leftrightarrow$	1, 0, 0
α	$\leftrightarrow$	0, 1, 0
α^2	$\leftrightarrow$	0, 0, 1
α^3	$\leftrightarrow$	1, 1, 0
α^4	$\leftrightarrow$	0, 1, 1
α^5	$\leftrightarrow$	1, 1, 1
α^6	$\leftrightarrow$	1, 0, 1

■

Given this sort of mapping, it is clear that the probability of occurrence of one incorrect code-word symbol α^x may differ from that of another incorrect symbol α^y. For example, suppose that the channel symbols in Example 10-4 are to be transmitted over a BSC with crossover probability p. A zero code-word symbol is sent as three consecutive channel zeroes. The probability that the incorrect symbol α arrives at the decoder at the other end of the channel is the probability that the first and third channel bits are correctly received, while the second is not ($p(1-p)^2$). The probability that α^3 arrives is the probability that the first two bits are incorrectly received while the third is correct ($p^2(1-p)$). Given our assumption that the all-zero code word is transmitted, the probability of reception for incorrect symbols becomes a monotonically decreasing function of the weight of the incorrect symbols. An exact performance analysis for nonbinary codes in such situations requires a weight enumerator that lists the number of times each nonzero symbol appears in the code words of a given weight. Such information is much harder to obtain than the simple weight distribution. For example, a complete enumerator has not yet been found for the RS codes. Since the compilation of a complete weight distribution by computer is currently impractical for most RS codes, an approximate analysis (used in conjunction with simulations) is necessary. In the following pages we develop an approximate analysis that bounds the performance of nonbinary codes.

The elements in the code-word symbol alphabet GF(2^m) are to be transmitted using a 2^b-ary constellation of channel symbols, where $b \leq m$. The channel symbols are transmitted across a noisy memoryless channel with probability of channel-symbol error p_{ce}. All incorrect *channel* symbols are assumed to occur with equal probability.[3] A code-word symbol error occurs whenever one or more of the code-word symbols' constituent channel symbols are incorrect. The probability of code-word symbol error is thus

$$p_e = 1 - (1 - p_{ce})^{m/b} \tag{10-16}$$

The **q^m-ary Uniform Discrete Symmetric Channel (UDSC)** shown in Figure 10-6 is adopted here. The probability that a symbol is correctly received is s, while

[3] This is an accurate assumption for most binary modulation formats and for nonbinary formats using orthogonal or simplex waveforms.

the probability that a *particular* incorrect symbol is received is p. The probability of code-word symbol error at the input to the decoder is thus $p_e = (1 - s) = (q^m - 1)p$.

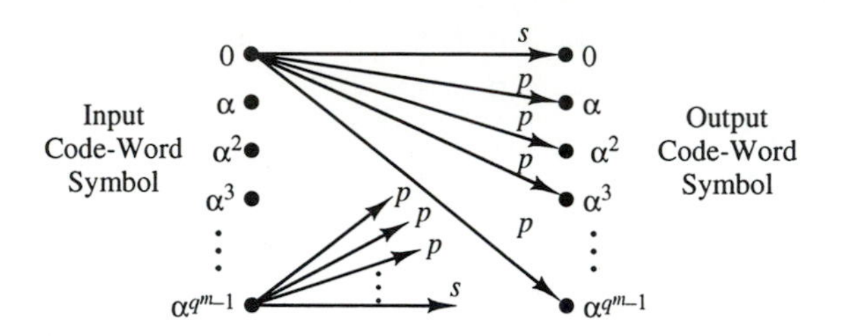

Figure 10-6. q^m-ary Uniform Discrete Symmetric Channel

10.2.1 Error Detection Performance

Given the UDSC model, we need only make a few simple modifications of the expressions in Section 10.1.1 to generalize them for the nonbinary case. Let **C** be a q^m-ary (n, k) code with minimum distance $d_{\min}$. The undetected and detected word error rates for **C** may be bounded as follows.

$$P_u(E) \le 1 - \left[\sum_{j=0}^{d_{\min}-1} \binom{n}{j} (1-s)^j s^{n-j}\right]$$
$$P_d(E) \le 1 - s^n \tag{10-17}$$

The exact values (given the assumed accuracy of the UDSC model) can be obtained if the weight distribution $\{A_j\}$ for **C** is known. A code word of weight j requires j specific incorrect symbols, each occurring with probability p.

$$P_u(E) = \sum_{j=d_{\min}}^{n} A_j p^j s^{n-j}$$
$$P_d(E) = 1 - s^n - P_u(E) \tag{10-18}$$

Bounds on the equivalent undetected and detected bit error rate can also be found. They assume the best- and worst-case impact on the $\log_2(q^m)$ bits comprising an incorrectly decoded code-word symbol.

$$P_u(E) \ge P_{u_b}(E) \ge \frac{1}{k \log_2(q^m)} P_u(E)$$
$$P_d(E) \ge P_{d_b}(E) \ge \frac{1}{k \log_2(q^m)} P_d(E) \tag{10-19}$$

10.2.2 Error Correction Performance

Given the UDSC channel assumption, the analysis for the decoder error rate for the nonbinary case is similar to that for the binary case (Eq. (10-12)) except that P^j_k, the probability that a received word is exactly Hamming distance k from a weight-j code word, is now

$$P^j_k = \sum_{r=0}^{k} \binom{j}{k-r}\binom{n-j}{r} p^{j-k+r}(1-p)^{k-r} s^{n-j-r}(1-s)^r \qquad (10\text{-}20)$$

The derivation of this expression proceeds as follows [Wic7]. It is assumed that the all-zero code word from an (n, k) q^m-ary linear code has been transmitted. Let $\mathbf{c}_j$ be a weight-j code word, where $j \neq 0$. The coordinates of $\mathbf{c}_j$ that contain nonzero symbols form a set Θ with cardinality j, while those containing the zero symbol form a set Ω with cardinality $(n - j)$. If a received pattern is to be within Hamming distance k of $\mathbf{c}_j$, then it must differ from $\mathbf{c}_j$ in exactly $(k - r)$ of the coordinates in Θ and r of the coordinates of Ω for some whole number r less than or equal to k.

First count the number of ways that a received word can differ from $\mathbf{c}_j$ in $(k - r)$ of the coordinates in Θ. The received word must contain $[j - (k - r)]$ coordinates with exactly the same nonzero value as in $\mathbf{c}_j$. There are

$$\binom{j}{j-(k-r)} = \binom{j}{k-r}$$

ways to pick these positions. Since the all-zero word was transmitted, each of these individual nonzero values occurs independently of the other coordinates with probability p. The remaining $(k - r)$ positions may contain any value except for the particular nonzero value in $\mathbf{c}_j$. Each of these independent events occurs with probability $(1 - p)$. The total probability of the received word's differing from $\mathbf{c}_j$ in exactly $(k - r)$ of the coordinates in Θ is thus

$$\binom{j}{k-r} p^{j-k+r}(1-p)^{k-r}$$

Now count the number of ways that a received word can differ from $\mathbf{c}_j$ in r of the coordinates in Ω. The received word must contain a nonzero value in r of the coordinates in Ω, while the remaining $(n - j - r)$ coordinates contain the zero symbol that was originally transmitted. The former independent events occur with probability $(1 - s)$, while the latter occur with probability s. Since there are

$$\binom{n-j}{r}$$

ways to pick the positions with the nonzero values, the total probability of the received word's differing from $\mathbf{c}_j$ in exactly $(k - r)$ of the coordinates in Ω is

$$\binom{n-j}{r}(1-s)^r s^{n-j-r}$$

Since Θ and Ω are disjoint, P_k^j is obtained by taking the product of the probabilities of the two required events and summing over all possible values of r, providing the desired expression.

$$\sum_{r=0}^{k}\left\{\left[\binom{j}{k-r}p^{j-k+r}(1-p)^{k-r}\right]\left[\binom{n-j}{r}(1-s)^r s^{n-j-r}\right]\right\}$$
$$= \sum_{r=0}^{k}\binom{j}{k-r}\binom{n-j}{r}p^{j-k+r}(1-p)^{k-r}s^{n-j-r}(1-s)^r$$

Note that for $q^m = 2$, we set $s = (1-p)$ and $(1-s) = p$ in Eq. (10-20) and obtain the binary case in Eq. (10-11).

The probability of decoder failure is also slightly changed from the binary case, becoming

$$P(F) = 1 - \left[\sum_{j=0}^{\lfloor(d_{\min}-1)/2\rfloor}\binom{n}{j}(1-s)^j s^{n-j}\right] - P(E) \tag{10-21}$$

Example 10-5—The Reliability Performance of (31, k) Reed-Solomon Codes

In this example we consider the performance of $(31,k)$ Reed-Solomon codes used in conjunction with coherent BPSK modulation over an AWGN channel ($q^m = 32$ and $q^b = 2$). The UDSC channel is used with the parameter p defined as follows.

$$p = \frac{p_e}{31} = \frac{1}{31}\left[1 - \left(1 - Q\left(\sqrt{\frac{2(k/31)E_b}{N_0}}\right)\right)^5\right]$$

The parameter s is then

$$s = 1 - p_e = \left(1 - Q\left(\sqrt{\frac{2(k/31)E_b}{N_0}}\right)\right)^5$$

Figures 10-7 and 10-8 show the decoder error and failure curves for RS codes with length 31 and various dimensions k.

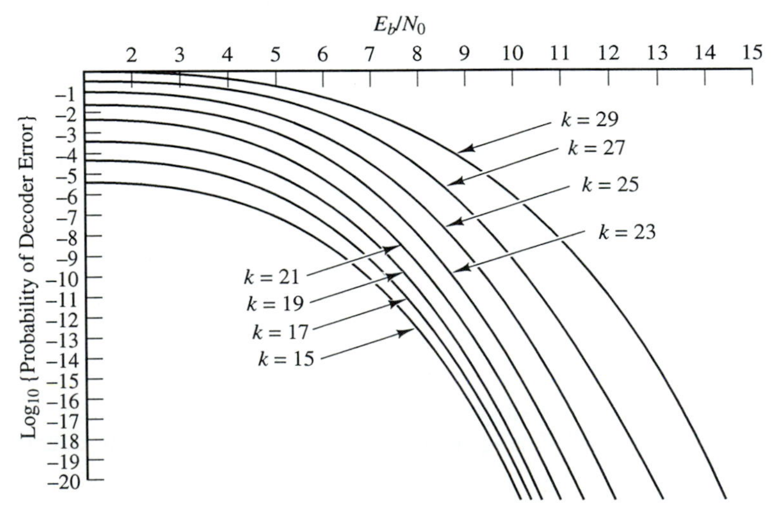

Figure 10-7. Decoder Error Rate for $(31,k)$ Reed-Solomon Codes, Code-Word Symbols Transmitted Using Coherent BPSK over an AWGN Channel

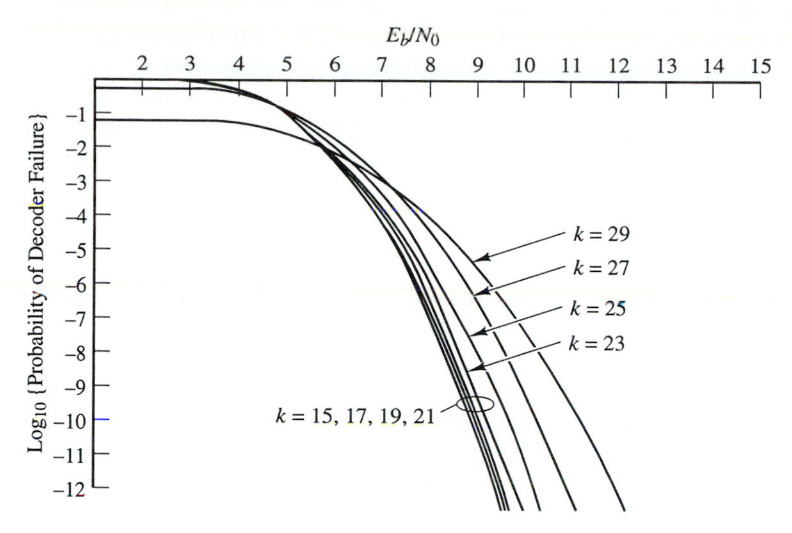

Figure 10-8. Decoder Failure Rate for $(31, k)$ Reed-Solomon Codes, Code-Word Symbols Transmitted Using Coherent BPSK over an AWGN Channel

It is interesting to note that the decoder error curves approach increasingly lower probabilities asymptotically as the received E_b/N_0 and code dimension decrease. This is due to the highly "imperfect" nature of RS codes. Given a completely random received vector, the probability that it will happen to fall into one of the decoding spheres established by the bounded-distance decoder becomes increasingly small with decreasing code dimension. This is made clear in Figure 10-8, in which the probability-of-decoder-failure curves for all but the highest code dimensions rapidly approach 1 as E_b/N_0 decreases. ■

10.3 ERASURE DECODING

In this section we analyze the performance of a bounded-distance decoder that is capable of erasure decoding. Let **C** be an (n, k) code with minimum distance $d_{\min}$. As shown in Chapter 9, the bounded-distance decoder can correctly decode all received words containing e errors and s erasures so long as e and s satisfy the constraint $(2e + s) < d_{\min}$. The decoding "sphere" surrounding each code word is thus defined as the collection of all n-tuples that differ from the code word in e errors and s erasures, where $(2e + s) < d_{\min}$.

The channel-symbol alphabet has cardinality q^b, while the code-word symbol alphabet has cardinality q^m. It is assumed that $m \geq b$, and thus each code-word symbol is composed of m/b consecutive channel symbols.[4] In this section it is assumed that the channel is memoryless, with channel symbols emerging from the

[4] In those cases in which $b > m$, single channel-symbol erasures cause multiple code-word symbol erasures, reducing, and in some cases eliminating, any benefit to be obtained through erasure decoding.

demodulator with probability of error p_{ce} and probability of erasure p_{cs}. The case for a channel with memory is treated in the next section.

As in the previous sections, the code-word symbol characteristics can be readily derived from those of the channel symbols. A code-word symbol is erased if one or more of the channel symbols of which it is composed is erased. The probability of channel-symbol erasure is thus

$$p_s = 1 - (1 - p_{cs})^{m/b} \tag{10-22}$$

A code-word symbol is in error if one or more of its constituent channel symbols is in error while none are erased.

$$p_e = (1 - p_{cs})^{m/b} - (1 - p_{ce} - p_{cs})^{m/b} \tag{10-23}$$

We shall once again assume that incorrect code-word symbols have equal probability of occurrence p. The UDSC channel model in Figure 10-6 can thus be extended to account for erasures as shown in Figure 10-9.

In this model the parameters $s = (1 - p_s - p_e)$ and $t = p_s$.

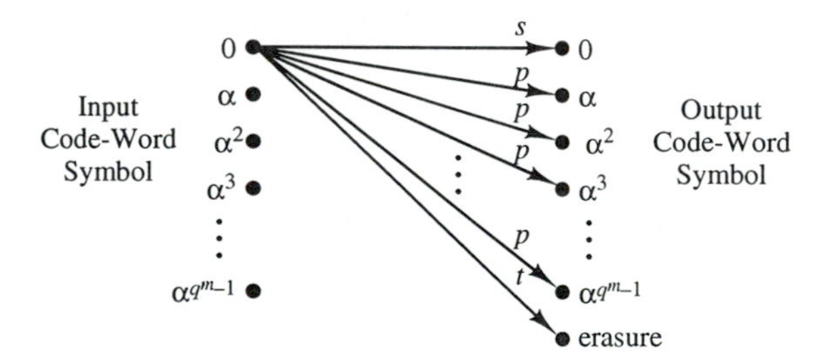

Figure 10-9. Model for the Uniform Discrete Symmetric Channel with Erasures (UDSC/E)

The next step in an exact performance analysis is the determination of $Q^j_{d_{\min}}$, the probability that a received word $\mathbf{r}$ falls into the decoding sphere surrounding a weight-j code word given that the all-zero code word was transmitted and assuming the UDSC/E model in Figure 10-9. The subscript $d_{\min}$ denotes the minimum distance of the code in use. After a bit of combinatorial effort, the following is obtained.

$$Q^j_{d_{\min}} = \sum_{v=0}^{\left\lfloor \frac{d_{\min}-1}{2} \right\rfloor} \sum_{w=0}^{d_{\min}-2v-1} \sum_{x=0}^{\left\lfloor \frac{d_{\min}-2v-w-1}{2} \right\rfloor} \sum_{y=0}^{d_{\min}-2v-w-2x-1} \sum_{z=0}^{\left\lfloor \frac{d_{\min}-2v-w-2x-y-1}{2} \right\rfloor} \tag{10-24}$$
$$\cdot \binom{n-j}{v}\binom{n-j-v}{w}\binom{j}{x}\binom{j-x}{y}\binom{j-x-y}{z} \cdot (q^m-1)^v (q^m-2)^x p^{j+v-y-z} t^{w+y} s^{n-j-v-w+z}$$

Equation (10-24) is derived as follows [Wic7]. Let $\mathbf{C}$ be an (n, k) code with minimum distance $d_{\min}$ for which there exists a bounded-distance erasure decoding algorithm. We wish to determine $Q^j_{d_{\min}}$, the probability that a received word $\mathbf{r}$ falls within the decoding sphere surrounding a code word $\mathbf{c}_i$ of weight j given the UDSC/E model in Figure 10-9.

Let Θ be the set containing the $(n - j)$ coordinates of $\mathbf{c}_i$ that contain zeros. Let Φ be the set containing the j coordinates of $\mathbf{c}_i$ that contain nonzero symbols. The desired probability expression can be obtained by allocating e errors and s erasures among the two sets of coordinates in all possible combinations within the constraint $(2e+s) < d_{\min}$.

Five distinct events must be accounted for in this derivation

- A Θ or Φ coordinate in $\mathbf{r}$ contains a zero symbol. This event occurs with probability s.
- A Θ or Φ coordinate in $\mathbf{r}$ contains an erasure. This event occurs with probability t.
- A Θ coordinate in $\mathbf{r}$ contains a nonzero, nonerased symbol. This event occurs with probability $(q^m - 1)p$.
- A Φ coordinate in $\mathbf{r}$ contains a nonzero, nonerased symbol that is different from the symbol at the same coordinate in $\mathbf{c}_i$. This event occurs with probability $(q^m - 2)p$.
- A Φ coordinate in $\mathbf{r}$ contains the same nonzero symbol as at the same coordinate in $\mathbf{c}_i$. This event occurs with probability p.

The allocation of errors and erasures in the expression is controlled by five counting variables as follows.

v = number of Θ coordinates for which $\mathbf{r}$ has a nonzero symbol

w = number of Θ coordinates for which $\mathbf{r}$ has an erasure

x = number of Φ coordinates in which $\mathbf{r}$ has a nonzero symbol other than the nonzero symbol in $\mathbf{c}_i$

y = number of Φ coordinates for which $\mathbf{r}$ has an erasure

z = number of Φ coordinates for which $\mathbf{r}$ has a zero

$Q^j_{d_{\min}}$ is computed by summing over all possible error/erasure patterns for the all-zero code word such that $(2v + w + 2x + y + 2z) < d_{\min}$.

$$Q^j_{d_{\min}} = \underbrace{\left\{ \sum_{v=0}^{\left\lfloor \frac{d_{\min}-1}{2} \right\rfloor} \binom{n-j}{v} [(q^m-1)p]^v \sum_{w=0}^{d_{\min}-2v-1} \binom{n-j-v}{w} t^w s^{n-j-v-w} \right\}}_{\Theta \text{ coordinates}}$$

$$\cdot \underbrace{\left\{ \sum_{x=0}^{\left\lfloor \frac{d_{\min}-2v-w-1}{2} \right\rfloor} \binom{j}{x} [(q^m-2)p]^x \sum_{y=0}^{d_{\min}-2v-w-2x-1} \binom{j-x}{y} t^y \sum_{z=0}^{\left\lfloor \frac{d_{\min}-2v-w-2x-y-1}{2} \right\rfloor} \binom{j-x-y}{z} s^z p^{j-x-y-z} \right\}}_{\Phi \text{ coordinates}}$$

$$= \sum_{v=0}^{\left\lfloor \frac{d_{\min}-1}{2} \right\rfloor} \sum_{w=0}^{d_{\min}-2v-1} \sum_{x=0}^{\left\lfloor \frac{d_{\min}-2v-w-1}{2} \right\rfloor} \sum_{y=0}^{d_{\min}-2v-w-2x-1} \sum_{z=0}^{\left\lfloor \frac{d_{\min}-2v-w-2x-y-1}{2} \right\rfloor}$$

$$\cdot \binom{n-j}{v}\binom{n-j-v}{w}\binom{j}{x}\binom{j-x}{y}\binom{j-x-y}{z} \cdot (q^m-1)^v (q^m-2)^x p^{j+v-y-z} t^{w+y} s^{n-j-v-w+z}$$

The probability of decoder error is obtained by weighting $Q^j_{d_{\min}}$ by the number of code words of weight j and summing over all nonzero j.

$$P(E) = \sum_{j=d_{\min}}^{n} A_j Q^j_{d_{\min}} \tag{10-25}$$

The probability of decoder failure is then

$$P(F) = 1 - P(E) - \sum_{\nu=0}^{\left\lfloor\frac{d_{\min}-1}{2}\right\rfloor} \sum_{w=0}^{d_{\min}-2\nu-1} \binom{n}{\nu}\binom{n-\nu}{w} s^{n-\nu-w} t^{w}(1-s-t)^{\nu} \tag{10-26}$$

10.4 FADING CHANNELS

The amplitude of the communication channel is defined as a Rayleigh random variable a with the probability density function

$$p_a(a) = 2ae^{-a^2} \tag{10-27}$$

The channel is also corrupted by additive white Gaussian noise with one-sided power spectral density N_0. It is assumed that channel phase variations due to multipath fading are detected and removed during the demodulation process. Methods for realizing such performance through pilot-tone techniques are discussed in [Dav] and [McG]. It is also assumed that channel fading is frequency-nonselective.

To determine the performance of nonbinary codes with erasure decoding over this channel, one must first select a modulation format and derive expressions for the probabilities of channel-symbol error and erasure. There are two basic methods for generating channel-symbol erasures that are distinguished by the existence or nonexistence of forward channel-amplitude side information at the receiver; both are considered in the following analysis.

Let the channel-symbol alphabet have cardinality q^b and the code-word symbol alphabet have cardinality q^m. Nonbinary codes provide a level of burst-error correction for slowly fading channels that is a function of the amount by which m exceeds b. The rationale is that a deep fade may affect several consecutive channel symbols while affecting only a few code-word symbols, thus allowing for the use of higher-rate codes. It follows that this burst-error-correcting capability is lost if the channel symbols are individually interleaved (as opposed to their being interleaved in clusters, each cluster corresponding to a code-word symbol).

The channel-symbol error and erasure expressions developed in this section are translated into code-word symbol error and erasure expressions for both the code-word symbol and channel-symbol interleaved channels. An example is provided to demonstrate that nonbinary codes provide better performance when used in conjunction with code-word symbol interleaving as opposed to channel-symbol interleaving.

10.4.1 Erasure Generation Without Side Information

If no channel-amplitude information is available, the received signal space is partitioned into several nonerasure (reliable) decision regions and an erasure (unreliable) decision region. The declaration of erasures by the receiver is then solely a function of the position of the received signal within the signal space. Consider the case of the 8-PSK modulation format. Figure 10-10 shows how the signal space is partitioned into eight nonerasure decision regions $\{\Lambda_0, \Lambda_1, \ldots, \Lambda_7\}$ and an erasure region Λ_s.

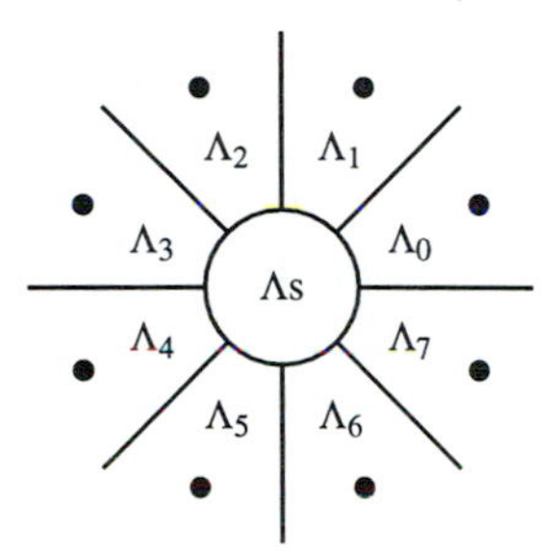

Figure 10-10. Decision Regions for the 8-PSK Modulation Format without Side Information [Sch]

In the general case an n-dimensional signal space Λ with $(2^b + 1)$ decision regions $\{\Lambda_s, \Lambda_0, \ldots, \Lambda_{2^b-1}\}$ is considered. Suppose that the channel symbol corresponding to the decision region Λ_i is transmitted. A conditional probability density function $p(\mathbf{z} \mid a, \Lambda_i)$ is derived for a fixed channel amplitude a and a received signal $\mathbf{z} = (z_0, z_1, \ldots, z_{n-1})$. This conditional pdf can be obtained by inverting the product of the characteristic functions for the Gaussian and Rayleigh processes defining the channel (e.g., the MPSK case is treated in Appendix 7A of [Pro]). The probability of channel-symbol error as a function of channel amplitude is then

$$p_{ce}(a) = \underset{i}{\mathrm{E}}\left\{\sum_{j \neq i}\left(\int_{\Lambda_j} p(\mathbf{z} \mid a, \Lambda_i)\, d\mathbf{z}\right)\right\} \tag{10-28}$$

where the expected-value operation is taken over all possible transmitted signals i corresponding to all reliable decision regions $\{\Lambda_i\}$. The probability of channel-symbol erasure as a function of a is

$$p_{cs}(a) = \underset{i}{E}\left\{\int_{\Lambda_s} p(\mathbf{z} \mid a, \Lambda_i)\, d\mathbf{z}\right\} \tag{10-29}$$

It is important to note that when side information is unavailable, erasures can be caused by the AWGN process as well as the channel fading process.

10.4.2 Erasure Generation with Side Information

If it is assumed that side information containing the exact value of the channel amplitude a is available, a simpler approach to erasure generation can be considered. Hagenauer and Lutz [Hag2] have examined the case in which erasure generation is based solely on the value of a, eliminating the impact of the AWGN process on the probability of channel and code-word symbol erasures. Let λ_s be the erasure threshold. A received channel symbol is declared an erasure any time the channel amplitude is less than λ_s. The probability of this occurring is

$$p_{cs}^{SI} = \int_0^{\lambda_s(a)} p_a(a)\, da \tag{10-30}$$

The derivation of an expression for the probability of channel-symbol error is also quite simple. Once the channel amplitude is known for a received channel symbol and it has been determined that the amplitude is not below the erasure threshold, the decision regions for the demodulator are scaled to match the channel amplitude, and the symbol decision is made. The 8-PSK decision regions for the side information case are shown in Figure 10-11. The signal scaling caused by the fading channel amplitude is indicated. Owing to the radial symmetry of the signal constellation in this example, the decision regions do not change with the value of a. This is not the case, however, with nonconstant envelope modulation formats (e.g., ASK and QAM).

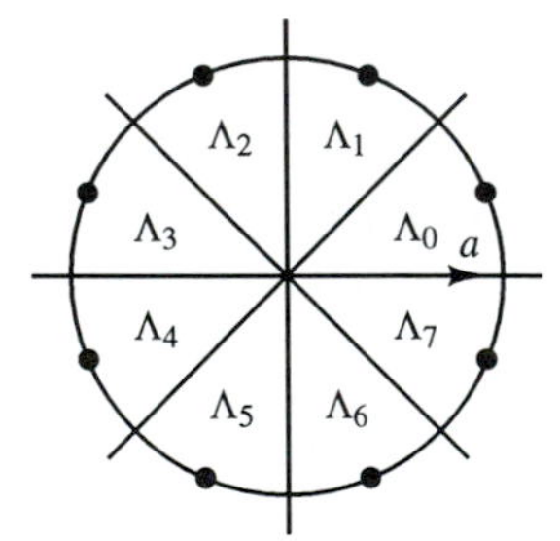

Figure 10-11. Decision Regions for the 8-PSK Modulation Format with Side Information

The probability of channel-symbol error for channel amplitude a and a given modulation format is obtained by taking the standard AWGN symbol error rate expression and weighting the signal energy by a^2. For example, the probability of bit error for coherent BPSK is

$$p_{ce}^{SI}(a) = Q\left(a\sqrt{\frac{2RE_b}{N_0}}\right) \tag{10-31}$$

where R is the rate of the code in use.

10.4.3 Code-Word Symbol Errors and Erasures on Code-Word Symbol Interleaved Slowly Fading Channels

In the literature frequent reference is made to "fast" and "slowly" fading channels. These terms can be understood only when taken in relation to the channel-symbol transmission rate. It is assumed here that a slowly fading channel exhibits fades whose duration exceeds the time required to transmit several channel symbols. A fade affecting one channel symbol is thus highly likely to affect temporally adjacent channel symbols. The techniques used to combat this correlative effect are one of the concerns of this section. Two interleaving techniques are considered: interleaving at the *channel*-symbol level and interleaving at the *code-word* symbol level.

The analysis of the code-word symbol interleaved slowly fading channel is predicated on the following assumptions.

- Channel amplitude is constant over any m/b consecutive channel symbols.
- The channel symbols are interleaved to infinite depth in clusters corresponding to individual code-word symbols. The noise processes affecting adjacent code-word symbols within a code word are thus uncorrelated.

Interleaving is used to eliminate the correlation of the noise/fading process affecting adjacent symbols in a received code word, but not that between adjacent channel symbols comprising a single received code-word symbol.

A code-word symbol erasure occurs whenever one or more of the m/b constituent channel symbols are declared to be erasures. The derivation of the code-word symbol erasure probabilities for both channel-symbol erasure generation mechanisms is straightforward. For the case where side information is not available and the channel amplitude is a constant a, the probability of there being at least one erased channel symbol among m/b channel symbols is

$$p_s(a) = 1 - (1 - p_{cs}(a))^{m/b} \tag{10-32}$$

The code-word symbol erasure probability p_s is then obtained through an expected-value operation using the probability density function for a. For the case with side information, the probability of code-word symbol erasure is determined solely by the value of a, which is assumed to be constant during the transmission of the code-word symbol. The probability of symbol erasure in this case is simply the probability that the value of a at any given moment is below the erasure threshold λ_s. The following expressions result.

$$p_s = \begin{cases} \int_0^\infty [1 - (1 - p_{cs}(a))^{m/b}]p_a(a)\,da & \text{no side information} \\ p_{cs}^{SI} & \text{side information} \end{cases} \tag{10-33}$$

A code-word symbol error occurs whenever the code-word symbol has not been declared an erasure and one or more of the constituent channel symbols is in error. For the case where side information is not available, the code-word symbol error probability for constant channel amplitude a is

$$p_e(a) = (1 - p_{cs}(a))^{m/b} - (1 - p_{ce}(a) - p_{cs}(a))^{m/b} \tag{10-34}$$

The code-word symbol error probability is then obtained through an expected-value operation. The corresponding result for the case with side information is similar, though the limits of integration rule out the possibility of a channel-symbol erasure, simplifying the integrand. The probability of code-word symbol error in both cases is then as shown in Eq. (10-35).

$$p_e = \begin{cases} \displaystyle\int_0^\infty [(1 - p_{cs}(a))^{m/b} - (1 - p_{ce}(a) - p_{cs}(a))^{m/b}]\, p_a(a)\, da & \text{no side information} \\ \displaystyle\int_{\lambda_s}^\infty [1 - (1 - p_{ce}^{SI}(a))^{m/b}]\, p_a(a)\, da & \text{side information} \end{cases} \tag{10-35}$$

10.4.4 Code-Word Symbol Errors and Erasures on Channel-Symbol Interleaved Slowly Fading Channels

The analytical ground rules are changed considerably for the case of channel-symbol interleaving. The following assumptions are made.

- The channel amplitude is constant over the time required to transmit one or more channel symbols.
- The channel symbols are interleaved to infinite depth. The noise processes affecting adjacent channel symbols are thus uncorrelated.

From the decoder's perspective, the channel-symbol interleaved channel appears to fade more rapidly than the code-word symbol interleaved channel, which in turn fades more rapidly than a stationary channel. It should be noted, however, that for both interleaved channels, *the physical propagation medium may be the same*: a Rayleigh fading channel whose fade duration exceeds the time required for the transmission of several channel symbols. It is the method used to format the coded information prior to transmission that differs between the code-word symbol and channel-symbol interleaved cases. Consider, for example, a mobile radio channel that is used for the transmission of data. Variations in vehicle velocity cause the physical channel fade frequency and duration to vary. In almost all applications, however, the fading is slow with respect to the channel-symbol transmission rate. Assuming sufficient depth, channel (code) symbol interleaving ensures that a given fade does not affect adjacent channel (code) symbols. The *effective* fade rate seen by the receiver after deinterleaving thus appears to be faster than the channel (code) symbol transmission rate.

The expected value operations of the previous section are performed at the channel-symbol level instead of the code-word symbol level for the case of the channel-symbol interleaved channel. Consider first the probability of channel-symbol erasure for the case without side information. The expected value of the channel erasure probability for constant a is computed using the probability density function for a as follows.

$$p_{cs} = \int_0^\infty p_{cs}(a) p_a(a)\, da \tag{10-36}$$

For the case with side information, the probability of channel-symbol erasure is again the probability that the channel amplitude during the bit transmission time is less than λ_s.

$$p_{cs}^{SI} = \int_0^{\lambda_s} p_a(a)\, da \tag{10-37}$$

In both cases the probability of code-word symbol erasure is the probability that, among m/b consecutive channel symbols, at least one symbol is erased. The probability of code-word symbol erasure for both cases is then

$$p_s = \begin{cases} 1 - (1 - p_{cs})^{m/b} & \text{no side information} \\ 1 - (1 - p_{cs}^{SI})^{m/b} & \text{side information} \end{cases} \tag{10-38}$$

The probability of channel-symbol error for the case without side information is the expected value of the channel error probability for fixed channel amplitude a.

$$p_{ce} = \int_0^{\infty} p_{ce}(a) p_a(a)\, da \tag{10-39}$$

The probability of code-word symbol error is then the probability that none of the channel symbols are erased and at least one channel symbol is in error. For the case with side information, the limits of integration are changed to reflect the erasure threshold.

$$p_{ce}^{SI} = \int_{\lambda_s}^{\infty} p_{ce}^{SI}(a) p_a(a)\, da \tag{10-40}$$

The probability of code-word symbol error for both cases is thus

$$p_e = \begin{cases} (1 - p_{cs})^{m/b} - (1 - p_{cs} - p_{ce})^{m/b} & \text{no side information} \\ (1 - p_{cs}^{SI})^{m/b} - (1 - p_{cs}^{SI} - p_{ce}^{SI})^{m/b} & \text{side information} \end{cases} \tag{10-41}$$

10.4.5 Comparing Channel-Symbol and Code-Word Symbol Interleaving

From the standpoint of a random-error-correcting nonbinary error control code, the probabilities of error and erasure for the symbols entering the decoder completely determine reliability performance. The following figures thus focus on code-word symbol error and erasure probabilities. These can in turn be translated directly into word error rates using the preceding development. It should be noted here, however, that the word error rate is a monotonically increasing function of the channel-symbol error rate. A reduction in the latter will thus ensure a reduction in the former.

Figure 10-12 compares the code-word symbol error probabilities for the channel and code-word symbol interleaved channels. These curves assume eight-bit code-word symbols transmitted using a coherent BPSK modem and a noncoherent BFSK modem. Both modems are assumed to have channel-amplitude side information. These curves indicate an increase in the probability of symbol error that corresponds to an effective reduction in E_b/N_0 of 1 to 2 dB when eight-bit code word symbols are interleaved bit-by-bit instead of symbol-by-symbol. In the examples it is seen that this reduction has a substantial impact on the decoder error probability for Reed-Solomon codes.

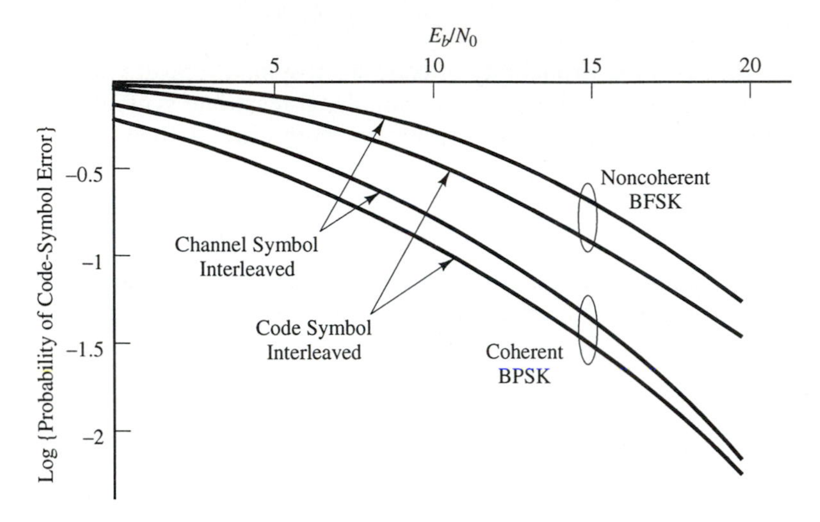

Figure 10-12. Probability-of-Symbol-Error Curves for Eight-Bit Code-Word Symbols ([Wic7], © 1992 IEEE)

Figure 10-13 compares the probabilities of code-word symbol erasure for the channel and code-word symbol interleaved channels as a function of the erasure threshold. As in Figure 10-12, the transmission of eight-bit code-word symbols over a binary modem with side information is assumed. The probability of code-word symbol erasure is substantially higher for the channel-symbol interleaved channel.

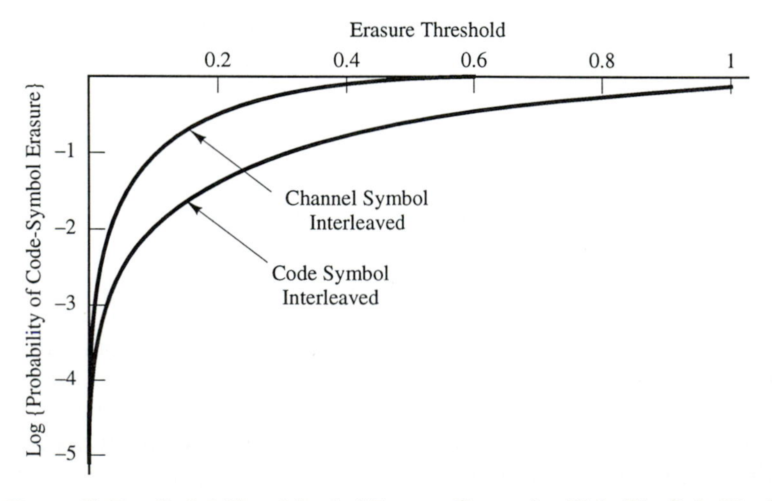

Figure 10-13. Probability-of-Symbol-Erasure Curves for Eight-Bit Code-Word Symbols ([Wic7], © 1992 IEEE)

Equation (10-33), (10-35), (10-38), and (10-41) can be used in deriving the performance of an arbitrary nonbinary code so long as appropriate reliability and throughput expressions as a function of code-word symbol erasure and error probabilities are known.

In the following examples the performance of a Reed-Solomon error control system with erasure decoding is examined for a coherent BPSK modem used over a code-word symbol interleaved slowly fading Rayleigh channel. The performance of the system over a channel-symbol interleaved channel is degraded because of the 1 to 2 dB decrease in the effective E_b/N_0 (see Fig. 10-12). It is assumed that side information is available for the declaration of erased channel symbols.

These examples consider the impact of a variation in the erasure threshold λ_s. In Figures 10-14, 10-15, and 10-16 the performance of a (16, 12) Reed-Solomon code is examined. Figure 10-14 shows the variation in the probability of word error $P(E)$ as the erasure threshold is increased. It is clear that a significant amount of improvement in reliability performance can be obtained through erasure decoding in the RS error control system at medium to high E_b/N_0, but very little improvement is obtained at smaller values of E_b/N_0.

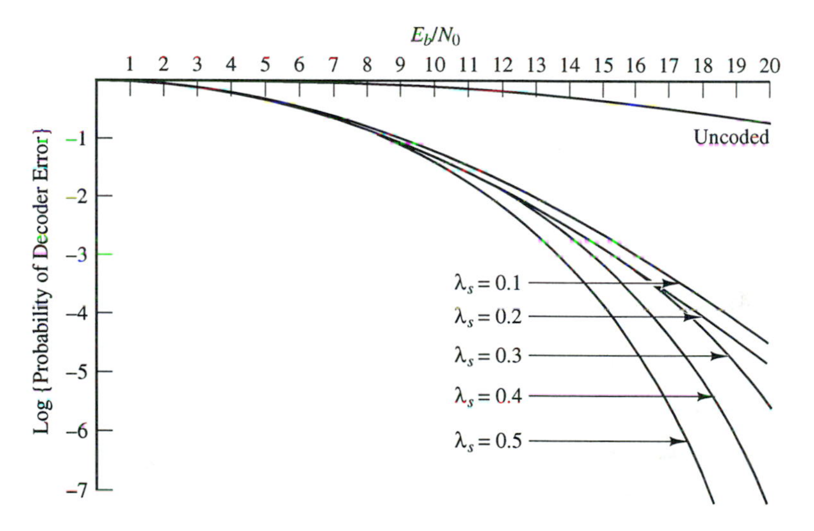

Figure 10-14. Probability of Word Error for a (16, 12) Reed-Solomon Code ([Wic7], © 1992 IEEE)

It should be noted that these word-error-rate curves and those that follow assume that a hybrid-ARQ protocol allows for an unlimited number of retransmission attempts. As a result, the word error rate in all cases approaches unity as the signal-to-noise ratio is reduced. The performance of a system that employs retry limits can be understood only through comparison of the word-error-rate curves with the throughput curves that follow. It is then seen that for extremely low signal-to-noise ratios the throughput is essentially zero, so that given a reasonable retry limit, it is highly improbable that erroneous words are accepted by the decoder.

Figure 10-15 shows the probability of failure $P(E)$ for the (16, 12) code as a function of E_b/N_0 and the erasure threshold λ_s. The floor effect is due to the fact that the probability of erasure is independent of the signal-to-noise ratio on the channel. After a certain point, any further reduction in the probability of symbol error is negated by the probability that the number of erasures alone will be sufficient to no longer meet the requirement $(2e + s) < d_{\min}$.

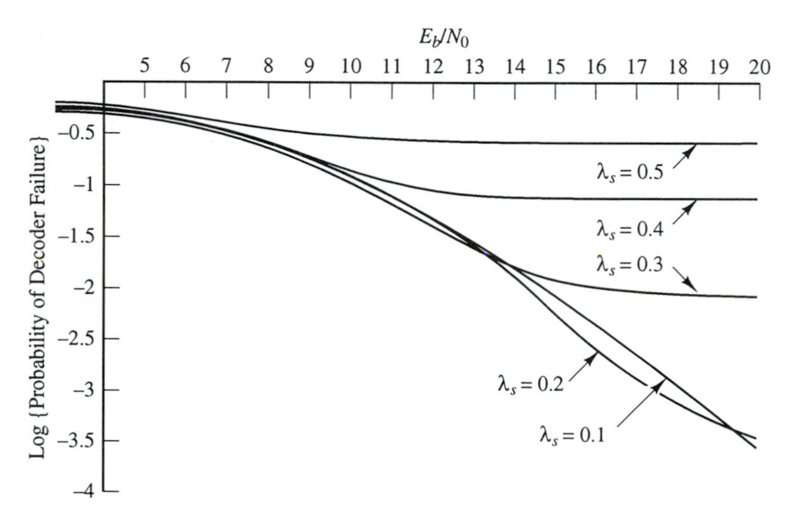

Figure 10-15. Probability of Word Failure for a (16, 12) Reed-Solomon Code ([Wic7], © 1992 IEEE)

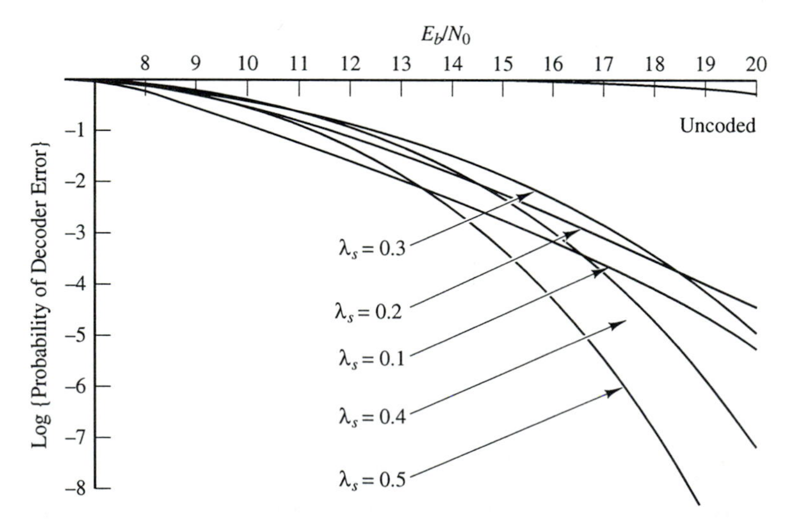

Figure 10-16. Probability of Decoder Error for a (64, 56) Reed-Solomon Code ([Wic7], © 1992 IEEE)

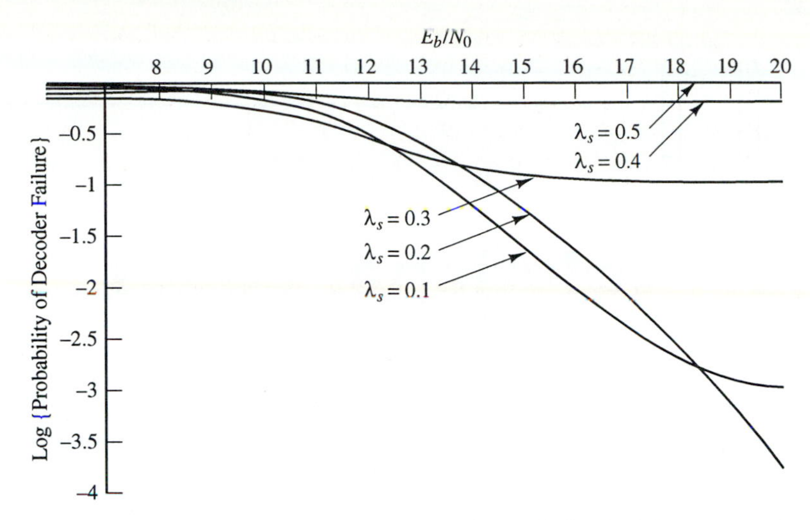

Figure 10-17. Probability of Decoder Failure for a (64, 56) Reed-Solomon Code ([Wic7], © 1992 IEEE)

Figures 10-16 and 10-17 show the performance of the (64, 56) Reed-Solomon code. Figure 10-16 shows the same asymptotic trends seen in Figure 10-14 but also shows that higher values of the erasure threshold can actually cause performance to degrade at low signal-to-noise ratios. The reason for this is seen in Figure 10-15; for the cases $\lambda_s = 0.4$ and 0.5 the probability of word error is being significantly increased by the large number of attempts required before the packet is accepted by the receiver. As E_b/N_0 increases, however, a point is reached beyond which the higher erasure thresholds provide better reliability performance than the lower erasure thresholds.

11

Convolutional Codes

Convolutional codes offer an approach to error control substantially different from that of block codes. A convolutional encoder converts the entire data stream, regardless of its length, into a single code word. Block encoders, on the other hand, segment the data stream into "blocks" of some fixed length k. These blocks are then mapped onto code words of some fixed length n. This fundamental difference in approach imparts a different nature to the design and evaluation of convolutional codes. Block codes are generally developed and analyzed through the use of algebraic/combinatorial techniques, while convolutional codes have been amenable almost solely to heuristic construction techniques.

Convolutional codes were first introduced by Elias in 1955 [Eli]. He showed that redundancy can be introduced into a data stream through the use of a linear shift register. He also showed that the resulting codes were very good when randomly chosen. This result was very interesting, for it correlated with Shannon's more theoretical work showing that there exist randomly selected codes that, on the average, provide arbitrarily high levels of reliability given data transmission at a rate less than the channel capacity [Sha1]. In 1961 Wozencraft and Reiffen described the first practical decoding algorithm for convolutional codes [Woz2]. This algorithm was the first of a class of "sequential algorithms" that provide fast, but suboptimal, decoding for convolutional codes. In 1963 Massey presented decoding algorithms for both block and convolutional codes that used a series of parity-check equations in a voting scheme [Mas1]. These "threshold" decoding algorithms (also called "majority-logic" decoding algorithms) were optimal for several classes of block codes (e.g., Reed-Muller codes), but, when applied to convolutional codes, were generally less powerful than the sequential algorithms.[1] Fano [Fan] and Jelinek [Jel] described modified sequential algorithms in 1963 and 1969, respectively, that improved on the performance of the Wozencraft-Reiffen algorithm, but were still suboptimal. In 1967 Viterbi discovered a third approach to decoding convolutional codes which he showed to be "asymptotically optimal" [Vit2]. Two years later Omura [Omu] showed that Viterbi's algorithm was a solution to the problem of finding the minimal-weight path through a weighted, directed graph. In 1973 Forney showed that what is now known as the Viterbi algorithm is actually a maximum-likelihood decoding algorithm for convolutional codes [For2].

In this chapter we consider the basic structure of convolutional codes. We begin by considering the encoder design and the various means of relating the output of the encoder to the input data stream. We then consider techniques for evaluating and comparing convolutional codes. An emphasis is placed on the graph-theoretic interpretation of convolutional codes, which leads to an elegant means for obtaining complete weight distributions.

In the two chapters that follow this one, we investigate the Viterbi algorithm and the various sequential decoding algorithms and derive appropriate performance models.

11.1 LINEAR CONVOLUTIONAL ENCODERS

Figure 11-1 shows a typical rate-1/2 linear convolutional encoder. The rate of this encoder is established by the fact that the encoder outputs two bits for every input bit. In general, an encoder with k inputs and n outputs is said to have rate k/n. For example, Figure 11-2 shows a rate-2/3 encoder.

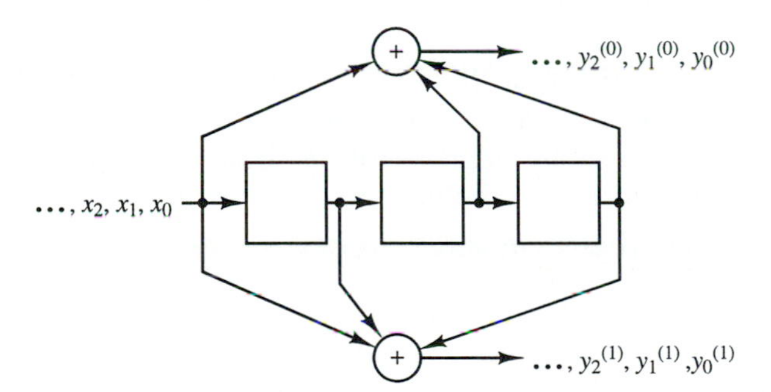

Figure 11-1. A Rate-1/2 Linear Convolutional Encoder

In Figure 11-1 the binary data stream $\mathbf{x} = (x_0, x_1, x_2, \ldots)$ is fed into a shift-register circuit consisting of a series of memory elements. With each successive input to the shift register, the values of the memory elements are tapped off and added

[1] Threshold decoding for convolutional codes does possess the benefit of an extremely fast shift-register implementation and is thus still a good choice for very high speed applications that need only a moderate amount of error control.

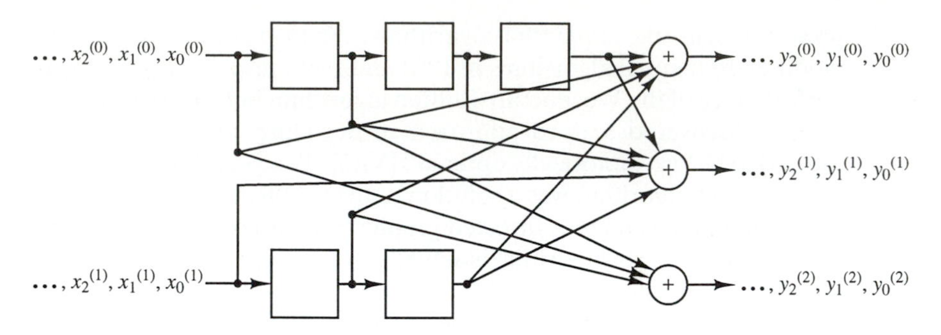

Figure 11-2. A Rate-2/3 Linear Convolutional Encoder

according to a fixed pattern, creating a pair of output coded data streams $\mathbf{y}^{(0)} = (y_0^{(0)}, y_1^{(0)}, y_2^{(0)}, \ldots)$ and $\mathbf{y}^{(1)} = (y_0^{(1)}, y_1^{(1)}, y_2^{(1)}, \ldots)$. These output streams can be multiplexed to create a single coded data stream $\mathbf{y} = (y_0^{(0)} y_0^{(1)}, y_1^{(0)} y_1^{(1)}, y_2^{(0)} y_2^{(1)}, \ldots)$. $\mathbf{y}$ is the convolutional code word. If there are multiple input streams, we can similarly refer to a single interleaved input stream $\mathbf{x} = (x_0^{(0)}, x_0^{(1)}, \ldots, x_0^{(k-1)}, x_1^{(0)}, x_1^{(1)}, \ldots, x_1^{(k-1)}, \ldots)$.

These encoders can be viewed in a number of different ways. For example, they may be considered as finite impulse response (FIR) digital filters or as finite state automata. Both of these approaches and their corresponding analytical tools yield some interesting insights into the structure of convolutional codes.

Forney [For3] has shown that all linear convolutional encoders are equivalent to a "minimal" encoder that is feedback-free. We may thus focus our attention on feedforward structures like those in the figures without excluding any of the linear codes. This structure does not, however, allow for the consideration of nonlinear convolutional codes.

Each element in the interleaved output stream $\mathbf{y}$ is a linear combination of the elements in the input streams $\mathbf{x}^{(0)}, \mathbf{x}^{(1)}, \ldots, \mathbf{x}^{(k-1)}$. For example, the output stream $\mathbf{y}^{(1)}$ in Figure 11-1 is computed from a single input stream $\mathbf{x}$. It is assumed that the shift-register contents are initialized to zero before the encoding process begins.

$$\begin{aligned}
y_0^{(1)} &= x_0^{(0)} + 0 + 0 \\
y_1^{(1)} &= x_1^{(0)} + x_0^{(0)} + 0 \\
y_2^{(1)} &= x_2^{(0)} + x_1^{(0)} + 0 \\
y_3^{(1)} &= x_3^{(0)} + x_2^{(0)} + x_0^{(0)} \\
y_4^{(1)} &= x_4^{(0)} + x_3^{(0)} + x_1^{(0)} \\
&\vdots
\end{aligned} \tag{11-1}$$

An arbitrary coordinate $y_j^{(1)}$ in the output stream $\mathbf{y}^{(1)}$ can be represented as follows.

$$y_i^{(1)} = x_i^{(0)} + x_{i-1}^{(0)} + x_{i-3}^{(0)} \tag{11-2}$$

Example 11-1

The rate-1/2 encoder in Figure 11-1 is used to encode the information sequence $\mathbf{x} = (10110)$. We obtain the following coded output sequences.

$$\mathbf{y}^{(0)} = (10001010)$$
$$\mathbf{y}^{(1)} = (11111110)$$

The convolutional code word corresponding to $\mathbf{x} = (10110)$ is then

$$\mathbf{y} = (11, 01, 01, 01, 11, 01, 11, 00)$$

The commas are used to bracket blocks of bits that are output at the same time. ■

Equation (11-2) shows that convolutional encoders like those in Figures 11-1 and 11-2 are linear, for if $\mathbf{y}_1$ and $\mathbf{y}_2$ are the code words corresponding to inputs $\mathbf{x}_1$ and $\mathbf{x}_2$, respectively, then $(\mathbf{y}_1 + \mathbf{y}_2)$ is the code word corresponding to the input $(\mathbf{x}_1 + \mathbf{x}_2)$. The linear structure of these codes allows for the use of some powerful techniques from linear systems theory.

An **impulse response** $\mathbf{g}_j^{(i)}$ is obtained for the ith output of an encoder by applying a single 1 at the jth input followed by a string of zeros. Strings of zeros are applied to all other inputs. The data stream $\mathbf{x}^{(j)} = \boldsymbol{\delta} = (10000\ldots)$ serves the same role as the Dirac delta function $\delta(x)$ in the analysis of continuous systems. The impulse responses for the encoder in Figure 11-1 are

$$\begin{aligned}\mathbf{g}^{(0)} &= (1011)\\ \mathbf{g}^{(1)} &= (1101)\end{aligned} \tag{11-3}$$

The encoder in Figure 11-2 has the following impulse responses. Note that each response is referenced to a single input and a single output.

$$\begin{aligned}\mathbf{g}_0^{(0)} &= (1001), & \mathbf{g}_1^{(0)} &= (0110)\\ \mathbf{g}_0^{(1)} &= (0111), & \mathbf{g}_1^{(1)} &= (1010)\\ \mathbf{g}_0^{(2)} &= (1100), & \mathbf{g}_1^{(2)} &= (0100)\end{aligned} \tag{11-4}$$

The impulse responses are often referred to as **generator sequences**, because their relationship to the code words generated by the corresponding convolutional encoder is similar to that between generator polynomials and code words in a cyclic code.

The generator sequences have been terminated at the point beyond which all of the output streams contain nothing but zeros. In this form it is clear that the generator sequences can be "read off" the encoder through an examination of the tap sequence.

Since there are three memory elements in the shift register in Figure 11-1, each bit in the input data stream can affect at most 4 bits, hence the length of the generator sequences in Eq. (11-3). In general, an input data stream $\mathbf{x}^{(i)}$ for a given convolutional encoder is fed into a shift register with m_i memory elements. The amount of memory in the encoder determines the extent to which an input bit directly affects the output data streams.

Definition 11-1—Constraint Length

The **constraint length** K of a convolutional code is the maximum number of bits in a single output stream that can be affected by any input bit.

As defined above, the constraint length of a convolutional code is the maximum number of taps on the shift registers in the encoder. This can be somewhat confusing, so in practice the constraint length is usually taken to be the length of the longest input shift register plus one.

$$K \triangleq 1 + \max_i m_i \tag{11-5}$$

It should be noted that the above definition is not in universal use. Some authors have adopted the understandable approach that constraint length should cover the number of bits in all output streams affected by a single input bit [Lin1], in which case the right-hand side of Eq. (11-5) is multiplied by n, the number of output coded data streams. But since (11-5) is in standard use in many military and industrial publications, we shall use it in this text.

The **total memory** M for a convolutional encoder is defined to be the total number of memory elements in the encoder.

$$M = \sum_{i=0}^{k-1} m_i \tag{11-6}$$

The total memory of an encoder has a strong impact on the complexity of the corresponding Viterbi decoder, as seen in the next chapter.

As a notational convenience we define the **maximal memory order** m to be the length of the longest input shift register. m is thus one less than the constraint length K.

In Example 11-1 the output code word is significantly longer than twice the length of the input. This is because the convolutional code word is not complete until the contents of the memory cells in the encoder have been "emptied" into the output coded data stream. These additional bits of redundancy reduce the effective rate of the code and are usually characterized in terms of a **fractional rate loss**. After the last input bit of an L-bit message enters the encoder, the shift registers must be clocked an additional m times before they have all emptied their contents into the output coded data stream. As there are n output streams, an additional nm bits have been added to the code word. We can thus compute the fractional rate loss for an L-bit input word as follows.

$$\gamma = \frac{R - R_{\text{effective}}}{R} = \left(\frac{k}{n}\right)^{-1}\left\{\left(\frac{k}{n}\right) - \left[\frac{L}{\left(\frac{n}{k}\right)L + nm}\right]\right\} = \frac{km}{L + km} \tag{11-7}$$

If the input is long relative to the total memory of the encoder, then the fractional rate loss is negligible.

Example 11-2—Fractional Rate Loss

The fractional rate loss for the convolutional code word in Example 11-1 is

$$\gamma = \frac{km}{L + km} = \frac{(1)(3)}{5 + (1)(3)} = \frac{3}{8} = 0.375$$

If the length of the input vector is increased to 12, then the fractional rate loss is reduced to 0.2. ■

Using the generator sequences developed above, Eq. (11-2) can be reexpressed in the following general form.

$$y_i^{(j)} = \sum_{l=0}^{m} x_{i-l} g_l^{(j)} \tag{11-8}$$

Equation (11-8) is the expression for the discrete convolution of a pair of sequences, hence the term "convolutional codes." Each coded output sequence $\mathbf{y}^{(i)}$ in a rate-$1/n$ code is the convolution of the input sequence $\mathbf{x}$ and the impulse response $\mathbf{g}^{(i)}$.

$$\mathbf{y}^{(i)} = \mathbf{x} * \mathbf{g}^{(i)} \tag{11-9}$$

Equation (11-8) can be further generalized for the k-input encoder.

$$y_j^{(i)} = \sum_{t=0}^{k-1} \left(\sum_{l=0}^{m} x_{j-l}^{(t)} g_{t,l}^{(i)} \right) \tag{11-10}$$

Equation (11-9) then generalizes to the following form.

$$\mathbf{y}^{(i)} = \sum_{t=0}^{k-1} (\mathbf{x}^{(t)} * \mathbf{g}_t^{(i)}) \tag{11-11}$$

Equation (11-8) can be reexpressed as a matrix multiplication operation, thus providing a generator matrix similar to that developed for block codes. The primary difference arises from the fact that the input sequence is not necessarily bounded in length, and thus the generator and parity check matrices for convolutional codes are semi-infinite.

For a rate-1/2 code the generator matrix is formed by interleaving the two generator sequences $\mathbf{g}^{(0)}$ and $\mathbf{g}^{(1)}$ as follows.

$$\mathbf{G} = \begin{bmatrix} g_0^{(0)} g_0^{(1)} & g_1^{(0)} g_1^{(1)} & g_2^{(0)} g_2^{(1)} & \cdots & g_m^{(0)} g_m^{(1)} & & & \mathbf{0} \\ & g_0^{(0)} g_0^{(1)} & g_1^{(0)} g_1^{(1)} & g_2^{(0)} g_2^{(1)} & \cdots & g_m^{(0)} g_m^{(1)} & & \\ & & g_0^{(0)} g_0^{(1)} & g_1^{(0)} g_1^{(1)} & g_2^{(0)} g_2^{(1)} & \cdots & g_m^{(0)} g_m^{(1)} & \\ & & & g_0^{(0)} g_0^{(1)} & g_1^{(0)} g_1^{(1)} & g_2^{(0)} g_2^{(1)} & \cdots & g_m^{(0)} g_m^{(1)} \\ \mathbf{0} & & & \ddots & \ddots & \ddots & \cdots & \ddots \end{bmatrix} \tag{11-12}$$

The code word $\mathbf{y}$ corresponding to an information sequence $\mathbf{x}$ is then obtained through matrix multiplication.

$$\mathbf{y} = \mathbf{xG} \tag{11-13}$$

Example 11-3—Convolutional Generator Matrices

The information sequence $\mathbf{x} = (1011)$ is to be encoded using the rate-1/2 encoder in Figure 11-1. The generator sequences for this encoder are $\mathbf{g}^{(0)} = (1011)$ and $\mathbf{g}^{(1)} = (1101)$. Since $\mathbf{x}$ has length 4, the general form in Eq. (11-12) is truncated so that it contains 4 rows. The code word corresponding to the message sequence $\mathbf{x} = (1011)$ is then

$$\mathbf{y} = \mathbf{xG} = (1011) \cdot \begin{bmatrix} 11 & 01 & 10 & 11 & 00 & 00 & 00 \\ 00 & 11 & 01 & 10 & 11 & 00 & 00 \\ 00 & 00 & 11 & 01 & 10 & 11 & 00 \\ 00 & 00 & 00 & 11 & 01 & 10 & 11 \end{bmatrix} = (11, 01, 01, 01, 11, 01, 11) \quad \blacksquare$$

Since the action of a convolutional encoder can be described as a discrete convolution operation, it follows that an appropriate transform will provide a simpler multiplicative representation for encoding. In this case we apply the delay transform, or ***D*-transform** [For3].

$$\begin{aligned} \mathbf{x}^{(i)} &= (x_0^{(i)}, x_1^{(i)}, x_2^{(i)}, \ldots) \leftrightarrow \mathbf{X}^{(i)}(D) = x_0^{(i)} + x_1^{(i)} D + x_2^{(i)} D^2 + \cdots \\ \mathbf{y}^{(i)} &= (y_0^{(i)}, y_1^{(i)}, y_2^{(i)}, \ldots) \leftrightarrow \mathbf{Y}^{(i)}(D) = y_0^{(i)} + y_1^{(i)} D + y_2^{(i)} D^2 + \cdots \\ \mathbf{g}_j^{(i)} &= (g_{j0}^{(i)}, g_{j1}^{(i)}, g_{j2}^{(i)}, \ldots) \leftrightarrow \mathbf{G}_j^{(i)}(D) = g_{j0}^{(i)} + g_{j1}^{(i)} D + g_{j2}^{(i)} D^2 + \cdots \end{aligned} \tag{11-14}$$

The indeterminate D is interpreted as a delay operator, with the exponent of D denoting the number of time units the coefficient is delayed with respect to the coefficient of the D^0 term. The delay transform is similar to the z-transform, but there is no corresponding interpretation of D as a number in the complex plane.

Using the D-transform, the encoding operation for a single input encoder can be represented as follows.

$$\mathbf{Y}^{(i)}(D) = \mathbf{X}(D)\mathbf{G}^{(i)}(D) \tag{11-15}$$

For a multiple input system, we obtain the following expression.

$$\mathbf{Y}^{(i)}(D) = \sum_{j=0}^{k-1} \mathbf{X}^{(j)}(D)\mathbf{G}_j^{(i)}(D) \tag{11-16}$$

Equation (11-16) can be represented as a matrix multiplication operation.

$$\mathbf{Y}^{(i)}(D) = [\mathbf{X}^{(0)}(D) \quad \mathbf{X}^{(1)}(D) \quad \cdots \quad \mathbf{X}^{(k-1)}(D)] \begin{bmatrix} \mathbf{G}_0^{(i)}(D) \\ \mathbf{G}_1^{(i)}(D) \\ \vdots \\ \mathbf{G}_{k-1}^{(i)}(D) \end{bmatrix} \tag{11-17}$$

Given the definition

$$\mathbf{Y}(D) = (\mathbf{Y}^{(0)}(D), \mathbf{Y}^{(1)}(D), \ldots, \mathbf{Y}^{(n-1)}(D))$$

we can obtain the following expression for the complete encoding operation.

$$\mathbf{Y}(D) = \mathbf{X}(D)\mathbf{G}(D) = [\mathbf{X}^{(0)}(D) \quad \mathbf{X}^{(1)}(D) \quad \cdots \quad \mathbf{X}^{(k-1)}(D)] \cdot \begin{bmatrix} \mathbf{G}_0^{(0)}(D) & \mathbf{G}_0^{(1)}(D) & \cdots & \mathbf{G}_0^{(n-1)}(D) \\ \mathbf{G}_1^{(0)}(D) & \mathbf{G}_1^{(1)}(D) & \cdots & \mathbf{G}_1^{(n-1)}(D) \\ \vdots & \vdots & \ddots & \vdots \\ \mathbf{G}_{k-1}^{(0)}(D) & \mathbf{G}_{k-1}^{(1)}(D) & \cdots & \mathbf{G}_{k-1}^{(n-1)}(D) \end{bmatrix} \tag{11-18}$$

The matrix $\mathbf{G}(D)$ is called the **transfer-function matrix** for the encoder.

Example 11-4—*D*-Transforms and Convolutional Encoding

The rate-2/3 encoder in Figure 11-2 is to be used to encode the message

$$\mathbf{x} = (11, 10, 11)$$

The message is separated into two data streams and the following D-transforms computed.

$$\mathbf{X}^{(0)}(D) = 1 + D + D^2$$

$$\mathbf{X}^{(1)}(D) = 1 + D^2$$

The D-transform of the output coded data streams is then computed as follows.

$$\mathbf{Y}(D) = [1 + D + D^2 \quad 1 + D^2]\begin{bmatrix} 1 + D^3 & D + D^2 + D^3 & 1 + D \\ D + D^2 & 1 + D^2 & D \end{bmatrix}$$

$$= [1 + D^5 \quad 1 + D + D^3 + D^4 + D^5 \quad 1 + D]$$

Inverting the transform we obtain

$$\mathbf{y}^{(0)} = (100001)$$

$$\mathbf{y}^{(1)} = (110111)$$

$$\mathbf{y}^{(2)} = (110000)$$

The output code word is then $\mathbf{y} = (111, 011, 000, 010, 010, 110)$. ■

A convolutional code is said to be **systematic** if, as in systematic block codes, the input data is reproduced unaltered in the output code word. A rate-1/2 systematic convolutional encoder is shown in Figure 11-3.

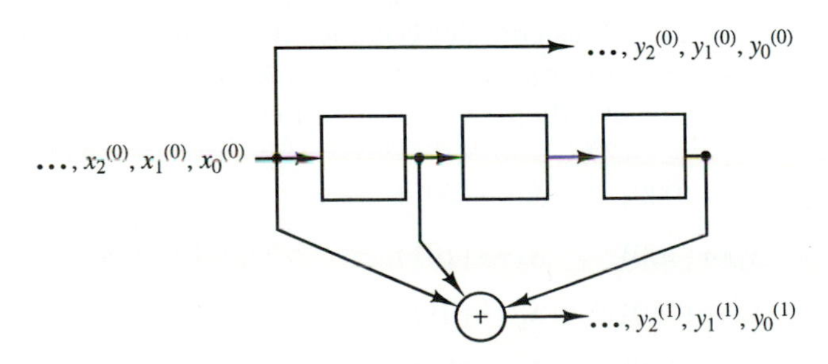

Figure 11-3. A Rate-1/2 Systematic Convolutional Encoder

In Figure 11-3 the output coded data stream $\mathbf{y}^{(0)}$ is identical to the input data stream $\mathbf{x}$. In a rate-k/n systematic encoder, k of the n output coded data streams $\mathbf{y}^{(i)}$ are identical to the k input streams $\mathbf{x}^{(i)}$. The corresponding transfer function matrix $\mathbf{G}(D)$ contains the $(k \times k)$ identity matrix $\mathbf{I}_k$ as a submatrix. It was noted earlier in the text that there exists a systematic encoding for any linear block code. This is not the case with convolutional codes, and it will be seen that the best convolutional codes are generally not systematic. In certain applications, however, systematic convolutional codes are preferred. For example, majority-logic decoding is substantially easier to implement with systematic convolutional codes than with nonsystematic codes.

Systematic codes also allow for easier inversion of the operation of the encoder. Nonsystematic codes require more complicated sequential circuits to perform the inversion, assuming that such circuits exist at all. Consider a rate-k/n code $\mathbf{C}$ with transfer-function matrix $\mathbf{G}(D)$. If the corresponding inversion circuit exists, it will have a transfer-function matrix of the form $\mathbf{G}^{-1}(D)$ such that $\mathbf{G}(D)\mathbf{G}^{-1}(D) = \mathbf{I}D^l$, where $\mathbf{I}$ is the $(k \times k)$ identity matrix and l is a nonnegative integer. Massey and Sain [Mas3] and Forney [For3] have conducted detailed investigations into the existence and design of inversion circuits for convolutional codes. Massey and Sain provide necessary and sufficient conditions for the existence of inversion circuits, which will be presented in the next section during the discussion of "catastrophic" codes.

11.2 THE STRUCTURAL PROPERTIES OF CONVOLUTIONAL CODES

The techniques used for analyzing and comparing block codes do not work well with convolutional codes. For example, how does the concept of minimum distance apply to an encoder that can generate code words of arbitrary length? We must begin anew with different tools and develop performance models that are appropriate for convolutional codes and their decoders. We apply two important graphical techniques to our analysis of convolutional codes: the state diagram and the trellis diagram.

11.2.1 State Diagrams

The convolutional encoder is a finite-state automaton. It contains memory elements whose contents determine a mapping between the next set of input bits and output bits. Consider the encoder in Fig. 11-1. This encoder contains three binary memory elements that can collectively assume any one of eight possible states. Designate these states $\{S_0, S_1, \ldots, S_7\}$ and associate them with the contents of the memory elements as shown below.

$$S_0 \leftrightarrow (000), \qquad S_4 \leftrightarrow (001)$$

$$S_1 \leftrightarrow (100), \qquad S_5 \leftrightarrow (101)$$

$$S_2 \leftrightarrow (010), \qquad S_6 \leftrightarrow (011)$$

$$S_3 \leftrightarrow (110), \qquad S_7 \leftrightarrow (111)$$

Note that the memory contents are the reverse of the customary binary representation of the state number. This causes the resulting state diagram to flow in a somewhat more intuitive manner.

As with most finite-state automata, the encoder can only move between states in a limited manner. Given a current state (XYZ), the next state can be either $(0XY)$ (corresponding to a zero input) or $(1XY)$ (corresponding to a one input). The **state diagram** in Figure 11-4 shows this constraint.

Each branch in the state diagram has a label of the form X/YY, where X is the input bit that causes the state transition and YY is the corresponding pair of output bits.

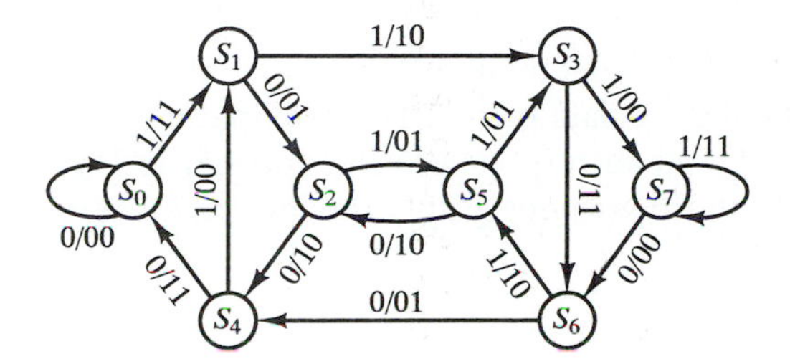

Figure 11-4. State Diagram for Encoder in Figure 11-1

We can better develop our understanding of convolutional codes by considering the encoder state diagram as a **weighted directed graph** [Liu]. A **graph** is a collection of nodes (states) that are interconnected in some manner by a series of branches. A graph is said to be **directed** if its branches have an associated direction. A graph is said to be **weighted** if the branch labels designate some quantity that accumulates as the graph is traversed. In the graph described by a convolutional encoder, the "accumulating quantities" are the weights (number of nonzero coordinates) of the information word and the corresponding code word.

A **path** through a graph is a sequence of nodes in which each pair of adjacent nodes are connected by a branch. If the graph is directed, the branches must point in the direction followed by the path. A path is said to be a **circuit** if it starts and stops at the same node. Finally, a circuit is said to be a **loop** if it does not enter any node more than once.

We shall henceforth refer to the encoder state diagram as the **encoder graph**. Since the encoding process begins and ends with the encoder memory in the all-zero state, every convolutional code word is associated with a circuit through the encoder graph that starts and stops at state S_0.

Example 11-5—Code Words and Paths

The information sequence $\mathbf{x} = (1011)$ is encoded using the rate-1/2 encoder in Figure 11-1. This information sequence designates the path $S_0, S_1, S_2, S_5, S_3, S_6, S_4, S_0$ through the encoder graph in Figure 11-4.

Once the information sequence has been completely input into the encoder, we find ourselves in state S_3. The circuit is completed by inputting zeros until the encoder returns to state S_0. The corresponding code word is thus (11, 01, 01, 01, 11, 01, 11). This result matches that in Example 11-3. ■

11.2.2 Catastrophic Convolutional Codes

The encoder graph is useful in both the description and detection of codes that can cause catastrophic error propagation. Such codes are called, appropriately enough, **catastrophic codes**.

Definition 11-2—Catastrophic Codes

A convolutional code is said to be **catastrophic** if its corresponding state diagram contains a circuit in which a nonzero input sequence corresponds to an all-zero output sequence.

With a catastrophic code, a small number of errors in the received code word can cause an unlimited number of data errors. For example, consider the encoder graph in Figure 11-5. This graph corresponds to the encoder in Figure 11-6. Note that the loop about state S_7 in the encoder graph provides an all-zero output for an input one.

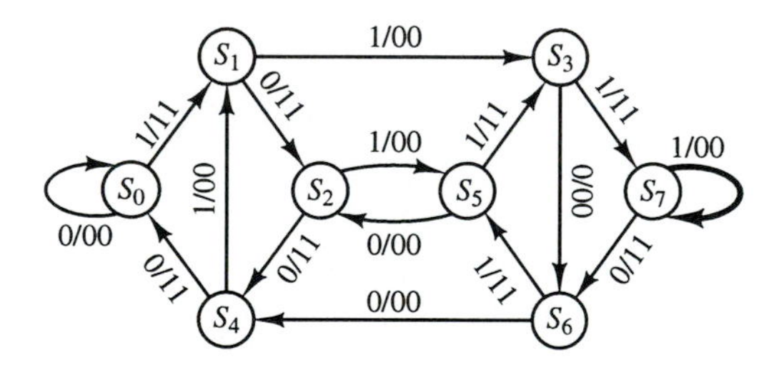

Figure 11-5. Encoder Graph for a Catastrophic Convolutional Code

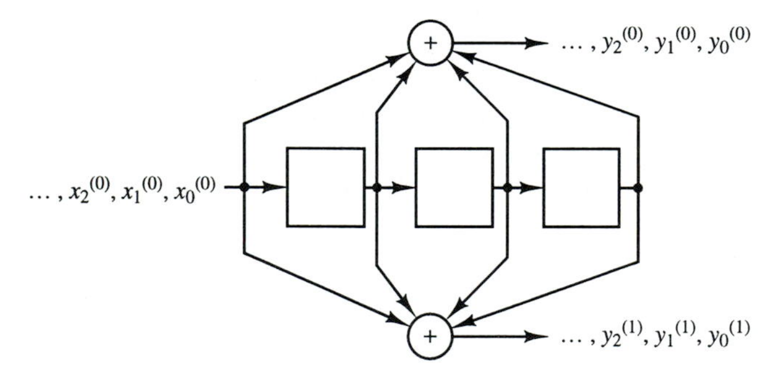

Figure 11-6. Catastrophic Encoder Corresponding to the Graph in Figure 11-5

Suppose that we wish to encode the data sequence $\mathbf{x}^{(0)} = (1, 1, 1, 1, 1, \ldots)$. The resulting code word is $\mathbf{y} = (11, 00, 11, 00, 00, 00, 00, \ldots)$. An information sequence with infinite weight thus corresponds to a code word with finite weight. If noise on the channel causes the received sequence to be, say, $\mathbf{y}' = (10, 00, 00, 00, 00, 00, \ldots)$, then a maximum-likelihood decoder will decide that the all-zero code word was transmitted. Since the all-zero code word corresponds to an all-zero information word, an infinite number of data errors results.

It is a simple matter to avoid the weight-zero loop about the all-ones state: one need only ensure that at least one of the output streams is formed through the summation of an odd number of taps. This expedient does not, however, ensure that the code will not be catastrophic. Consider the encoder graph in Figure 11-7 and its corresponding encoder in Figure 11-8. The state sequence $S_0, S_1, S_3, S_6, S_5, S_3, S_6, S_5, \ldots$ corresponds to a nonzero input sequence of the form $(1,1,0,1,1,0,1,1,\ldots)$ and an output sequence of the form $(01,10,00,00,00,00,00,00,\ldots)$.

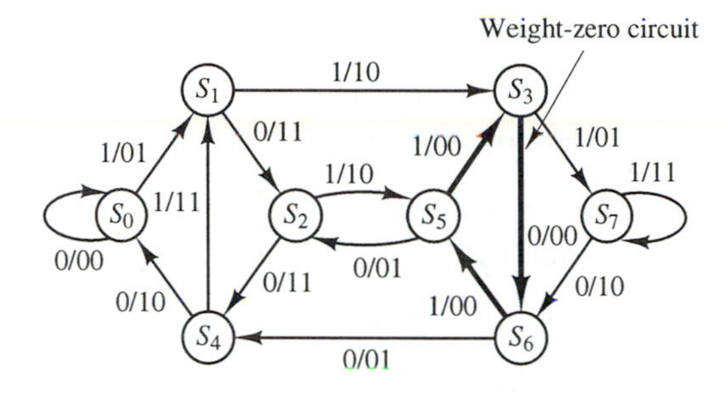

Figure 11-7. Encoder Graph for a Catastrophic Convolutional Code

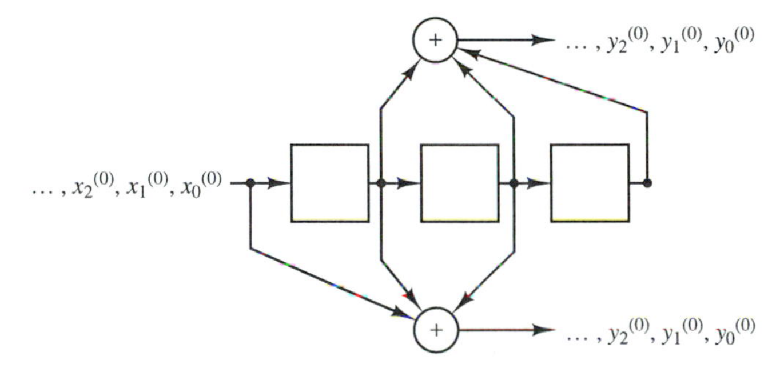

Figure 11-8. Catastrophic Encoder Corresponding to the Graph in Figure 11-5

In [Mas3] Massey and Sain provide a set of necessary and sufficient conditions for a convolutional code to be noncatastrophic. They begin by showing that a code is catastrophic if and only if there does not exist a feedforward sequential circuit that inverts the operation of the encoder. They then proceed to develop a set of necessary and sufficient conditions for the existence of an inverting circuit. This result is summarized in the following theorem.

Theorem 11-1—Catastrophic Codes [Mas3]

1. **Let C be a rate-$1/n$ convolutional code with transfer-function matrix $\mathbf{G}(D)$ whose generator sequences have the transforms $\{\mathbf{G}^{(0)}(D), \mathbf{G}^{(1)}(D), \ldots, \mathbf{G}^{(n-1)}(D)\}$. C is not catastrophic if and only if**

$$\mathrm{GCD}(\mathbf{G}^{(0)}(D), \mathbf{G}^{(1)}(D), \ldots, \mathbf{G}^{(n-1)}(D)) = D^l$$

   for some nonnegative integer l.

2. **Let C be a rate-k/n convolutional code with transfer-function matrix $\mathbf{G}(D)$. C is not catastrophic if and only if**

$$\mathrm{GCD}\left[\Delta_i(D), i = 1, 2, \ldots, \binom{n}{k}\right] = D^l$$

for some nonnegative integer l, where $\Delta_i(D)$ is the determinant of the ith $k \times k$ submatrix of $\mathbf{G}(D)$.

Proof. See [Mas3].

Example 11-6—Massey-Sain Conditions for Catastrophic Codes

The encoder graph in Figure 11-7 shows that the encoder in Figure 11-8 generates a catastrophic code. The generator sequences for the encoder are as follows.

$$\mathbf{g}^{(0)} = (0111)$$

$$\mathbf{g}^{(1)} = (1110)$$

The corresponding D-transforms are easily obtained.

$$\mathbf{G}^{(0)} = D + D^2 + D^3$$

$$\mathbf{G}^{(1)} = 1 + D + D^2$$

It follows that $\text{GCD}[\mathbf{G}^{(0)}, \mathbf{G}^{(1)}] = 1 + D + D^2 \neq D^l$ for any integer l, and the code is thus catastrophic by Theorem 11-1. ■

Though it may not be clear from the preceding theorem, catastrophic codes are relatively infrequent and thus do not pose a serious obstacle to computer searches for good convolutional codes. Forney [For3] has shown that precisely $1/(2^n - 1)$ of all convolutional codes of rate $1/n$ and a given constraint length are catastrophic.

11.2.3 Graphs and Weight Enumerators

When convolutional encoders are associated with weighted directed graphs, the way is cleared for the application of some very powerful analysis techniques. In this section we introduce Mason's formula for computing the **generating function** of a graph. Generating functions are combinatorial devices that convey information about a process in the form of the coefficients and exponents of the terms in an algebraic series. In this section we apply Mason's formula to a modified encoder graph to obtain a generating function, or **weight enumerator**, of the form

$$T(X, Y) = \sum_{i=1}^{\infty} \sum_{j=1}^{\infty} a_{i,j} X^i Y^j \tag{11-19}$$

where $a_{i,j}$ is the number of code words in the convolutional code with weight i that correspond to information sequences of weight j.

In Chapter 15 Mason's formula is used to determine the performance of retransmission request systems that can be described using weighted directed graphs. The following discussion is thus kept as general as possible.

In Chapter 4 of [Mas], Mason and Zimmermann present a very nice tutorial on what are commonly called **linear signal flow graphs**. Linear signal flow graphs are weighted directed graphs that have been constructed to represent series of simultaneous equations. The nodes in these graphs represent the variables in the equations, while the branch labels denote dependencies among the variables, as shown in Figure 11-9.

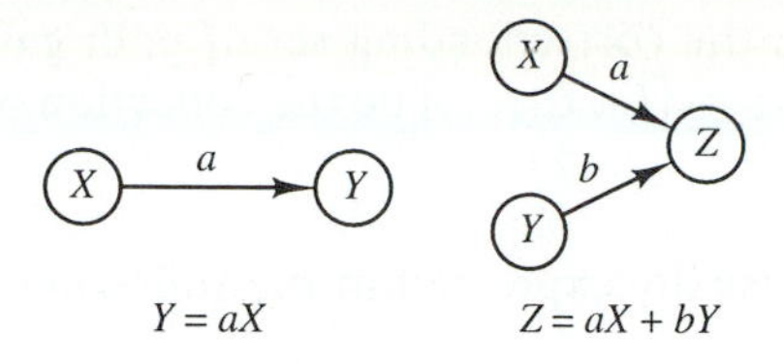

Figure 11-9. Nodal Relationships in a Linear Signal Flow Graph

The signal flow graph for a convolutional encoder is obtained by modifying the branch labels of the encoder graph and splitting the state S_0 into a starting state (a **source node**) and a stopping state (a **sink node**). For a given branch label **Y/X** in the encoder graph we substitute the label $X^i Y^j$, where j is the weight of the input vector **Y** (the number of nonzero coordinates) and i is the weight of the output vector **X**. Figure 11-10 is the signal flow graph corresponding to the encoder graph in Figure 11-4.

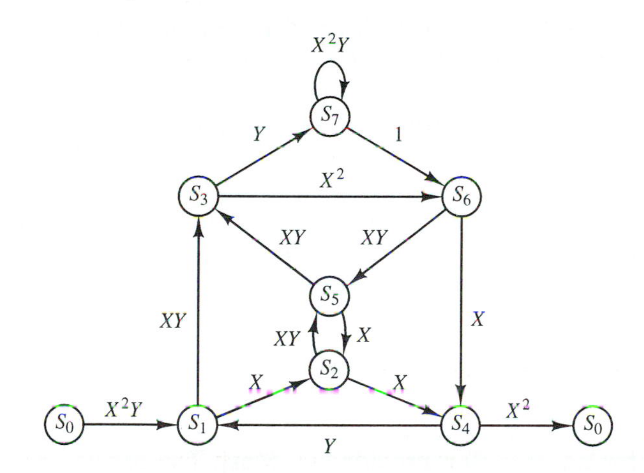

Figure 11-10. The Encoder Signal Flow Graph for the Encoder in Figure 11-1

Consider the path $S_0, S_1, S_3, S_6, S_4, S_0$, which corresponds to the information word (11000) and the code word (11, 10, 11, 01, 11). The weights on the corresponding branches are X^2Y, XY, X^2, X, X^2, which when multiplied together provide the path gain X^8Y^2. The weight of the information word is the power of Y (2), while the weight of the code word is the power of X (8).

Mason's formula computes the source-to-sink **transmission** T of a signal flow graph. This is defined as the "signal" appearing at the sink node per unit signal input at the source node [Mas]. For the encoder signal flow graph, the transmission is thus a sum of the path gains for all paths that start and stop at state S_0.

Before proceeding directly to a statement of Mason's formula, a few definitions are in order. Recall that a **loop** is a circuit that does not enter any state more than once. A **forward loop** is a loop that starts and stops at state S_0. For example, the sequence S_0, S_1, S_2, S_4, S_0 is a forward loop, while the sequence $S_0, S_1, S_2, S_5, S_2, S_4, S_0$ is not. A set of loops is said to be **nontouching** if they do not share a common vertex. The sequences S_0, S_1, S_2, S_4, S_0 and S_3, S_6, S_5, S_3 are nontouching loops.

Let $K = \{K_1, K_2, \ldots\}$ be the set of all forward loops in the signal flow graph and let $F = \{F_1, F_2, \ldots\}$ be the corresponding set of path gains (F_1 is the path gain for the loop K_1, etc.). Let $L = \{L_1, L_2, \ldots\}$ be the collection of all loops in the graph that do not contain the vertex S_0 and $C = \{C_1, C_2, \ldots\}$ the corresponding set of path gains.

Mason's formula is usually expressed in the following form [Mas].

$$T(X, Y) = \frac{\sum_l F_l \Delta_l}{\Delta} \tag{11-20}$$

The **graph determinant Δ** is computed as follows.

$$\Delta = 1 - \sum_{L_l} C_l + \sum_{(L_l, L_m)} C_l C_m - \sum_{(L_l, L_m, L_n)} C_l C_m C_n + \cdots \tag{11-21}$$

The first summation in Eq. (11-21) is over all indices i for the loops $\{L_i\} = L$. The second summation is taken for all pairs of indices $\{(l, m)\}$ such that (L_l, L_m) is a pair of nontouching loops in L. The third summation is taken for all triplets of indices $\{(l, m, n)\}$ such that (L_l, L_m, L_n) forms a triplet of nontouching loops in L—the process continuing until a sufficiently large collection of nontouching loops in L cannot be found.

The computation of Δ_i, the **cofactor of the path $\mathbf{K}_i$**, is identical to that for Δ in Eq. (11-21), except that all loops touching K_i are deleted from the summations. For example, the second summation is over all pairs of indices $\{(l, m)\}$ such that the triplet of loops (K_i, L_l, L_m) is nontouching.

$$\Delta_i = 1 - \sum_{(K_i, L_l)} C_l + \sum_{(K_i, L_l, L_m)} C_l C_m - \sum_{(K_i, L_l, L_m, L_n)} C_l C_m C_n + \cdots \tag{11-22}$$

A sketch of a proof for Eq. (11-20) is provided in [Mas]. The development in [Mas] is highly constructive and may prove interesting to the reader whose interest in Mason's formula extends beyond the rather restricted applications of the formula made in this text. As Mason and Zimmermann note, the applications of Eq. (11-20) are practically unbounded.

In the encoder graph in Figure 11-10 the set K consists of seven forward loops [Lin1].

K_1: $S_0, S_1, S_3, S_7, S_6, S_5, S_2, S_4, S_0$ $F_1 = X^8 Y^4$
K_2: $S_0, S_1, S_3, S_7, S_6, S_4, S_0$ $F_2 = X^6 Y^3$
K_3: $S_0, S_1, S_3, S_6, S_5, S_2, S_4, S_0$ $F_3 = X^{10} Y^3$
K_4: $S_0, S_1, S_3, S_6, S_4, S_0$ $F_4 = X^8 Y^2$
K_5: $S_0, S_1, S_2, S_5, S_3, S_7, S_6, S_4, S_0$ $F_5 = X^8 Y^4$
K_6: $S_0, S_1, S_2, S_5, S_3, S_6, S_4, S_0$ $F_6 = X^{10} Y^3$
K_7: S_0, S_1, S_2, S_4, S_0 $F_7 = X^6 Y$

The set L consists of eleven loops.

$$
\begin{array}{lll}
L_1: & S_1, S_3, S_7, S_6, S_5, S_2, S_4, S_1 & C_1 = X^4 Y^4 \\
L_2: & S_1, S_3, S_7, S_6, S_4, S_1 & C_2 = X^2 Y^3 \\
L_3: & S_1, S_3, S_6, S_5, S_2, S_4, S_1 & C_3 = X^6 Y^3 \\
L_4: & S_1, S_3, S_6, S_4, S_1 & C_4 = X^4 Y^2 \\
L_5: & S_1, S_2, S_5, S_3, S_7, S_6, S_4, S_1 & C_5 = X^4 Y^4 \\
L_6: & S_1, S_2, S_5, S_3, S_6, S_4, S_1 & C_6 = X^6 Y^3 \\
L_7: & S_1, S_2, S_4, S_1 & C_7 = X^2 Y \\
L_8: & S_2, S_5, S_2 & C_8 = X^2 Y \\
L_9: & S_3, S_7, S_6, S_5, S_3 & C_9 = X^2 Y^3 \\
L_{10}: & S_3, S_6, S_5, S_3 & C_{10} = X^4 Y^2 \\
L_{11}: & S_7, S_7 & C_{11} = X^2 Y
\end{array}
$$

There are ten pairs of nontouching loops in L.

$$
\begin{array}{ll}
(L_2, L_8) & C_2 C_8 = X^4 Y^4 \\
(L_3, L_{11}) & C_3 C_{11} = X^8 Y^4 \\
(L_4, L_8) & C_4 C_8 = X^6 Y^3 \\
(L_4, L_{11}) & C_4 C_{11} = X^6 Y^3 \\
(L_6, L_{11}) & C_6 C_{11} = X^8 Y^4 \\
(L_7, L_9) & C_7 C_9 = X^4 Y^4 \\
(L_7, L_{10}) & C_7 C_{10} = X^6 Y^3 \\
(L_7, L_{11}) & C_7 C_{11} = X^4 Y^2 \\
(L_8, L_{11}) & C_8 C_{11} = X^4 Y^2 \\
(L_{10}, L_{11}) & C_{10} C_{11} = X^6 Y^3
\end{array}
$$

There are two triplets of nontouching loops in L.

$$
\begin{array}{ll}
(L_4, L_8, L_{11}) & C_4 C_8 C_{11} = X^8 Y^4 \\
(L_7, L_{10}, L_{11}) & C_7 C_{10} C_{11} = X^8 Y^4
\end{array}
$$

There are no sets of four or more nontouching loops in L.

The graph determinant is then

$$
\begin{aligned}
\Delta = 1 &- (X^4 Y^4 + X^2 Y^3 + X^6 Y^3 + X^4 Y^2 + X^4 Y^4 + X^6 Y^3 + X^2 Y + X^2 Y \\
&+ X^2 Y^3 + X^4 Y^2 + X^2 Y) + (X^4 Y^4 + X^8 Y^4 + X^6 Y^3 + X^6 Y^3 + X^8 Y^4 \\
&+ X^4 Y^4 + X^6 Y^3 + X^4 Y^2 + X^4 Y^2 + X^6 Y^3) - (X^8 Y^4 + X^8 Y^4) \\
= 1 &- 2X^2 Y^3 - 3X^2 Y + 2X^6 Y^3
\end{aligned}
$$

The computation of the cofactors of the forward paths proceeds as follows. There are no loops that do not contain vertices traversed by forward paths K_1 and K_5. Δ_1 and Δ_5 are thus equal to 1.

$$\Delta_1 = \Delta_5 = 1$$

Forward paths K_3 and K_6 do not traverse vertex S_7 and thus do not touch loop L_{11}. K_3 and K_6 touch all other loops, so we have the following.

$$\Delta_3 = \Delta_6 = 1 - C_{11} = 1 - X^2Y$$

Forward path K_2 does not touch L_8 but touches all other loops.

$$\Delta_2 = 1 - C_8 = 1 - X^2Y$$

Forward path K_4 is shown in Figure 11-11. We see that loops L_8 and L_{11} (shaded) do not touch K_4, and furthermore, L_8 and L_{11} are themselves a pair of nontouching loops.

$$\Delta_4 = 1 - (C_8 + C_{11}) + C_8C_{11} = 1 - 2X^2Y + X^4Y^2$$

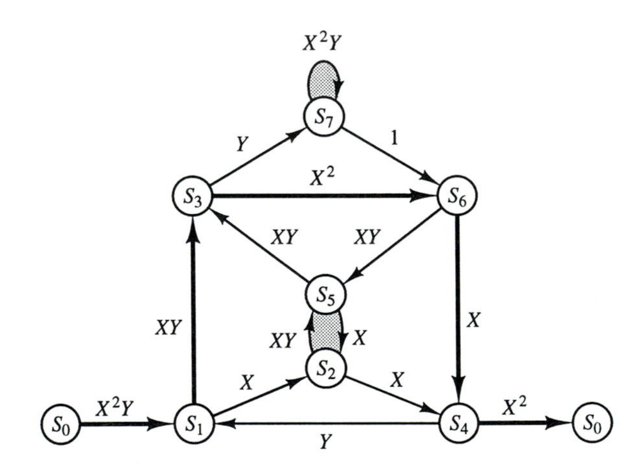

Figure 11-11. Forward path K_4 and the loops not touching K_4

Forward path K_7 is shown in Figure 11-12. The shaded area in the figure highlights the states that are not traversed by K_7 (states S_3, S_5, S_6, and S_7). K_7 thus does not touch the loops L_9, L_{10}, L_{11}, of which L_{10} and L_{11} form a nontouching pair.

$$\Delta_7 = 1 - (C_9 + C_{10} + C_{11}) + (C_{10}C_{11}) = 1 - (X^2Y^3 + X^4Y^2 + X^2Y) + (X^6Y^3)$$

The gain formula thus provides us with the following.

$$T(X,Y) = \frac{\begin{pmatrix} X^8Y^4(1) + X^6Y^3(1 - X^2Y) + X^{10}Y^3(1 - X^2Y) + X^8Y^2(1 - 2X^2Y + X^4Y^2) \\ + X^8Y^4(1) + X^{10}Y^3(1 - X^2Y) + X^6Y(1 - X^2Y^3 - X^4Y^2 - X^2Y + X^6Y^3) \end{pmatrix}}{1 - 2X^2Y^3 - 3X^2Y + 2X^6Y^3}$$

$$= \frac{X^6Y^3 - X^{10}Y^3 + X^6Y}{1 - 2X^2Y^3 - 3X^2Y + 2X^6Y^3} \qquad (11\text{-}23)$$

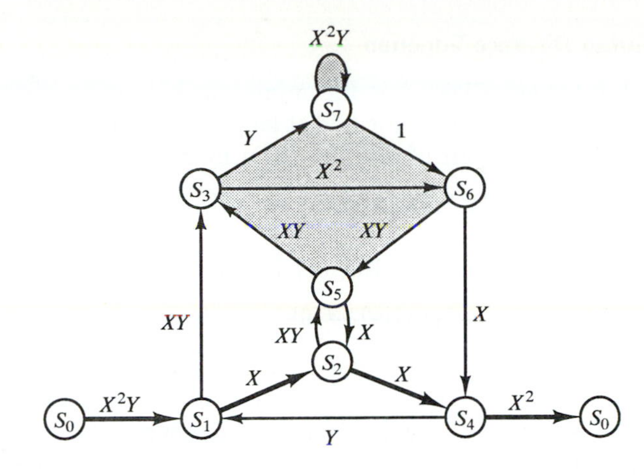

Figure 11-12. Forward path K_7 and the loops not touching K_7

When Eq. (11-23) is expanded through long division, we obtain the desired information.

$$\begin{aligned} T(X,Y) &= X^6Y + X^6Y^3 + 3X^8Y^2 + 5X^8Y^4 + 2X^8Y^6 + 8X^{10}Y^3 \\ &\quad + 21X^{10}Y^5 + 16X^{10}Y^7 + 4X^{10}Y^9 + \cdots \qquad (11\text{-}24) \\ &= X^6(Y+Y^3) + X^8(3Y^2+5Y^4+2Y^6) + X^{10}(8Y^3+21Y^5+16Y^7+4Y^9) + \cdots \end{aligned}$$

This generating function tells us that there are two code words of weight 6: one corresponding to an information sequence of weight 1 (X^6Y), the other corresponding to an information sequence of weight 3 (X^6Y^3). There are then ten code words of weight 8, and 49 of weight 10, and so on.

Unfortunately Mason's formula becomes computationally intractable rather quickly as the constraint length of the subject code increases. It is still possible, however, to accurately predict the performance of the more complex codes using the performance measures in the following section.

11.2.4 Convolutional-Code Performance Measures

There are several different performance measures that can be used to compare convolutional codes. The appropriateness of a given measure depends on the decoding technique that is to be used in the application. Three performance measures are considered here: the *column distance function*, *minimum distance*, *and minimum free distance*.

The **column distance function** is particularly useful in evaluating the performance of sequential decoders (see Chapter 13). Consider an (n,k) code $\mathbf{C}$ with constraint length K. Let the input sequence $\mathbf{x}$ and the output sequence $\mathbf{y}$ for an encoder for $\mathbf{C}$ be truncated at length i. These truncated sequences can be represented as follows.

$$[\mathbf{x}]_i = (x_0^{(0)}, x_0^{(1)}, \ldots, x_0^{(k-1)}, x_1^{(0)}, x_1^{(1)}, \ldots, x_1^{(k-1)}, \ldots, x_{i-1}^{(0)}, x_{i-1}^{(1)}, \ldots, x_{i-1}^{(k-1)})$$

$$[\mathbf{y}]_i = (y_0^{(0)}, y_0^{(1)}, \ldots, y_0^{(n-1)}, y_1^{(0)}, y_1^{(1)}, \ldots, y_1^{(n-1)}, \ldots, y_{i-1}^{(0)}, y_{i-1}^{(1)}, \ldots, y_{i-1}^{(n-1)})$$

Definition 11-3—The Column Distance Function

The **column distance function (CDF)** $\boldsymbol{d_i}$ is the minimum Hamming distance between all pairs of output sequences truncated at length i given that the input sequences corresponding to the pair of outputs differ in the first k-bit block.

$$d_i \triangleq \min\{d([\mathbf{y}']_i, [\mathbf{y}'']_i) \mid [\mathbf{x}']_1 \neq [\mathbf{x}'']_1\} \tag{11-25}$$

If the convolutional code is linear, Eq. (11-25) can be re-expressed as

$$d_i = \min\{w([\mathbf{y}]_i) \mid [\mathbf{x}]_1 \neq \mathbf{0}\} \tag{11-26}$$

where $w(\mathbf{y})$ is the Hamming weight of the vector $\mathbf{y}$.

The term "column distance" is derived from the relationship of the CDF to the expression of a code word as a product of the information sequence and the subject code's generator matrix $\mathbf{G}$ [Lin1]. If the information sequence is truncated to length ik bits, then the correspondingly truncated output sequences (and thus d_i) depend only on the first ni columns of $\mathbf{G}$.

The CDF for the code in Figure 11-1 is graphed in Figure 11-13.

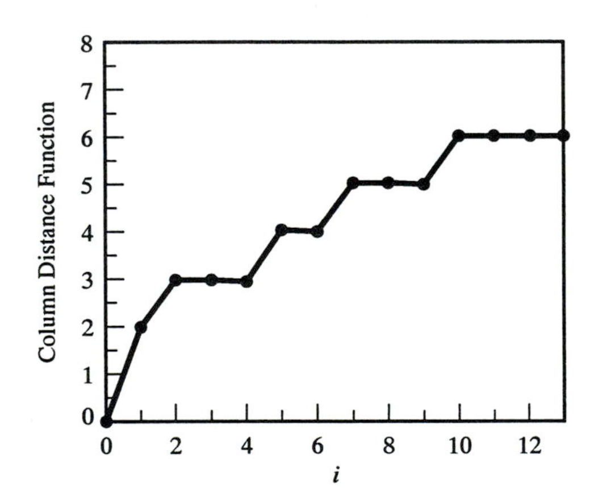

Figure 11-13. Column Distance Function for the Encoder in Figure 11-1

The early decoding techniques developed for convolutional codes (e.g., threshold decoding [Mas1]) used only the first nK bits of a received word to decode each received bit, where n is the number of output streams produced by the encoder and K is the constraint length of the code. Under these circumstances the error correcting capability of a convolutional code is a strong function of the Hamming distance between code words over their first nK bits corresponding to a nonzero input sequence. For this distance, coding theorists adopted the term **minimum distance** from the world of block coding.

Definition 11-4—The Minimum Distance of a Convolutional Code

The **minimum distance** $\mathbf{d_{min}}$ of an (n, k) convolutional code with constraint length K is the CDF d_i evaluated at $i = K$.

The minimum distance for the code defined by the encoder in Figure 11-1 is thus $d_{\min} = d_4 = 3$.

The Viterbi decoder, on the other hand, uses the entire convolutional code word to decode a single bit. In this case we are interested in the Hamming distance between all complete convolutional code words (i.e., those output sequences corresponding to a circuit that starts and stops at the state S_0 in the encoder graph).

Definition 11-5—The Minimum Free Distance of a Convolutional Code

The **minimum free distance**, denoted d_{free}, is the minimum Hamming distance between all pairs of complete convolutional code words.

$$\begin{aligned} d_{\text{free}} &\triangleq \min\{d(\mathbf{y}', \mathbf{y}'') \mid \mathbf{y}' \neq \mathbf{y}''\} \\ &= \min\{w(\mathbf{y}) \mid \mathbf{y} \neq 0\} \end{aligned} \tag{11-27}$$

The code defined by the encoder in Figure 11-1 has $d_{\text{free}} = 6$, as verified by the weight enumerator in Eq. (11-24). Note that the minimum *free* distance of this code is twice its minimum distance. A decoding technique that makes use of the entire code word as opposed to the first nK bits thus provides significantly better performance.

For noncatastrophic codes, the CDF approaches the minimum free distance as i increases. This can be shown by noting that the CDF is monotonically nondecreasing and that noncatastrophic codes have no zero-weight loops except that about the state S_0.

$$\lim_{n \to \infty} d_i = d_{\text{free}} \qquad \text{(for noncatastrophic codes)} \tag{11-28}$$

The catastrophic code defined by the encoder graph in Figure 11-7 provides an excellent example of a case in which Eq. (11-28) does not hold. Though the minimum free distance for this code is 6, d_i never rises above 2.

Costello [Cos] and Forney [For4] have developed a series of bounds on the minimum free distance of convolutional codes. The bounds differ, depending on whether the codes under consideration are systematic or not, and imply a rather interesting result. Unlike linear block codes, nonsystematic convolutional codes offer a higher minimum free distance than systematic codes of comparable constraint length and rate.

TABLE 11-1 Maximum d_{free} for rate-1/3 convolutional codes

Constraint length K	Systematic maximum d_{free}	Nonsystematic maximum d_{free}
2	5	5
3	6	8
4	8	10
5	9	12
6	10	13
7	12	15

TABLE 11-2 Maximum d_{free} for rate-1/2 convolutional codes

Constraint length K	Systematic maximum d_{free}	Nonsystematic maximum d_{free}
2	3	3
3	4	5
4	4	6
5	5	7
6	6	8
7	6	10

11.3 THE BEST-KNOWN CONVOLUTIONAL CODES

The following tables list by constraint length nonsystematic rate-1/4, rate-1/3, and rate-1/2 convolutional codes with maximal minimum free distance. These codes were found by computer search. The code generators are written in octal; for example, the generator sequences for the best rate-1/3 code with $m = 3$ are $\mathbf{g}_0 = (101100)$, $\mathbf{g}_1 = (110100)$, and $\mathbf{g}_2 = (111100)$, which are written in the chart as (54), (64), and (74). Zeros are padded onto the end of the sequences so that their total length is a multiple of three.

TABLE 11-3 Rate-1/4 convolutional codes with maximal minimum free distance [Lin1][2]

K	$\mathbf{g}^{(0)}$	$\mathbf{g}^{(1)}$	$\mathbf{g}^{(2)}$	$\mathbf{g}^{(3)}$	d_{free}
3	5	7	7	7	10
4	54	64	64	74	13
5	52	56	66	76	16
6	53	67	71	75	18
7	564	564	634	714	20
8	472	572	626	736	22
9	463	535	733	745	24
10	4474	5724	7154	7254	27
11	4656	4726	5562	6372	29
12	4767	5723	6265	7455	32
13	44624	52374	66754	73534	33
14	42226	46372	73256	73276	36

[2] Tables 11-3 through 11-7 are reprinted (in slightly modified form), with permission, from S. Lin and D. J. Costello, Jr., *Error Control Coding: Fundamentals and Applications*, pp. 330–331 (Englewood Cliffs, NJ: Prentice Hall, 1983).

TABLE 11-4 Rate-1/3 convolutional codes with maximal minimum free distance [Lin1]

K	$\mathbf{g}^{(0)}$	$\mathbf{g}^{(1)}$	$\mathbf{g}^{(2)}$	d_{free}
3	5	7	7	8
4	54	64	74	10
5	52	66	76	12
6	47	53	75	13
7	554	624	764	15
8	452	662	756	16
9	557	663	711	18
10	4474	5724	7154	20
11	4726	5562	6372	22
12	4767	5723	6265	24
13	42554	43364	77304	24
14	43512	73542	76266	26

TABLE 11-5 Rate-1/2 convolutional codes with maximal minimum free distance [Lin1]

K	$\mathbf{g}^{(0)}$	$\mathbf{g}^{(1)}$	d_{free}
3	5	7	5
4	64	74	6
5	46	72	7
6	65	57	8
7	554	744	10
8	712	476	10
9	561	753	12
10	4734	6624	12
11	4672	7542	14
12	4335	5723	15
13	42554	77304	16
14	43572	56246	16
15	56721	61713	18
16	447254	627324	19
17	716502	514576	20

16-95

G(561) = 101110001
G(753) = 111101011

TABLE 11-6 Rate-2/3 convolutional codes with maximal minimum free distance [Lin1]

K	M	$\mathbf{g}_0^{(0)}$ $\mathbf{g}_1^{(0)}$	$\mathbf{g}_0^{(1)}$ $\mathbf{g}_1^{(1)}$	$\mathbf{g}_0^{(2)}$ $\mathbf{g}_1^{(2)}$	d_{free}
2	2	6	2	6	3
		2	4	4	
3	3	4	2	6	4
		1	4	7	
3	4	7	1	4	5
		2	5	7	
4	5	60	30	70	6
		14	40	74	
4	6	64	30	64	7
		30	64	74	
5	7	60	34	54	8
		16	46	74	
5	8	64	12	52	8
		26	66	44	
6	9	52	06	74	9
		05	70	53	
6	10	63	15	46	10
		32	65	61	

TABLE 11-7 Rate-3/4 convolutional codes with maximal minimum free distance [Lin1]

K	M	$\mathbf{g}_0^{(0)}$ $\mathbf{g}_1^{(0)}$ $\mathbf{g}_2^{(0)}$	$\mathbf{g}_0^{(1)}$ $\mathbf{g}_1^{(1)}$ $\mathbf{g}_2^{(1)}$	$\mathbf{g}_0^{(2)}$ $\mathbf{g}_1^{(2)}$ $\mathbf{g}_2^{(2)}$	$\mathbf{g}_0^{(3)}$ $\mathbf{g}_1^{(3)}$ $\mathbf{g}_2^{(3)}$	d_{free}
2	3	4	4	4	4	4
		0	6	2	4	
		0	2	5	5	
3	5	6	2	2	6	5
		1	6	0	7	
		0	2	5	5	
3	6	6	1	0	7	6
		3	4	1	6	
		2	3	7	4	
4	8	70	30	20	40	7
		14	50	00	54	
		04	10	74	40	
4	9	40	14	34	60	8
		04	64	20	70	
		34	00	60	64	

11.4 TRANSPARENT CONVOLUTIONAL CODES, RECEIVER PHASE OFFSET, AND DIFFERENTIAL ENCODING

Digital receivers commonly use phase-locked-loops to generate an estimate of the phase of the received signal [Gag1]. This phase estimate is used to bias a voltage-controlled oscillator, which generates a sinusoid that is used to demodulate the signal. Since not all sequences of signals are valid coded sequences, an offset in this phase estimate has a visible impact within the demodulator and the decoder. Many of the more powerful digital receivers feed back information from the decoder to aid in the estimation of the carrier phase. The resulting PLL is commonly called a **decision-directed PLL**.

For a variety of reasons, the phase estimate may not be accurate. For example, phase offset can result from ambient channel noise, jamming, or the deterioration (or poor selection) of components within the receiver. If the phase offset is so large that long, distinct code words become indistinguishable, then the information from the decoder becomes useless. The phase estimate then proceeds on a random walk and the receiver loses synchronization.

Synchronization is regained whenever the random carrier phase estimate causes the demodulated signal to resemble a valid signal sequence. It is thus desirable that the coded signal sequences show as much symmetry under phase rotations as possible. The problem of phase ambiguity that arises from this symmetry is handled through the use of **differential encoding** and **transparent convolutional codes**.

We assume here that the modulation format is binary; the coded sequences may thus, if properly designed, be symmetric under 180° phase rotations. A convolutional code that allows for such symmetry is called a **transparent code**. A code is **transparent** if given any code word **c**, the complement of **c** is also a code word. It is sufficient that the code be linear and contain the all-ones code word. This is equivalent to having an odd number of taps summed together at every decoder output. A 180° change in receiver phase thus converts a valid code word into another valid code word—the Viterbi decoder performs its function as if the phase were correct.

The 180° phase ambiguity that results from code transparency can be removed through differential encoding. Figure 11-14 shows a differential encoder used in conjunction with a transparent convolutional code. The differential encoder outputs

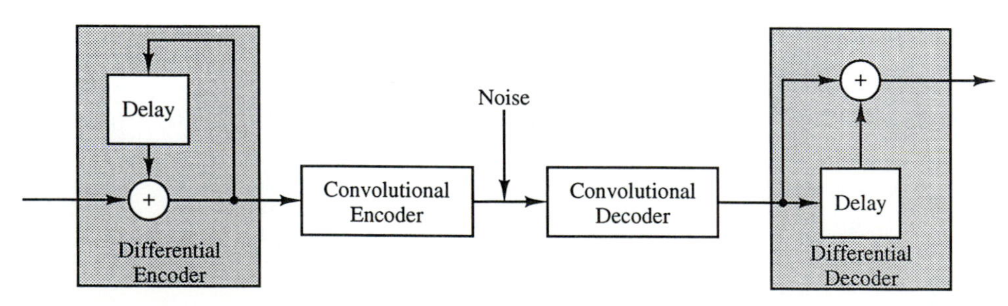

Figure 11-14. Differential encoding

the difference between each successive pair of data bits. Phase changes at the receiver thus cause a momentary disruption of two bit errors, but all following bits are decoded correctly. Since a single bit error in the differentially encoded data causes two errors at the output, there is a slight (0.1 to 0.2 dB) loss in receiver performance [Cla].

Techniques for obtaining higher levels of phase invariance in convolutionally encoded systems employing nonbinary signaling are discussed in [Wei1] and [Wei2].

PROBLEMS

Convolutional encoders

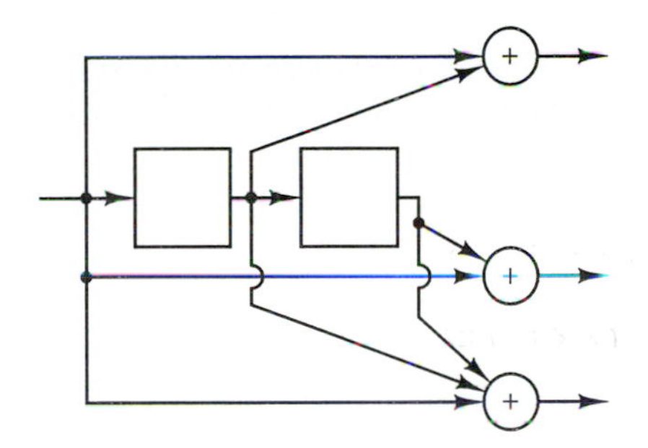

Figure 11-15. A rate-1/3 convolutional encoder

1. Prove that the convolutional encoder in Figure 11-15 generates a linear code.
2. **(a)** Find the impulse response for the encoder in Figure 11-15.
   (b) Find the transfer function matrix for the encoder in Figure 11-15.
   (c) Use the transfer function matrix to determine the code word associated with the input sequence $\mathbf{x} = (11101)$.

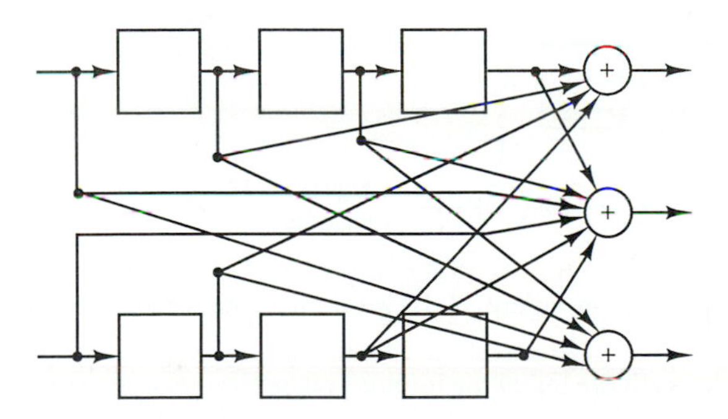

Figure 11-16. A rate-2/3 convolutional encoder

3. **(a)** Find the impulse response for the encoder in Figure 11-16.
   (b) Find the transfer function matrix for the encoder in Figure 11-16.
   (c) Use the transfer function matrix to determine the code word associated with the input sequence $\mathbf{x} = (11, 10, 01)$.

4. Construct an encoder for the convolutional code represented by the following transfer function matrix.

$$\mathbf{G}(D) = \begin{bmatrix} 1 + D + D^2 & 1 + D + D^3 & 1 + D^2 + D^3 \\ 1 + D^3 & 1 + D^2 & 1 + D + D^3 \end{bmatrix}$$

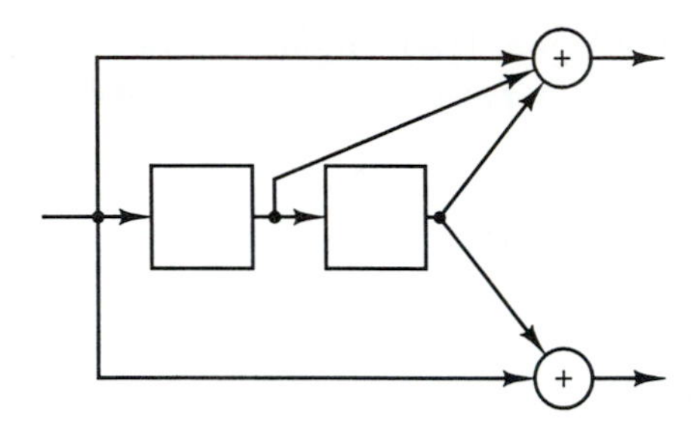

Figure 11-17. A rate-1/2 convolutional encoder

5. For all noncatastrophic convolutional codes there are inversion circuits that accept code words and output the associated information sequence (thereby inverting the operation of the encoder). Design an inversion circuit for the encoder in Figure 11-17.
6. Construct a generalized transfer function matrix for a rate-k/n systematic convolutional encoder.

Structural properties

7. Construct a state diagram for the encoder in Figure 11-15.
8. Compute the weight enumerator for the code generated by the encoder in Figure 11-15. What is the minimum free distance of this code?
9. Describe what happens when Mason's formula is applied to a catastrophic convolutional code to obtain a weight enumerator.

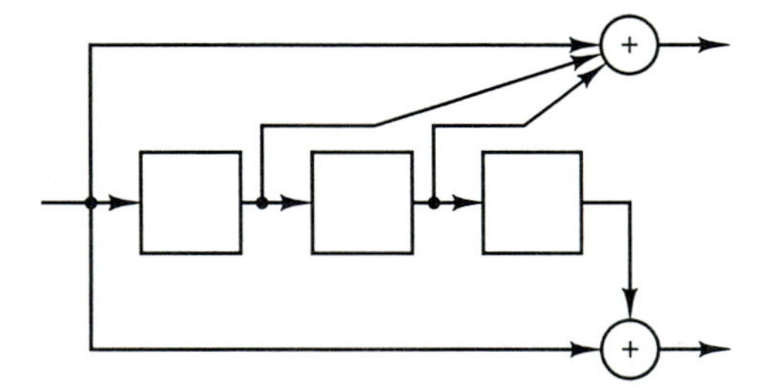

Figure 11-18. A rate-1/2 convolutional encoder

10. Determine whether the encoder in Figure 11-18 generates a catastrophic convolutional code.
11. **(a)** Prove that the encoder with the following impulse responses generates a catastrophic convolutional code.

$$\mathbf{g}^{(0)} = (0111)$$
$$\mathbf{g}^{(1)} = (1001)$$

(b) Find an infinite weight input sequence **x** that corresponds to a finite weight output sequence **y** for this encoder.

12. (*Easy*) Can a convolutional code be catastrophic if it can be implemented using a systematic encoder? Explain your answer.

13. (*Difficult*) Prove that an inversion circuit cannot exist for a convolutional code that is catastrophic.

The column distance function

14. Prove that $\lim_{i \to \infty} d_i = d_{\text{free}}$ for all noncatastrophic convolutional codes.

12

The Viterbi Decoding Algorithm

The Viterbi algorithm is named after the gentleman who proposed the algorithm in 1967 as an "asymptotically optimum" approach to the decoding of convolutional codes in memoryless noise [Vit2]. The algorithm had already been known for at least ten years in various forms in the field of operations research. For example, in 1957 Minty presented a solution to the "shortest-route" problem that is an exact analog of the Viterbi algorithm [Min]. But since Minty's presentation focuses on the manipulation of a physical model constructed from bits of string, it is only possible through the clarity of hindsight to see that Viterbi's and Minty's algorithms are the same solution to the same problem.

In 1973 and 1974, Forney showed that the Viterbi algorithm provides both a maximum-likelihood and a maximum a posteriori (MAP) decoding algorithm for convolutional codes [For2, For4]. He went on to demonstrate that, in its most general form, the Viterbi algorithm is a solution to the problem of "MAP estimation of the state sequence of a finite-state discrete-time Markov process observed in memoryless noise" [For2]. Given this general form, the algorithm can be applied to a host of problems encountered in the design of communication systems (e.g., the resolution of intersymbol interference).

In 1979, Cain, Clark, and Geist [Cai] showed that Viterbi decoders for high-rate codes can be substantially simplified through the application of puncturing. In 1988 Hagenauer [Hag3] extended the work of Cain et al. by developing rate-compatible punctured convolutional (RCPC) codes. RCPC codes can be used as the foundation for highly efficient adaptive rate error control systems.

In this chapter we begin by introducing the Viterbi algorithm as a maximum likelihood decoding algorithm for convolutional codes. Several decoding examples are provided, including one that demonstrates soft-decision decoding. In the sections that follow, the performance of the Viterbi algorithm is determined under a variety of conditions. The final sections focus on the implementation of the algorithm and related practical issues, including punctured convolutional and RCPC codes.

12.1 TRELLIS DIAGRAMS

A **trellis diagram** is an extension of a convolutional code's state diagram that explicitly shows the passage of time. For example, consider the encoder in Figure 12-1. This rate-1/3 encoder has two memory cells, so the associated state diagram (Figure 12-2) has four states. In Figure 12-3 the state diagram is extended in time to form a trellis diagram. The branches of the trellis diagram are labeled with the output bits corresponding to the associated state transitions. We use the notation jk to denote the branch moving from state S_j to state S_k in the trellis. The corresponding output bits are denoted $\mathbf{y}_{jk} = (y_{jk,0}, y_{jk,1}, \ldots, y_{jk,(n-1)})$. It is assumed that the convolutional code is time invariant, so that $\mathbf{y}_{jk}$ is constant.

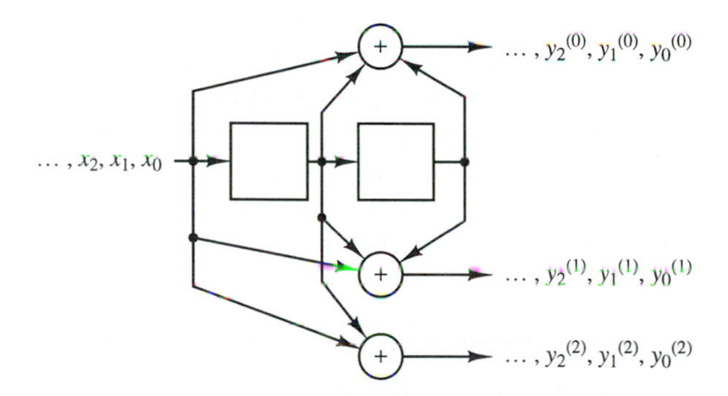

Figure 12-1. Encoder for a rate-1/3 convolutional code

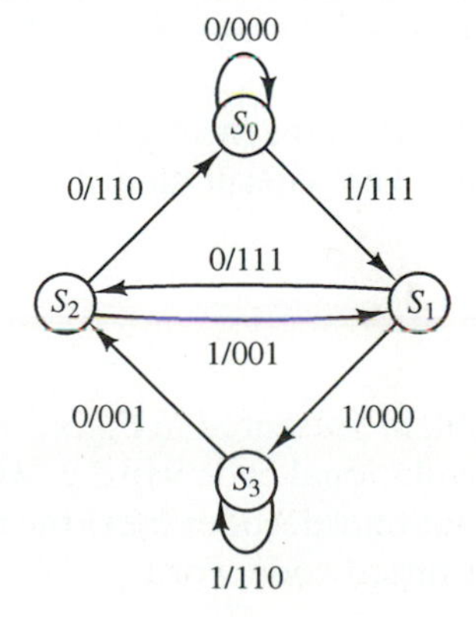

Figure 12-2. State diagram for encoder in Figure 12-1

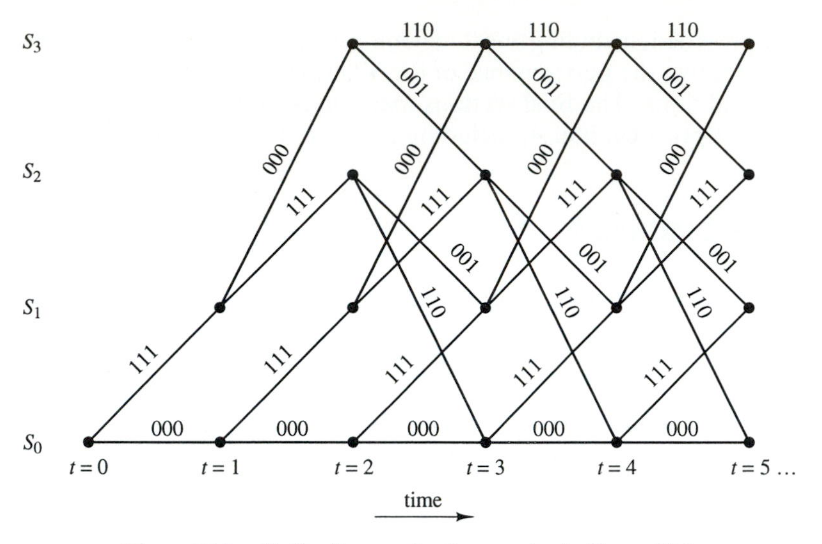

Figure 12-3. Trellis diagram for the encoder in Figure 12-1

Every code word in a convolutional code is associated with a unique path, starting and stopping at state S_0, through the associated trellis diagram. The trellis structure allows us to perform some simple counting exercises that lead to some useful results. Consider a general (n, k) binary convolutional encoder with total memory M and maximal memory order m. The associated trellis diagram has 2^M nodes at each stage, or time increment t. There are 2^k branches leaving each node, one branch for each possible combination of input values. After time $t = m$, there are also 2^k branches entering each node. It is assumed that after the input sequence has been entered into the encoder, m state transitions are necessary to return the encoder to state S_0. Given an input sequence of kL bits, the trellis diagram must have $L + m$ stages, the first and last stages starting and stopping, respectively, in state S_0. There are thus 2^{kL} distinct paths through the general trellis, each corresponding to a convolutional code word of length $n(L + m)$. For example, the length-3 input sequence $\mathbf{x} = (011)$ is shown in Figure 12-4 to correspond to a five-branch path associated with the $3(3 + 2) = 15$-bit convolutional code word $\mathbf{y} = (000, 111, 000, 001, 110)$.

Note that in Figure 12-4 all paths through the trellis intersect all other possible paths at one or more nodes. The Viterbi algorithm takes great advantage of this.

12.2 THE VITERBI ALGORITHM

Consider the decoding problem presented in Figure 12-5. An information sequence $\mathbf{x}$ is encoded to form a convolutional code word $\mathbf{y}$, which is then transmitted across a noisy channel. The convolutional decoder takes the received vector $\mathbf{r}$ and generates an estimate $\mathbf{y}'$ of the transmitted code word.

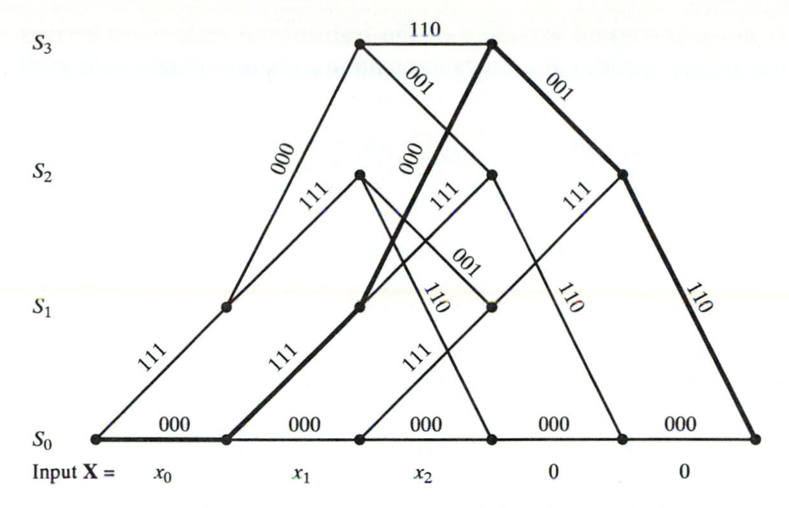

Figure 12-4. Trellis diagram for inputs of length 3 to the encoder in Figure 12-1

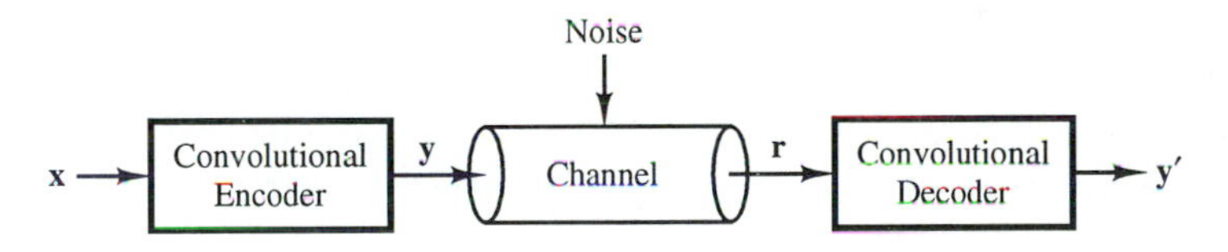

Figure 12-5. The convolutional decoding problem

The **maximum likelihood (ML) decoder** selects, by definition, the estimate $\mathbf{y}'$ that maximizes the probability $p(\mathbf{r} \mid \mathbf{y}')$, while the **maximum a posteriori (MAP) decoder** selects the estimate that maximizes $p(\mathbf{y}' \mid \mathbf{r})$. If the distribution of the source words $\{\mathbf{x}\}$ is uniform, then the two decoders are identical; in general, they can be related by Bayes' rule [Pap].

$$p(\mathbf{r} \mid \mathbf{y})p(\mathbf{y}) = p(\mathbf{y} \mid \mathbf{r})p(\mathbf{r}) \qquad \text{(Bayes' rule)} \tag{12-1}$$

The development of the ML decoder is pursued in this section.

A rate k/n convolutional encoder takes k input bits and generates n output bits with each shift of its internal registers. Suppose that we have an input sequence $\mathbf{x}$ composed of L k-bit blocks.

$$\mathbf{x} = (x_0^{(0)}, x_0^{(1)}, \ldots, x_0^{(k-1)}, x_1^{(0)}, x_1^{(1)}, \ldots, x_1^{(k-1)}, \ldots, x_{L-1}^{(k-1)})$$

The output sequence $\mathbf{y}$ will consist of L n-bit blocks (one for each input block) as well as m additional blocks, where m is the length of the longest shift register in the encoder.

$$\mathbf{y} = (y_0^{(0)}, y_0^{(1)}, \ldots, y_0^{(n-1)}, y_1^{(0)}, y_1^{(1)}, \ldots, y_1^{(n-1)}, \ldots, y_{L+m-1}^{(n-1)})$$

A noise-corrupted version **r** of the transmitted code word arrives at the receiver, where the decoder generates a maximum likelihood estimate $\mathbf{y}'$ of the transmitted sequence. **r** and $\mathbf{y}'$ have the following form.

$$\mathbf{r} = (r_0^{(0)}, r_0^{(1)}, \ldots, r_0^{(n-1)}, r_1^{(0)}, r_1^{(1)}, \ldots, r_1^{(n-1)}, \ldots, r_{L+m-1}^{(n-1)})$$
$$\mathbf{y}' = (y_0'^{(0)}, y_0'^{(1)}, \ldots, y_0'^{(n-1)}, y_1'^{(0)}, y_1'^{(1)}, \ldots, y_1'^{(n-1)}, \ldots, y_{L+m-1}'^{(n-1)})$$

A few assumptions about the channel need to be made to facilitate the analysis. We assume that the channel is **memoryless**, i.e., that the noise process affecting a given bit in the received word **r** is independent of the noise process affecting all of the other received bits. Since the probability of joint, independent events is simply the product of the probabilities of the individual events [Pap], it follows that

$$\begin{aligned} p(\mathbf{r}\,|\,\mathbf{y}') &= \prod_{i=0}^{L+m-1} [p(r_i^{(0)}\,|\,y_i'^{(0)})p(r_i^{(1)}\,|\,y_i'^{(1)}) \cdots p(r_i^{(n-1)}\,|\,y_i'^{(n-1)})] \\ &= \prod_{i=0}^{L+m-1} \left(\prod_{j=0}^{n-1} p(r_i^{(j)}\,|\,y_i'^{(j)}) \right) \end{aligned} \tag{12-2}$$

There are two sets of product indices, one corresponding to the block numbers (subscripts) and the other corresponding to bits within the blocks (superscripts). Equation (12-2) is sometimes called the **likelihood function** for $\mathbf{y}'$ [Vit3]. Since logarithms are monotonically increasing, the estimate that maximizes $p(\mathbf{r}\,|\,\mathbf{y}')$ is also the estimate that maximizes $\log p(\mathbf{r}\,|\,\mathbf{y}')$. By taking the logarithm of each side of Eq. (12-2) we obtain the **log likelihood function**

$$\log p(\mathbf{r}\,|\,\mathbf{y}') = \sum_{i=0}^{L+m-1} \left(\sum_{j=0}^{n-1} \log p(r_i^{(j)}\,|\,y_i'^{(j)}) \right) \tag{12-3}$$

In hardware implementations of the Viterbi decoder, the summands in Eq. (12-3) are usually converted to a more easily manipulated form called the **bit metrics**.

$$M(r_i^{(j)}\,|\,y_i'^{(j)}) = a[\log p(r_i^{(j)}\,|\,y_i'^{(j)}) + b] \tag{12-4}$$

a and b are chosen such that the bit metrics are small positive integers that can be easily manipulated by digital logic circuits. The **path metric** for a code word $\mathbf{y}'$ is then computed as follows.

$$M(\mathbf{r}\,|\,\mathbf{y}') = \sum_{i=0}^{L+m-1} \left(\sum_{j=0}^{n-1} M(r_i^{(j)}\,|\,y_i'^{(j)}) \right) \tag{12-5}$$

If a in Eq. (12-4) is positive and real, while b is simply real, then the code word $\mathbf{y}'$ that maximizes $p(\mathbf{r}\,|\,\mathbf{y}')$ also maximizes $M(\mathbf{r}\,|\,\mathbf{y}')$. a may also be chosen to be negative, in which case $\mathbf{y}'$ is selected so that it *minimizes* $M(\mathbf{r}\,|\,\mathbf{y}')$. Later on we will discuss some good reasons for choosing to minimize path metrics.

At times it is useful for us to focus on the contribution made to the path metric by a single block of **r** and $\mathbf{y}'$. Recall that a single block corresponds to a single branch in the trellis. The **kth branch metric** for a code word $\mathbf{y}'$ is defined as the sum of the bit metrics for the kth block of $\mathbf{r}$ given $\mathbf{y}'$.

$$M(\mathbf{r}_k \mid \mathbf{y}'_k) = \sum_{j=0}^{n-1} M(r_k^{(j)} \mid y_k'^{(j)}) \tag{12-6}$$

The **kth partial path metric** for a path is obtained by summing the branch metrics for the first k branches that the path traverses.

$$M^k(\mathbf{r} \mid \mathbf{y}') = \sum_{i=0}^{k-1} M(\mathbf{r}_i \mid \mathbf{y}'_i) = \sum_{i=0}^{k-1} \left(\sum_{j=0}^{n-1} M(r_i^{(j)} \mid y_i'^{(j)}) \right) \tag{12-7}$$

In the Viterbi algorithm we use our trellis diagrams for the computation of path metrics. Recall that we have labeled the branches of our trellis with the output bits corresponding to a particular input to the encoder and the encoder's current state. Now assume that we have a received sequence $\mathbf{r}$. $\mathbf{r}$ is written at the bottom of the trellis one block at a time, with each block corresponding to a trellis stage. For example, if we are working with the code whose trellis is shown in Figure 12-3, the beginning of the trellis would be as shown in the accompanying drawing.

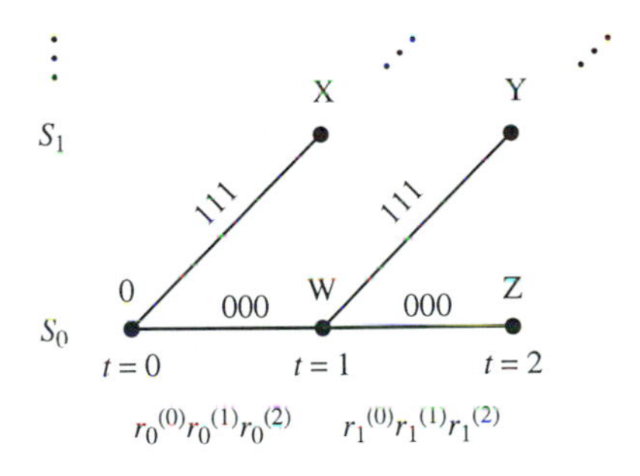

In the Viterbi algorithm, each node in the trellis is assigned a number. This number is the partial path metric of the path that starts at state S_0 at time $t = 0$ and terminates at that node. For example, in the accompanying drawing the label Y corresponds to the two-branch path that terminates at state S_1 at time $t = 2$. Given that the output bits corresponding to this path consist of three zeros followed by three ones, we have

$$\begin{aligned} \mathrm{Y} &= M^2(\mathbf{r} \mid \mathbf{y}' = (000, 111, \ldots)) \\ &= W + M(\mathbf{r}_1 \mid \mathbf{y}'_1 = (111)) \\ &= (M(r_0^{(0)} \mid 0) + M(r_0^{(1)} \mid 0) + M(r_0^{(2)} \mid 0)) + (M(r_1^{(0)} \mid 1) + M(r_1^{(1)} \mid 1) + M(r_1^{(2)} \mid 1)) \end{aligned}$$

The assignment of numbers to trellis nodes is routine until we reach the point in the trellis in which more than one path enters each node. In this case we choose as the node label the "best" partial path metric among the metrics for all of the entering paths (the best partial path metric may be either the largest or smallest, depending on how we selected a and b in Eq. (12-4)). The path with the best metric is the **survivor**, while the other entering paths are nonsurvivors. If the best metric is shared by two or more paths, the survivor is selected from among the best paths at random. Consider the accompanying diagram and assume that our path metrics have been designed so that the path that has the minimum path metric is the maximum-likelihood code word. The label Z is then determined as follows.

$$Z = \min\{[X + M(r_t^{(0)}\,|\,0) + M(r_t^{(1)}\,|\,0) + M(r_t^{(2)}\,|\,1)],$$
$$[Y + M(r_t^{(0)}\,|\,1) + M(r_t^{(0)}\,|\,1) + M(r_t^{(0)}\,|\,1)]\}$$

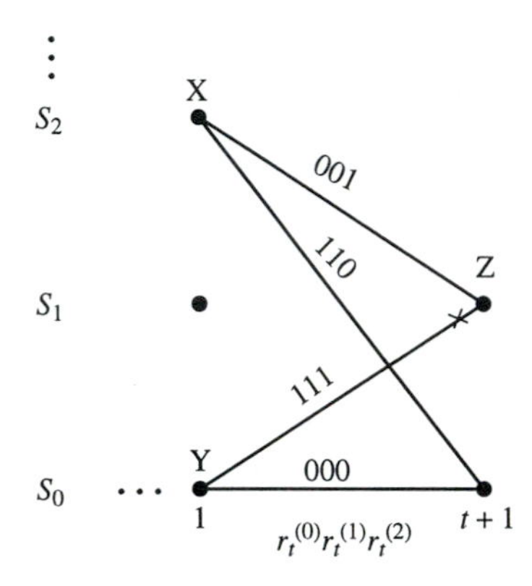

Suppose that the path from S_2 survives at S_1 at time $t + 1$. The nonsurviving path can be denoted pictorially by a cross mark, as shown in the diagram. The selection of survivors lies at the heart of the Viterbi algorithm, and insures, as we shall see, that the algorithm terminates with the ML path.

The algorithm terminates when all of the nodes in the trellis have been labeled and their entering survivors determined. We then go to the last node in the trellis (state S_0 at time $L + m$) and trace back through the trellis. At any given node, we can only continue backward on a path that survived upon entry into that node. Since each node has only one entering survivor, our trace-back operation always yields a unique path. This path is the ML path.

We can now summarize the algorithm.

The Viterbi algorithm

Let the node corresponding to state S_j at time t be denoted $S_{j,t}$. Each node in the trellis is to be assigned a value $V(S_{j,t})$. The node values are computed in the following manner.

1. Set $V(S_{0,0}) = 0$ and $t = 1$.
2. At time t, compute the partial path metrics for all paths entering each node.
3. Set $V(S_{k,t})$ equal to the best partial path metric entering the node correspond-

ing to state S_k at time t. Ties can be broken by the flip of a coin. The nonsurviving branches are deleted from the trellis or indicated by a cross mark.

4. If $t < L + m$, increment t and return to step 2.

Once all node values have been computed, start at state S_0, time $t = L + m$, and follow the surviving branches backward through the trellis. The path thus defined is unique and corresponds to the ML code word.

Theorem 12-1

The path selected by the Viterbi decoder is the maximum likelihood path.

Proof. If the ML path **y** is not selected by the decoder, then at some time t the partial ML path does not survive when compared to some other partial path **z**. If the remainder of **y** after time t is appended to **z**, then the resulting path must have a better path metric than the ML path **y**. This contradicts the assumption that **y** was the ML path. **QED**

12.2.1 Hard-Decision Decoding

In hard-decision decoding each received signal is examined and a "hard" decision made as to whether the signal represents a transmitted zero or a one. These decisions form the input to the Viterbi decoder. If the channel is assumed to be memoryless, then from the decoder's perspective it can be modeled as shown in Figure 12-6 below. The individual branches are labeled with the likelihood functions. This channel model is commonly called the **binary memoryless channel** model.

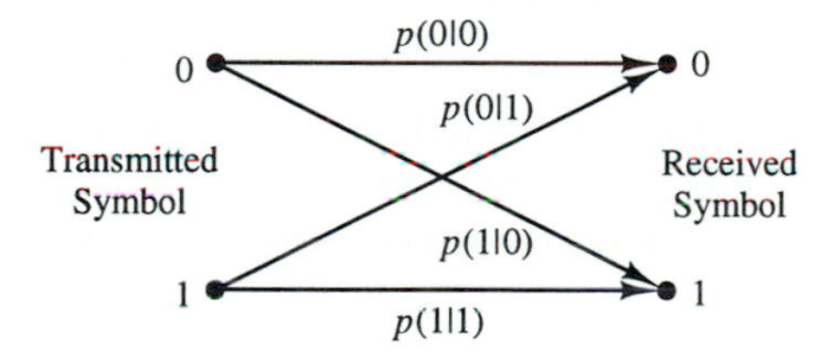

Figure 12-6. Binary memoryless channel model

The first step in defining the bit metrics for this channel is the compilation of the likelihood functions in a table. These conditional probabilities are then converted into log likelihood functions. Finally we convert the log likelihood functions into bit metrics using Eq. (12-4).

Example 12-1—Bit Metrics for Binary Memoryless Channels

Consider a binary optical channel that utilizes on-off keying (OOK) [Gag1]. In such a system a one is transmitted as a burst of photons during the signaling interval, while a zero is transmitted by leaving the interval empty. Bit decisions are made by photodetectors in the receiver. It is much more likely, in most applications, that a one becomes a zero than a zero becomes a one. The following likelihood functions are typical for a highly dispersive channel.

$p(r_i^{(j)} \mid y_i^{(j)})$	$y_i^{(j)} =$ 0	1
$r_i^{(j)} = 0$	0.9999	0.01
$r_i^{(j)} = 1$	0.0001	0.99

These are in turn converted into log likelihood functions.

$\log_2 p(r_i^{(j)} \mid y_i^{(j)})$	$y_i^{(j)} =$ 0	1
$r_i^{(j)} = 0$	-1.443×10^{-4}	-6.644
$r_i^{(j)} = 1$	-13.288	-1.450×10^{-2}

The goal in selecting values for a and b in Eq. (12-4) is the conversion of the numbers in the foregoing table into a set of integers that can be easily manipulated by digital circuitry. If a is set to $-1/6.6$, b is set to 0, and we round to the nearest tenth, the following set of bit metrics is obtained.

$M(r_i^{(j)} \mid y_i^{(j)})$	$y_i^{(j)} = 0$	1
$r_i^{(j)} = 0$	0	1
$r_i^{(j)} = 1$	2	0

These simple bit metrics are suboptimal, but the loss of performance is very small. Given this set of bit metrics, the surviving paths are those paths with the minimum partial path metric at each node. ■

If the probability of bit error is independent of the value of the transmitted bit, then the channel is said to be a **binary symmetric channel (BSC)** (see Figure 12-7).

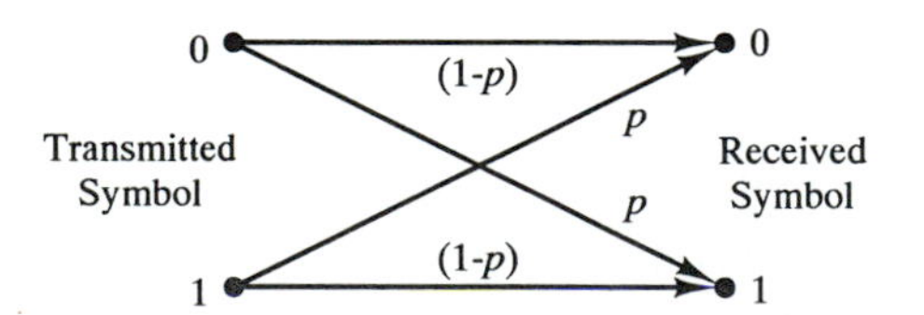

Figure 12-7. Binary symmetric channel model

We can assume that the crossover probability p for the BSC is less than or equal to 1/2, for if it were greater, we could simply relabel the zeros and ones at the receiving end, converting the effective value of p to a value less than 1/2.

If a and b in Eq. (12-4) are set to $[\log_2 p - \log_2(1-p)]^{-1}$ and $-\log_2(1-p)$, respectively, the bit metrics are independent of the value of the crossover probability p.

$$M(r_i^{(j)} \mid y_i^{(j)}) = \frac{1}{[\log_2 p - \log_2(1-p)]}[\log_2 p(r_i^{(j)} \mid y_i^{(j)}) - \log_2(1-p)]$$

$M(r_i^{(j)} \mid y_i^{(j)})$	$r_i^{(j)} = 0$	1
$y_i^{(j)} = 0$	0	1
$y_i^{(j)} = 1$	1	0

For the BSC case, the path metric for a code word $\mathbf{y}$ given a received word $\mathbf{r}$ is simply the Hamming distance $d(\mathbf{r}, \mathbf{y})$. The surviving paths are those paths with the minimum partial path metric at each node.

On the other hand, we can set a equal to $[\log_2(1-p) - \log_2 p]$ and b to $-\log_2 p$. We then get the following bit metrics.

$M(r_i^{(j)} \mid y_i^{(j)})$	$r_i^{(j)} = 0$	1
$y_i^{(j)} = 0$	1	0
$y_i^{(j)} = 1$	0	1

Given this set of bit metrics, the surviving paths are those paths with the maximum partial path metric at each node.

Example 12-2—Hard-Decision Viterbi Decoding for the BSC

The encoder in Figure 12-1 encodes the sequence $\mathbf{x} = (110101)$, generating the code word

$$\mathbf{y} = (111, 000, 001, 001, 111, 001, 111, 110)$$

$\mathbf{y}$ is transmitted over a noisy binary symmetric channel, and the received word

$$\mathbf{r} = (1\bar{0}1, \bar{1}00, 001, 0\bar{1}1, 111, \bar{1}01, 111, 110)$$

emerges from the receiver detection circuit and is sent to the Viterbi decoder. Erroneous bits have been overstruck in the expression above. The second set of bit metrics developed above is used. The surviving paths are thus those that have the highest partial path metrics. Figure 12-8 shows the results of the decoding operation. Each node in the trellis is labeled with its value as computed by the Viterbi algorithm. Nonsurviving paths are denoted by dashed lines. Several ties occur during decoding (note, for example, the evaluation of S_3 at times $t = 3$ and $t = 5$); however, the resolution of these ties has no impact on the path selected by the decoder. The ML path, denoted by the thick solid lines, is obtained by starting at state $S_{0,8}$ and tracing back along the surviving paths. The corresponding code word is the code word that was transmitted—the Viterbi decoder has corrected four errors in the received word. ■

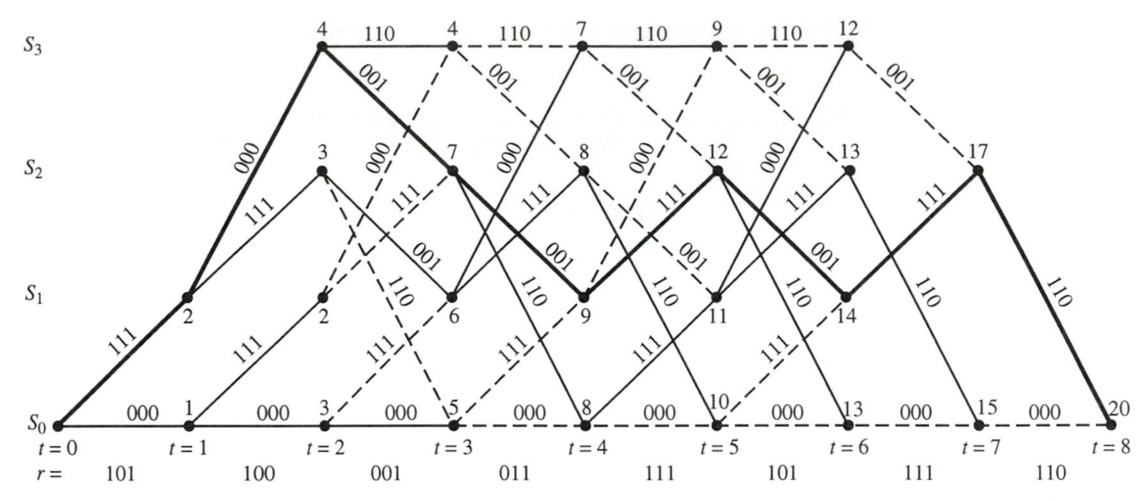

Figure 12-8. Hard-decision Viterbi decoding, Example 12-2

Example 12-3—More Hard-Decision Viterbi Decoding for the BSC

We start with the same transmitted code word as in the previous example, but this time we add three additional bit errors at the end of the received word, creating a four-error burst:

$$\mathbf{y} = (111, 000, 001, 001, 111, 001, 111, 110)$$

becomes

$$\mathbf{r} = (1\overline{0}1, \overline{1}00, 001, 0\overline{1}1, 11\overline{0}, \overline{110}, 111, 110).$$

Figure 12-9 shows that the ML path is not the transmitted path. At node $S_{1,6}$, the transmitted path (the bold dotted line entering the node) does not survive when its partial path metric is compared to that of the ML path. "Error events" of this type play a fundamental role in the analysis of the performance of the Viterbi decoder (see Section 12.3).

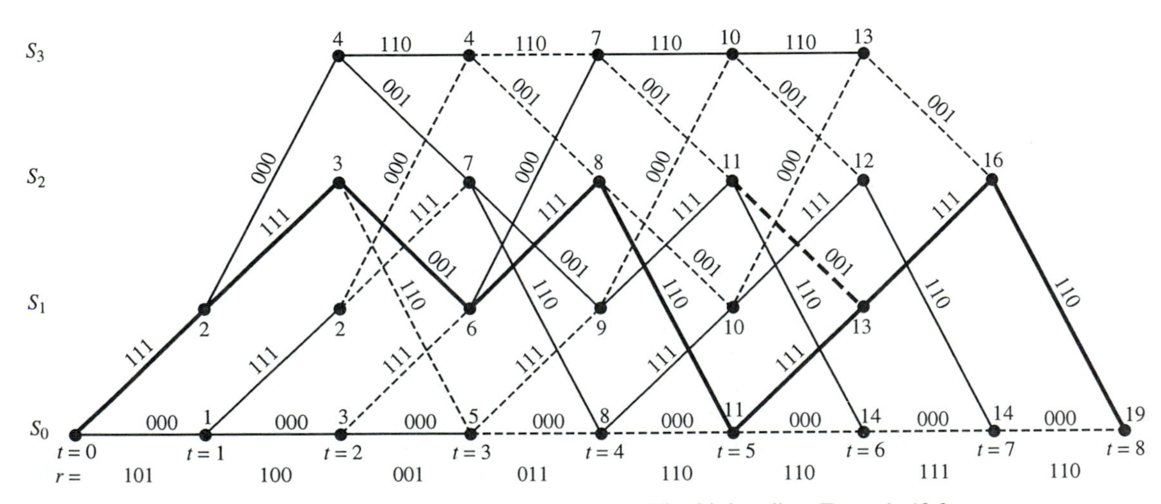

Figure 12-9. Hard-decision Viterbi decoding, Example 12-3

The ML word in this example is

$$\mathbf{y}' = (111, \overline{1}\overline{1}\overline{1}, 001, \overline{1}11, 11\overline{0}, \overline{1}\overline{1}1, 111, 110)$$

Note that the Viterbi decoder has found a code word that contains most of the erroneous bits in the received word caused by the burst error. The corresponding data word is $\mathbf{x}' = (101001)$ as opposed to the actual word (110101), so the burst of four errors in the received word has caused three bit errors in the received data. ■

12.2.2 Soft Decision Decoding

In soft-decision decoding the receiver takes advantage of "side information" generated by the receiver bit decision circuitry. Rather than simply assign a zero or a one to each received, noisy binary signal, a more flexible approach is taken through the use of multibit quantization. Four or more decision regions are established, ranging from a "strong-one" decision to a "strong-zero" decision. Intermediate values are given to signals for which the decision is less clear. On an additive white Gaussian noise channel, soft-decision decoding can provide an increase in coding gain of 2 to 3 dB over the hard-decision Viterbi decoder. A performance improvement exceeding 9 dB is possible on fading channels.

Consider the case of a BPSK modem (see Section 1-1) operating over an additive white Gaussian noise (AWGN) channel. Let E_b be the received energy per code-word bit and let the one-sided noise spectral density be N_0 W/Hz. Let the transmitted bits $\{y_i^{(j)}\}$ take the values ± 1, so that the received signals can be represented as Gaussian random variables with mean $y_i^{(j)}\sqrt{E_b}$ and variance $N_0/2$. The likelihood functions then take on the standard form for a Gaussian probability density function with nonzero mean.

$$p(r_i^{(j)} \mid y_i^{(j)}) = \frac{1}{\sqrt{\pi N_0}} e^{-(r_i^{(j)} - y_i^{(j)}\sqrt{E_b})^2/N_0} \tag{12-8}$$

The log likelihood function is simplified by noting that $(y_i^{(j)})^2 = 1$, regardless of the value of $y_i^{(j)}$. All terms that are not a function of $\mathbf{y}$ are collected into a pair of constants C_1 and C_2 [Vit3].

$$\begin{aligned}
\log p(\mathbf{r} \mid \mathbf{y}) &= \sum_{i=0}^{L-1} \left(\sum_{j=0}^{n-1} \log p(r_i^{(j)} \mid y_i^{(j)}) \right) \\
&= \sum_{i=0}^{L-1} \left\{ \sum_{j=0}^{n-1} \left[-\frac{(r_i^{(j)} - y_i^{(j)}\sqrt{E_b})^2}{N_0} - \log \sqrt{\pi N_0} \right] \right\} \\
&= \frac{-1}{N_0} \sum_{i=0}^{L-1} \left\{ \sum_{j=0}^{n-1} (r_i^{(j)} - y_i^{(j)}\sqrt{E_b})^2 \right\} - \frac{Ln}{2} \log \pi N_0 \\
&= \frac{-1}{N_0} \sum_{i=0}^{L-1} \left\{ \sum_{j=0}^{n-1} [r_i^{(j)2} - 2r_i^{(j)} y_i^{(j)}\sqrt{E_b} + y_i^{(j)2} E_b] \right\} - \frac{Ln}{2} \log \pi N_0 \\
&= C_1 \sum_{i=0}^{L-1} \left\{ \sum_{j=0}^{n-1} r_i^{(j)} y_i^{(j)} \right\} + C_2 \\
&= C_1(\mathbf{r} \cdot \mathbf{y}) + C_2
\end{aligned} \tag{12-9}$$

The path metrics for the AWGN channel are thus the inner product of the received word and the code words. The individual bit metrics are

$$M(r_i^{(j)} \mid y_i^{(j)}) = r_i^{(j)} y_i^{(j)} \tag{12-10}$$

The minimization of the path metric in Eq. (12-9) is equivalent to finding the code word **y** that is the closest to **r** in terms of Euclidean distance. This provides an interesting contrast to the metric for the BSC, which minimizes Hamming distance.

The foregoing analysis assumes that the receiver is capable of processing real numbers with infinite precision. In the practical world of digital circuitry, however, we must at some point quantize the received signal and sacrifice some of the power of the Viterbi decoder. It will be seen, however, that the loss of coding gain through quantization to as few as eight levels is surprisingly small.

In Figure 12-10 a **discrete symmetric channel** is shown in which the receiver assigns one of four values to each received signal. The underlined zero and one indicate the reception of a clear, strong signal, while the nonunderlined pair denote the reception of a weaker signal.

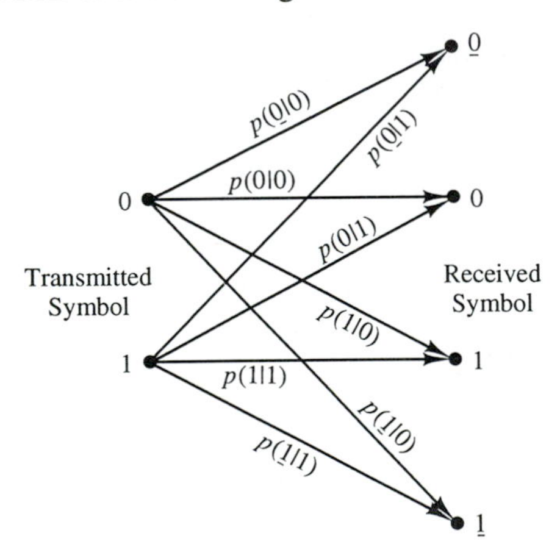

Figure 12-10. A discrete symmetric channel model

The bit decisions in a hard-decision receiver are made through the use of a hard limiter. In soft-decision receivers this role is played by a multiple-bit analog-to-digital converter (ADC). For example, the model in Figure 12-10 implies the use of a 2-bit ADC in the decision circuitry.

Decoding is performed almost exactly as shown for the hard-decision case, the only difference being the increased number (and resolution) of the bit metrics.

Example 12-4—Soft-Decision Viterbi Decoding

Consider the following values for the conditional probabilities in Figure 12-10.

$p(r \mid y)$	$r = \underline{0}$	0	1	$\underline{1}$
$y = 0$	0.50	0.32	0.13	0.05
$y = 1$	0.05	0.13	0.32	0.50

They provide the following log likelihood functions.

$\log_2 p(r \mid y)$	$r = \underline{0}$	0	1	$\underline{1}$
$y = 0$	-1.00	-1.64	-2.94	-4.32
$y = 1$	-4.32	-2.94	-1.64	-1.00

Using the expression below, we obtain a set of bit metrics that can be easily implemented in digital hardware.

$$M(r \mid y) = 1.5[\log_2 p(r \mid y) - \log_2 (0.05)]$$

$M(r \mid y)$	$r = \underline{0}$	0	1	$\underline{1}$
$y = 0$	5	4	2	0
$y = 1$	0	2	4	5

As in Example 12-2, we assume that the code word

$$\mathbf{y} = (111, 000, 001, 001, 111, 001, 111, 110)$$

is transmitted and corrupted by an error pattern that includes a burst of errors toward the end of the code word. However, this time we have 2 bits of soft decision information available in the decoder.

$$\mathbf{r} = (\underline{1}\overline{0}1, \underline{\overline{1}00}, 0\underline{01}, \underline{0}\overline{1}1, \underline{11}\overline{0}, \overline{110}, 1\underline{11}, 1\underline{1}0)$$

In Figure 12-11, we see that, with the added soft-decision information, we are able to correct the error burst that caused decoded data errors in the second of the hard-decision examples. ■

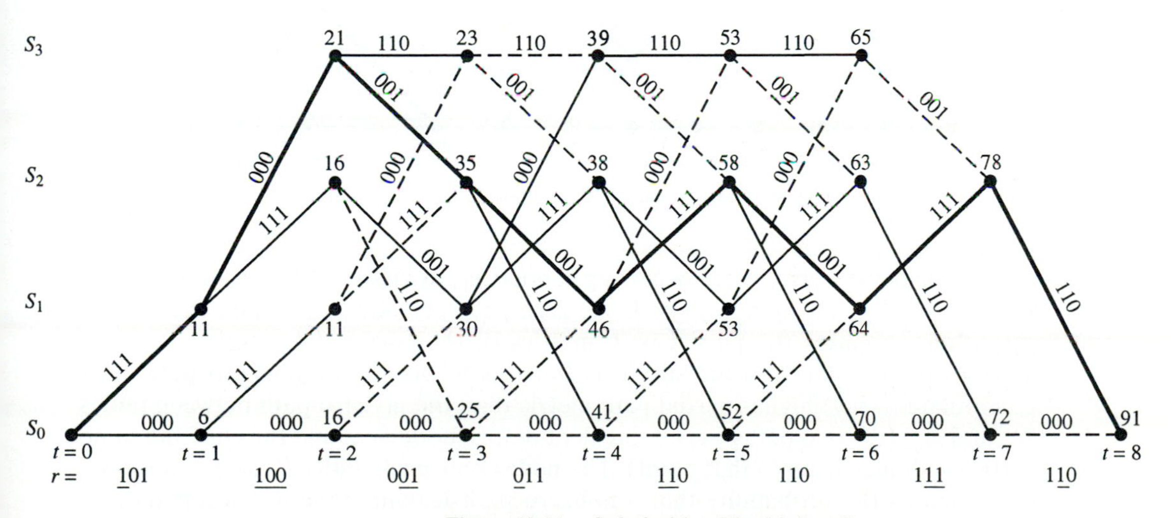

Figure 12-11. Soft-decision Viterbi decoding

12.3 PERFORMANCE ANALYSIS

In discussing the performance of block codes, we focus on the development of expressions for the word error rate. This is defined as the number of times per received word that the decoder picks a code word other than that which was transmitted. These decoder-error-rate expressions are in turn translated into bit-error-rate expressions. In the realm of convolutional codes, however, word error rate is practically meaningless. Communication between two parties using convolutional codes consists of the exchange of a single code word of unbounded length. The word error rate may thus approach unity as the length of the convolutional code word increases, while the bit error rate remains at an acceptably low level. Furthermore, when we discuss implementation issues (see Section 12.5), we will see that the decoder actually releases information bit estimates *before* the entire convolutional code word has been received. For these and other reasons, the analysis of the performance of convolutional codes focuses solely on the **instantaneous bit error rate** (henceforth referred to simply as the **bit error rate**). This is defined as the average number of erroneous information bits emerging from the decoder per information bit decoded. In this section upper and lower bounds on decoded bit error rate are derived for several channels of interest.

We begin by assuming that the code is linear and that the all-zero code word has been transmitted. An **error event** occurs at a node $S_{0,t}$ whenever a nonzero path leaves that node and, upon remerging with the all-zero path at a later node, is declared the survivor. Several such error events are shown in Figure 12-12. The path marked in bold lines is the ML path selected by the Viterbi decoder.

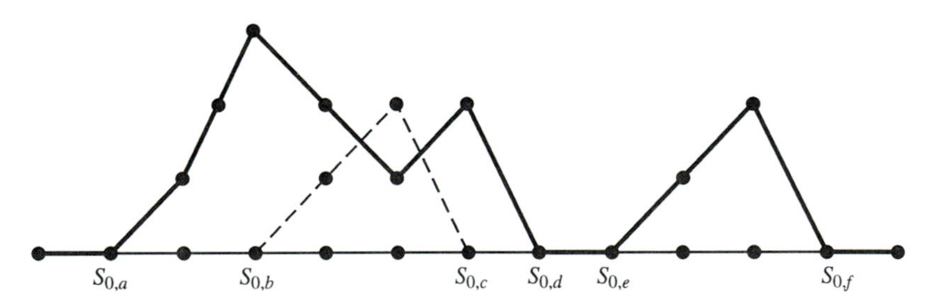

Figure 12-12. Error events at nodes $S_{0,a}$, $S_{0,b}$, and $S_{0,e}$

Consider the error event occurring at node $S_{0,a}$ in Figure 12-12. A nonzero path leaves node $S_{0,a}$ and remerges with the zero path at node $S_{0,d}$. When the Viterbi algorithm computes the partial path metrics of the two paths entering $S_{0,d}$, the nonzero path is declared the survivor. For this to occur, the nonzero path must accumulate a (say) higher partial path metric than the all-zero path between nodes $S_{0,a}$ and $S_{0,d}$ (the two paths were coincident before node $S_{0,a}$, and thus had the same partial path metric up to that point). The **node error probability** $P_e(S_{0,j})$ at node $S_{0,j}$ is defined as the probability that a nonzero path leaving node $S_{0,j}$ accumulates a higher partial path metric than the all-zero path before remerging with the all-zero path. Note that this definition allows for error events that may not be included in the ML path. Consider the error event at node $S_{0,b}$ in Figure 12-12. A nonzero path leaves $S_{0,b}$ and remerges with the all-zero path at $S_{0,c}$ and is declared the survivor. This path is not included in the ML path, however, because it does not survive the comparison with another nonzero path at node $S_{0,d}$. The node error at $S_{0,b}$ thus has no impact on the output of the decoder. We obtain an upper bound on bit error rate, however, by assuming that all node errors cause information bit errors at the decoder output.

Let $Y'(S_{0,j}) = \{\mathbf{y}'\}$ be the set of all nonzero paths $\mathbf{y}'$ diverging from node $S_{0,j}$. The amount by which the partial path metric of the all-zero path is exceeded by that of $\mathbf{y}'$ when the latter remerges with the all-zero path is denoted $\Delta M(\mathbf{y}', \mathbf{0})$. The previous development provides the following bound on the node error probability $P_e(S_{0,j})$.

$$P_e(S_{0,j}) \leq P\left\{\bigcup_{\mathbf{y}' \in Y'(S_{0,j})} [\Delta M(\mathbf{y}', \mathbf{0}) \geq 0]\right\} \tag{12-11}$$

Equation (12-11) is made more tractable through the use of the union bound.

Theorem 12-2—The Union Bound

Let $\{E_1, E_2, \ldots, E_n\}$ be events in a probability space S with probability measure $P(\cdot)$. It follows that the sum of the probabilities of the individual events is greater than or equal to the probability of the union of the events.

$$P(E_1) + P(E_2) + \cdots + P(E_n) \geq P(E_1 \cup E_2 \cup \cdots \cup E_n)$$

Proof. See [Pap].

The union bound converts Eq. (12-11) to the following.

$$P_e(S_{0,j}) \leq \sum_{\mathbf{y}' \in Y'(S_{0,j})} P\{\Delta M(\mathbf{y}', \mathbf{0}) \geq 0\} \tag{12-12}$$

The bracketed term in the summation of Eq. (12-12) is the **pairwise error probability** for the nonzero path $\mathbf{y}'$ and the all-zero path $\mathbf{0}$. The pairwise error probabilities are readily calculated for a number of channels.

12.3.1 General Memoryless Channels

Since the channel is memoryless, the branch metric computations for a noise-corrupted received word are treated as independent events. It follows that the difference in path metrics $\Delta M(\mathbf{y}', \mathbf{0})$ is only a function of those bits for which $\mathbf{y}'$ and $\mathbf{0}$ have different values. The number of such bits is, of course, the weight of the path

$\mathbf{y}'$. Let P_d be the probability that a weight-d path $\mathbf{y}'$ has a higher metric than the all-zero path, which is assumed to have been transmitted.

$$\begin{aligned} P_d &= P\{M(\mathbf{r}\,|\,\mathbf{y}') \geq M(\mathbf{r}\,|\,\mathbf{0})\} \\ &= P\left\{\sum_{i=0}^{L-1}\left(\sum_{j=0}^{n-1} \log p(r_i^{(j)}\,|\,y_i'^{(j)})\right) \geq \sum_{i=0}^{L-1}\left(\sum_{j=0}^{n-1} \log p(r_i^{(j)}\,|\,0)\right)\right\} \\ &= P\left\{\sum_{i=1}^{d} \log p(r_i\,|\,1) \geq \sum_{i=1}^{d} \log p(r_i\,|\,0)\right\} \\ &= P\left\{\sum_{i=1}^{d} \log\left[\frac{p(r_i\,|\,1)}{p(r_i\,|\,0)}\right] \geq 0\right\} \\ &= P\left\{\prod_{i=1}^{d} \frac{p(r_i\,|\,1)}{p(r_i\,|\,0)} \geq 1\right\} \end{aligned} \tag{12-13}$$

where $\{r_1, r_2, \ldots, r_d\}$ are received signals at the d coordinates at which $\mathbf{y}'$ is nonzero. Eq. (12-13) can be reexpressed as

$$P_d = \sum_{r \in R'} \left[\prod_{i=1}^{d} p(r_i\,|\,0)\right] \tag{12-14}$$

where R' is the set of all distinct vectors $\mathbf{r} = (r_1, r_2, \ldots, r_d)$ such that

$$\prod_{i=1}^{d} \frac{p(r_i\,|\,1)}{p(r_i\,|\,0)} \geq 1 \tag{12-15}$$

Since the left-hand side of Eq. (12-15) is greater than or equal to one, the same must hold for any power s of the left-hand side, where $0 < s < 1$. Equation (12-14) is thus made into an upper bound as follows.

$$\begin{aligned} P_d &< \min_{0<s<1} \sum_{r \in R'} \left\{\prod_{i=1}^{d} p(r_i\,|\,0)\left[\frac{p(r_i\,|\,1)}{p(r_i\,|\,0)}\right]^s\right\} \\ &= \min_{0<s<1} \sum_{r \in R'} \left\{\prod_{i=1}^{d} p(r_i\,|\,0)^{1-s}\, p(r_i\,|\,1)^s\right\} \end{aligned} \tag{12-16}$$

The summands in Eq. (12-16) are the products of powers of nonnegative values, so they themselves must be nonnegative, regardless of the value of r_i. We can thus loosen the bound, but facilitate its computation, by substituting for R' the set R of all possible received d-tuples.

$$P_d < \min_{0<s<1} \sum_{r \in R} \left\{\prod_{i=1}^{d} p(r_i\,|\,0)^{1-s}\, p(r_i\,|\,1)^s\right\} \tag{12-17}$$

The bound is optimized by selecting the value of s that minimizes the RHS.

The order of the summation and product can be reversed by breaking the

d-dimensional summation up into d one-dimensional summations over the individual coordinate positions $\{r_1, r_2, \ldots, r_d\}$.

$$\begin{aligned} P_d &< \min_{0<s<1} \sum_{r_1} \sum_{r_2} \cdots \sum_{r_d} \left\{ \prod_{i=1}^{d} p(r_i \mid 0)^{1-s} p(r_i \mid 1)^s \right\} \\ &= \min_{0<s<1} \prod_{i=1}^{d} \left[\sum_{r_i} p(r_i \mid 0)^{1-s} p(r_i \mid 1)^s \right] \end{aligned} \tag{12-18}$$

Equation (12-18) is the **Chernoff bound** on the pairwise error probability. If the channel is symmetric, $s = 1/2$ optimizes the bound, providing the **Bhattacharyya bound** in Eq. (12-19).

$$P_d < \prod_{i=1}^{d} \left[\sum_{r_i} \sqrt{p(r_i \mid 0) p(r_i \mid 1)} \right] \tag{12-19}$$

At this point the analysis becomes highly channel dependent, so an example is in order. We consider an asymmetric binary memoryless channel. In this case there are only two possible values for the r_i in Eq. (12-18), so we obtain the following.

$$\begin{aligned} P_d &< \min_{0<s<1} \prod_{i=1}^{d} [p(0 \mid 0)^{1-s} p(0 \mid 1)^s + p(1 \mid 0)^{1-s} p(1 \mid 1)^s] \\ &= \prod_{i=1}^{d} \min_{0<s<1} [p(0 \mid 0)^{1-s} p(0 \mid 1)^s + p(1 \mid 0)^{1-s} p(1 \mid 1)^s] \\ &= \{\min_{0<s<1} [p(0 \mid 0)^{1-s} p(0 \mid 1)^s + p(1 \mid 0)^{1-s} p(1 \mid 1)^s]\}^d \\ &= Z^d \end{aligned} \tag{12-20}$$

The form of the last expression in Eq. (12-20) is particularly useful, for it allows us to use the closed-form weight enumerator in the bound on node error probability in Eq. (12-12).

The weight enumerator $T(X, Y)$ for a convolutional code (see Section 11.2.3) lists all paths through the encoder graph that leave the all-zero state and later remerge with the all-zero state exactly once. If we set Y equal to one, the resulting expression $T(X, 1) = T(X)$ has the form

$$T(X) = a_1 X + a_2 X^2 + a_3 X^3 + \cdots + a_d X^d + \cdots$$

where a_d is the number of nonzero paths with weight d leaving the all-zero state. The probability of node error is then bounded above by

$$P_e(S_{0,j}) \leq \sum_{d=1}^{\infty} a_d P_d \tag{12-21}$$

The bound on P_d is then inserted into Eq. (12-21).

$$\begin{aligned} P_e(S_{0,j}) &\leq \sum_{d=1}^{\infty} a_d Z^d \\ &= T(X)\big|_{X=Z} \end{aligned} \tag{12-22}$$

Since the RHS of Eq. (12-22) is not a function of j, it serves as an upper bound on node error probability P_e for any node in the trellis.

Each of the nonzero paths leaving the node $S_{0,j}$ is associated with a nonzero information word. The weight of this information word is the number of bit errors that will occur should that path be chosen as the ML code word over the correct all-zero path. Once again the weight enumerator provides the desired information. The full weight enumerator has the form

$$T(X, Y) = a_{1,1}XY + a_{1,2}XY^2 + a_{1,3}XY^3 + \cdots + a_{j,k}X^jY^k + \cdots$$

where $a_{j,k}$ is the number of code words of weight j associated with information sequences of weight k. If we take the partial derivative of $T(X, Y)$ with respect to Y and set Y equal to 1, we obtain an expression of the following form. Note that the exponents for the Y terms now weight the coefficients of the weight enumerator, giving us an expected-value expression.

$$\begin{aligned}\left.\frac{\partial T(X, Y)}{\partial Y}\right|_{Y=1} &= a_{1,1}X + 2a_{1,2}X + 3a_{1,3}X + \cdots + ka_{j,k}X^j + \cdots \\ &= \left(\sum_{l=1}^{\infty} la_{1,l}\right)X + \left(\sum_{l=1}^{\infty} la_{2,l}\right)X^2 + \cdots + \left(\sum_{l=1}^{\infty} la_{j,l}\right)X^j + \cdots \\ &= b_1X + b_2X^2 + b_3X^3 + \cdots + b_jX^j + \cdots\end{aligned} \tag{12-23}$$

The $\{b_j\}$ are the total number of nonzero information bits associated with code words of weight j. Let $\#_b(S_{0,j})$ be the number of information-bit errors caused by a node error at $S_{0,t}$. The expected value of $\#_b(S_{0,j})$ is bounded above by evaluating X in Eq. (12-23) at Z.

$$E[\#_b(S_{0,j})] \leq \left.\frac{\partial T(X, Y)}{\partial Y}\right|_{X=Z,\, Y=1} \tag{12-24}$$

Since the RHS of Eq. (12-24) is independent of t, it serves as an upper bound for the expected number of information-bit errors caused by a node error at any node.

At each stage in the trellis, the decoder is acting on a single output block that is associated with a single input block. Equation (12-24) provides an upper bound on the number of bit errors caused by this action. The upper bound on the instantaneous bit error rate follows.

$$P_b \leq \frac{1}{k}\left.\frac{\partial T(X, Y)}{\partial Y}\right|_{X=Z,\, Y=1} \tag{12-25}$$

For more complex convolutional codes, it is extremely difficult to obtain the weight enumerator. In such cases an estimate of the performance can be obtained based on knowledge of the paths whose weight is equal to the minimum free distance d_{free}. Let $b_{d_{\text{free}}}$ be the number of nonzero information bits associated with code words of weight d_{free}.

$$P_b \approx \frac{1}{k} b_{d_{\text{free}}} Z^{d_{\text{free}}} \tag{12-26}$$

Example 12-5—An Asymmetric Memoryless Channel

As in Example 12-1, we assume the following likelihood functions.

$p(r_i^{(j)} \mid y_i^{(j)})$	$y_i^{(j)} =$ 0	1
$r_i^{(j)} = 0$	0.9999	0.01
$r_i^{(j)} = 1$	0.0001	0.99

The expression

$$Z = \min_{0<s<1} [0.9999^{1-s} 0.01^s + 0.0001^{1-s} 0.99^s]$$

takes on the minimum value $Z = 0.0875372$ at $s = 0.617017$.

The rate-1/2 constraint length-4 encoder in Figure 11-1 was shown in Eq. (11-23) to have a weight enumerator of the form

$$T(X, Y) = \frac{X^6 Y^3 - X^{10} Y^3 + X^6 Y}{1 - 2X^2 Y^3 - 3X^2 Y + 2X^6 Y^3}$$

Taking the partial derivative with respect to Y and setting Y to 1, we obtain

$$\left.\frac{\partial T(X, Y)}{\partial Y}\right|_{Y=1} = \frac{4X^6 - 3X^{10}}{1 - 5X^2 + 2X^6} - \frac{(6X^6 - 9X^2)(2X^6 - X^{10})}{(1 - 5X^2 + 2X^6)^2}$$

Substituting for X, we obtain the upper bound on bit error rate.

$$P_b < \frac{1}{k}\left.\frac{\partial T(X, Y)}{\partial Y}\right|_{X=Z,\, Y=1} = \left.\frac{\partial T(X, Y)}{\partial Y}\right|_{X=0.0875372,\, Y=1} = 1.94 \times 10^{-6}$$

Equation (12-26) provides the following estimate of the performance. Recall that there are two paths of weight $d_{\text{free}} = 6$: one associated with a weight-one information sequence, the other with a sequence of weight three. $b_{d_{\text{free}}}$ is thus 4.

$$P_b \approx b_{d_{\text{free}}} Z^{d_{\text{free}}} = 4(0.0875372)^6 = 1.80 \times 10^{-6}$$ ■

12.3.2 Binary Symmetric Channels

For the BSC, the pairwise error probability P_d is the probability that half or more of d transmitted bits are in error. Assuming a BSC crossover probability p, P_d is obtained through a simple combinatorial exercise.

$$P_d = \begin{cases} \displaystyle\sum_{k=(d+1)/2}^{d} \binom{d}{k} p^k (1-p)^{d-k}, & d \text{ odd} \\ \displaystyle\frac{1}{2}\binom{d}{d/2} p^{d/2}(1-p)^{d-d/2} + \sum_{k=d/2+1}^{d} \binom{d}{k} p^k (1-p)^{d-k}, & d \text{ even} \end{cases} \qquad (12\text{-}27)$$

This expression can be simplified for the case in which d is odd in the following manner (it can be shown that the bound holds for even values of d as well). It is assumed that the BSC crossover probability p is less than or equal to 1/2.

$$\begin{aligned} P_d &= \sum_{k=(d+1)/2}^{d} \binom{d}{k} p^k (1-p)^{d-k} \\ &< \sum_{k=(d+1)/2}^{d} \binom{d}{k} p^{d/2}(1-p)^{d/2} \\ &= p^{d/2}(1-p)^{d/2} \sum_{k=(d+1)/2}^{d} \binom{d}{k} \end{aligned} \tag{12-28}$$

The bound on P_d in Eq. (12-28) is tighter for the BSC than the **Bhattacharyya bound** in Eq. (12-19). Equation (12-28) can be converted into the Bhattacharyya bound by extending the limits of the summation.

$$\begin{aligned} P_d &< p^{d/2}(1-p)^{d/2} \sum_{k=(d+1)/2}^{d} \binom{d}{k} \\ &< p^{d/2}(1-p)^{d/2} \sum_{k=0}^{d} \binom{d}{k} \\ &= 2^d p^{d/2}(1-p)^{d/2} \\ &= (2\sqrt{p(1-p)})^d \end{aligned} \tag{12-29}$$

A lower bound on the bit error rate for the BSC case is obtained through a rather interesting ploy. It is assumed that the decoder is inhabited by a magic (and very small) genie [Vit2]. This digital djin examines the received word and offers the decoder two code words from which to choose: the actual transmitted code word $\mathbf{y}$ and a second code word $\mathbf{y}_{d_{\text{free}}}$ that is distance d_{free} from $\mathbf{y}$. Since an unassisted decoder would have to decide between these two code words as well as all of the other code words, the bit error rate of the genie-assisted decoder provides a lower bound on the bit error rate for an unassisted decoder.

$$\begin{aligned} P_b &\geq \frac{1}{k} P_{d_{\text{free}}} \\ &= \begin{cases} \dfrac{1}{k} \displaystyle\sum_{k=(d_{\text{free}}+1)/2}^{d_{\text{free}}} \binom{d_{\text{free}}}{k} p^k (1-p)^{d_{\text{free}}-k}, & d \text{ odd} \\ \dfrac{1}{2k} \dbinom{d_{\text{free}}}{d_{\text{free}}/2} p^{d_{\text{free}}/2}(1-p)^{d_{\text{free}}/2} + \dfrac{1}{k} \displaystyle\sum_{k=d_{\text{free}}/2+1}^{d_{\text{free}}} \binom{d_{\text{free}}}{k} p^k (1-p)^{d_{\text{free}}-k}, & d \text{ even} \end{cases} \end{aligned} \tag{12-30}$$

Example 12-6—BSC Performance Example

We use the rate-1/2, constraint-length-4 encoder from Example 12-5. Equation (12-25) and the Bhattacharyya bound in Eq. (12-29) are used to compute an upper bound on the bit error rate for this code over a BSC with crossover probability p.

$$P_b < \frac{1}{k} \left. \frac{\partial T(X,Y)}{\partial Y} \right|_{X=Z,\, Y=1} = \left. \frac{\partial T(X,Y)}{\partial Y} \right|_{X=2\sqrt{p(1-p)},\, Y=1}$$

A lower bound is obtained through the use of Eq. (12-30). The code in this example has a minimum free distance of 6.

$$P_b \geq \frac{1}{2}\binom{6}{3}p^3(1-p)^3 + \sum_{k=4}^{6}\binom{6}{k}p^k(1-p)^{6-k}$$
$$= 10p^3(1-p)^3 + 15p^4(1-p)^2 + 6p^5(1-p) + p^6$$

Equation (12-26) provides an estimate on the performance that falls between the two bounds.

$$P_b \approx b_{d_{\text{free}}}[2\sqrt{p(1-p)}]^{d_{\text{free}}} = 256(p(1-p))^3$$

The two bounds and the approximation are plotted as a function of p in the graph in Figure 12-13. The bit error rate for the uncoded case is included for reference. The bounds are not very tight, for they differ by over an order of magnitude over the range of the plot. The approximation lies on top of the upper bound except for large values of p. As p gets large, the union bound begins to explode, and the upper bound becomes increasingly less tight. ■

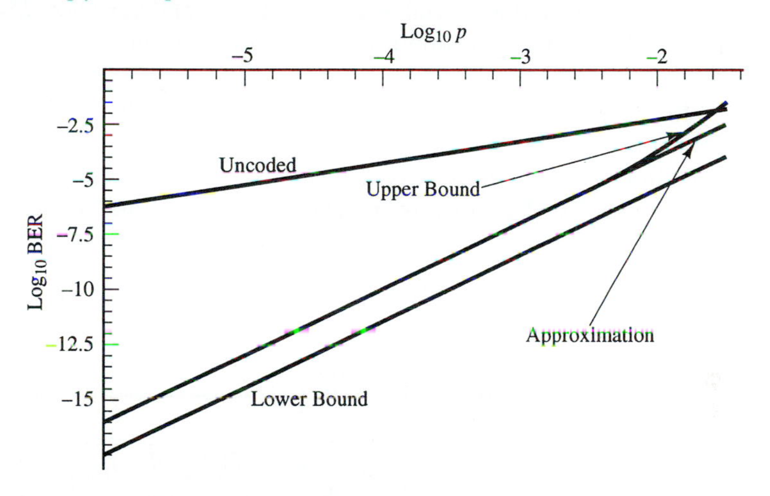

Figure 12-13. Upper and lower bounds on the performance of a (2, 1, 3) code over the binary symmetric channel

12.3.3 Additive White Gaussian Noise Channels

For the AWGN analysis we assume the use of the inner-product bit metric in Eq. (12-10). It follows that the signals sent to the decoder are assumed to be unquantized. The resulting performance analysis provides a good approximation, however, for a practical receiver using an analog-to-digital converter with three or more bits of precision.

Suppose that a BPSK modem is used for communication across an AWGN channel with one-sided noise spectral density N_0. As in the BSC case, we need to compute the probability P_d that a path containing exactly d ones has a higher metric than the all-zero path, given that the latter path is transmitted. To facilitate the analysis, the transmitted bits are assumed to be ± 1, with the all-zero path being transmitted as a sequence of -1s. The pairwise error probability is the probability that the d bits where another path has 1s are received, as a whole, more positive than negative.

$$P_d = P\left\{\sum_{j=0}^{d-1} r_i^{(j)} \geq -\sum_{j=0}^{d-1} r_i^{(j)}\right\} = P\left\{\sum_{j=0}^{d-1} r_i^{(j)} \geq 0\right\} \tag{12-31}$$

Let E_b be the received signal energy per code-word bit. If a string of -1s is transmitted, then the $r_i^{(j)}$ are Gaussian random variables with mean $\mu = -\sqrt{E_b}$ and variance $\sigma^2 = N_0/2$. The probability density function for $r_i^{(j)}$ is then

$$p(r_i^{(j)} \mid y_i^{(j)} = -1) = \frac{1}{\sqrt{\pi N_0}} e^{-(r_i^{(j)}+\sqrt{E_b})^2/N_0} \tag{12-32}$$

The last expression in Eq. (12-31) is the sum of k independent, identically distributed, Gaussian random variables with mean μ and variance σ^2. The sum is thus itself a Gaussian random variable with mean $d\mu$ and variance $dN_0/2$ [Pap]. P_d can thus be evaluated using a simple Gaussian error integral.

$$P_d = P\left\{\sum_{j=0}^{d-1} r_i^{(j)} \geq 0\right\} = \frac{1}{\sqrt{\pi\, dN_0}} \int_0^\infty e^{-(r+d\sqrt{E_b})^2/dN_0}\, dr = Q\left(\sqrt{\frac{2dE_b}{N_0}}\right) \tag{12-33}$$

As in the earlier cases, it is convenient to reexpress Eq. (12-33) in terms of a constant raised to the dth power. This is done by making use of the following inequality [Vit3].

$$Q(\sqrt{x+y}) \leq Q(\sqrt{x})e^{-y/2} \tag{12-34}$$

Equation (12-33) then becomes

$$\begin{aligned} P_d &= Q\left(\sqrt{\frac{2dE_b}{N_0}}\right) \\ &= Q\left(\sqrt{\frac{2d_{\text{free}}E_b}{N_0} + \frac{2(d-d_{\text{free}})E_b}{N_0}}\right) \\ &\leq Q\left(\sqrt{\frac{2d_{\text{free}}E_b}{N_0}}\right) e^{-(d-d_{\text{free}})E_b/N_0} \\ &= e^{d_{\text{free}}E_b/N_0}\, Q\left(\sqrt{\frac{2d_{\text{free}}E_b}{N_0}}\right)(e^{-E_b/N_0})^d \end{aligned} \tag{12-35}$$

The node error probability is then

$$P_e(S_{0,j}) \leq e^{d_{\text{free}}E_b/N_0}\, Q\left(\sqrt{\frac{2d_{\text{free}}E_b}{N_0}}\right) T(x)\big|_{X=e^{-E_b/N_0}} \tag{12-36}$$

and the instantaneous bit error rate is

$$P_b \leq \frac{1}{k} e^{d_{\text{free}}E_b/N_0}\, Q\left(\sqrt{\frac{2d_{\text{free}}E_b}{N_0}}\right) \frac{\partial T(X,Y)}{\partial Y}\bigg|_{X=e^{-E_b/N_0},\, Y=1} \tag{12-37}$$

The genie-assisted decoder (Section 12.3.2) can be used to derive a lower bound on the bit-error-rate performance of the Viterbi decoder over the AWGN channel.

$$P_b \geq \frac{1}{k} P_{d_{\text{free}}} = \frac{1}{k} Q\left(\sqrt{\frac{2d_{d_{\text{free}}} E_b}{N_0}}\right) \tag{12-38}$$

Example 12-7—Soft-Decision vs. Hard-Decision Decoding

In this example we compare the performance of a hard-decision decoder and an unquantized soft-decision decoder given the same code and the same channel conditions. The rate-1/2, constraint-length-4 encoder from Examples 12-5 and 12-6 is used once again. The code words are BPSK modulated and transmitted over an AWGN channel.

In the hard-decision case the received signal is converted to a sequence of zeros and ones before it is sent to the decoder. The decoder thus "sees" a BSC whose crossover probability p is related to the received bit-energy to noise-spectral-density ratio (E_b/N_0) given by the expression

$$p = Q\left(\sqrt{\frac{2E_b}{N_0}}\right), \qquad \text{where } Q(x) = \frac{1}{\sqrt{2\pi}} \int_x^\infty e^{-y^2/2}\, dy$$

The bounds are then derived in the same manner as in Example 12-6.

In the AWGN case the received signals are sent directly to the decoder without being quantized. The resulting upper and lower bounds on the bit error rate at the output of the decoder are provided by Eqs. (12-37) and (12-38), respectively.

In order to determine the absolute coding gain provided by the two decoders, we need to compare their performance to that of an uncoded system. A fair comparison requires that we take into account the redundancy in the coded data stream. It is assumed that the same amount of energy is allocated to the transmission of the coded and uncoded data streams; the received energy E_s per information bit is thus constant, while the received energy E_b per code word bit varies with code rate R. The two are related by the expression $E_b = RE_s$. Figure 12-14 thus allows for a fair comparison by expressing performance as a function of E_s/N_0. To achieve an output BER of 1×10^{-6}, the soft-decision decoder requires a 6-dB E_s/N_0, while the uncoded system requires 10.5 dB. The soft-decision system thus provides a 4.5 dB coding gain.

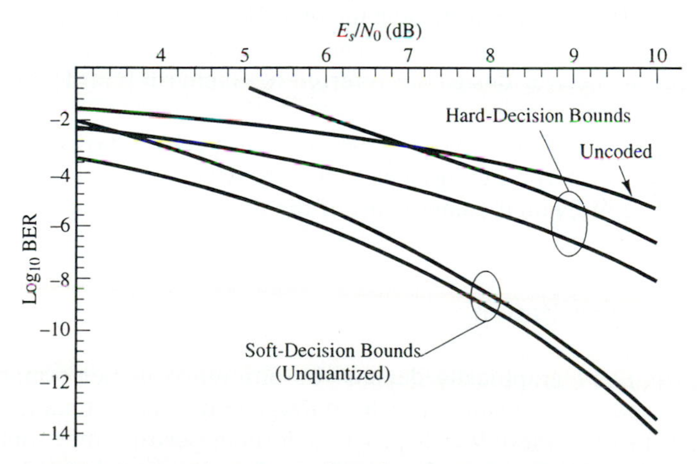

Figure 12-14. The performance of a (2, 1, 3) code with hard-decision and unquantized soft-decision decoding over an AWGN channel

These curves also show that soft-decision decoding offers an approximately 3 dB increase in coding gain over hard-decision decoding for the $(2, 1, 3)$ code. The results are similar for most other convolutional codes when used for error control over an AWGN channel. The performance difference increases dramatically, however, when the channel is Rician or Rayleigh fading. ■

12.4 THE DESIGN AND IMPLEMENTATION OF THE VITERBI DECODER

A real Viterbi decoder differs from the ideal image thus far considered in that it must contend with several constraints imposed by the real world. For example:

- Arbitrarily long decoding delays cannot be tolerated in most applications. The decoder therefore outputs decoded information bits before the entire encoded message has been received.
- If the decoder is to be implemented using digital logic, incoming analog signals must be quantized by an analog-to-digital converter.
- The decoder will frequently be brought on line in the middle of a transmission, and will thus not know where one n-bit block ends and the next begins. Some form of block synchronization is necessary.

Figure 12-15 shows a block diagram for a Viterbi decoder that satisfies the above constraints. For completeness, a few of the relevant parts of the receiver front end have been included as well.

The input to the receiver is a continuous stream of analog, modulated signals. The primary tasks of the receiver are the recovery of the carrier and bit timing so that the individual received data bits can be removed from the carrier and separated from one another in an efficient manner. Both tasks are generally performed through the use of phase-locked loops. Those readers interested in pursuing this interesting aspect of receiver design are referred to Gagliardi [Gag1] and Lindsey and Simon [Lin].

The analog baseband signals are sent to the Viterbi decoder. The decoder is provided with a bit timing clock so that the decoder can distinguish between separate bits. The individual components of the decoder are treated in the subsections that follow.

12.4.1 Quantization

Figure 12-14 graphically depicts the difference in performance between an unquantized soft-decision and a hard-decision decoder. It stands to reason that b-bit quantization, where $b > 1$, provides decoder performance somewhere between the two extremes. Fortunately a number of researchers have discovered that even a

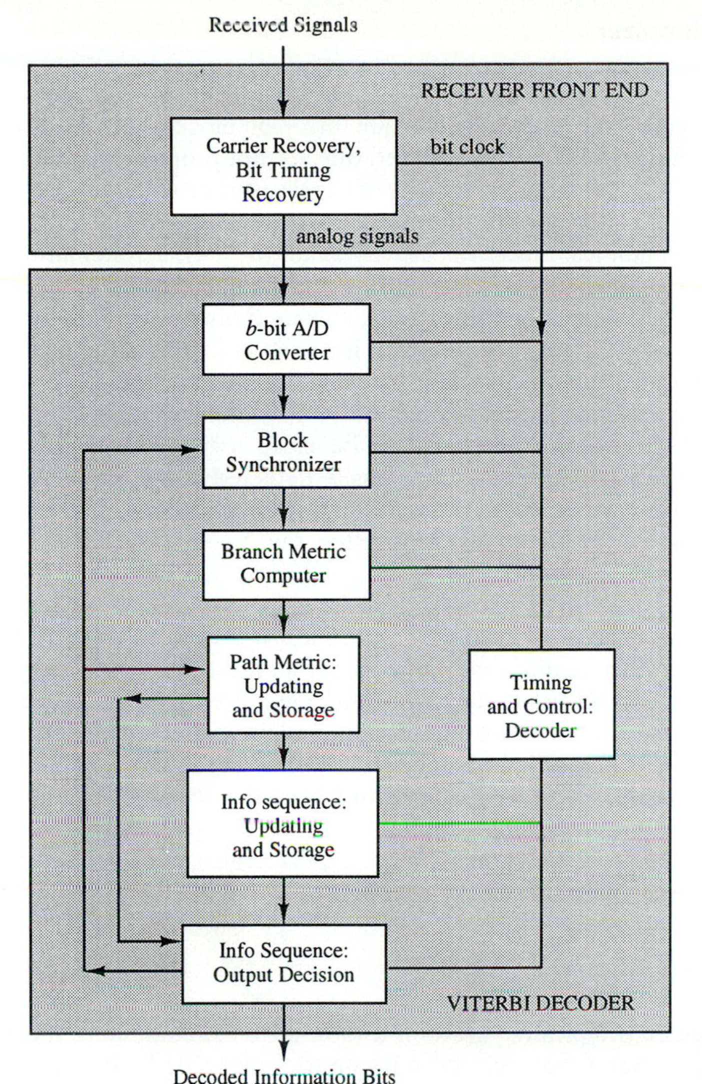

Figure 12-15. Block diagram of a practical Viterbi decoder

coarse quantization to 8 levels ($b = 3$) introduces only a slight reduction in performance with respect to the unquantized case. For example, Heller and Jacobs [Hel] conducted a simulation study that showed that an 8-level quantization resulted in only a 0.25-dB reduction in coding gain for a number of codes. The additional complexity implicit in the selection of a higher-resolution analog-to-digital converter is thus, in many cases, not balanced by a significant improvement in decoder performance. Coarser quantization levels show a more substantial decline in performance, however, with 0.9- and 2.2-dB reductions for 4- and 2-level quantization, respectively [Cai].

12.4.2 The Block Synchronizer

In order to compute the branch metrics at any given point in time, the Viterbi decoder must be able to segment the received bit stream into n-bit blocks, each block corresponding to a stage in the trellis. If the received bits are not properly divided up, the results are disastrous. Fortunately the nature of the disaster provides its own resolution. There are n distinct positions in the received data stream at which the decoder can draw block boundaries. Since n is usually small ($n \leq 4$), a systematic search of the positions can be efficiently conducted. If a guess at block synchronization is correct, one or two partial path metrics will be substantially lower than the others at each stage in the trellis after a few constraint lengths of branch metric computations (it is assumed that the ML path has the smallest metric). If a guess is wrong, the data stream is misaligned in the decoder and is essentially random. The partial path metrics for the survivors thus tend to be close together and very high—there is no dominant path. This situation is easily detected and used as an "out-of-sync" indicator. A simple thresholding circuit can be designed that compares the minimum partial path metric to a fixed threshold after, say, 5 constraint lengths of the received bit stream have been decoded.

The block synchronizer can also help resolve polarity ambiguity in the receiver demodulator. In many digital receivers, the phase-locked loops used to recover the carrier phase have several stable equilibrium points. Each point assigns ones and zeros to the received signals in a different manner. If the linear convolutional code in use does not have the all-ones word as a code word, then given a code word $\mathbf{c}$ we know that its complement $(\mathbf{c} + \mathbf{1})$ cannot be a valid code word. It follows that if the receiver switches its zeros and ones, a valid code word becomes invalid, and the Viterbi decoder acts as if it is out of synchronization. If the receiver phase-recovery loop has r equilibrium points, the block/phase synchronizer must contend with nr-fold ambiguity.

12.4.3 The Branch Metric Computer

The branch metric computer is typically based on a look-up table containing the various bit metrics. The computer looks up the n bit metrics associated with each branch and sums them to obtain the branch metric. The result is passed along to the path metric updating and storage unit. The design of the branch metric computer is made much simpler if the channel is symmetric. Note that the second row of the bit metric table in Example 12-4 (shown again below) is simply a reversed image of the first row; only one-half of the table need be placed in memory. It should also be noted that the same look-up function is performed n times per branch for each of 2^{Mk} branches per stage in the trellis. An extremely fast decoder may thus be designed using $n2^{Mk}$ look-up table circuits, or a simple decoder can be designed that uses the same look-up table $n2^{Mk}$ times.

The size of the look-up table implementation is, of course, a function of the numbers that are stored in it. The following table from Example 12-4 is a bit inefficient in that it requires 3 bits for the storage of 4 numbers.

$M(r\|y)$	$r = \underline{0}$	0	1	$\underline{1}$
$y = 0$	5	4	2	0
$y = 1$	0	2	4	5

In this case the branch metric computer can be simplified through the use of the following modified bit metrics, at the expense of a small loss in performance. Two bits are required for the implementation of this table.

$M(r\|y)$	$r = \underline{0}$	0	1	$\underline{1}$
$y = 0$	3	2	1	0
$y = 1$	0	1	2	3

12.4.4 Path Metric Updating and Storage

The path metric updating and storage unit takes the branch metrics computed by the branch metric computer and computes the partial path metrics at each node in the trellis. The surviving path at each node is identified, and the information-sequence updating and storage unit notified accordingly. These functions are performed most efficiently by breaking the trellis up into a number of identical elements. For example, the trellis diagram for a rate-$1/n$ convolutional code can be broken up into elements containing a pair of origin and destination states and four interconnecting branches, as shown in Figure 12-16. Since the entire trellis is multiple images of the same simple element, we can design a single circuit that can be used repeatedly in our decoder. Figure 12-17 shows such a circuit for a rate-$1/n$ code. Since the basic functions performed by this circuit are adding, comparing, and selecting, the circuit is commonly called an add-compare-select (ACS) circuit. The ACS circuit in Figure 12-17 is used twice to perform the functions associated with the trellis element in Figure 12-16. The entire decoder can be based on one such ACS, resulting in a very slow, but low-cost, implementation. On the other hand, a separate ACS circuit can be dedicated to every element in the trellis, resulting in a fast, massively parallel implementation. Given a rate-$1/n$ code with total memory M, $L2^M$ ACS operations

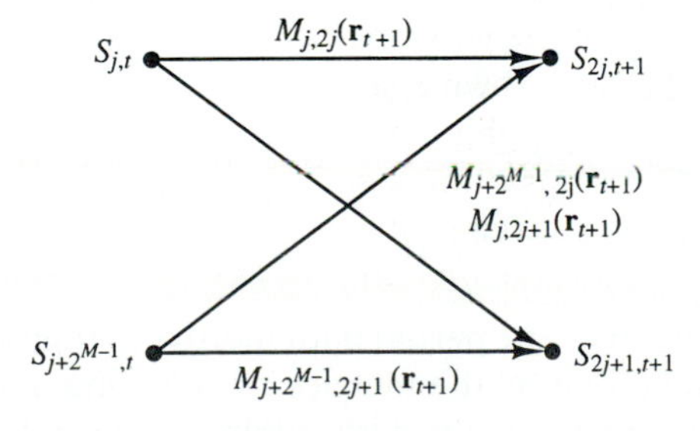

Figure 12-16. The basic trellis element for a rate-$1/n$ convolutional code

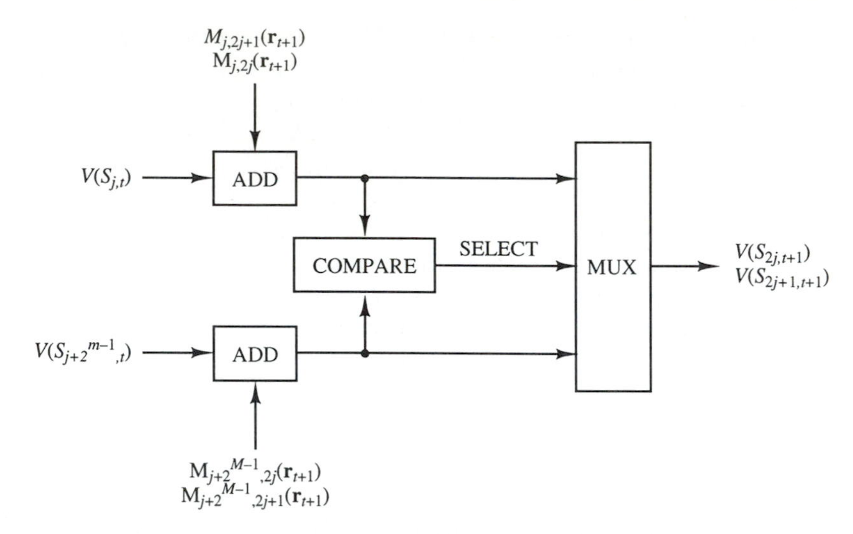

Figure 12-17. An Add-Compare-Select (ACS) circuit for a rate-$1/n$ convolutional code

are required to decode a received sequence of length L (assuming the encoder can start and stop in any state, otherwise the number of operations is slightly smaller). The ACS circuit is thus the primary functional unit of the Viterbi decoder; its design and utilization are the principal factors in determining its complexity and speed.

For a rate-k/n convolutional code, a single trellis element contains two sets of 2^k states interconnected by 2^{2k} branches. For example, Figure 12-18 shows the basic element for a code with $k = 2$. Since four branches enter each node, a four-way comparison of partial path metrics must be made to determine the surviving path. The complexity of the ACS circuit thus increases exponentially with k. This problem is avoided through the use of **punctured convolutional codes**, which are discussed in Section 12.5.

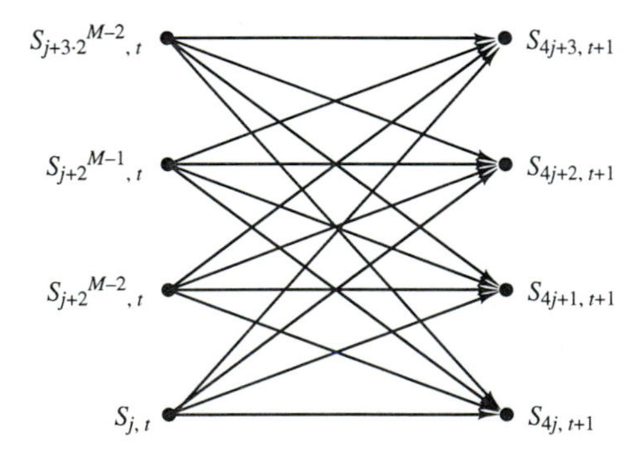

Figure 12-18. Basic trellis element for a rate-$2/n$ convolutional code

Since the decoder must store the partial path metrics in digital memory, we can limit the range of values taken on by the metrics by selecting a and b in Equation (12-4) so that the survivors are the paths with minimum partial path metrics.

12.4.5 Information Sequence Updating and Storage

The information sequence and storage unit is responsible for keeping track of the information bits associated with the surviving paths designated by the path metric updating and storage unit. There are two basic design approaches: register exchange and trace back. In both techniques a shift register is associated with every trellis node throughout the decoding operation. In register exchange, the register for a given node at a given time contains the information bits associated with the surviving partial path that terminates at that node. Consider Figure 12-19. Here we have the contents of the registers for the decoding example in Figure 12-8, assuming a register-exchange implementation. Only the surviving paths have been drawn to make the exchange operations clearer. As the decoding operations proceed, the contents of the registers in the bank are updated and exchanged as dictated by the surviving branches.

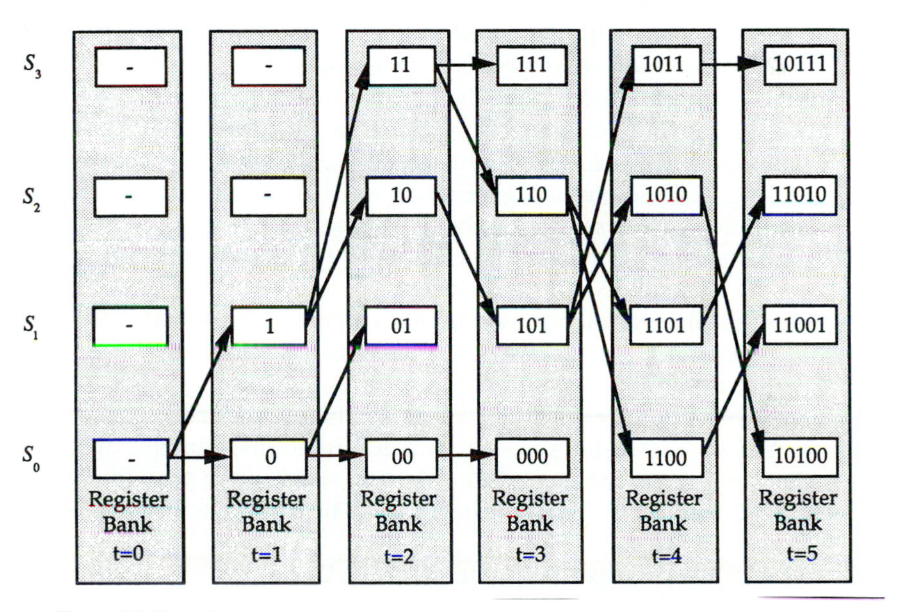

Figure 12-19. Register contents for a register-exchange decoding operation (connections follow the example shown in Figure 12-8)

The number of bits that a register must be capable of storing is a function of the **decoding depth**. A detailed discussion of this topic must wait until the section on information output decisions, but at this point we can note that at some point during decoding the decoder can begin to output information bits. For example, the decoder may be designed such that the information bits associated with a surviving branch at time t can be released when the decoder begins operation on the branches at time $t + \Gamma$. Γ is the decoding depth, and is usually set to be five to ten times the constraint length of the code, as we shall see. Given a fixed decoding depth Γ, the registers need only be of length $k\Gamma$; once the register is full (this occurs at time $t = \Gamma$), the oldest bits in the register are output as new bits are entered. The registers are thus FIFOs of fixed length. The information-sequence output device is responsible for deciding which of the 2^M register outputs is selected as the decoder output at each branch.

Note that the registers must be capable of sending and receiving strings of bits to and from two other registers. In a fast decoder all of the exchanging must take place simultaneously, leading to a hardware-intensive implementation. From a functional standpoint, however, the register-exchange approach is quite simple and may thus be preferred for applications calling for a relatively slow decoder.

In the trace-back method each state is once again assigned a register, but the contents of the registers do not move back and forth. Each state's register contains the past history of the surviving branches entering that state. Information bits are obtained by "tracing" back through the trellis as dictated by this connection history.

In Section 11.2.1 the states in the state diagram (and therefore in the trellis) were associated with the encoder shift-register contents. For example, state S_2 in a two-state encoder corresponds to the encoder shift-register contents 01. In general, we note that a state S_{XY} can be preceded only by states S_{Y0} or S_{Y1}. A zero or one may thus be used to uniquely designate the surviving branch entering a given state. For a rate-k/n code, k bits are necessary to designate the surviving branch. We can now make sense out of Figure 12-20, which shows the equivalent trace-back implementation for the register-exchange example in Figure 12-19. Note that the contents of the register associated with each state are not exchanged with other registers, but simply updated by one bit with each successive stage of decoding. The decoder outputs information bits by selecting a register at time t and tracing back as dictated by the connection history. Only those information bits associated with branches before time $t - \Gamma$ are output.

As with the register-exchange technique, the trace-back registers must be of length $k\Gamma$. Trace-back decoder implementations, though, are significantly faster than register exchange because they do not require that the register contents be moved about. On the other hand, trace back is somewhat more complicated.

12.4.6 Information-Sequence Output Decisions

If a decoding depth Γ is assumed, then at a given time t the decoder needs to release an n-bit information block decoded at time $t - \Gamma$. This raises a problem, for if at time t the entire received word has not yet been processed, there are still 2^M surviving paths from which to choose. There are three basic design solutions.

- Output the $(t - \Gamma)$th information block on a randomly selected surviving path.
- Output the $(t - \Gamma)$th information block on the surviving path with the best metric.
- Output the information block that occurs most often at time $t - \Gamma$ on the 2^M surviving paths.

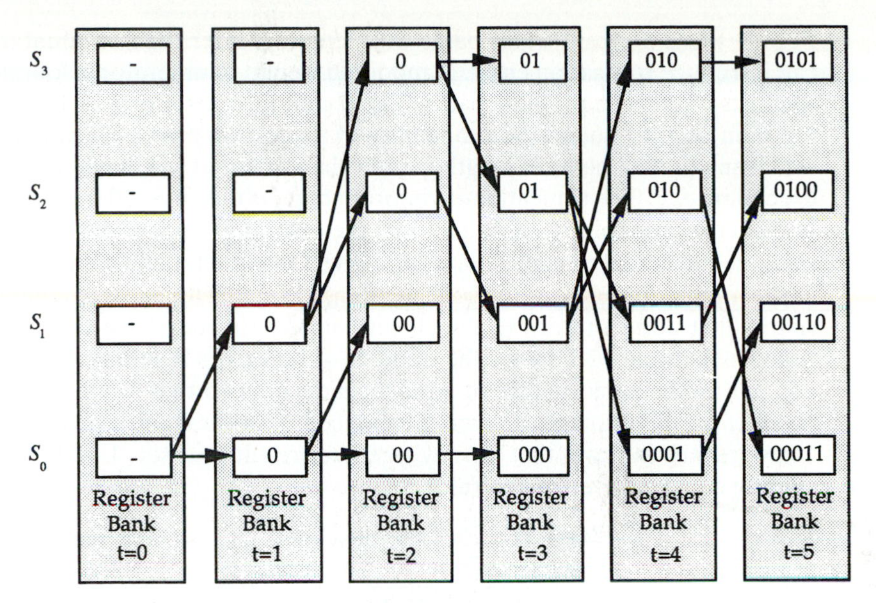

Figure 12-20. Register contents for a trace-back decoding operation (connections follow the example shown in Figure 12-8)

The first choice is clearly the simplest to implement but offers the worst performance. The second choice requires that 2^M partial path metrics be sorted to determine their maximum values at each stage of the decoding operation. Finally, it should be noted that all three choices allow for a decoded information sequence that does not correspond to any valid complete path through the trellis.

The difference in performance offered between the three choices is actually quite small if the decoding depth Γ is sufficiently large. As the 2^M surviving paths are traced back through the trellis, they tend to merge. Eventually a point is reached beyond which the 2^M survivors coincide. Clearly the information blocks from this point back to the beginning of the trellis will be associated with the maximum-likelihood code word. It is also clear that the choice of output-decision algorithm at time t is moot if all of the paths have merged at time $t - \Gamma$.

An information-sequence error that occurs as a result of the release of information bits before the entire received word has been decoded is called a **truncation error**. Forney [For4, App. B] has shown through ensemble coding arguments that the probability of truncation error decreases exponentially with decoding depth Γ. At low signal-to-noise ratios, his results show that the probability of truncation error is negligible for $\Gamma \geq 5.8m$, where m is the maximal memory order (the length of the longest shift register in the encoder).

Hemmati and Costello [Hem] have taken a more direct approach, using transfer-function techniques to bound the probability of truncation error for specific codes. Their analysis can be briefly summarized without doing it too much injustice. Assume that the all-zero code word has been transmitted. Given a decoder depth

Γ, only those nonzero trellis paths with length greater than or equal to Γ branches can introduce truncation error. A modified weight enumerator is found for the code that lists only these problem paths. The techniques described earlier in this chapter are then used to bound the probability of truncation error. Hemmati and Costello set Γ such that all paths of length greater than or equal to Γ have code-word weights exceeding d_{free}. The truncation-error probability is then a small fraction of the overall node-error probability. Table 12-1 lists the resulting decoder depths for a few of the rate-1/2 codes that maximize d_{free} for a given maximal memory order. d_Γ is defined as the minimum weight of the code words associated with paths of length Γ or greater (paths that do not return to the all-zero state until they have traversed Γ or more branches). Table 12-2 shows the number of bit errors resulting from truncation by comparing the performance of a practical decoder to that of an ideal ML decoder. In all cases the number of errors introduced by truncation is proportionally small. These results indicate that Forney's general result that sets Γ at $5.8m$ is a good, if slightly conservative, design rule.

TABLE 12-1 Decoder depths for rate-1/2 code with maximal free distance codes ([Hem], © 1977 IEEE)

Maximal memory order m	Code generators (octal)	d_{free}	Γ	d_Γ
2	5, 7	5	8	6
3	64, 74	6	10	7
4	46, 72	7	15	8
5	65, 57	8	19	9
6	554, 744	10	27	11
7	712, 476	10	28	11
8	561, 753	12	33	13

TABLE 12-2 The number of bit errors made while decoding 50,000 bits given the maximal free distance, rate 1/2, $m = 3$ code and a binary symmetric channel ([Hem], © 1977 IEEE)

Crossover probability p	$\Gamma = 8$	$\Gamma = 10$	$\Gamma = \infty$ (ML)
.0098	4	2	2
.0176	35	27	25
.0254	91	69	66
.0332	184	142	115
.0371	255	191	154

12.4.7 Initializing the Decoder

Digital receivers generally need a small amount of time to acquire an incoming signal. As noted earlier, this acquisition process includes carrier-phase and bit-timing synchronization. The Viterbi decoder introduces an additional requirement for block synchronization. By the time all of these preliminary operations have been concluded, the decoder has almost certainly missed the first few blocks of the transmitted code word. The decoder must thus begin its operation in midstream. Surprisingly this has virtually no impact on the decoder's performance beyond the first few constraint lengths of the received word. Assume that the all-zero code word has been transmitted. Since the decoder is beginning operation in midstream, branch metrics are computed for branches leaving all 2^M states at decoder time $t = 0$. We define as an **initialization error** any error resulting from the all-zero path failing to survive a comparison to a nonzero path that starts at a nonzero state at $t = 0$ and has not previously merged with the all-zero path. The problem paths at time t are thus those nonzero paths of length greater than or equal to t. The initialization problem is the dual of the truncation problem [For4] and is thus subject to the analysis presented in the previous section. It follows that at time $t = \Gamma$ and beyond, the probability of initialization error is negligible. As a result, no special initialization procedure is necessary for the Viterbi decoder in most applications.

12.5 PUNCTURED CONVOLUTIONAL CODES

In Section 12.4.4 it was seen that the complexity of the ACS circuits for a rate-k/n convolutional code increases exponentially with k. As the ACS circuit is the fundamental unit of the Viterbi decoder, the use of standard high-rate convolutional codes in conjunction with Viterbi decoding becomes problematic. This complexity issue can be avoided, however, through the use of punctured convolutional codes.

In Definition 4-11, puncturing was defined as the systematic deletion of one or more parity coordinates in every code word in a code. Cain et al. [Cai] applied this concept to convolutional codes in a very natural way: given a fixed encoder structure, higher-rate codes are achieved by periodically deleting bits from one or more of the encoder output streams. For example, let $\mathbf{C}$ be a rate-1/2 convolutional code, and let $\mathbf{x}$ be the information sequence corresponding to a code word $\mathbf{y} \in \mathbf{C}$.

$$\mathbf{x} = (x_0, x_1, x_2, \ldots)$$
$$\mathbf{y} = (y_0^{(0)} y_0^{(1)}, y_1^{(0)} y_1^{(1)}, y_2^{(0)} y_2^{(1)}, y_3^{(0)} y_3^{(1)}, y_4^{(0)} y_4^{(1)}, \ldots)$$

If every fourth bit of $\mathbf{y}$ is deleted, the resulting punctured code word $\mathbf{y}_P$ has the form

$$\mathbf{y}_P = (y_0^{(0)} y_0^{(1)}, y_1^{(0)} E, y_2^{(0)} y_2^{(1)}, y_3^{(0)} E, y_4^{(0)} y_4^{(1)}, \ldots)$$

E's have been inserted to mark the location of the deleted bits, though nothing is actually transmitted for these bits. Since $\mathbf{y}_P$ has three code-word bits for every two information bits, $\mathbf{y}_P$ is a code word in a rate-2/3 punctured code $\mathbf{C}_P$. If the receiver inserts erasures at the points where bits have been punctured, the decoder for $\mathbf{C}$ can be used to decode code words from $\mathbf{C}_P$. A simpler rate-1/2 decoder may thus be used instead of a more complicated rate-2/3 decoder.

Example 12-8—A Punctured Convolutional Code

Consider the rate-1/2 convolutional code with generator sequences (101) and (111). A rate-2/3 punctured code is created by deleting every other bit in the second output stream $\mathbf{y}^{(1)}$, as shown on following page.

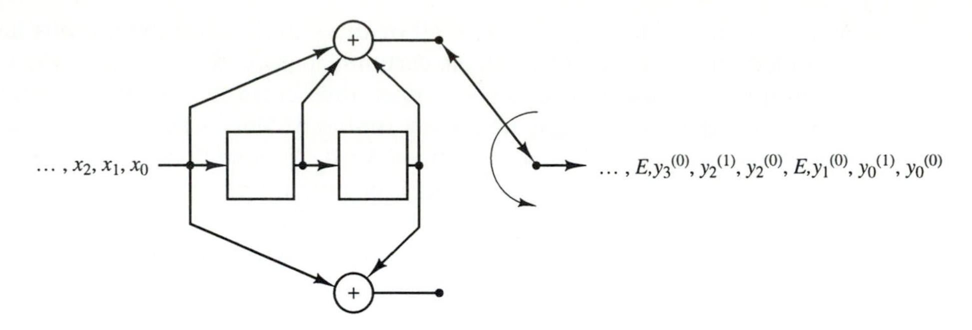

The erasures create a small timing/buffering problem, for one output is generating bits at twice the rate of the other output. This problem is avoided by adopting a standard rate-2/3 encoder design. The encoder below generates the same code as that above, but does not require the erasure of any of the output bits.

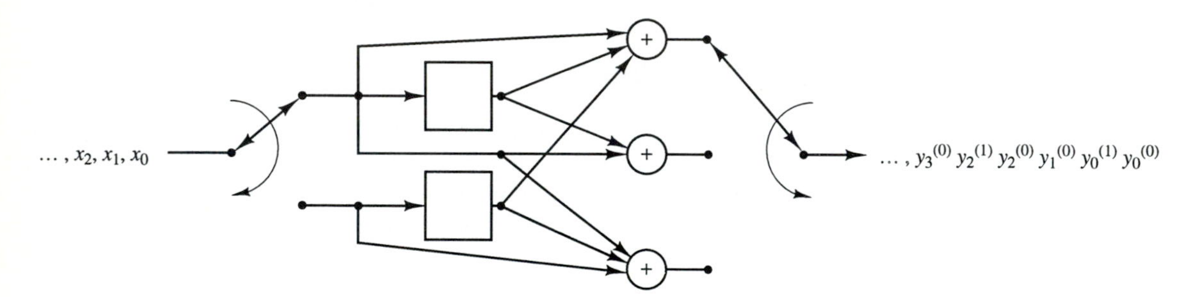

The trellis stages for a standard rate-2/3 code have the form shown below. Note that four partial path metrics must be compared at the input to each node. Three pairwise comparisons are thus necessary to pick out the surviving path.

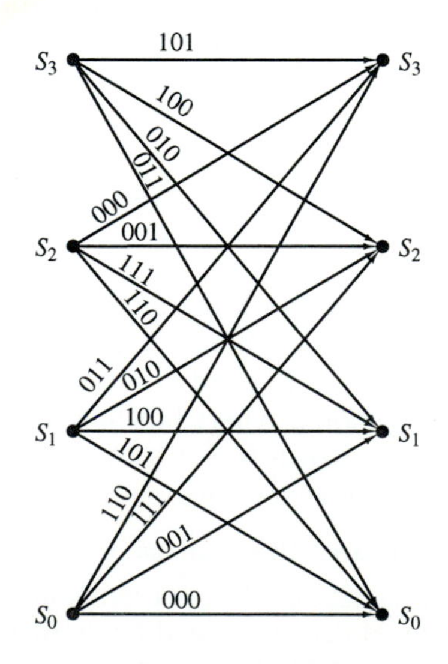

A rate-1/2 decoder that has been modified for the punctured code has the following basic trellis element.

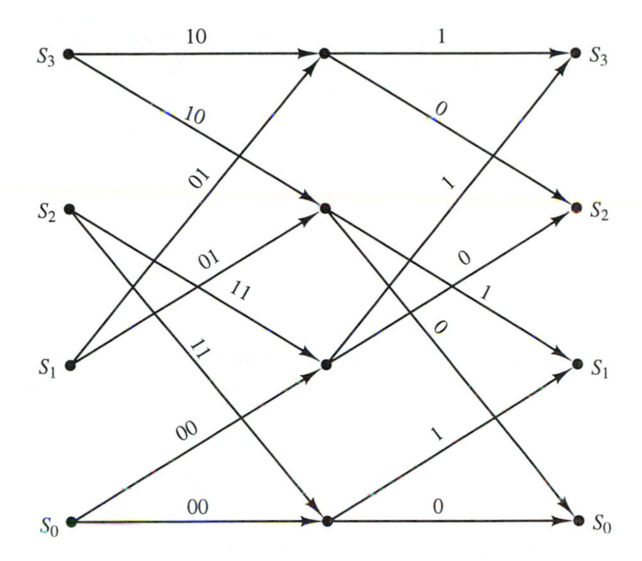

The two trellis elements correspond to the same code, but the second element breaks up each received block, associating two bits with one branch and the third bit with the next branch. The benefit to the second approach lies in the fact that it requires only one binary comparison at the input to each state. A Viterbi decoder based on the second element thus performs two binary comparisons per received block of three bits, while an implementation based on the first element requires three comparisons per block. ■

Not all convolutional codes of rate $n/(n+1)$ can be obtained by puncturing a rate-1/2 code; however, in most cases the best punctured code of a given rate attains the highest free distance of any code at that rate. (See Tables 12-3 and 12-4.)

TABLE 12-3 The best rate-3/4 punctured convolutional codes (based on minimum free distance) ([Cai], © 1979 IEEE)

Constraint length K	Code generators (octal)	d_{free}
3	5, 7, 5, 7	3
4	64, 74, 64, 74	4
5	62, 56, 46, 46	4
6	61, 47, 65, 65	5
7	724, 534, 474, 754	6

TABLE 12-4 The best rate-2/3 punctured convolutional codes (based on minimum free distance) ([Cai], © 1979 IEEE)

Constraint length K	Code generators (octal)	d_{free}
3	7, 5, 7	3
4	44, 74, 64	4
5	62, 56, 52	5
6	61, 53, 57	6
7	714, 564, 714	6
8	676, 522, 676	8

Cain et al [Cai] emphasize the benefits of puncturing in reducing the complexity of decoders for high-rate codes. Puncturing also allows for the generation of codes of various rates using the same encoder. In Example 12-8 a rate-2/3 code was obtained by deleting one out of every four bits emerging from a rate-1/2 encoder. The same encoder can be used to generate a rate-4/5 code by deleting three out of every eight output bits. If the two bits deleted in creating the rate-2/3 code are among the three bits deleted to form the rate-4/5 code, then the rate-4/5 code words are embedded in the rate-2/3 code words. In other words, as shown below, the bits in the rate-4/5 code word are all included in the rate-2/3 code word, which is in turn a part of a rate-1/2 code word. Such codes are said to be **rate compatible** [Hag3].

$$\mathbf{y} = (y_0^{(0)}, y_0^{(1)}, y_1^{(0)} y_1^{(1)}, y_2^{(0)} y_2^{(1)}, y_3^{(0)} y_3^{(1)}, y_4^{(0)} y_4^{(1)}, \ldots)$$

$$\mathbf{y}_{P,2/3} = (y_0^{(0)} y_0^{(1)}, y_1^{(0)} E, y_2^{(0)} y_2^{(1)}, y_3^{(0)} E, y_4^{(0)} y_4^{(1)}, \ldots)$$

$$\mathbf{y}_{P,4/5} = (y_0^{(0)} y_0^{(1)}, y_1^{(0)} E, y_2^{(0)} E, y_3^{(0)} E, y_4^{(0)} y_4^{(1)}, \ldots)$$

Definition 12-1—Rate-Compatible Punctured Convolutional Codes

A family of punctured codes are **rate compatible** if the code-word bits from the higher-rate codes are embedded in the lower-rate codes.

Rate-compatible punctured convolutional (RCPC) codes provide an efficient means of implementing a variable-rate error control system using a single encoder/decoder pair. There are two basic ways to achieve rate adaptivity.

- Design the error control system so that it offers two or more different code rates. The system uses channel-state information to decode which of the various codes should be used at any given moment. The channel-state information can be obtained through a variety of means, including side information generated by the Viterbi decoder [Har3].
- On a system that allows for retransmission requests, the initial transmissions consist of information that has been encoded using a high-rate punctured code.

The punctured bits are transmitted in response to subsequent retransmission requests. When they are combined with the initial packet, the resulting code word has more error control capability than either of the packets alone. An extended chain of RCPC codes allows for the combination of multiple packets, thus forming a multirate system.

Tables 12-5 and 12-6 (on pages 328 and 329) describe a family of 13 RCPC codes based on a rate-1/4 mother code. The **puncturing matrices** show which of the bits in the $M = 4$ output streams are to be punctured prior to transmission. These matrices operate on a **period** of $P = 8$ bits at a time. A zero is used to indicate deleted bits. By varying M and P, we can obtain different sets of RCPC codes for different mother codes. Table 12-5 lists b_d, the total weight of the information blocks associated with code words of weight d, for each of the 13 codes. Table 12-6 lists a_d, the number of code words of weight d.

Figure 12-21 shows an encoder that implements three codes from the family of RCPC codes listed in Tables 12-5 and 12-6. These three codes have rates 8/9, 2/3, and 1/2 and are generated using the same rate-1/2 encoder. Each of the four output streams is punctured according to the puncturing table associated with the selected rate. Bits associated with 1s in the table are passed through to the encoder output, while those associated with 0s are punctured. The puncturing tables thus show that only the top two output streams are used by the two higher-rate codes, while the lower rate code uses all of the bits from all four output streams.

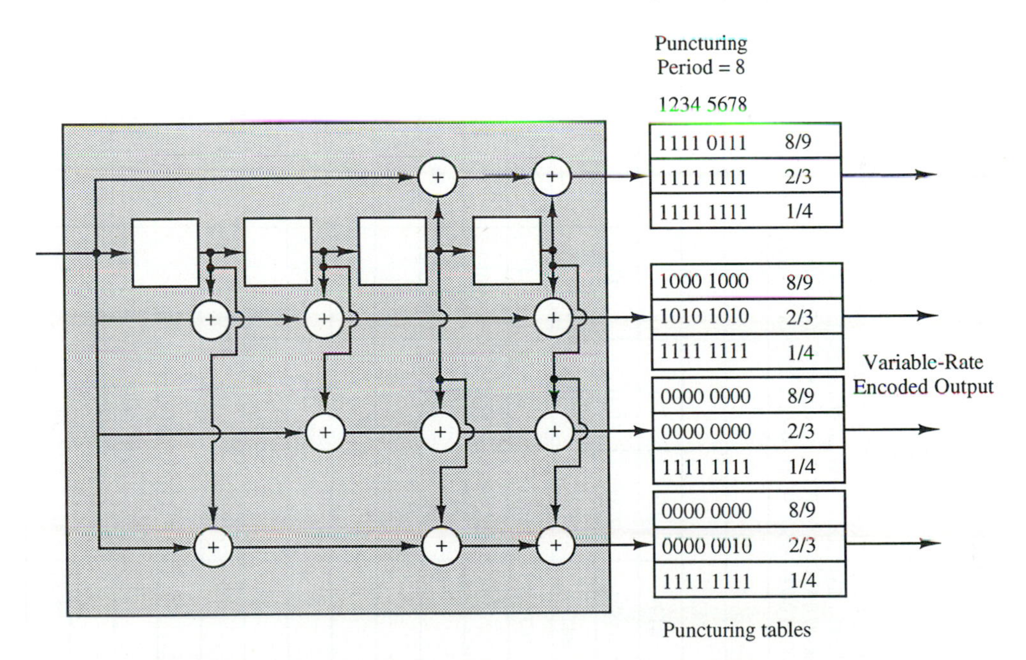

Figure 12-21. A three-rate convolutional encoder based on RCPC codes in Tables 12-5 and 12-6

TABLE 12-5 Puncturing table and b_d for a set of RCPC codes with $M = 4$ and $P = 8$ ([Hag3], © 1988 IEEE)

Rate	8/9	8/10 4/5	8/12 2/3	8/14 4/7	8/16 1/2	8/18 4/9	8/20 2/5	8/22 4/11	8/24 1/3	8/26 4/13	8/28 2/7	8/30 4/15	8/32 1/4
Puncturing Matrices	1111 0111 1000 1000 0000 0000 0000 0000	1111 1111 1000 1000 0000 0000 0000 0000	1111 1111 1010 1010 0000 0000 0000 0000	1111 1111 1110 1110 0000 0000 0000 0000	1111 1111 1111 1111 0000 0000 0000 0000	1111 1111 1111 1111 1000 1000 0000 0000	1111 1111 1111 1111 1100 1100 0000 0000	1111 1111 1111 1111 1110 1110 0000 0000	1111 1111 1111 1111 1111 1111 0000 0000	1111 1111 1111 1111 1111 1111 1000 1000	1111 1111 1111 1111 1111 1111 1010 1010	1111 1111 1111 1111 1111 1111 1110 1110	1111 1111 1111 1111 1111 1111 1111 1111
d = 2	1												
3	242	42											
4	4199	274	4										
5	63521	2688	0	2									
6	885318	21692	496	62									
7	11678199	154684	0	144	32	2							
8		1103894	10884	350	96	36	2						
9			0	2006	160	60	34	10					
10				5394	576	82	28	8					
11					1800	354	66	36	8	2			
12					4000	856	226	72	48	4			
13							354	114	72	56	20	2	
14								228	48	40	20	16	
15									104	38	36	38	32
16									256	104	24	0	16
17											56	18	0
18											184	74	32
19													48
20													96

Mother Code Generator Sequences
10011
11101
10111
11011

This table contains the first six values of b_d for a set of RCPC codes with $M = 4$ and $P = 8$.

TABLE 12-6 Puncturing table and a_d for a set of RCPC codes with $M = 4$ and $P = 8$ ([Hag3], © 1988 IEEE)

Rate	8/9	8/10 4/5	8/12 2/3	8/14 4/7	8/16 1/2	8/18 4/9	8/20 2/5	8/22 4/11	8/24 1/3	8/26 4/13	8/28 2/7	8/30 4/15	8/32 1/4
Puncturing Matrices	1111 0111 1000 1000 0000 0000 0000 0000	1111 1111 1000 1000 0000 0000 0000 0000	1111 1111 1010 1010 0000 0000 0000 0000	1111 1111 1110 1110 0000 0000 0000 0000	1111 1111 1110 1111 0000 0000 0000 0000	1111 1111 1111 1111 1000 1000 0000 0000	1111 1111 1111 1111 1100 1100 0000 0000	1111 1111 1111 1111 1110 1110 0000 0000	1111 1111 1111 1111 1111 1111 0000 0000	1111 1111 1111 1111 1111 1111 1000 1000	1111 1111 1111 1111 1111 1111 1010 1010	1111 1111 1111 1111 1111 1111 1110 1110	1111 1111 1111 1111 1111 1111 1111 1111
d ↓ 2	1												
3	30	8											
4	327	40	4										
5	3493	274	0	2									
6	37729	1686	108	18									
7	406015	9842	0	32	16	2							
8		59406	1380	74	24	14	2						
9			0	308	32	18	14	4					
10				696	80	22	10	6					
11					296	76	20	14	8	2			
12					544	164	54	22	16	4			
13							78	30	24	20	12	2	
14								52	16	14	12	8	
15									24	12	12	16	16
16									64	30	8	0	8
17											16	14	0
18											52	24	8
19													16
20													32

Mother Code Generator Sequences
10011
11101
10111
11011

This table contains the first six values of a_d for a set of RCPC codes with $M = 4$ and $P = 8$.

Figure 12-22 shows that the three codes offer varying levels of error control. The graph assumes an AWGN channel with unquantized Viterbi decoding. The higher-rate codes have lower overhead requirements because they require the insertion of fewer redundant bits into the transmitted bit stream. On the other hand, the lower-rate codes provide more coding gain. Let us suppose that the error control system is required to maintain a 1×10^{-6} bit error rate. The graph shows that the variable-rate encoder provides sufficient error control at received signal-to-noise ratios as low as 6 dB. Furthermore, should the channel improve, the variable-rate encoder can reduce the overhead rate from 3 bits out of every 4 to as low as 1 in every 9, while still maintaining the desired level of data reliability.

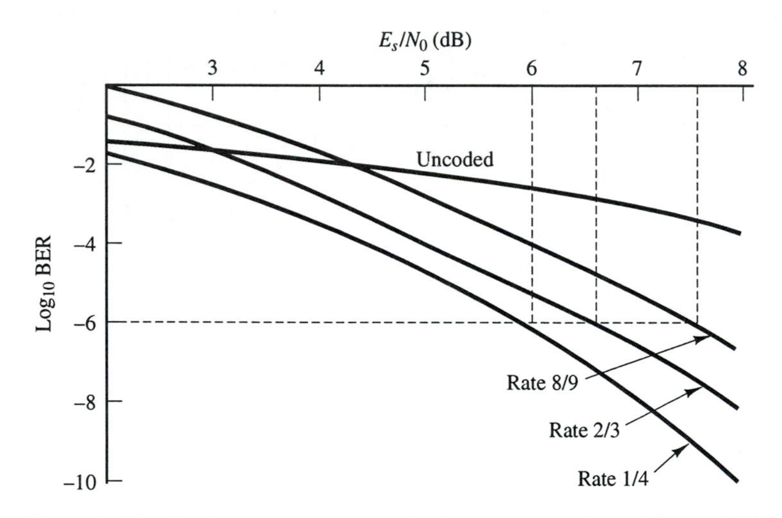

Figure 12-22. Performance curves for the three-rate encoder in Figure 12-21

PROBLEMS

1. Draw the first three stages of the trellis diagram corresponding to the following encoder.

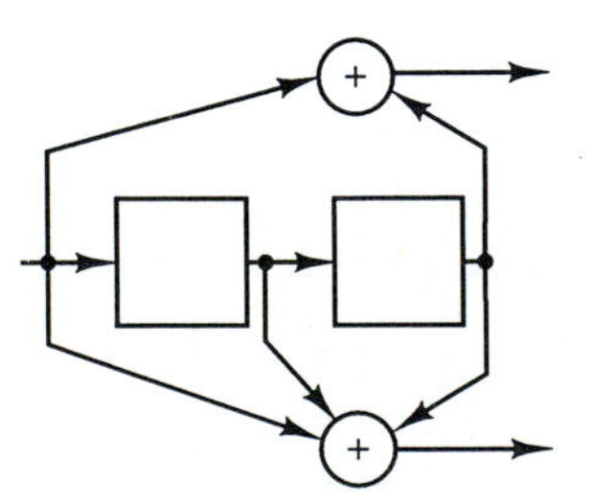

2. Draw the first three stages of the trellis diagram corresponding to the following encoder.

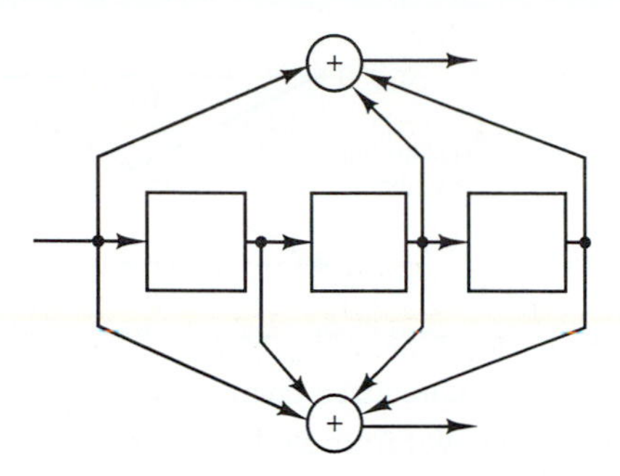

3. Compute an efficient set of bit metrics for the following channel.

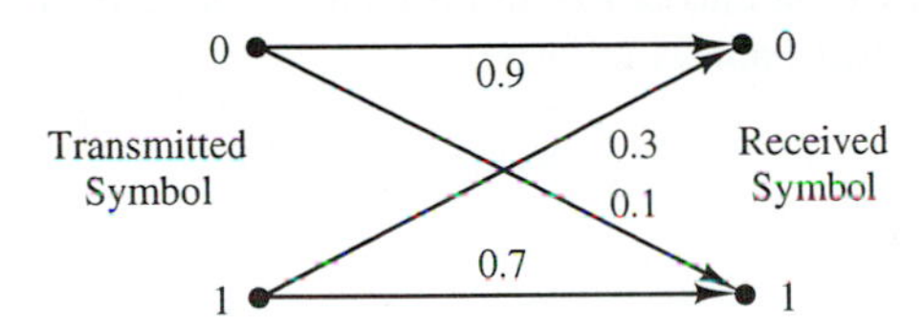

4. Compute an efficient set of bit metrics for the following channel.

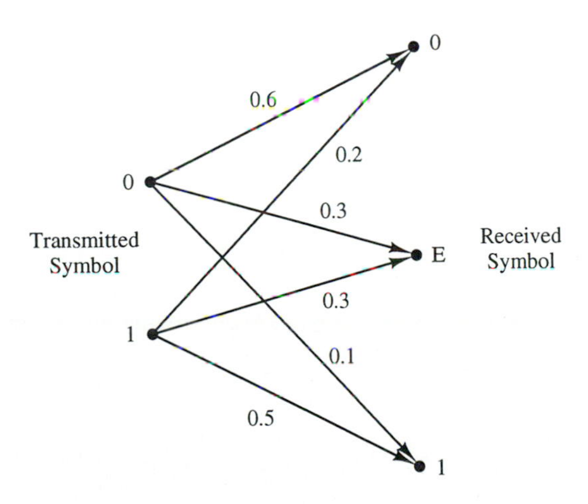

5. Compute the upper and lower bounds on the node-error probability and the bit-error-rate performance for a Viterbi decoder used in conjunction with a code with the following weight enumerator. The channel is binary symmetric with $p = 0.01$. Assume that $k = 1$.

$$T(X, Y) = \frac{X^5 Y}{1 - 2XY}$$

6. Determine the performance of the Viterbi decoder used in conjunction with the code from Problem 5 and the following channel.

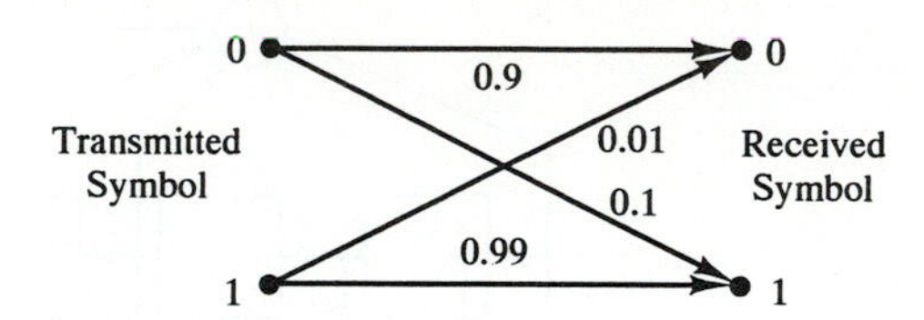

7. Determine the performance of the Viterbi decoder used in conjunction with the code from Problem 5 and an unquantized channel with $E_b/N_0 = 6$ dB.

8. Assume that the encoder from Problem 1 is used over a binary symmetric channel. Find the maximum-likelihood code word corresponding to the following received sequence.

$$\mathbf{r} = (10, 11, 11, 11, 01, 00)$$

9. Assume that the encoder from Problem 1 is used over a symmetric memoryless channel. The bit metrics are as follows.

$M(r \mid y)$	$r = \underline{0}$	0	1	$\underline{1}$
$y = 0$	0	1	3	6
$y = 1$	6	3	1	0

Find the maximum-likelihood code word corresponding to the following received sequence.

$$\mathbf{r} = (1\underline{0}, 01, \underline{1}\underline{0}, \underline{1}\underline{1}, 00, \underline{1}\underline{0}, \underline{1}1, 0\underline{0})$$

13

The Sequential Decoding Algorithms

At the beginning of Chapter 11 it was noted that Wozencraft and Reiffen described the first practical decoding algorithm for convolutional codes in 1961 [Woz2]. This algorithm was the first of a class of "sequential algorithms" that provide fast, but suboptimal, decoding for convolutional codes. In 1963 Fano presented an improved sequential decoding algorithm which now bears his name. In 1966 Zigangirov described several sequential algorithms [Zig]; unfortunately his work did not receive the immediate attention in the West that it deserved, in part because the original manuscript was in Russian. One of Zigangirov's schemes was rediscovered independently by Jelinek in 1969 [Jel1]. The Zigangirov-Jelinek (ZJ) or "stack" algorithm has since seen a great deal of application. The popularity of the sequential algorithms decreased substantially with the development of the Viterbi algorithm in 1967. The Viterbi algorithm is a maximum-likelihood algorithm, while the sequential algorithms are not. For a given code, the performance of the sequential algorithms will lag that of the Viterbi algorithm (though, in some cases, only by a negligible amount). However, the sequential algorithms still remain the better choice for many applications. The principal reason is that the complexity of the Viterbi decoder increases exponentially with the constraint length of the code, while the complexity of the sequential algorithms increases only linearly. Sequential algorithms can thus be used quite efficiently with constraint-length-50 codes, while the Viterbi systems are typically limited to constraint lengths of 10 or less.

In this chapter we examine the Fano and stack algorithms in some detail, comparing their various advantages and disadvantages with respect to each other and in comparison to the Viterbi algorithm. We then proceed to a performance analysis.

13.1 THE FANO ALGORITHM

It is quite easy to follow the intuitive path that resulted in the various sequential decoding algorithms of 1961 and 1963. Though the fine details required some brilliant engineering insights, the "big picture" for the sequential algorithms was and still is immediately obvious. We begin by considering the encoding process. Consider the $(2, 1, 3)$ convolutional encoder in Figure 11-1. If we restrict ourselves to inputs of a fixed length, say 4, then all 16 of the possible input words can be mapped to their associated code words through the use of the tree diagram in Figure 13-1. The encoding process begins at the "origin" node on the left side of the tree. We then move to the right through the branching portion of the tree, with each branching decision based on the value of an input bit. The individual branches are labeled with the output bits corresponding to the input bit. Once all of the input bits have entered the encoder, the encoding operation is concluded by inputting zeros until the encoder state returns to zero. This last action is represented by the nonbranching portion of the tree. The encoding process terminates at a "leaf" node, which corresponds to a unique convolutional code word.

The tree also provides a convenient code representation for the decoder. Consider a word $\mathbf{r}$ taken from the output of a noise-free channel. To determine the associated information sequence, we need only match up one block of received bits at a time with the labels on successive tree branches. At each branch point we simply follow the path whose n-bit label matches the appropriate portion of the received word. The various branching decisions determine the information bits associated with the received code word.

If the channel is not noise-free, the decoding process becomes a bit more complicated. We start at the origin and follow the branches that seem to provide the best match to the noisy data. If, after making a few branch decisions, we find that the received word and the branch labels are not matching well further along the path, we have to back up and try a different route. This simple approach lies at the heart of all sequential decoding algorithms. We need only define a metric for comparing the various paths given a particular received word, and then select a systematic means for determining when to proceed along a path and when to back up and try a different route.

We begin by selecting a metric. In the development of the Viterbi algorithm, a maximum likelihood decoder is obtained by using metrics that are directly proportional to the bit-likelihood functions. The sequential approach presents a different situation, however. All of the comparisons in the Viterbi algorithm are between partial paths of the same length. In a sequential algorithm, on the other hand, we often want to compare paths of differing length when deciding whether to continue through the tree or to back up and try a different branch. We cannot use the bit-likelihood functions by themselves, because the resulting partial path metrics are biased against longer paths. As a simple example of this problem, consider the Viterbi metrics for the binary symmetric channel. The partial path metric is the Hamming distance between a given path and the received word. If bit errors occur

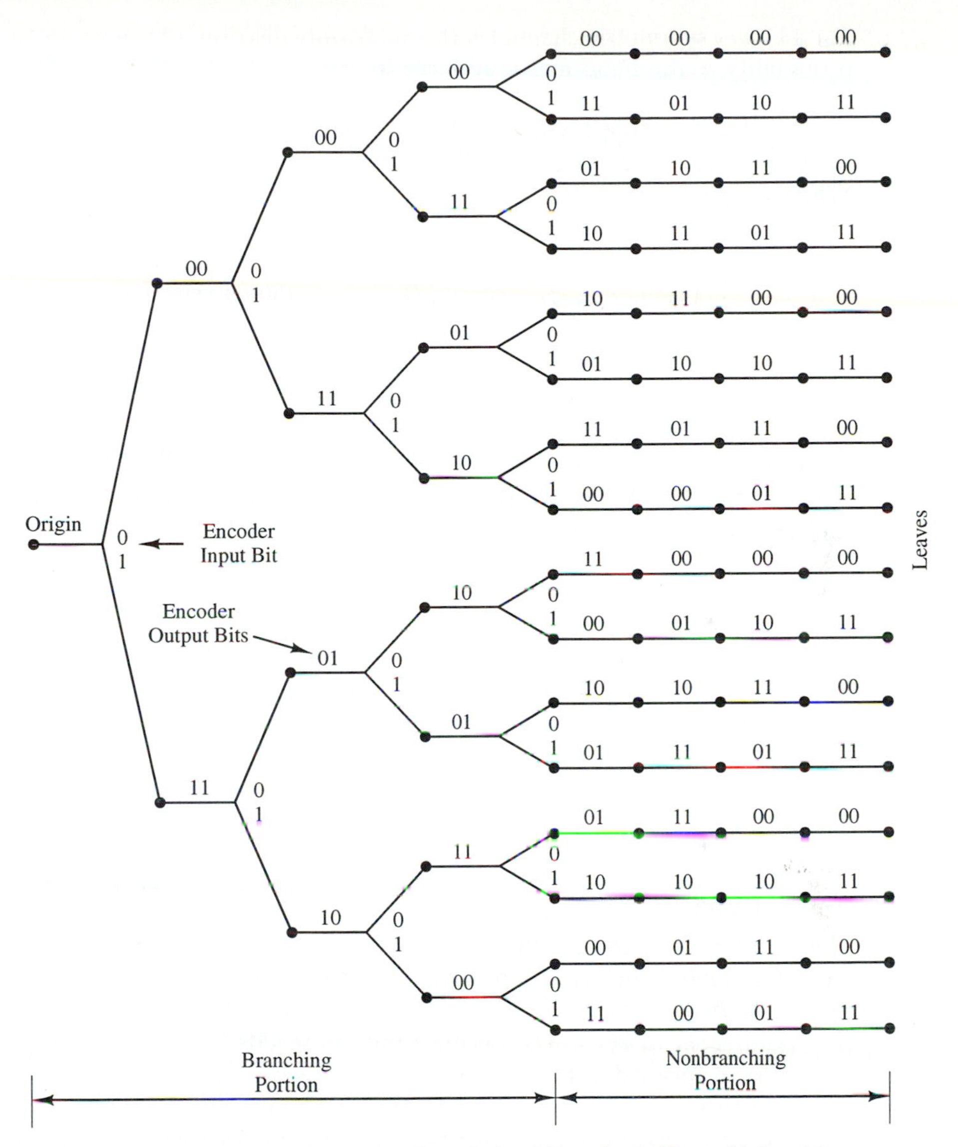

Figure 13-1. Tree Diagram for Convolutional Encoder in Figure 11-1 and Input Blocks of Length 4

at some fixed rate, then a longer path will generally have more errors and thus a higher (worse) metric than a shorter path.

Fano suggested that the following bit metric be used for sequential decoding [Fan]. Note that R is the rate of the code in use, $y_i^{(j)}$ is the ith bit of the jth transmitted block, and $r_i^{(j)}$ is the corresponding received bit.

$$M(r_i^{(j)} \mid y_i^{(j)}) = \log_2\left[\frac{P(r_i^{(j)} \mid y_i^{(j)})}{P(r_i^{(j)})}\right] - R \tag{13-1}$$

For a binary symmetric channel with a uniformly distributed source and a crossover probability p, the Fano metric reduces to

$$M(r_i^{(j)} | y_i^{(j)}) = \begin{cases} \log_2 2(1-p) - R, & r_i^{(j)} = y_i^{(j)} \\ \log_2 2p - R, & r_i^{(j)} \neq y_i^{(j)} \end{cases} \tag{13-2}$$

When a partial path metric is computed, a "length-bias" term is generated which is a function of the length of the partial path, as shown below for k branches.

$$\begin{aligned} M^k(\mathbf{r} | \mathbf{y}) &= \sum_{i=0}^{k-1} \sum_{j=0}^{n-1} M(r_i^{(j)} | y_i^{(j)}) = \sum_{i=0}^{k-1} \sum_{j=0}^{n-1} \left\{ \log_2 \left[\frac{P(r_i^{(j)} | y_i^{(j)})}{P(r_i^{(j)})} \right] - R \right\} \\ &= \sum_{i=0}^{k-1} \sum_{j=0}^{n-1} \log_2 P(r_i^{(j)} | y_i^{(j)}) + \underbrace{\left[\sum_{i=0}^{k-1} \sum_{j=0}^{n-1} \log_2 \frac{1}{P(r_i^{(j)})} - nkR \right]}_{\text{length-bias term}} \end{aligned} \tag{13-3}$$

For $R \leq 1$ and $P(r_i^{(j)}) \leq 1/2$, the length-bias term can be shown to be positive. For the BSC with a uniformly distributed source, Eq. (13-3) reduces to

$$M^k(\mathbf{r} | \mathbf{y}) = \sum_{i=0}^{k-1} \sum_{j=0}^{n-1} \log_2 P(r_i^{(j)} | y_i^{(j)}) + \underbrace{nk(1-R)}_{\text{length-bias term}} \tag{13-4}$$

The length-bias term is a linear function of the length of the path.

Fano's original selection of this metric was based on a heuristic argument, and on occasion other researchers/designers have used other metrics. Massey has since shown, however, that the Fano metric causes the decoder to always extend the most likely path based on the information currently available to the decoder [Mas5]. It is thus usually (but not always) a good choice.

We now proceed to a description of the Fano sequential decoding algorithm. A flow chart is provided in Figure 13-2. The decoder's movement through the tree is dictated by the partial path metrics for the paths ending at the various tree nodes and a threshold which is varied during decoding. At the beginning of the decoding operation the position of the decoder is placed at the root of the tree and the decoder threshold is set to zero. The decoder then proceeds to compute the partial path metric for a forward node. Generally there are two or more forward nodes from which to choose at each node in the branching portion of the tree (one for each possible input symbol value). The first forward node tested at any position is usually the node whose output block is closest in Hamming distance to the corresponding received block. If this node proves to be a bad choice, the other forward nodes may later be tested in order of their corresponding input values (lexicographic order), without regard to the Hamming distance between the received block and the branch labels. This simplifies the implementation of the decoder, as we shall see momentarily.

If the partial path metric for the forward node tested exceeds the threshold, then the decoder moves its position to the forward node (this movement is represented by P^+ in the flow chart). If this new position is a leaf, decoding is complete.

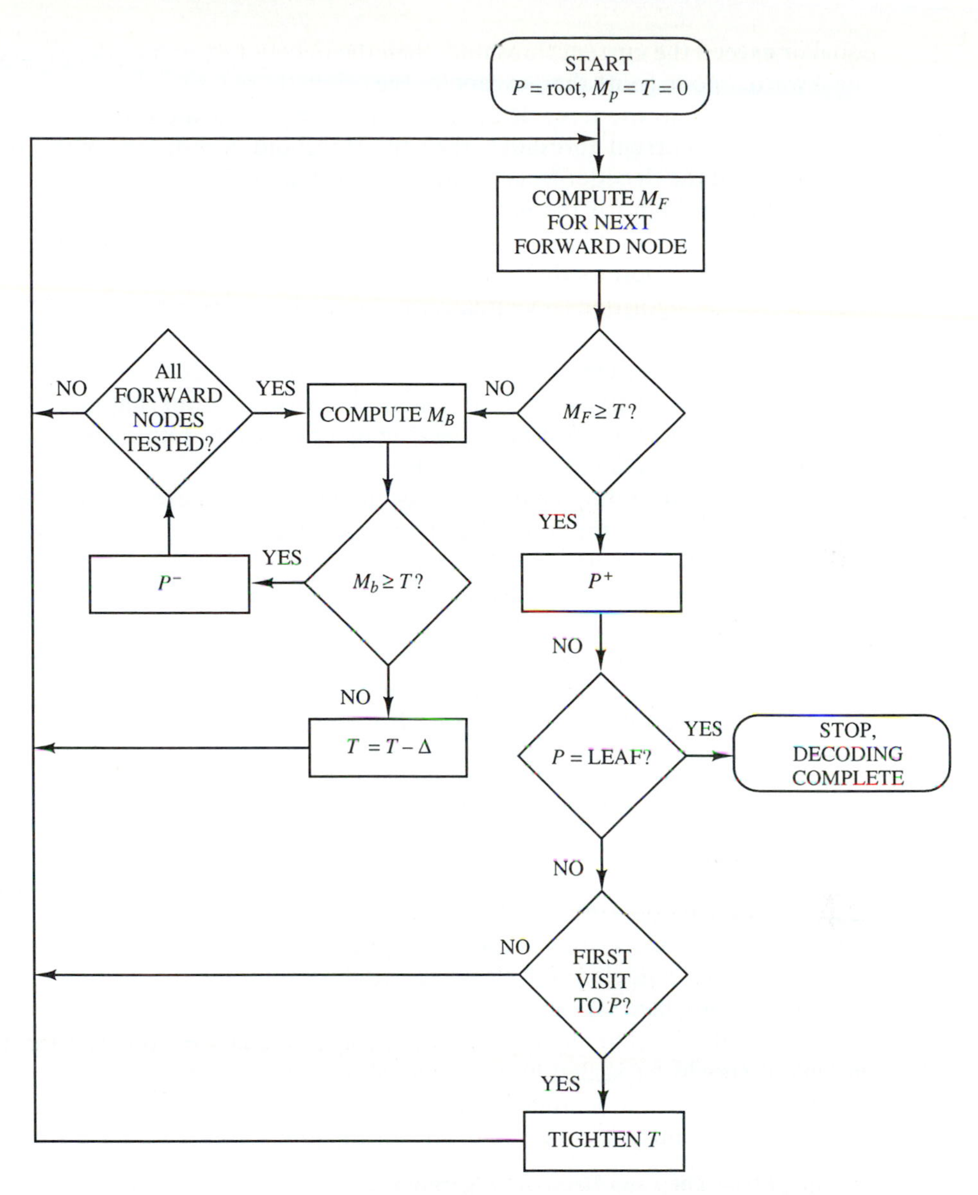

Figure 13-2. Flow Chart for Fano Sequential Decoding Algorithm

Otherwise, if this is the decoder's first visit to the current node, the threshold is "tightened." This tightening involves raising the threshold by integer multiples of the threshold increment Δ to the highest possible value less than or equal to the partial path metric associated with the decoder's current position. The decoder then continues by testing a forward node referenced to the new position.

If the next forward node to be tested has a partial path metric that does not equal or exceed the current threshold, then the decoder examines the preceding, or "backward," node (note that, except for the origin there is exactly one node preceding a given node in the tree). If the partial path metric at the backward node does not exceed the current threshold, then the threshold is reduced by the threshold increment and the decoder tests the next forward node. This event occurs when the threshold is too high for the number of errors corrupting the received word. The lowering of the threshold also insures that no node is ever tested twice at the same threshold value. There is thus no chance of an infinite loop occurring during the execution of the algorithm—eventually the threshold is lowered to a point that allows the decoder to move all the way through the tree to a leaf.

If the partial path metric of the backward node exceeds the threshold, then the decoder backs up (denoted by P^-). If all nodes forward from that point have been tested, then the decoder considers backing up once again. Otherwise, the decoder tests one of the remaining forward nodes.

The node preceding the origin is assumed to have a metric of $-\infty$. This forces the algorithm to lower the threshold and return to the testing of the forward paths.

The threshold increment Δ determines the number of node computations performed by the Fano metric for a given received word. The larger the increment, in general, the fewer the number of computations performed. Unfortunately this increase in decoder speed is obtained at the expense of reliability. The setting of Δ establishes a balance between the processing rate of the decoder and the output bit error rate. The rationale is as follows. If the decoding algorithm is to select the maximum-likelihood path, the threshold must eventually be lowered below the metric for this path. The more rapidly this lowering is performed, the more rapidly the path is accepted. If the threshold is lowered by too large an increment, however, several other paths through the tree may also become acceptable to the decoder. The decoder may thus select an incorrect path, resulting in bit errors at the output of the decoder. The greater the number of acceptable paths, the greater the average number of bit errors at the output of the decoder. Lin and Costello provide the following design guideline for choosing the threshold increment [Lin1]. If the bit metrics are not scaled, then the threshold increment should be between 2 and 8. If the metrics are scaled to make them more easily implemented in hardware, then the increment should be scaled by an identical amount. In any design, however, the selection of a particular increment should be made only after an extensive simulation study that takes into account the details of the application.

Example 13-1—The Fano Decoding Algorithm

The rate-1/2 code associated with the tree in Figure 13-1 is used to encode the information sequence $\mathbf{x} = (1100)$, resulting in the code word

$$\mathbf{y} = 11, 10, 11, 01, 11, 00, 00$$

The code word is transmitted across a binary symmetric channel with crossover probability $p = 0.125$. The received sequence is

$$\mathbf{r} = 11, 1\underline{1}, 11, 01, 11, 00, \underline{1}0$$

The erroneous bits are underlined. The decoder bit metrics are computed as follows.

$$M(r_i^{(j)} \mid y_i^{(j)}) = \begin{cases} \log_2 2(0.875) - \frac{1}{2} = 0.31, & r_i^{(j)} = y_i^{(j)} \\ \log_2 2(0.125) - \frac{1}{2} = -2.5, & r_i^{(j)} \neq y_i^{(j)} \end{cases}$$

The metrics are scaled by 1/0.31 to obtain integer metrics.

$M(r_i^{(j)} \mid y_i^{(j)})$	$y_i^{(j)} =$ 0	1
$r_i^{(j)} = 0$	1	−8
$r_i^{(j)} = 1$	−8	1

The threshold increment range [2, 8] is scaled by 1/0.31 to give a range of approximately [6, 26]. $\Delta = 10$ is selected for the example, and decoding proceeds as follows. The position P of the decoder is denoted by the binary input sequence corresponding to the current node in the tree. X is used to represent the root. The reader can note the movement of the decoder back and forth through the tree by following the variation in the length of the representation for the decoder position.

STAGE	***P***	**COMPUTATION AND RESULT**	**METRIC *M***	**THRESHOLD *T***
1	X	$M_F = 2$	0	0
	1	$M_F \geq T \Rightarrow P^+$	2	0
	1	First visit	2	0
2	1	$M_F = -5$	2	0
	1	$M_B = 0$	2	0
	X	$M_B \geq T \Rightarrow P^-$	0	0
	X	All forward nodes not yet tested	0	0
3	X	$M_F = -16$	0	0
	X	$M_B = -\infty$	0	0
	X	$M_B < T \Rightarrow T = T - \Delta$	0	−10
4	X	$M_F = 2$	0	−10
	1	$M_F \geq T \Rightarrow P^+$	2	−10
	1	Not first visit	2	−10
5	1	$M_F = -5$	2	−10
	10	$M_F \geq T \Rightarrow P^+$	−5	−10
	10	First visit $\Rightarrow$ Tighten T	−5	−10
6	10	$M_F = -12$	−5	−10
	10	$M_B = 2$	−5	−10
	1	$M_B \geq T \Rightarrow P^-$	2	−10
7	1	$M_F = -5$	2	−10
	11	$M_F \geq T \Rightarrow P^+$	−5	−10
	11	First visit $\Rightarrow$ Tighten T	−5	−10

STAGE	*P*	COMPUTATION AND RESULT	METRIC *M*	THRESHOLD *T*
8	11	$M_F = -3$	-5	-10
	110	$M_F \geq T \Rightarrow P^+$	-3	-10
	110	First visit $\Rightarrow$ Tighten T	-3	-10
9	110	$M_F = -1$	-3	-10
	1100	$M_F \geq T \Rightarrow P^+$	-1	-10
	1100	First visit $\Rightarrow$ Tighten T	-1	-10
10	1100	$M_F = 1$	-1	-10
	11000	$M_F \geq T \Rightarrow P^+$	1	-10
	11000	First visit $\Rightarrow$ Tighten T	1	0
11	11000	$M_F = 3$	1	0
	110000	$M_F \geq T \Rightarrow P^+$	3	0
	110000	First visit $\Rightarrow$ Tighten T	3	0
12	110000	$M_F = -4$	3	0
	110000	$M_B = 1$	3	0
	11000	$M_B \geq T \Rightarrow P^-$	1	0
	11000	All forward nodes tested, $M_B = -1 < T \Rightarrow T = T - \Delta$	1	-10
13	11000	$M_F = 3$	1	-10
	110000	$M_F \geq T \Rightarrow P^+$	3	-10
	110000	Not first visit	3	-10
14	110000	$M_F = -4$	3	-10
	1100000	$M_F \geq T \Rightarrow P^+$	-4	-10
	1100000	$P =$ leaf $\Rightarrow$ STOP	-4	-10

This example contains a number of events typical of decoding with the Fano decoding algorithm. At stage 1 the decoder moves forward without any trouble. At the next stage, however, a single bit error causes the partial path metric for the forward path at $P = 1$ ($M_F = -5$) to be below the current threshold (0). The metric of the previous node is above the threshold, so the decoder backs up to the root, $P = X$. The decoder then tests the next forward node ($P = 0$). The metric there (-16) is well below the threshold, so the decoder does not move forward. The decoder looks back once again, but all metrics preceding the origin are assumed to be $-\infty$. The decoder is thus forced to lower the threshold by the increment Δ and then return to the original forward path. Now that the threshold has been lowered, the decoder is willing to proceed down the path starting at $P = 10$. It immediately discovers that this is a bad path and backs up to $P = 1$. When the decoder proceeds down the path starting at $P = 11$, it is able to proceed for some time without any reversal. The bit error in the final block causes the decoder to back up once again, but once the threshold is lowered, decoding terminates at the correct leaf. ■

The most obvious characteristic of the Fano algorithm is that its execution time varies with the number of errors in the received word. This can be both a benefit and a disadvantage, depending on the situation. Recall that the Viterbi algorithm performed 2^M add-compare-select operations for each received block of bits. Most of these operations are wasted if the received word is error free. The sequential decoder, on the other hand, needs to perform only one metric computation per received block of an error-free word, moving straight from the origin of the tree to the appropriate leaf. Under moderately noisy conditions, the number of nodes visited by the Fano algorithm increases, but on the average the number of computations is still less than that performed by the Viterbi algorithm. Under extremely noisy conditions, however, the Fano algorithm is forced to perform more computations than the Viterbi decoder.

The most important benefit of the sequential algorithms as opposed to the Viterbi algorithms is their ability to use codes with very long constraint lengths. Recall that the complexity of the Viterbi algorithm increases exponentially with the constraint length of the code in use. Constraint lengths of the order of 10 stretch the limits of what is feasible in Viterbi decoder design. Unfortunately the performance of the Viterbi decoder is proportional to the minimum free distance of the code, which is in turn proportional to the code's constraint length (see Tables 11-3 through 11-7). The complexity of the Fano algorithm, on the other hand, varies linearly with the constraint length (the length of the nonbranching portion of the tree is equal to the constraint length). Constraint-length-48 codes have been used to achieve extremely high levels of reliability performance with sequential decoders without introducing an enormous amount of complexity [For8].

Finally, another principal benefit of the Fano algorithm in comparison to the stack or Viterbi algorithm is its parsimonious use of memory. The various partial path metrics are computed as needed, so they need not be stored. In most implementations memory is used only to buffer the received word, the decoded information sequence, and a "trial" code word, as shown in Figure 13-3.

As the decoder progresses through the tree, the encoder is used to generate the corresponding output blocks for comparison with the received word. The control logic monitors comparisons between the partial path metrics and the threshold. Depending on the results of these comparisons, the buffers are shifted back and forth as the decoder moves back and forth through the tree.

As with the Viterbi decoder, we often need to output information bits before the received word has been completely processed. A maximum depth may thus be established for the information-sequence buffer. Once this depth has been exceeded, the information bits furthest back in the tree are sent to the data sink. The decoder is prevented from backing up beyond this point by setting the partial path metric for all preceding nodes to $-\infty$.

The most serious constraint in the design of a Fano or stack decoder is the size of the input buffer. In many applications the received data arrives in a continuous stream at some rate R_S symbols per second. If the decoder is to move back and forth through the tree, it must be able to buffer incoming blocks until they are needed. It must also be able to process blocks faster than they arrive, otherwise buffer overflow is inevitable. The *speed factor* of the decoder is defined as the ratio of the number of branches the decoder can process in the time it takes one branch to arrive at the decoder input. As an example, Forney and Bower [For8] extensively tested

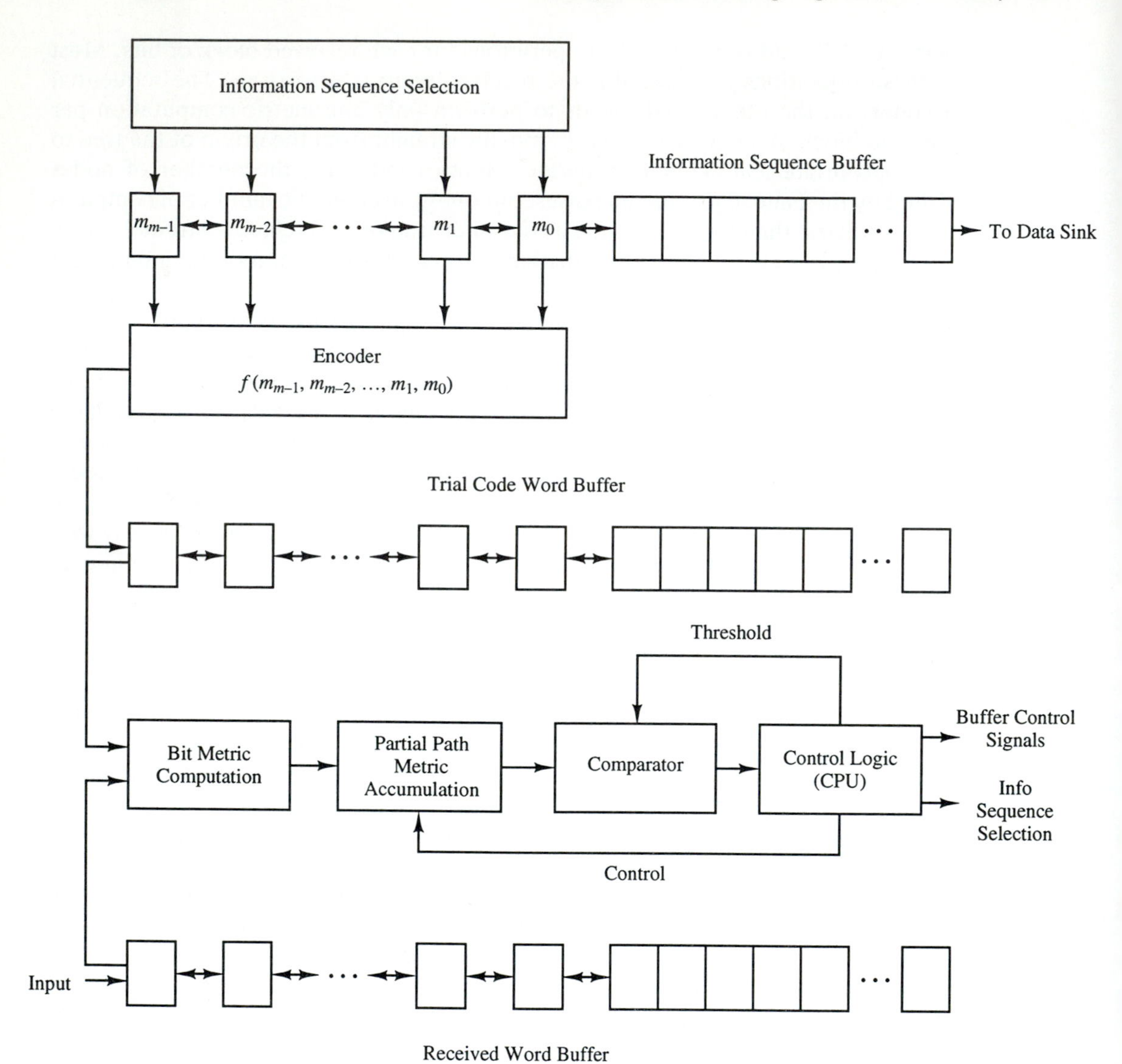

Figure 13-3. Block Diagram for a Fano Sequential Decoder

a 1971-technology decoder at speed factors from 2.66 to 13.3. Much higher speed factors are possible, depending on the application.

Given a fixed speed factor and buffer size, buffer overflow due to excessively noisy received blocks remains a possibility. For this reason the encoded data is frequently broken up into frames of fixed length. The framing is achieved by periodically inserting strings of zeros into the data prior to encoding. This initializes the encoder at the end of each frame. Buffer overflows are then handled through the declaration of erasures. When the capacity of the input buffer is exceeded, the current frame is erased. The decoder then resynchronizes at the beginning of the next frame and continues decoding. Erased frames can be resolved in a number of ways.

- Erasure correction through the use of an outer code with erasure correction capability (e.g., a Reed-Solomon code). An interleaver (see Section 16.1) should be placed between the inner and outer codes.
- Erasure concealment through muting or interpolation (very useful in applications in which the source is highly redundant, as in speech).
- Retransmission requests.

Given that extremely long constraint lengths can be used with the Fano or stack algorithms, the probability of a decoder error is extremely small. The erasure rate thus becomes the dominant performance statistic, with rates of the order of 10^{-3} being common. Such a rate in a retransmission request system will have a negligible impact on throughput. The use of sequential decoders in retransmission request schemes has been considered by a number of authors, including Khan et al. [Kha], Drukarev and Costello [Dru2], and Kallel and Haccoun [Kal1], [Kal2].

13.2 THE STACK ALGORITHM

Fano's algorithm moves through the tree in a very limited manner. With each iteration of the algorithm, the decoder position can move only one node to the left or right (i.e., one node into or out of the tree). Nodes are thus often revisited (though the threshold is always lower with each additional visit), and their partial path metrics recomputed. In the Zigangirov-Jelinek (ZJ) algorithm no node is visited more than once. The nodes with the best partial path metrics are stored in memory, and only the node with the best metric is extended. The ZJ algorithm is usually called the **stack algorithm**, because the stack provides a convenient model for the operation of the algorithm.

Consider the flow chart in Figure 13-4. The decoder begins at the origin of the tree and computes the metrics for ALL succeeding forward nodes. The positions and metrics for these nodes are stored in a stack, whose contents are then arranged in decreasing order of metric (the forward node with the best metric is thus on top).[1] The decoder then proceeds to the position at the top of the stack and computes the metrics for all nodes that are immediate successors of that point. These new values are then added to the stack, which is then reordered. This process continues until the position at the top of the stack corresponds to a leaf in the tree. Example 13-2 shows the result when the stack algorithm is applied to the received word from Example 13-1.

[1] The stack paradigm, though convenient, is thus not entirely accurate. In the decoding operation the contents of the "stack" must be accessed at all levels and repeatedly reordered.

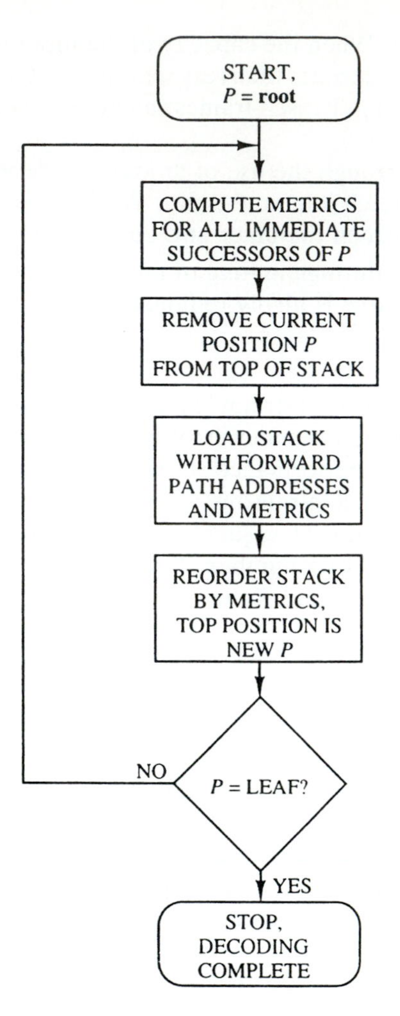

Figure 13-4. Flow Chart for the Stack Algorithm

Example 13-2—The Stack Decoding Algorithm

Reconsider Example 13-1, this time using the stack decoding algorithm. The rate-1/2 code associated with the tree in Figure 13-1 is used to encode the information sequence $\mathbf{x} = (1100)$, resulting in the code word

$$\mathbf{y} = 11, 10, 11, 01, 11, 00, 00$$

The code word is transmitted across a binary symmetric channel with crossover probability $p = 0.125$. The received sequence (with errors underlined) is

$$\mathbf{r} = 11, 1\underline{1}, 11, 01, 11, 00, \underline{1}0$$

The bit metrics are once again

$M(r_i^{(j)} \mid y_i^{(j)})$	$y_i^{(j)} = \;\; 0$	1
$r_i^{(j)} = 0$	1	-8
$r_i^{(j)} = 1$	-8	1

The stack algorithm then proceeds as follows. At each stage the contents of the stack are listed in order with the best tree position on top.

1. 1 +2
0 −16

2. 10 −5
11 −5
0 −16

3. 11 −5
100 −12
101 −12
0 −16

4. 110 −3
100 −12
101 −12
0 −16
111 −21

5. 1100 −1
100 −12
101 −12
0 −16
1101 −19
111 −21

6. 11000 +1
100 −12
101 −12
0 −16
1101 −19
111 −21

7. 110000 +3
100 −12
101 −12
0 −16
1101 −19
111 −21

8. 1100000 −4
100 −12
101 −12
0 −16
1101 −19
111 −21

1100000 is a leaf ⇒ STOP

The stack algorithm takes slightly more than half the time required by the Fano algorithm. In particular, note that where the Fano algorithm was forced to back up when it encountered the erroneous bit in the last received block, the stack algorithm "knows" that there are no better choices among the previously visited nodes, so it forges on ahead to the end without hesitation. One should also note, however, that even in this simple example the stack algorithm was keeping track of six nodes and their associated partial path metrics after eight iterations. ■

The stack algorithm has several characteristics that differentiate it from the Fano algorithm. We have noted already that no node is ever visited more than once. The decoder can jump between nodes that are not adjacent, but it may never move to a node that is a predecessor (i.e., connected by a direct backward path) of the current position. This flexibility reduces the number of operations, relative to the Fano algorithm, required to decode a given received word.

On the other hand, the stack algorithm introduces an additional memory dependency that is not shared by the Fano algorithm. Suppose that the code in use is an (n, k, m) convolutional code and that the position on the top of the stack is in the branching portion of the tree. When the stack algorithm removes the top position from the stack and computes the metric for the forward nodes, k entries are added to the stack. When the top position is in the nonbranching portion of the tree, the size of the stack is not changed (one position is removed from the stack and its single successor added). If we assume that the length of the information sequence is large with respect to the code constraint length (i.e., the asymptotic rate loss is negligible), then the number of entries that must be held by the stack can be approximated by $k \cdot$ (# nodes visited).

Stack overflow is not as dramatic a problem as buffer overflow. If the stack capacity is exceeded, then the bottom entry in the stack can be thrown out with only a slight reduction in performance.

As the stack increases in size (1000 entries is typical), the reordering operation takes on increasing significance. Jelinek [Jel1] has suggested a scheme called the "stack-bucket" algorithm that obviates the need for stack reordering. The stack is divided into partitions (buckets), with each partition assigned a range of metric values. When the stack algorithm computes the metric for a node in the tree, that node is stored in the top position of the partition in which its metric falls. The path in the top position of the nonempty partition with the highest metric range is always selected for extension. Reordering is thus no longer necessary, though we run the risk that the path selected for extension is not the best path, merely one in the best bucket. If the partitions are fine enough, however, and noise conditions are not too severe, the stack-bucket algorithm gives the same performance as the regular stack algorithm while reducing the complexity substantially [Jel1].

13.3 PERFORMANCE ANALYSIS FOR SEQUENTIAL DECODERS

Sequential decoders are examined in the literature on the basis of individual codes and on the basis of all possible codes as an ensemble. The latter approach is originally due to Shannon, who used it to prove the noisy channel coding theorem (Theorem 1-1). The basic idea is ingenious in its simplicity. Assume a code whose code words have been selected completely at random, then determine the performance of this code in a given system. We can now take the average performance over all possible codes. Since the resulting performance measures are "average," we are guaranteed at least one real code that equals or exceeds this performance level. This "random coding" argument is similar to that used in the proof of Gilbert's bound (Theorem 4-2) and in its asymptotic version (Theorem 4-6).

Aside from these two theorems, we have thus far studiously avoided the use of random coding arguments. Though of great interest to the information theorist, their importance in practical applications is not usually obvious. When it comes to sequential decoders, however, random coding arguments generate a series of results that provide an excellent basis for the modeling of the performance of real systems. The following summary presents the basic results. The reader who would like a more detailed exposition is referred to Jacobs and Berlekamp [Jac], Jelinek [Jel2], and Forney [For5].

The most important performance parameter for a sequential decoder is the probability of erasure. Suppose that the input buffer has enough room for the received blocks associated with B branches in the tree. Let μ be the speed factor for the decoder. It follows that an erasure occurs whenever the decoding of a given branch requires more than μB operations. In the random coding analysis we consider a branch to be decoded whenever the sequential algorithm moves beyond it and does not return to that branch. This is best expressed quantitatively through the use of "incorrect subtrees" [Lin1], as shown in Figure 13-5. The correct path through the tree in this figure is shown as a bold line. At each branch point an incorrect subtree is defined as all portions of the tree extending beyond the incorrect branch. Given this definition, we can then say that a given branch has been decoded once the decoder has performed its last computation in the incorrect subtree with respect to that branch.

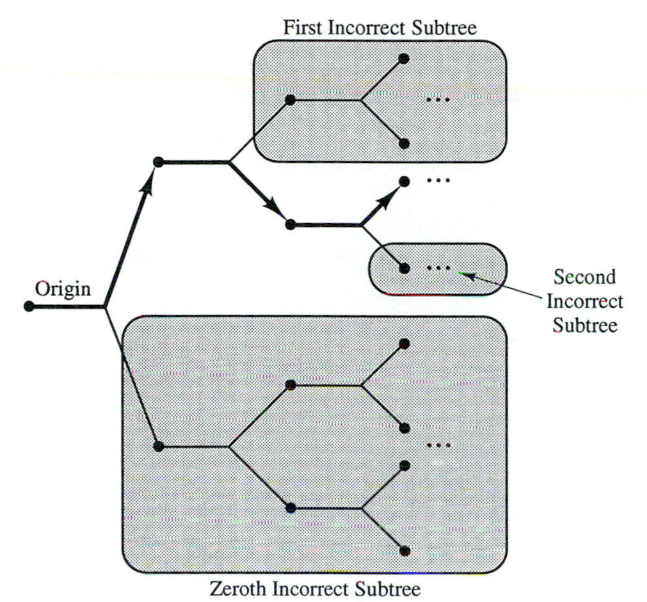

Figure 13-5. Incorrect Subtrees

The random coding argument proceeds by examining the question of how much time the decoder spends in an incorrect subtree for the ensemble of all convolutional codes. Let C_j be the number of computations performed by the decoder in the jth incorrect subtree. The average distribution on C_j can be shown to have the following form [Lin1].

$$P\{C_j \geq \eta\} \approx A\eta^{-\rho}, \qquad 0 < \rho < \infty, \quad 0 \leq j \leq L-1 \tag{13-5}$$

This is a *Pareto* distribution. The parameter ρ is the **Pareto exponent**, while the parameter A is a constant whose value depends on the sequential decoding algorithm in use. The Pareto exponent can be related to the code rate and channel conditions through the use of the **Gallager function**. Consider a discrete memoryless channel whose transition probabilities are defined by the conditional probabilities $P(j\,|\,k)$. The Gallager function for this channel is defined as follows.

$$E_0(\rho) = \rho - \log_2\left\{\frac{1}{2}\sum_j \left[P(j\,|\,0)^{1/1+\rho} + P(j\,|\,1)^{1/1+\rho}\right]^{1+\rho}\right\} \tag{13-6}$$

For a binary symmetric channel (BSC) with crossover probability ϵ, Eq. (13-6) reduces to the following.

$$E_0(\rho) = \rho - \log_2\left[\epsilon^{1/1+\rho} + (1-\epsilon)^{1/1+\rho}\right]^{1+\rho} \tag{13-7}$$

The rate R of the code in use is then related to the Pareto exponent by the following expression.

$$R = \frac{E_0(\rho)}{\rho}, \qquad 0 < R < C \tag{13-8}$$

The upper limit C on R in Eq. (13-8) is the channel capacity described in Theorem 1-1. If we substitute Eq. (13-7) into Eq. (13-8), the Pareto exponent and the crossover probability for the BSC can be graphed for fixed code rates, as shown in Fig. 13-6. This figure shows that, for fixed channel conditions, the Pareto exponent increases as the code rate decreases. In Eq. (13-5) we see that an increase in the Pareto exponent causes the average number of computations in a given incorrect subtree to be reduced.

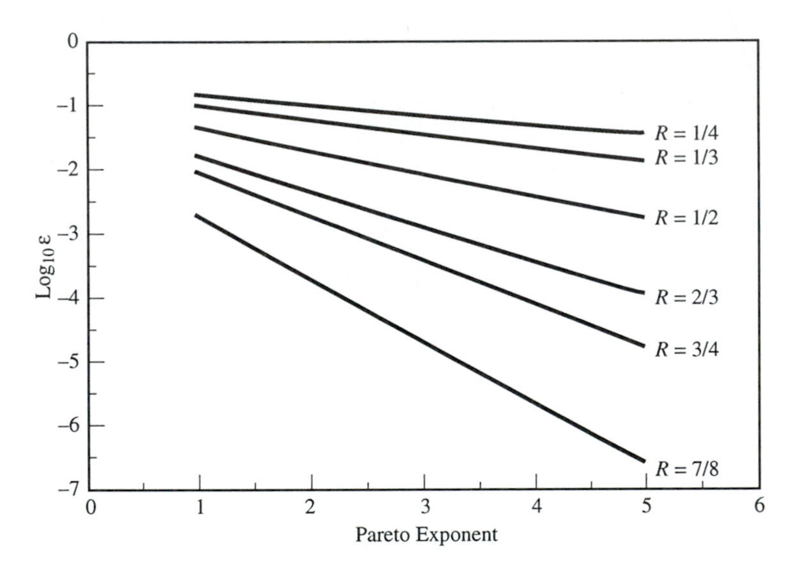

Figure 13-6. The Pareto exponent as a function of code rate and BSC crossover probability

One of the more interesting results to emerge from the random coding analysis is that the expected number of computations in a given incorrect subtree is unbounded if $\rho \leq 1$. Setting $\rho = 1$ and substituting into Eq. (13-8), we obtain an expression for the maximum code rate at which decoding can be completed in a finite amount of time.

$$\rho > 1 \Rightarrow R = \frac{E_0(\rho)}{\rho} < E_0(1) \triangleq R_0 \tag{13-9}$$

R_0 is commonly called the **computational cutoff rate**. It is often treated as a general parameter for a communication channel, though its meaning outside of the context of sequential decoding is not always clear.

Having considered a few of the properties of the Pareto exponent, we can now apply it to the problem at hand. Erasures occur whenever the time spent in a given incorrect subtree exceeds the time required to fill the receiver buffer with incoming bits. Given a speed factor μ and a buffer that can hold B branches, an erasure occurs whenever more than μB computations are performed in an incorrect subtree. If we assume that the branch processing times are independently distributed, the probability of erasure can be approximated by

$$P_s = \sum_{i=1}^{L} P\{C_i \geq \mu B\} \approx LA(\mu B)^{-\rho} \tag{13-10}$$

Chevillat and Costello found bounds on the computational behavior of sequential decoders for specific codes [Che]. Given a BSC with crossover probability ϵ and an (n,k,m) code with column distance function d_l, it can be shown that

$$P\{C_j \geq \eta\} < \sigma n_d e^{-\mu d_l + \phi l} \tag{13-11}$$

where σ, μ, and ϕ are functions of ϵ, R, and the decoder implementation. n_d is the number of code words of length $(l+1)$ branches with weight d_l. The index l is determined by the relation

$$l \triangleq \lfloor \log_{2^k} \eta \rfloor \tag{13-12}$$

where k is the number of input bits per output block. Finally, for Eq. (13-11) to hold, R must satisfy

$$R < 1 + 2\epsilon \log_2 \epsilon + (1 - 2\epsilon) \log_2 (1 - \epsilon) \tag{13-13}$$

The important thing to note about Eq. (13-11) is that it shows a strong relationship between the probability of erasure and the column distance function (CDF—see Definition (11-3)). If the erasure rates are to be low, then Eq. (13-11) should be a rapidly decreasing function of η. This in turn indicates a requirement that the CDF increase rapidly as a function of its index l (a typical CDF is shown in Figure 11-13).

Let code $\mathbf{C}_0$ have distance profile $d_0^0, d_1^0, d_2^0, \ldots$ and code $\mathbf{C}_1$ have distance profile $d_0^1, d_1^1, d_2^1, \ldots$. We say that code $\mathbf{C}_0$ has a superior distance profile [Lin1] if

$$d_i^0 \begin{cases} = d_i^1, & i = 0, 1, \ldots, l-1 \\ > d_i^1, & i = l \end{cases}$$

Note that this definition of superiority emphasizes only the first l values. If Eq. (13-11) is rapidly decreasing, the values of the CDF beyond the index l will have little impact.

Definition 13-1—Optimum Distance Profile

A code is said to be **ODP**, or to have an **optimum distance profile**, if its CDF is superior or equal to that of all codes of the same memory order.

Several ODP codes with long constraint lengths are listed in the next section.

We finally turn to the question of reliability performance. Since a sequential decoder is not a maximum likelihood decoder, one expects its performance for a given code to lag behind that of the Viterbi algorithm.[2] If the parameters of the sequential algorithm in use are carefully selected, however, this disparity in performance is not substantial. The lower bounds on performance developed in Chapter 12 for the Viterbi decoder may thus be used as approximations for the reliability performance of the sequential decoder. It follows that the reliability performance of the sequential decoders is a function of the minimum free distance of the codes in use. In any event, given that the computational complexity of the sequential decoder is only linearly related to the constraint length, one need only increase the constraint length to improve reliability performance to a desired level.

13.4 GOOD CODES FOR USE WITH SEQUENTIAL DECODERS

Table 13-1 contains several rate-1/2 systematic ODP codes. The generator sequences are listed in octal form along with each code's minimum free distance. Table 13-2 contains several rate-1/2 nonsystematic ODP codes. The minimum free distances for the codes are provided once again—note that the nonsystematic codes significantly outperform the systematic codes. Specific codes that have been discussed in the literature in some detail are noted with a reference. Tables 13-3 through 13-6 contain ODP codes for other commonly used rates.

Table 13-5 lists some systematic rate-2/3 ODP codes. It is assumed that the encoder implementation has minimal memory. Figure 13-7 shows a typical example. For all of the codes listed in Table 13-5, the constraint length K is thus one more than the total memory of the encoder (i.e., $K = m + 1$).

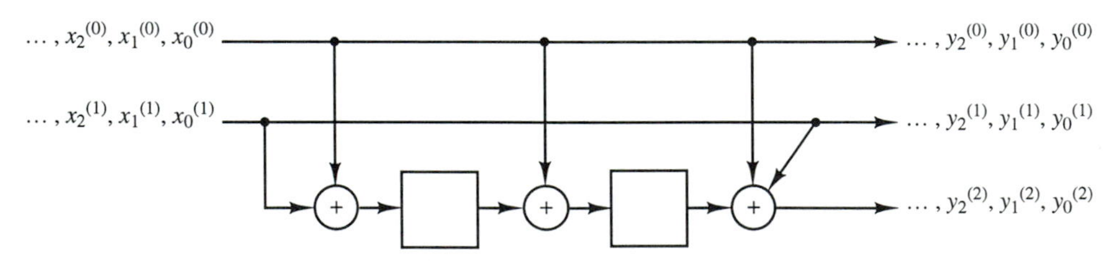

Figure 13-7. A rate-2/3 systematic code with minimal memory ($M = 2, K = 3$), $\mathbf{g}_1^{(2)} = (101) = 5$ (octal), $\mathbf{g}_0^{(2)} = (111) = 7$

[2] Forney [For5] and Viterbi and Omura [Vit1] use random coding arguments to show that this is true only when the code rate is below the cutoff rate.

TABLE 13-1 Rate-1/2 systematic codes with optimum distance profiles [Lin1][3]

K[4]	$\mathbf{g}^{(1)}$ (octal)	d_{free}
2	6	3
3	7	4
4	64	4
5	72	5
6	73	6
7	734	6
8	714	6
9	715	7
10	7154	8
11	7152	8
12	7153	9
13	67114	9
14	67114	9
15	67115	10
16	714474	10
17	671166	12
18	671166	12
19	6711454	12
20	7144616	12
21	7144761	12
22	67114544	12
23	71446166	14
24	67114543	14
25	714461654	15
26	671145536	15
27	671151433	16
28	7144760524	16
29	6711454306	16
30	7144760535	18
31	71446162654	≥16
32	67114543066	18
33	71447605247	≥18
34	714461626554	≥18
35	714461625306	≥18
36 [Joh]	714461626555	≥19

[3] Tables 13-1 through 13-6 are reprinted (in slightly modified form), with permission, from S. Lin and D. J. Costello, Jr., *Error Control Coding: Fundamentals and Applications*, pp. 374–378 (Englewood Cliffs, NJ: Prentice Hall, 1983).

[4] Recall that the constraint length K is defined as one plus the length of the longest shift register in the encoder (Eq. (11-5)).

TABLE 13-2 Rate-1/2 nonsystematic codes with optimum distance profiles [Lin1]

K	$\mathbf{g}^{(0)}$	$\mathbf{g}^{(1)}$	d_{free}
2	6	4	3
3	7	5	5
4	74	54	6
5	62	56	7
6	75	55	8
7	634	564	10
8	626	572	10
9	751	557	12
10	7664	5714	12
11	7512	5562	14
12	6643	5175	14
13	63374	47244	15
14	45332	77136	16
15	65231	43677	17
16	517604	664134	18
17	717066	522702	19
18	506477	673711	20
19	5653664	7746714	21
20	4305226	6574374	22
21	6567413	5322305	22
22	67520654	50371444	24
23	67132702	50516146	24
24	55346125	75744143	25

TABLE 13-3 Rate-1/3 systematic codes with optimum distance profiles [Lin1]

K	$\mathbf{g}^{(1)}$	$\mathbf{g}^{(2)}$	d_{free}
2	6	6	5
3	5	7	6
4	64	74	8
5	56	72	9
6	57	73	10
7	564	754	12
8	626	736	12
9	531	676	13
10	5314	6764	15
11	5312	6766	16
12	5312	6766	16
13	65304	71274	17
14	65306	71276	18
15	65305	71273	19
16	653764	712614	20
17	514112	732374	20
18	653761	712611	22
19	6530574	7127304	24
20	5141132	7323756	24
21	6530547	7127375	26
22	65376164	71261060	26
23	51445036	73251266	26
24	65305477	71273753	28

TABLE 13-4 Rate-1/3 nonsystematic codes with optimum distance profiles [Lin1]

K	$\mathbf{g}^{(0)}$	$\mathbf{g}^{(1)}$	$\mathbf{g}^{(2)}$	d_{free}
2	4	6	6	5
3	5	7	7	8
4	54	64	74	10
5	52	66	76	12
6	47	53	75	13
7	—	—	—	—
8	516	552	656	16

TABLE 13-5 Rate-2/3 systematic codes with optimum distance profiles [Lin1]

K	$\mathbf{g}_0^{(2)}$	$\mathbf{g}_1^{(2)}$	d_{free}
2	4	6	2
3	5	7	3
4	54	64	4
5	56	62	4
6	57	63	5
7	554	704	5
8	664	742	6
9	665	743	6
10	5734	6370	6
11	5736	6322	7
12	5736	6323	8
13	66414	74334	8
14	57372	63226	8
15	57371	63225	8
16	664150	743314	8
17	664072	743346	10
18	573713	632255	10
19	6640344	7431024	10
20	5514632	7023726	10
21	5514633	7023725	11
22	57361424	63235074	12
23	66415416	74311464	11
24	66415417	74311465	12

TABLE 13-6 Rate-2/3 nonsystematic codes with optimum distance profiles [Lin1]

K	M	$\mathbf{g}_0^{(0)}$ / $\mathbf{g}_1^{(0)}$	$\mathbf{g}_0^{(1)}$ / $\mathbf{g}_1^{(1)}$	$\mathbf{g}_0^{(2)}$ / $\mathbf{g}_1^{(2)}$	d_{free}
3	3	6	2	4	4
		1	4	7	
3	4	6	3	7	5
		1	5	5	
4	5	60	30	70	6
		34	74	40	
4	6	50	24	54	6
		24	70	54	
5	7	54	30	64	7
		00	46	66	
5	8	64	12	52	8
		26	66	44	
6	9	54	16	66	8
		25	71	60	
6	10	53	23	51	9
		36	53	67	
7	11	710	260	670	10
		320	404	714	
9	12	740	260	520	10
		367	414	515	
9	13	710	260	670	11
		140	545	533	
8	14	676	046	704	12
		256	470	442	
9	15	722	054	642	12
		302	457	435	
10	16	7640	2460	7560	12
		0724	5164	4260	
10	17	5330	3250	5340	13
		0600	7650	5434	
10	18	6734	1734	4330	14
		1574	5140	7014	
11	19	5044	3570	4734	14
		1024	5712	5622	
11	20	7030	3452	7566	14
		0012	6756	5100	
12	21	6562	2316	4160	15
		0431	4454	7225	
13	22	57720	12140	63260	16
		15244	70044	47730	
13	23	51630	25240	42050	16
		05460	61234	44334	

Though it is not known to be ODP, the following rate-1/2 systematic code can be considered for high-reliability applications. Forney and Bower provide a thorough examination of an implementation of this code in [For8].

K	$\mathbf{g}^{(1)}$	d_{free}
49	71547370131746504	18 at $m = 32$

The following rate-1/2 nonsystematic code has been considered for use in space applications [Mas6]. It is also not known to be ODP.

K	$\mathbf{g}^{(0)}$	$\mathbf{g}^{(1)}$	d_{free}
47	5335336767373553	7335336767373553	23 at $m = 31$

PROBLEMS

1. Show that Eq. (13-1) reduces to Eq. (13-2) for a binary symmetric channel (BSC) with a uniformly distributed source.
2. Show that the length bias term in Eq. (13-3) is always positive for $R < 1$ and $P(r_i^{(j)}) \leq 1/2$.
3. Compute the Fano metric given a rate-1/2 code used over a binary symmetric channel with the following crossover probabilities. Convert these metrics to integer form.
   (a) $p = 0.25$ **(b)** $p = 0.07$ **(c)** $p = 0.001$
4. Determine the Fano metric for the binary asymmetric channel shown below. Assume that 0s are transmitted 60% of the time (i.e., the source output is not uniformly distributed).

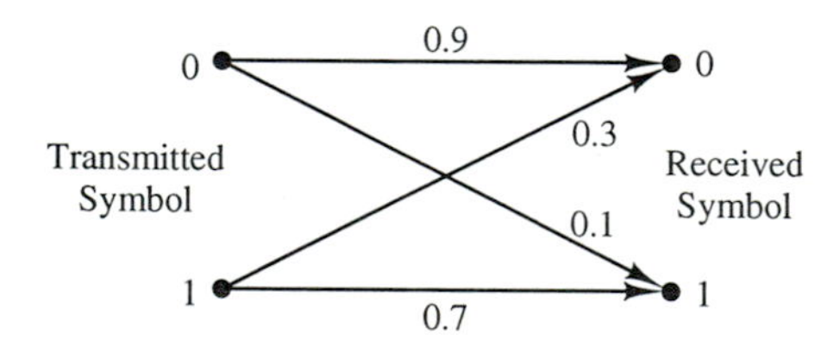

5. Draw a tree diagram for length-4 inputs to a rate-1/2 code with generator sequences $G_0(D) = 1 + D + D^2$ and $G_1(D) = 1 + D^2$. Determine the information sequences corresponding to the following code words.
   (a) $\mathbf{c}_1 = 00, 11, 10, 00, 10, 11$
   (b) $\mathbf{c}_2 = 11, 01, 01, 11, 00, 00$
   (c) $\mathbf{c}_3 = 00, 11, 10, 00, 10, 11$
6. A rate-1/2 code with generator sequences $G_1(D) = 1 + D + D^2$ and $G_2(D) = 1 + D^2$ is used for error control on a BSC with crossover probability $p = 0.1$. Use the Fano decoding algorithm to decode the received word

$$\mathbf{r} = 01, 10, 11, 00, 00, 00$$

Perform the decoding operation twice, the first time with $\Delta = 1$ (before normalization), the second time with $\Delta = 4$. Compare the number of operations required and the reliability of the results.

7. Assuming the same code and channel as in Problem 6, use the stack decoding algorithm to decode the received word

$$\mathbf{r} = 01, 10, 11, 00, 00, 00$$

Compare the decoded sequence and the number of operations required to the results from Problem 6.

8. Assuming the same code and channel as in Problem 6 and received words containing k blocks of 2 bits, what is the maximum amount of memory required to maintain the "stack" in the stack decoding algorithm?

9. A rate-1/2 code with generator sequences $G_1(D) = 1 + D^2 + D^3$ and $G_2(D) = 1 + D + D^3$ (see Figure 13-1) is used for error control on a BSC with crossover probability $p = 0.2$. Use the stack decoding algorithm to decode the following received words.

(a) $\mathbf{r}_1 = 11, 11, 10, 00, 00, 01, 11$
(b) $\mathbf{r}_2 = 01, 00, 11, 01, 11, 01, 00$
(c) $\mathbf{r}_3 = 11, 01, 10, 10, 10, 11, 00$

14

Trellis Coded Modulation

The treatment of coding and modulation as inseparable parts of a single system was first suggested by Massey [Mas7] in 1974. Two years later Ungerboeck presented a feasible means for integrating coding and modulation through what is now called "trellis coded modulation" (TCM) [Ung1]. At that point TCM began its phenomenally rapid move from theoretical concept to worldwide application.

TCM uses convolutional codes and multidimensional signal constellations to provide reliable, high-data-rate communication over bandwidth-limited channels. The most immediate application for TCM was in the area of digital data transmission over standard telephone lines. The telephone channel has a bandwidth of roughly 2700 Hz. In the early 1960s, many engineers felt that 2400 bps was the maximum data rate that would ever be supported on this channel. By 1970 the estimated ceiling on data rate had been raised to 9600 bps, with many exceedingly fine engineers believing that it would never go any higher [For6]. But of course, TCM has demolished this ceiling as surely as the 2400-bps ceiling was demolished before it. CCITT[1] Study Group XVII has recently approved a 9.6/14.4-Kbps trellis coded modem for use with both intercomputer and FAX communication [McC], [IBM1]. The 19.2-Kbps modem is already a reality, and the corresponding standards are not far behind. We should also note that TCM has had a substantial impact on other bandwidth-limited applications, particularly in cellular mobile ratio and satellite communications [Big2].

We begin this chapter by considering one- and two-dimensional signaling constellations and the trade-off between spectral efficiency and power efficiency. In

[1] Comité Consultatif International de Télegraphique et Téléphonique.

the next section we examine one- and two-dimensional trellis coded modulation (TCM) systems. The emphasis is placed on three key concepts: expanded signal constellations, signal constellation partitioning, and the selection of partitions by convolutional encoders. We then briefly consider the multiple trellis coded modulation (MTCM) systems, in which multiple-signal sequences are assigned to individual trellis branches. This is followed by a discussion of decoding by the Viterbi decoder and a presentation of various performance analysis techniques. The chapter concludes with a discussion of implementation considerations and a look at the CCITT 9.6/14.4 Kbps standard.

14.1 *M*-ARY SIGNALING

In Chapter 1 the basic subsystems comprising a digital communication system were discussed (see Figures 1-1 and 1-3). Such systems are "digital" in the sense that they transmit information in the form of discrete symbols. In a binary system (e.g., binary phase-shift keying (BPSK) and binary frequency-shift keying (BFSK)) each of the transmitted symbols is assigned one of two possible values. The Nyquist bandwidth for a communication signal is equal to the rate at which the symbols are transmitted. If the symbols are binary, an R_S bit-per-second data rate requires an R_S-Hz Nyquist bandwidth. The **spectral efficiency** of a system is defined to be the number of bits per second transmitted per 1 Hz of bandwidth. The binary system mentioned above thus has a spectral efficiency of 1 bit/sec/Hz.

The binary signal constellation is generally represented by a pair of points on a number line. The position of a signal on the line is proportional to its magnitude and its relative phase or, equivalently, to the square root of the transmitted or received signal energy. Figure 14-1 shows the signal diagram for the BPSK constellation. Normalized, integer labels are frequently used in such diagrams; they preserve the relative differences in signal magnitude but are less cumbersome than absolute received or transmitted signal magnitudes.

Figure 14-1. BPSK Signal Constellation

When we need to compute performance figures given some received noise energy, the labels in the signal constellation diagram are weighted by $\sqrt{E_s}$, the square root of the average received signal energy. The symbol error rate is computed by determining the probability that the channel noise causes the received signal to be closer to one of the incorrect signals than to the signal that was actually transmitted.

For example, consider an additive white Gaussian noise channel with two-sided spectral density $N_0/2$ at the input to the receiver. On such a channel the noise can be represented by a Gaussian random variable n that is added to the transmitted signal; if the signal x has been transmitted, the receiver will see a signal of the form $r = x + n$. If all signals are equally likely to be transmitted, the maximum likelihood receiver selects the signal that is closest in Euclidean distance to the received signal.

Consider the BPSK constellation and assume that +1 has been transmitted. The bit error rate (BER) is then computed as follows.

$$\begin{aligned}
\text{BER}_{BPSK} &= P\{\|\sqrt{E_b} - r\| \geq \|-\sqrt{E_b} - r\|\} \\
&= P\{\|\sqrt{E_b} - r\| - \|-\sqrt{E_b} - r\| \geq 0\} \\
&= P\left\{n \geq \frac{\sqrt{E_b} - (-\sqrt{E_b})}{2}\right\} \qquad (14\text{-}1) \\
&= Q\left(\sqrt{\frac{2E_b}{N_0}}\right), \qquad \text{where } Q(x) = \frac{1}{\sqrt{2\pi}}\int_x^\infty e^{-y^2/2}\,dy
\end{aligned}$$

In many applications the desired data rate (in bits per second) far exceeds the available bandwidth (in hertz). In such cases it is necessary to increase the spectral efficiency of the communication system. This is done by increasing the size of the signaling constellation; the symbol transmission rate (and thus the Nyquist bandwidth) remains the same, but the number of possible values taken on by each symbol is increased.

The most obvious expansion of the constellation in Figure 14-1 is amplitude modulation (AM),[2] as exemplified by the 8-AM constellation in Figure 14-2. The 8-AM constellation provides 3 bits/sec/Hz (double-sideband AM), three times the spectral efficiency of the BPSK signal.

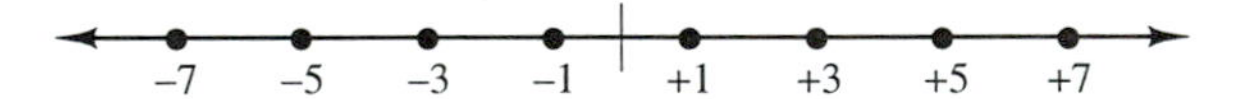

Figure 14-2. 8-AM Signal Constellation

At high signal-to-noise ratios (SNRs), the most likely error events are those in which the transmitted symbol is confused with one of its nearest neighbors. Let the minimum Euclidean distance ($d_{\min}$) for a constellation be the shortest distance between any pair of distinct signals. In Figure 14-2 we have $d_{\min} = 2$. If we assume that there are always two nearest neighbors, the symbol error rate for AM on an AWGN channel can be easily approximated. Let E_s be the average received signal energy.

$$P(E)_{AM} \approx 2Q\left(\sqrt{\frac{2E_s d_{\min}^2}{N_0}}\right) \qquad (14\text{-}2)$$

The amount of energy required to transmit one of the 8-AM signals in Figure 14-2 is proportional to the square of the distance of the signal from the origin. If the distance between adjacent signals in Figure 14-2 is to be the same at the receiver input as that in Figure 14-1, then the average energy per transmitted symbol is significantly more for 8-AM than it is for BPSK. Let the normalized average received energy for the BPSK case be $S_1 = 1$. The letter "S" denotes normalized average received energy, while the subscript denotes the log (to the base 2) of the cardinality

[2] Also called amplitude-shift keying (ASK), digital AM is simply the discrete version of the familiar analog amplitude modulation.

of the signal constellation. For a 2^m-ary AM constellation with the same minimum free distance as BPSK, the normalized average energy is

$$S_m = \frac{(4^m - 1)}{3} \tag{14-3}$$

8-AM thus requires $S_3 = 21.0$ times (13.22 dB) more energy than BPSK to maintain the same minimum distance. Equation (14-3) can also be used to derive the additional energy required for each additional bit/sec/Hz.

$$S_{m+1} = 4S_m + 1 \tag{14-4}$$

From Eq. (14-4) it is clear that, asymptotically, each additional bit/sec/Hz requires 4 times (6 dB) more average energy per transmitted symbol. This provides a rough estimate of the price to be paid for the additional spectral efficiency.

The AM constellations discussed above are one-dimensional. We can obtain better performance at the expense of a little additional complexity by moving to two-dimensional, or "quadrature," amplitude modulation (QAM). Figure 14-3 shows a set of commonly used rectangular QAM constellations, while Fig. 14-4 shows some *M*-ary phase-shift keying (MPSK) constellations.[3] Both types have their respective advantages and disadvantages. The rectangular constellations provide better minimum-distance versus average-energy performance but are subject to distortion when they pass through nonlinear devices (e.g., traveling-wave tubes and other amplifiers that are operated in saturation). In MPSK constellations the minimum distances are relatively small, but the modulated signals have a constant envelope and are thus not distorted by channel nonlinearities.

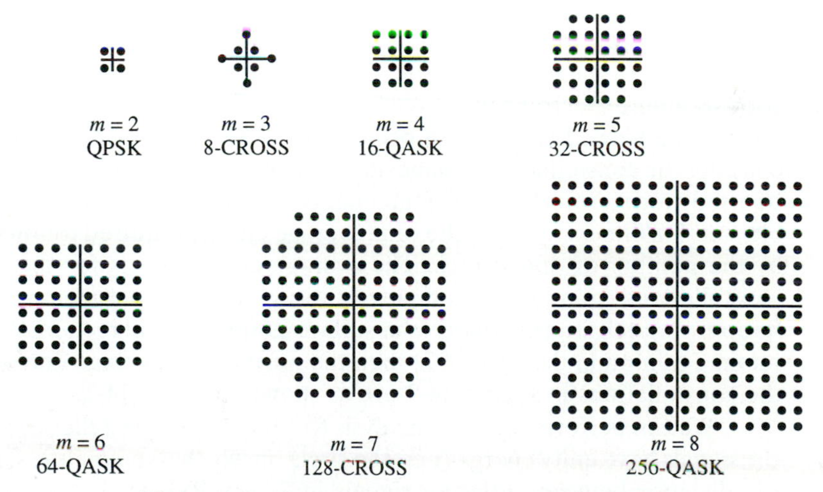

Figure 14-3. Several Rectangular Signal Constellations

[3] MPSK signals can be parameterized in terms of a single variable (the phase), so it can be argued that they are one-dimensional constellations. However, most practical MPSK receiver designs project the received signal onto a pair of orthogonal signals before making a phase estimate, thus treating MPSK as a form of QAM.

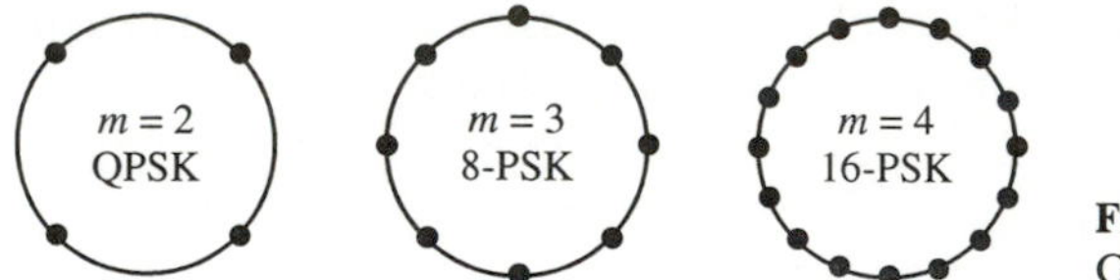

Figure 14-4. A Few MPSK Signal Constellations

Two-dimensional constellations can be implemented by modulating a pair of orthogonal carriers, as shown in Figure 14-5. m bits are taken from the source and mapped onto an ordered pair (X, Y). These two coordinates are then used to modulate a pair of orthogonal carriers, which when added form a single complex signal z. The modulated sine and cosine are orthogonal and do not interfere with each other on a linear channel. Two-dimensional rectangular constellations can thus be viewed as a pair of orthogonal one-dimensional constellations.

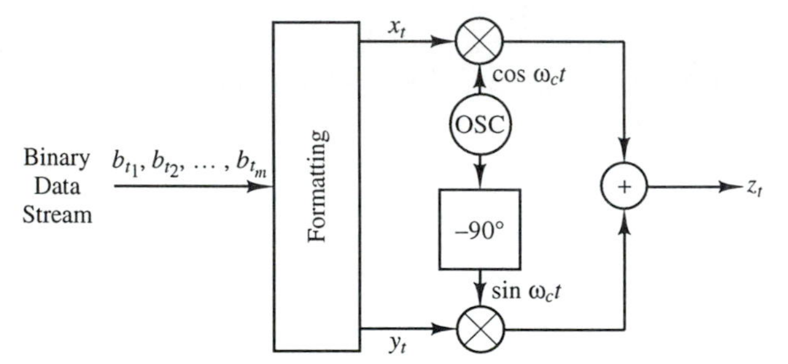

Figure 14-5. QAM Modulator

Now take a closer look at 64-QASK. Since there are 64 possible signals that can be sent in each transmission interval, each signal represents 6 bits, and the corresponding spectral density is 6 bits/sec/Hz. The constellation is perfectly square and can be viewed as a pair of orthogonal 8-AM constellations. We can thus also consider the constellation as achieving 3 bits/sec/Hz/dimension. Equation (14-3) can be used to determine the average energy S required per dimension to achieve m bits/sec/Hz/dimension, and the total average energy required follows by multiplying by the number of dimensions. The average energy levels for the various CROSS constellations (m odd) are computed through more direct means but are found to be appropriately intermediate values. If we assume that all rectangular constellations retain a minimum distance of 2, then the normalized average energy S required by the constellations in Figure 14-3 are as shown in Table 14-1.

We can perform a similar analysis for the MPSK constellations. Assuming that the signals have unity energy (i.e., the circle on which the signals rest has unit radius), the distance between adjacent signals in 2^m-ary PSK is $d_{\min} = 2\sin(\pi/2^m)$. If the average signal energy is adjusted to ensure that the minimum distance is always 2, then we obtain the results in Table 14-2.

Clearly MPSK is a much less efficient means of obtaining higher spectral efficiency. Note, for example, that 64-PSK requires an additional 10 dB of energy to obtain the same $d_{\min}$/average energy performance as rectangular 64-QASK. Table

TABLE 14-1 Average energy requirements for the 2^m-ary rectangular constellations in Figure 14-3 ([For6], © 1984 IEEE)

m	Number of points	S	(dB)
1	2	1	0.0
2	4	2	3.0
3	8	5.5	7.4
4	16	10	10.0
5	32	20	13.0
6	64	42	16.2
7	128	82	19.1
8	256	170	22.3

TABLE 14-2 Average energy requirements for the 2^m-ary PSK constellations in Figure 14-4

m	Number of points	S	(dB)
1	2	1	0.0
2	4	2	3.0
3	8	6.8	8.3
4	16	26.3	14.2
5	32	104.1	20.2
6	64	415.3	26.2
7	128	1660.4	32.3

14-2 also shows that, asymptotically, each additional bit/sec/Hz requires 6 dB of additional energy (the same as for one-dimensional AM), while the rectangular constellations require an average increase of only 3 dB.

14.2 ONE- AND TWO-DIMENSIONAL TRELLIS CODED MODULATION

The block and convolutional codes discussed in this text have all provided coding gain through the insertion of redundant bits. Redundancy makes some symbol patterns invalid, thus allowing the receiver to detect, and in some cases correct, errors caused by noise on the channel. Redundancy is frequently measured in terms of the code rate R, the ratio of the number of information bits to the total number of transmitted bits (see Definition 4-1). If the coded system is to retain the same information transmission rate as the uncoded system, then the symbol transmission rate R_S (and thus the Nyquist bandwidth) for the coded system must be R^{-1} times that of the uncoded system. In a bandwidth-limited system this presents a problem, for by definition the additional bandwidth required for conventional error control is unavailable. One alternative is to reduce the information rate to make room for the redundant bits, but in many applications this is either extremely difficult or impossible. Trellis coded modulation provides a powerful, additional alternative. Ungerboeck introduced the redundancy required for error control without increasing the signal bandwidth through a three-step procedure [Ung1]. Suppose that the uncoded system uses a 2^m-ary signal constellation (m bits/sec/Hz). Ungerboeck encoding is performed as follows.

1. Add one bit of redundancy to every m source bits.
2. Expand the signal constellation from 2^m to 2^{m+1} signals.
3. Use the $(m + 1)$-bit encoded source blocks to select signals from the expanded constellation.

The symbol transmission rate for the coded system is the same as that of the uncoded system, so the Nyquist bandwidth is not increased. The coded signal has redundancy, however, because a 2^{m+1}-ary constellation is being used to transmit information at a rate of m bits/sec/Hz.

The genius of Ungerboeck's system lies in the manner by which the m information bits are mapped onto the 2^{m+1} signals in the expanded constellation. This mapping is achieved through **set partitioning**. Most practical signal constellations can be partitioned in a systematic manner to form a series of smaller subconstellations. If this partitioning is done properly, the resulting subconstellations have a higher minimum distance than their "parent" constellation. Figure 14-6 shows the partition tree for the 32-CROSS constellation. The parent constellation (labeled A0) is assumed to have $d_{\min} = \Delta$. A0 is partitioned into B0 and B1, two constellations with $d_{\min} = \sqrt{2}\Delta$. Each of these constellations is in turn partitioned to form C0, C1, C2, and C3, and so on. Each partitioning results in subconstellations with a minimum distance that is $\sqrt{2}$ times that of its immediate predecessor in the tree. At the νth level of the tree (the top of the tree is level 0) there are thus 2^{ν} subconstellations with $d_{\min} = (\sqrt{2})^{\nu}\Delta$.

The goal of constellation partitioning: each partition should produce subconstellations with increased minimum distance.

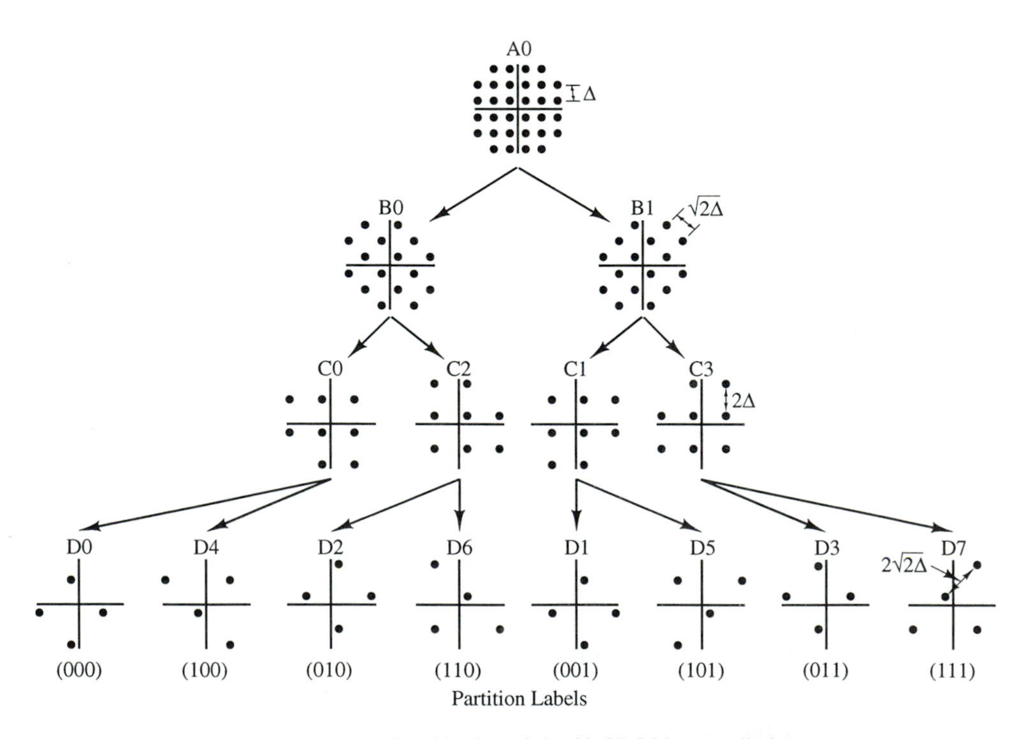

Figure 14-6. A Partitioning of the 32-CROSS constellation

Note that each of the partitions at the bottom of Figure 14-6 has a binary label. The labels have a natural interpretation as a tree address: each coordinate corresponds to a movement downward in the tree, with leftward branches denoted by a 0, and rightward branches by a 1. We shall see that these labels provide a convenient means for partition selection by the output of a convolutional encoder.

Figure 14-7 shows how partitioning can be applied to MPSK constellations. In this case the 16-PSK constellation is partitioned into a pair of 8-BPSK constellations. The average symbol energy for all of these constellations is assumed to be unity. Note that the minimum distance increases more rapidly with each successive partition of 16-PSK than with 32-CROSS.

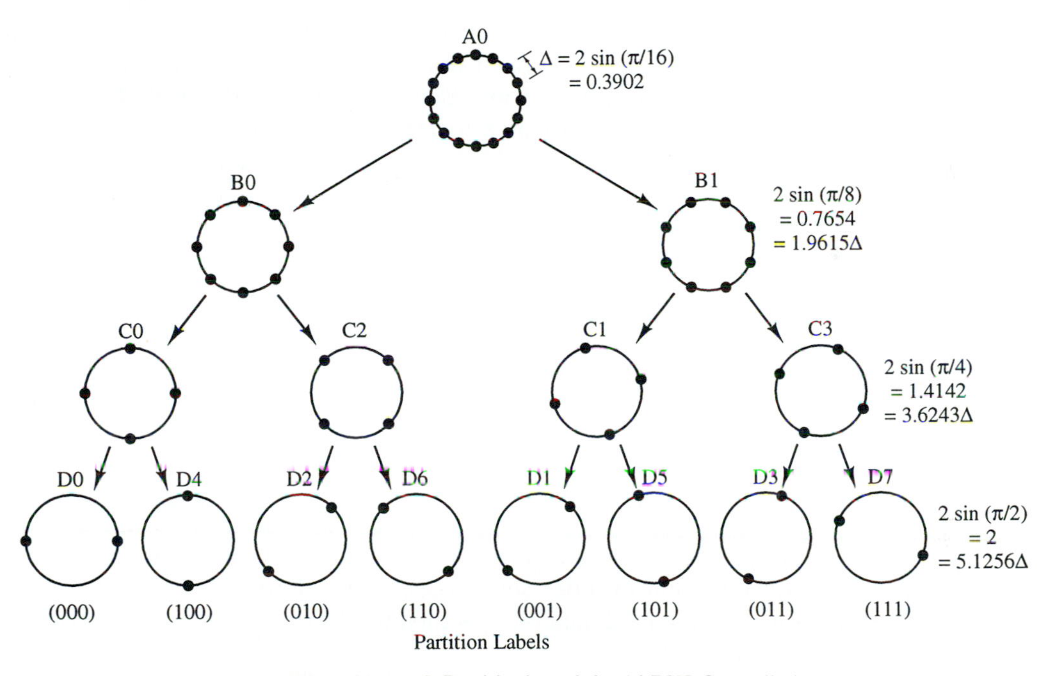

Figure 14-7. A Partitioning of the 16-PSK Constellation

An Ungerboeck encoder [Ung2] is shown in Figure 14-8. m information bits $x_1, x_2, \ldots, x_m$ are taken from the data stream for mapping onto a signal from the 2^{m+1}-ary constellation. k of the m information bits are encoded using a rate $k/(k + 1)$ convolutional encoder. The resulting $(k + 1)$ bits are used to select one of the 2^{k+1} partitions of the 2^{m+1}-ary signal constellation at the $(k + 1)$st level of the constellation's partition tree. The remaining $(m - k)$ information bits are then used to select a signal z within the designated partition.

The convolutional encoder within the Ungerboeck encoder imparts a trellis structure on the allowed signal sequences. The encoder is thus referred to as a **trellis coded modulation system** (TCM).

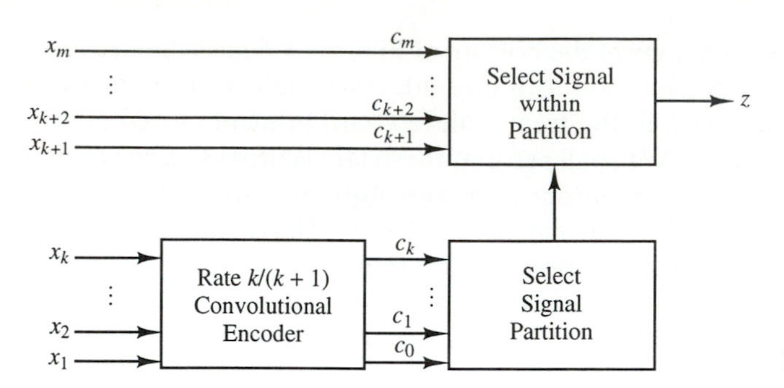

Figure 14-8. An Ungerboeck Encoder

To understand the impact of the trellis structure upon the symbol sequences generated by the Ungerboeck encoder, it is best to begin with a detailed example and then proceed to the general conclusions. We use one of Ungerboeck's simpler examples [Ung3], which compares uncoded QPSK to a coded 8-PSK system. The uncoded and coded systems provide a spectral efficiency of 2 bits/sec/Hz. The two constellations are shown in Figure 14-9. A rate-1/2 encoder is to be used, so the 8-PSK constellation is partitioned into four subconstellations {C0, C1, C2, C3}. Each subconstellation is assigned a label that corresponds to a unique two-bit output from the convolutional encoder.

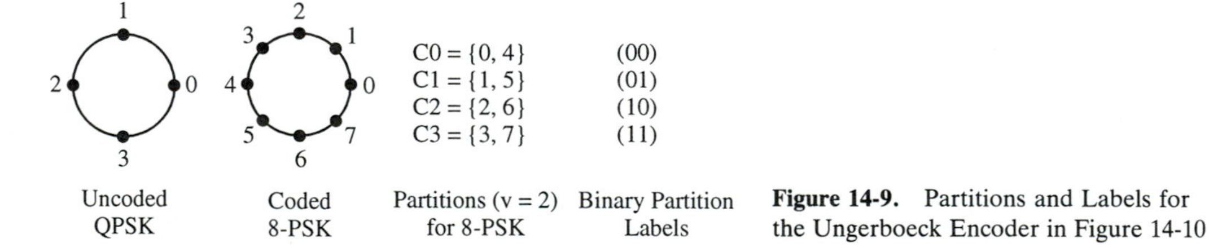

Figure 14-9. Partitions and Labels for the Ungerboeck Encoder in Figure 14-10

The encoder is shown in Figure 14-10. The output of a four-state (constraint-length-3) convolutional encoder selects one of the four partitions of the 8-PSK constellation. The uncoded data bit x_1 is then used to choose one of the two signals in the selected partition.

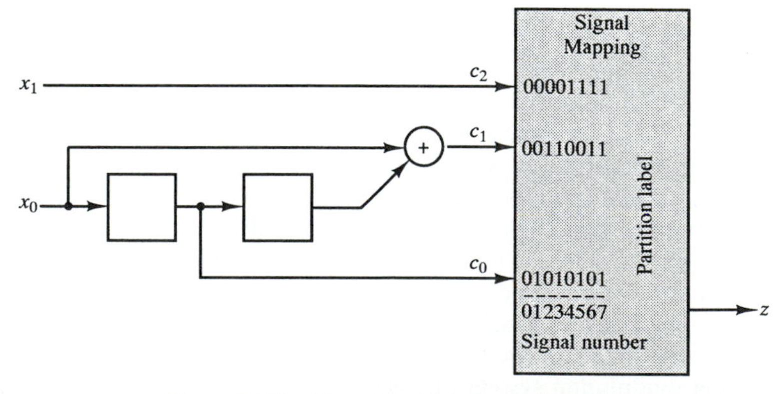

Figure 14-10. Four-state Ungerboeck Encoder

Consider the uncoded and coded cases through the use of trellis diagrams. In the uncoded case, two source bits are taken at a time and mapped onto one of the four signals in the QPSK constellation. As there is no memory in this system, any of the four signals can be selected at any symbol transmission time slot. The uncoded system can thus be viewed as a one-state trellis, as shown in Figure 14-11.

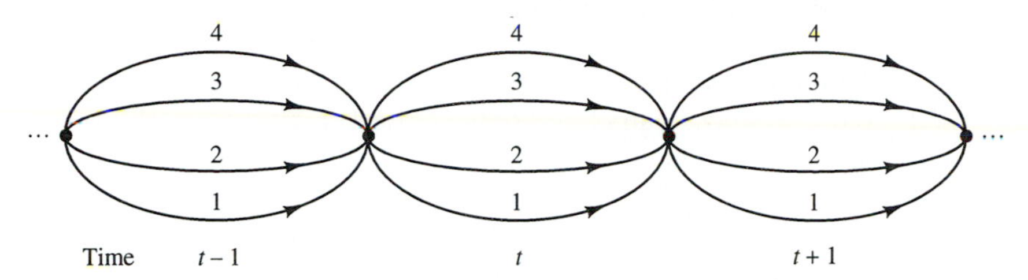

Figure 14-11. Trellis for Uncoded Transmission

The probability of symbol error for transmission over a noisy channel is a function of the minimum Euclidean distance between pairs of distinct signal sequences. The minimum Euclidean distance between a pair of valid, distinct sequences is called the **minimum free distance** (d_{free}). In the uncoded case two valid sequences of symbols may differ in only one coordinate, and any values can be taken by that coordinate. The minimum free distance between any pair of uncoded sequences in our example is thus the minimum distance for the QPSK constellation: $d_{\text{free/uncoded}} = \sqrt{2}$.

In the coded case we have a more interesting situation. An examination of the encoder in Figure 14-10 shows that the valid sequences can be described by the four-state trellis diagram in Figure 14-12. The states of the trellis diagram correspond to the contents of the memory elements in the convolutional encoder within the Ungerboeck encoder. The labels on the branches are the signal partitions from which signals may be selected for transmission, given the associated state transition. For example, when the convolutional encoder moves from state S_0 (00) to state S_1 (01), only a signal from partition C2 may be selected for transmission.

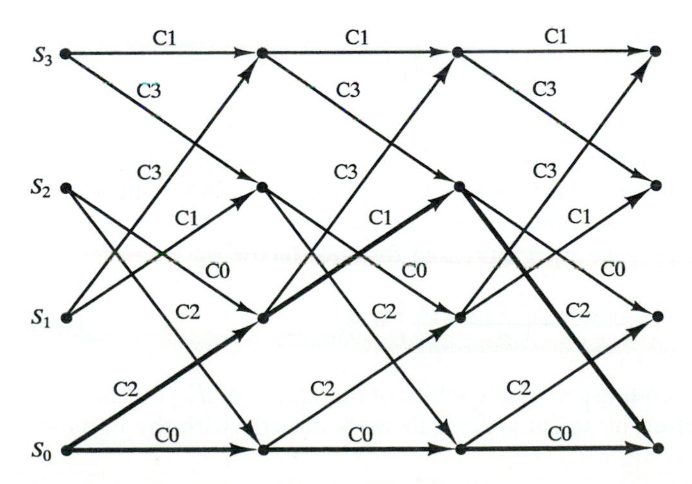

Figure 14-12. Trellis Diagram for Ungerboeck Encoder in Figure 14-10

Since there is more than one signal associated with each branch in the trellis, we can have signal sequences that differ by a single **parallel transition**. For example, the two signal sequences $Z_1 = 00000\ldots$ and $Z_2 = 04000\ldots$ are both perfectly valid and differ in only one coordinate. The minimum free distance for parallel transitions is the minimum distance for the partitions labeling each branch. For the encoder under consideration, the minimum free distance for parallel transitions is the smallest minimum distance among C0, C1, C2, C3. Thus $d_{\text{free}_{\text{parallel}}} = 2$.

Though parallel transitions differ in only one branch, they do not necessarily denote the sequence pairs that are the closest. Figure 14-12 shows one way in which two nonparallel sequences can differ in three consecutive branches. This pair of sequences is one of the closest nonparallel pairs of sequences for this encoder. To determine the minimum distance between sequences following these branches, we apply the familiar Euclidean distance metric. Let $d_{\min}(\text{CA}, \text{CB})$ be the minimum distance between signals in partition CA and signals in partition CB.

$$\begin{aligned} d_{\text{free}_{\text{nonparallel}}} &= \sqrt{d^2_{\min}(\text{C0}, \text{C2}) + d^2_{\min}(\text{C0}, \text{C3}) + d^2_{\min}(\text{C0}, \text{C2})} \\ &= \sqrt{2 + \left(2 \sin \frac{\pi}{8}\right)^2 + 2} \\ &= 2.14 \end{aligned}$$

The minimum free distance for the encoder is the minimum of $d_{\text{free}_{\text{parallel}}}$ and $d_{\text{free}_{\text{nonparallel}}}$. In this case the parallel transitions provide for the closest sequences and the minimum free distance for the coded system is $d_{\text{free/coded}} = 2$.

The performance improvement provided by the coded system relative to the uncoded is usually measured in terms of the **asymptotic coding gain** γ. Let S_{uncoded} be the normalized average received energy for the uncoded case and S_{coded} the normalized average received energy for the coded case. The asymptotic coding gain is then defined as follows.

$$\gamma = \frac{\left(\dfrac{S_{\text{uncoded}}}{d^2_{\text{free/uncoded}}}\right)}{\left(\dfrac{S_{\text{coded}}}{d^2_{\text{free/coded}}}\right)} \tag{14-5}$$

Equation (14-5) can be reexpressed as shown below to illustrate the impact on coding gain of the increased energy required for the expanded constellation and the increased minimum free distance provided by the redundancy. The additional energy is represented by a **constellation expansion factor** γ_C, while the increased minimum distance is represented by a **increased distance factor** γ_D.

$$\gamma = \left(\frac{S_{\text{uncoded}}}{S_{\text{coded}}}\right) \cdot \left(\frac{d^2_{\text{free/coded}}}{d^2_{\text{free/uncoded}}}\right) = \gamma_C \cdot \gamma_D \tag{14-6}$$

Since the asymptotic coding gain is a function of the square of d_{free} for the coded and uncoded case, it is often more convenient to work directly with the minimum squared Euclidean free distance. In the coded 8-PSK example the uncoded and coded constellations have the same average energy (assumed to be unity). The minimum squared free distance for the uncoded case is 2, while that for the coded case is 4. The asymptotic coding gain is thus $\gamma = 2$ (3.01 dB).

The trellis in Figure 14-12 reflects some heuristic design rules whose goal is to maximize d^2_{free} in a straightforward manner. They were presented by Ungerboeck in [Ung3] and are thus referred to here as Ungerboeck design rules.

Ungerboeck design rules

- Signals in the same, lowest partition in the partition tree are assigned to parallel transitions.
- Signals in the preceding partition are assigned to transitions that start or stop in the same state.
- All signals are used equally often.

The first of the above rules maximizes the distance between signals assigned to parallel transitions. The second rule maximizes the distance between nonparallel paths at their first and last branches (i.e., the points at which they diverge and later merge).

The asymptotic coding gain for the 8-PSK Ungerboeck encoder can be increased by increasing the memory and the rate of the convolutional encoder, as demonstrated in Figure 14-13. A rate-2/3 encoder is used to select from among the third-level partitions of 8-PSK, each of which contains only a single signal. There is thus no need for an uncoded information bit to complete the signal selection process, and the corresponding trellis has no parallel transitions, as shown in Figure 14-14. Note how the Ungerboeck design rules have been followed in assigning signals to trellis branches.

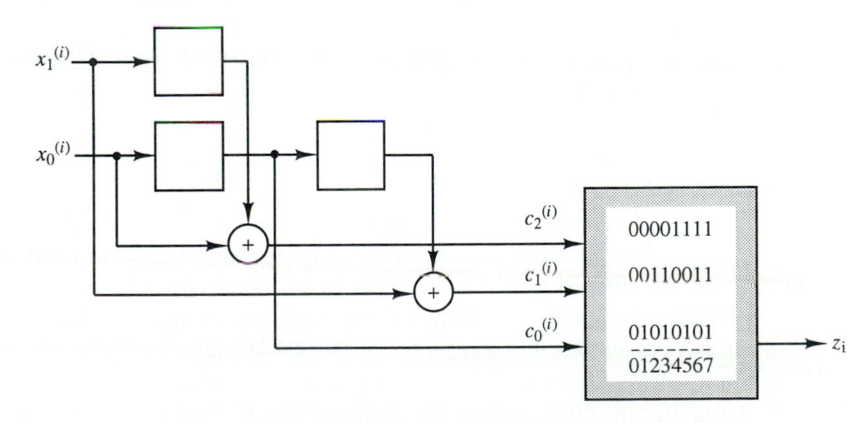

Figure 14-13. Eight-state Ungerboeck Encoder [Ung2]

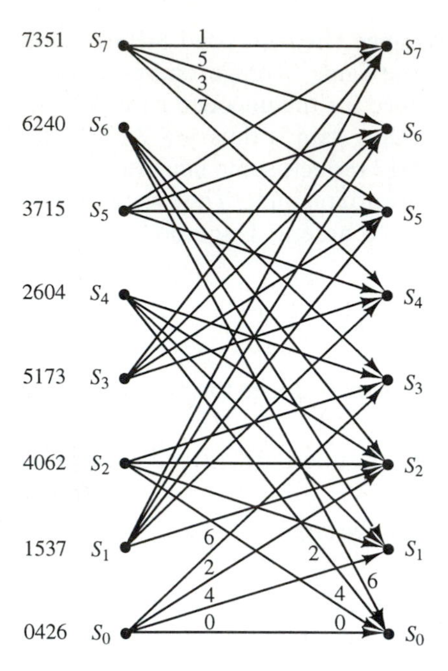

Figure 14-14. Trellis Diagram for Ungerboeck Encoder in Figure 14-13

Since there are no parallel transitions in this example, we need only examine nonparallel paths to determine the minimum free distance of the code. In this case the pair of trellis paths $S_0 \rightarrow S_3 \rightarrow S_6 \rightarrow S_0$ and $S_0 \rightarrow S_0 \rightarrow S_0 \rightarrow S_0$ correspond to the closest signal sequences. The sequences are 0, 0, 0 and 6, 7, 6, respectively, and are separated by a squared Euclidean distance of 4.58. The coding gain is thus

$$\gamma = \left(\frac{S_{\text{uncoded}}}{S_{\text{coded}}}\right) \cdot \left(\frac{d^2_{\text{free/coded}}}{d^2_{\text{free/uncoded}}}\right) = \left(\frac{1}{1}\right)\left(\frac{4.58}{2}\right) = 2.29 \quad (3.60 \text{ dB}) \qquad (14\text{-}7)$$

The extra memory cell in the encoder has increased the coding gain by approximately 0.6 dB. Unfortunately the complexity of the trellis (and thus the decoder) is substantially increased by the change from a rate-1/2 to a rate-2/3 convolutional code. Further increases in the constraint length of the convolutional code have a moderate effect on the asymptotic coding gain for the 8-PSK Ungerboeck encoder, as shown in Table 14-3. The generator sequences for the convolutional encoder are listed in octal form. The encoder is assumed to have the systematic feedback form shown in Figure 14-15.

Table 14-4 contains a series of Ungerboeck codes for QAM constellations. Convolutional codes with total memory ranging from 2 to 9 are used in conjunction with 8, 16, 32, and 64 QAM to provide from 2 to 6 dB of asymptotic coding gain. As with Table 14-3, the convolutional encoders are assumed to have the systematic, canonical form shown in Figure 14-15. The table includes coding gains for systems with spectral efficiencies of 2, 3, 4, and 5 bits/sec/Hz.

Consider the construction of a trellis coded 32-CROSS system using Table 14-4. The system provides a spectral efficiency of 4 bits/sec/Hz. The asymptotic coding

TABLE 14-3 Ungerboeck encoders based on 8-PSK modulation ([Ung2], © 1982 IEEE)

M	k	Generator sequences $\mathbf{h}^{(0)}, \mathbf{h}^{(1)}, \mathbf{h}^{(2)}$	Gain with respect to uncoded QPSK	
2	1	5, 2	2.000	(3.0 dB)
3	2	11, 02, 04	2.293	(3.6 dB)
4	2	23, 04, 16	2.586	(4.1 dB)
5	2	45, 16, 34	2.879	(4.6 dB)
6	2	105, 036, 074	3.000	(4.8 dB)
7	2	203, 014, 016	3.172	(5.0 dB)
8	2	405, 250, 176	3.465	(5.4 dB)
9	2	1007, 0164, 0260	3.758	(5.7 dB)

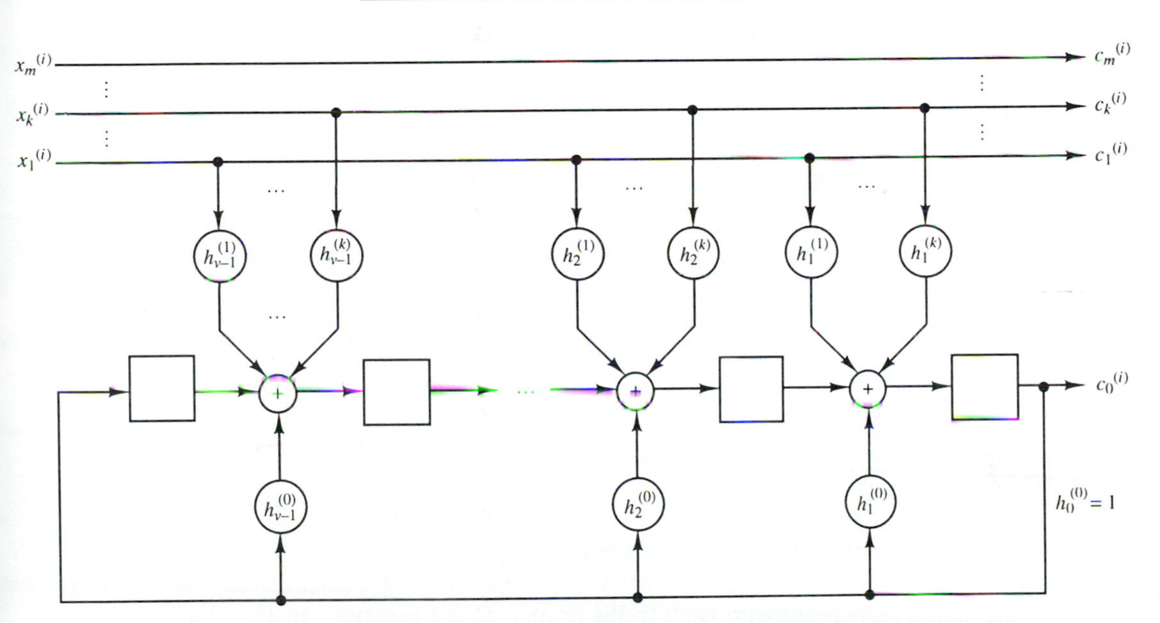

Figure 14-15. Systematic, Canonical, 2^{ν}-State Convolutional Encoder with Feedback ([Ung2], © 1982 IEEE)

gains in the column labeled "$\gamma_{32/16}$" tell us the coding gains we can expect relative to an uncoded 16-QAM system as a function of the complexity in the convolutional encoder. Assuming that we need 3.8 dB of coding gain, we select the encoder in the second row ($M = 3$). The octal generator sequences 11, 02, and 04 describe the convolutional encoder in Figure 14-16.

The convolutional encoder in Figure 14-16 has $k + 1 = 3$ coded output bits: $c_0^{(i)}$, $c_1^{(i)}$, and $c_2^{(i)}$. They are used to select one of the eight partitions of the 32-CROSS constellation shown at the bottom of Figure 14-6. The remaining two uncoded bits, $c_3^{(i)}$ and $c_4^{(i)}$, are used to select one of the four signals in the selected partition.

TABLE 14-4 Ungerboeck encoders based on rectangular QAM modulation ([Ung2], © 1982 IEEE)

M	k	Generator sequences $\mathbf{h}^{(0)}, \mathbf{h}^{(1)}, \mathbf{h}^{(2)}$	$\gamma_{\text{coded 8/uncoded 4}}$	$\gamma_{16/8}$	$\gamma_{32/16}$	$\gamma_{64/32}$
2	1	5, 2	2.0 dB	3.0 dB	2.8 dB	3.0 dB
3	2	11, 02, 04	3.0 dB	4.0 dB	3.8 dB	4.0 dB
4	2	23, 04, 16	3.8 dB	4.8 dB	4.6 dB	4.8 dB
5	2	41, 06, 10	3.8 dB	4.8 dB	4.6 dB	4.8 dB
6	2	101, 016, 034	4.5 dB	5.4 dB	5.2 dB	5.4 dB
7	2	203, 014, 042	5.1 dB	6.0 dB	5.8 dB	6.0 dB
8	2	401, 056, 354	5.1 dB	6.0 dB	5.8 dB	6.0 dB
9	2	1001, 0346, 0510	5.6 dB			

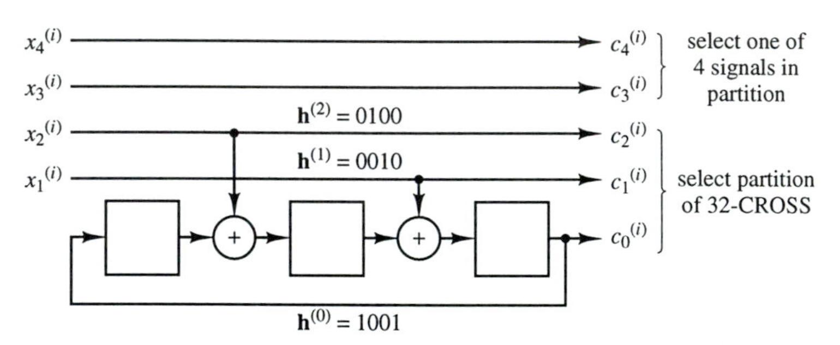

Figure 14-16. Convolutional Encoder for Trellis Coded 32-CROSS in Table 14-4

14.3 MULTIPLE TCM

In conventional TCM, the contents of the convolutional encoder (and thus the state of the TCM encoder) change as each signal is selected for transmission. There is thus one signal associated with each trellis branch at the receiver. In 1988 Divsalar and Simon [Div1] showed that it is possible to achieve a significant increase in coding gain relative to conventional TCM (almost 2 dB in one example) by increasing the number of signals transmitted for each trellis branch. They called the resulting system **multiple trellis coded modulation (MTCM)**. In this section we briefly examine the two-state MTCM systems that they investigated in their 1988 paper.

Consider the simple TCM system in Figure 14-17. A rate-1/2 convolutional encoder with a single memory element is used to select signals from a QPSK constellation. The corresponding two-state trellis is shown in the same figure. The QPSK signals are labeled as in the uncoded constellation in Figure 14-9.

Since there are no parallel transitions, the minimum free distance for this code corresponds to nonparallel pairs of two-branch paths, as shown by the dotted lines in Figure 14-17. Let $d^2(x, y)$ denote the squared distance between signals x and y.

Assuming QPSK signals with unity energy, d^2_{free} for this system is

$$\begin{aligned} d^2_{\text{free}_{\text{TCM}}} &= d^2(0,2) + d^2(0,1) \\ &= 4 + 4\sin^2\left(\frac{\pi}{4}\right) \\ &= 6 \end{aligned}$$

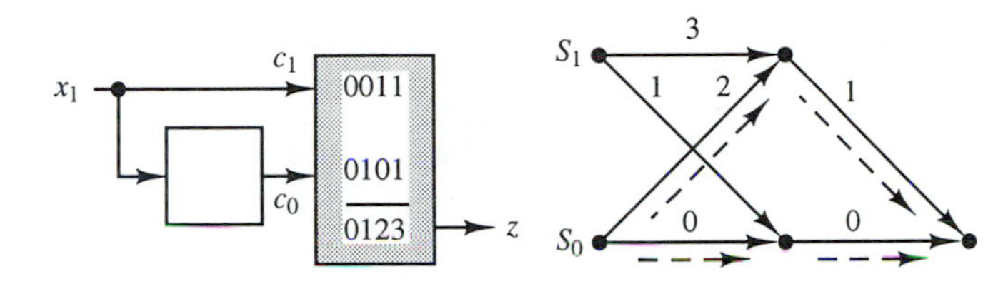

Figure 14-17. Two-State Encoder and Trellis

The encoder in Figure 14-17 outputs a signal from the partition $\{0, 2\}$ when the encoder is in state S_0 and a signal from $\{1, 3\}$ when the encoder is in state S_1. We now allow the encoder to select $r = 2$ signals (with replacement) from these respective partitions for each trellis branch. For each trellis state there are now four possible two-signal sequences to be distributed among the two branches leaving the state. Consider state S_0. The four two-signal sequences $\{0, 0; 0, 2; 2, 0; 2, 2\}$ can be viewed as the vertices of a square.

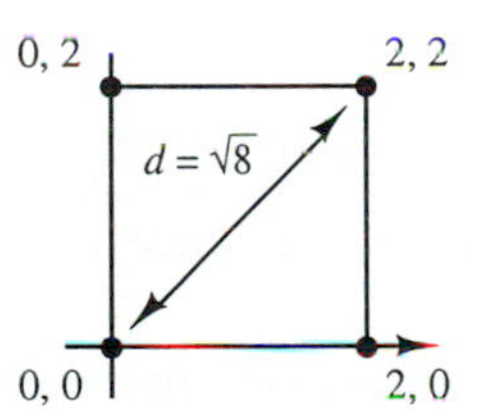

Figure 14-18. Two-Signal Sequences Leaving State S_0

Since there are four possible two-signal sequences, but only two successor states for S_0, we are clearly going to have parallel transitions. The two-signal sequences are partitioned into pairs such that the intersequence distance in each pair is maximized. Figure 14-19 shows the resulting MTCM trellis.

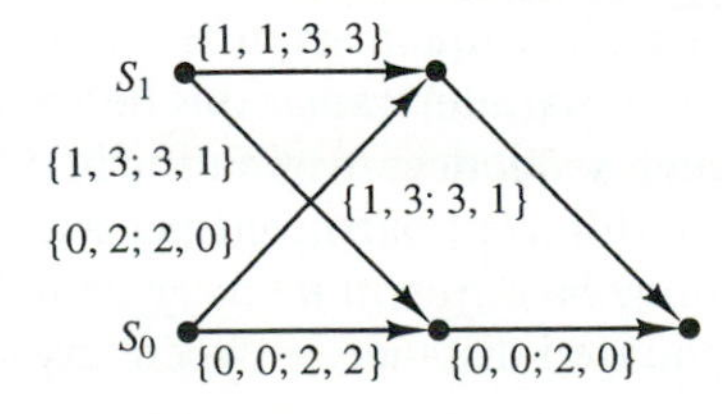

Figure 14-19. Two-state, MTCM trellis with $r = 2$

The minimum squared free distance for parallel transitions is

$$\begin{aligned} d^2_{\text{free}_{\text{parallel}}} &= d^2(0,2) + d^2(0,2) \\ &= 2^2 + 2^2 \\ &= 8 \end{aligned}$$

while that for two branch paths is

$$\begin{aligned} d^2_{\text{free}_{\text{nonparallel}}} &= d^2(0,0) + d^2(0,2) + d^2(1,0) + d^2(3,0) \\ &= 0^2 + 2^2 + 4\sin^2\left(\frac{\pi}{4}\right) + 4\sin^2\left(\frac{\pi}{4}\right) \\ &= 8 \end{aligned}$$

d^2_{free} for the MTCM system is thus 8, while d^2_{min} for the uncoded (BPSK) constellation is 4. The MTCM system provides a coding gain of 8/4 = 2 (3.01 dB). We have obtained an increase in coding gain of 8/6 = 1.33 (1.25 dB) relative to the conventional TCM system in Figure 14-17. Given the simplicity of the example, this is a substantial improvement.

The preceding system can be generalized to provide a design procedure for two-state systems in the following manner [Div1]. Suppose that a two-state convolutional code is used with a constellation of 2^{m+1} signals. Each of the two trellis states thus has an associated partition containing 2^m signals. If each trellis branch is to correspond to a sequence of r signals, we must allocate 2^{mr} sequences between the two branches leaving each trellis state. This is done in the following manner.

- Transitions to the same state ($S_0 \rightarrow S_0$ or $S_1 \rightarrow S_1$) are assigned 2^{mr-1} sequences such that, in any pair of sequences, the two sequences differ in at least two coordinates.
- The remaining 2^{mr-1} sequences are assigned to the transitions to differing states ($S_0 \rightarrow S_1$ or $S_1 \rightarrow S_0$).

The minimum squared free distance of the resulting MTCM system is determined by three different minimum intersequence distances. Let d^2 be the minimum squared distance for the entire constellation of 2^{m+1} signals to be used in the system. Let d'^2 be the minimum squared distance for the partitions assigned to each of the two trellis states. The minimum squared distance for parallel transitions between like states is then $2d'^2$ (by design the sequences differ in at least two coordinates). The minimum squared distance for parallel transitions between unlike states is d'^2. There remains the distance between sequences originating in different states, but terminating in the same state. Since different partitions are assigned to different states, the sequences in such a pair *must be different in every coordinate*. Given a sequence of r symbols, the minimum squared distance between any such pair of sequences must be rd^2. We thus have the following situation for an MTCM system with r signals per branch.

$$d^2_{\text{free}_{\text{parallel}}} = 2d'^2$$
$$d^2_{\text{free}_{\text{nonparallel}}} = d'^2 + rd^2 \tag{14-8}$$
$$d^2_{\text{free}} = \min\left(d^2_{\text{free}_{\text{parallel}}}, d^2_{\text{free}_{\text{nonparallel}}}\right)$$

If the coded constellation is MPSK, where $M = 2^{m+1}$, we have the following values for the various distances.

$$d^2 = 4\sin^2\left(\frac{\pi}{2^{m+1}}\right)$$
$$d'^2 = 4\sin^2\left(\frac{\pi}{2^m}\right) \tag{14-9}$$

Equation (14-8) gives the following d^2_{free} for a two-state MPSK MTCM system.

$$d^2_{\text{free}} = \min\left(8\sin^2\left(\frac{\pi}{2^m}\right), 4\sin^2\left(\frac{\pi}{2^m}\right) + 4r\sin^2\left(\frac{\pi}{2^{m+1}}\right)\right) \tag{14-10}$$

Divsalar and Simon [Div1] have shown that the minimum value of r for which Eq. (14-10) is maximized is $r = \lceil 2[1 + \cos(\pi/2^m)]\rceil \leq 4$.

14.4 DECODING AND PERFORMANCE ANALYSIS

In this section we first examine the basic decoder operation and then proceed to derive upper and lower bounds on performance. As with convolutional codes, much of the analysis makes use of generating functions and signal flow graphs.

14.4.1 Decoding with the Viterbi Decoder

The decoding of trellis codes is very similar to that of convolutional codes. In fact, after a certain point the two are identical, for the trellis structure imparted on the transmitted sequence by the convolutional encoder within the TCM system allows for the use of the Viterbi decoder. The vast majority of the work on trellis codes has thus far focused on decoding with the Viterbi decoder, though sequential decoders can certainly be used and may have their own set of advantages and disadvantages. In this section we consider only the Viterbi decoder.

In general the TCM trellis assigns a subconstellation of signals to each branch in the trellis. Decoding is performed in two steps.

1. At each branch in the trellis, the decoder compares the received signal to each of the signals allowed for that branch. The identity of the allowed signal closest to the received signal is saved in memory, and the branch is labeled with a metric proportional to the distance between the two signals.

2. The Viterbi algorithm is then applied to the trellis, with surviving partial paths corresponding to partial signal sequences that are closest to the received sequence. The maximum likelihood path is then the complete signal sequence closest in Euclidean distance to the received sequence.

The design and implementation of this encoder is an extension of the material in Section 12.4.

14.4.2 An Upper Bound on Node Error Probability and Bit Error Rate for the AWGN Channel

Given the important part played by the Viterbi decoder, it naturally follows that the performance analysis for TCM systems is similar to that for convolutional codes. There are, however, two basic differences for which we must account: TCM systems allow for parallel transitions and are, in general, nonlinear codes. We begin the analysis of the performance of the decoder by reviewing the notation, and then proceed to the development of upper bounds on the node error probability and the bit error rate.

The Ungerboeck encoder (recall Figure 14-8) performs two mapping operations. A block of m information bits $\mathbf{x}_i = x_1^{(i)}, x_2^{(i)}, \ldots, x_m^{(i)}$ is first mapped onto an $(m+1)$-bit "label" $\mathbf{c}_i = c_0^{(i)}, c_1^{(i)}, \ldots, c_m^{(i)}$. The label $\mathbf{c}_i$ is then mapped onto a signal $\mathbf{z}_i$. The mapping $\mathbf{x}_i \rightarrow \mathbf{c}_i$ is performed by a convolutional encoder, which may be linear. The second mapping, $\mathbf{c}_i \rightarrow \mathbf{z}_i$, which we shall denote by $f(\mathbf{c}_i) = \mathbf{z}_i$, is generally nonlinear, thus making the overall TCM system nonlinear. This notation extends naturally to cover entire sequences. A sequence of information blocks $\mathbf{X} = \mathbf{x}_1, \mathbf{x}_2, \mathbf{x}_3, \ldots$ is encoded onto a sequence of labels $\mathbf{C} = \mathbf{c}_1, \mathbf{c}_2, \mathbf{c}_3, \ldots$, which is then mapped onto a sequence of signals $\mathbf{Z} = \mathbf{z}_1, \mathbf{z}_2, \mathbf{z}_3, \ldots$.

As with the Viterbi decoding of convolutional codes (see Section 12.3), we are concerned with the characterization of node error events. A **node error** is said to occur whenever a nonzero path leaves a given node and, upon remerging with the correct path at a later node, is declared the survivor. Once again we pursue the problem by applying the union bound and focusing on pairwise error probabilities. Let P_e be the node error probability. Let $\mathbf{Z}$ be the transmitted sequence and let $\mathbf{Z}_L$ be a sequence that diverges from $\mathbf{Z}$ at some point in the trellis and remerges after exactly L branches. $P(\mathbf{Z} \rightarrow \mathbf{Z}_L)$ is the pairwise error probability for the sequences $\mathbf{Z}$ and $\mathbf{Z}_L$. The union bound provides the following upper bound on the node error probability P_e. Note that, because the code may be nonlinear, we must sum over all possible transmitted sequences $\mathbf{Z}$.

$$P_e \leq \sum_{L=1}^{\infty} \sum_{\mathbf{Z}} \sum_{\mathbf{Z}_L} P(\mathbf{Z}) P(\mathbf{Z} \rightarrow \mathbf{Z}_L) \tag{14-11}$$

It is usually convenient to reexpress Eq. (14-11) in terms of the signal-label sequences

{**C**} [Big1]. Since the mapping f is one-to-one, the resulting expressions are unambiguous.

$$
\begin{aligned}
P_e &\leq \sum_{L=1}^{\infty} \sum_{\mathbf{C}} \sum_{\mathbf{C}_L} P\{\mathbf{C}\} P\{\mathbf{C} \rightarrow \mathbf{C}_L\} \\
&= \sum_{L=1}^{\infty} \sum_{\mathbf{C}} P\{\mathbf{C}\} \sum_{\mathbf{E}_L \neq 0} P\{\mathbf{C} \rightarrow \mathbf{C} \oplus \mathbf{E}_L\}
\end{aligned}
\tag{14-12}
$$

We can now focus on sequences of binary labels. $\mathbf{E}_L$ is the label sequence for an error event of length L. To compute the probability $P\{f(\mathbf{C}) \rightarrow f(\mathbf{C} \oplus \mathbf{E}_L)\}$, we need to make some assumptions about the channel. We assume here that the channel is perturbed by additive white Gaussian noise with two-sided noise spectral density $N_0/2E_S$; the Euclidean distances between sequences are thus normalized. Given a received sequence $\mathbf{R}$, we obtain the following.

$$
\begin{aligned}
P(\mathbf{C} \rightarrow \mathbf{C} \oplus \mathbf{E}_L) &= P(\|f(\mathbf{C}) - \mathbf{R}\| \geq \|f(\mathbf{C} \oplus \mathbf{E}_L) - \mathbf{R}\|) \\
&= P(\|f(\mathbf{C}) - \mathbf{R}\| - \|f(\mathbf{C} \oplus \mathbf{E}_L) - \mathbf{R}\| \geq 0) \\
&= P\left\{ n \geq \frac{\|f(\mathbf{C}) - f(\mathbf{C} \oplus \mathbf{E}_L)\|}{2} \right\} \\
&= Q\left\{ \frac{\|f(\mathbf{C}) - f(\mathbf{C} \oplus \mathbf{E}_L)\|}{\sqrt{2N_0/E_s}} \right\}
\end{aligned}
\tag{14-13}
$$

where n is a one-dimensional Gaussian random variable. The computation of the expression in Eq. (14-13) can be simplified through the use of the upper bound

$$
Q(\sqrt{x + y}) \leq Q(\sqrt{x}) e^{-y/2} \tag{14-14}
$$

Proceeding as in Section 12.3.3, we obtain

$$
\begin{aligned}
&Q\left(\frac{\|f(\mathbf{C}) - f(\mathbf{C} \oplus \mathbf{E}_L)\|}{\sqrt{2N_0/E_s}} \right) \\
&\qquad = Q\left(\sqrt{\frac{d_{\text{free}}^2 + [\|f(\mathbf{C}) - f(\mathbf{C} \oplus \mathbf{E}_L)\|^2 - d_{\text{free}}^2]}{2N_0/E_s}} \right) \\
&\qquad \leq Q\left(\sqrt{\frac{d_{\text{free}}^2 E_s}{2N_0}} \right) \cdot \exp\left\{ \frac{d_{\text{free}}^2 - \|f(\mathbf{C}) - f(\mathbf{C} \oplus \mathbf{E}_L)\|^2}{4N_0/E_s} \right\} \\
&\qquad = Q\left(\sqrt{\frac{d_{\text{free}}^2 E_s}{2N_0}} \right) \cdot \exp\left(\frac{d_{\text{free}}^2 E_s}{4N_0} \right) \cdot \exp\left\{ -\frac{E_s \|f(\mathbf{C}) - f(\mathbf{C} \oplus \mathbf{E}_L)\|^2}{4N_0} \right\}
\end{aligned}
\tag{14-15}
$$

We now have the rather cumbersome expression

$$
\begin{aligned}
P_e \leq Q\left(\sqrt{\frac{d_{\text{free}}^2 E_s}{2N_0}} \right) &\cdot \exp\left(\frac{d_{\text{free}}^2 E_s}{4N_0} \right) \\
&\cdot \sum_{L=1}^{\infty} \sum_{\mathbf{C}} P(\mathbf{C}) \sum_{\mathbf{E}_L \neq 0} \exp\left\{ -\frac{E_s \|f(\mathbf{C}) - f(\mathbf{C} \oplus \mathbf{E}_L)\|^2}{4N_0} \right\}
\end{aligned}
\tag{14-16}
$$

as an upper bound on the node error rate. As with convolutional codes, the remaining summations can be removed through the use of a generating function for all normalized squared distances $\|f(\mathbf{C}) - f(\mathbf{C} \oplus \mathbf{E}_L)\|^2$. Our goal is to construct a function $T(D)$ of the form

$$T(D) = \sum_{L=1}^{\infty} \sum_{\mathbf{C}} P(\mathbf{C}) \sum_{\mathbf{E}_L} D^{\|f(\mathbf{C}) - f(\mathbf{C} \oplus \mathbf{E}_L)\|^2} \tag{14-17}$$

Given such a function, an upper bound on the node error rate for an AWGN channel follows immediately.

$$P_e \leq Q\left(\sqrt{\frac{d_{\text{free}}^2 E_s}{2N_0}}\right) \cdot \exp\left(\frac{d_{\text{free}}^2 E_s}{4N_0}\right) \cdot T(D)\big|_{D=\exp(-E_s/4N_0)} \tag{14-18}$$

The derivation of the generator function is discussed in the next section. The derivation can be quite involved, and rapidly becomes intractable as the number of states in the TCM system increases. Fortunately the node error rate can be approximated quite accurately without the use of a generating function at all if the signal-to-noise ratio is sufficiently high. We simply assume that only the closest sequences to the transmitted sequence have any significant probability of being selected by the decoder. Let $N(d_{\text{free}})$ be the average number of sequences that are distance d_{free} from the transmitted sequence. The probability of node error is then approximated by

$$P_e \approx N(d_{\text{free}}) Q\left(\sqrt{\frac{d_{\text{free}}^2 E_s}{2N_0}}\right) \tag{14-19}$$

As was the case for the Viterbi decoding of convolutional codes in Chapter 12, the node-error-rate expression in Eq. (14-18) can be converted into a bit-error-rate expression through the use of an augmented generating function $T(D, I)$, where the exponents of the I terms denote the number of information-bit errors associated with each error event. Note that each node in the decoding trellis represents the processing of m information bits.

$$P_b \leq \frac{1}{m} Q\left(\sqrt{\frac{d_{\text{free}}^2 E_s}{2N_0}}\right) \cdot \exp\left(\frac{d_{\text{free}}^2 E_s}{4N_0}\right) \cdot \frac{\partial T(D, I)}{\partial I}\bigg|_{D=\exp(-E_s/4N_0),\, I=1} \tag{14-20}$$

The approximation in Eq. (14-19) can be similarly modified. Let $b_{d_{\text{free}}}$ be the total number of information bit errors associated with erroneous paths that are distance d_{free} from the transmitted path, averaged over all possible transmitted paths. We then have

$$P_b \approx \frac{b_{d_{\text{free}}}}{m} Q\left(\sqrt{\frac{d_{\text{free}}^2 E_s}{2N_0}}\right) \tag{14-21}$$

14.4.3 The Squared-Distance Generating Function

The squared-distance generating function is obtained as follows. Each error sequence can be written as a sequence of L $(m + 1)$-bit error blocks: $\mathbf{E}_L = \mathbf{e}_1, \mathbf{e}_2, \mathbf{e}_3, \ldots, \mathbf{e}_L$.

If we assume that the source output is uniformly distributed, then all 2^m m-bit input blocks are equally likely to occur. It follows that the TCM encoder is in any one of the 2^ν trellis states at any given point in time with equal probability.

Each trellis branch from some state p to some state q is assigned one or more signals, all of which can be denoted by the label $\mathbf{c}_{p \to q}$. We can represent the squared distances associated with the transition from state p to state q and some error block $\mathbf{e}_i$ in the following manner [Big1].

$$[\mathbf{G}(\mathbf{e}_i)]_{p,q} = \frac{1}{2^m} \sum_{\mathbf{c}_{p \to q}} D^{\|f(\mathbf{c}_{p \to q}) - f(\mathbf{c}_{p \to q} \oplus \mathbf{e}_i)\|^2} \tag{14-22}$$

The summation in Eq. (14-22) allows for parallel transitions.

A $(2^\nu \times 2^\nu)$ matrix $\mathbf{G}(\mathbf{e}_i)$ is then constructed by selecting each summation $[\mathbf{G}(\mathbf{e}_i)]_{p,q}$ as the element in the pth row and qth column. The squared distances associated with the error sequence $\mathbf{E}_L = \mathbf{e}_1, \mathbf{e}_2, \mathbf{e}_3, \ldots, \mathbf{e}_L$ are then obtained in the form of a matrix $\mathbf{G}(\mathbf{E}_L)$ by taking the product of the appropriate matrices $\mathbf{G}(\mathbf{e}_i)$.

$$\mathbf{G}(\mathbf{E}_L) = \prod_{i=1}^{L} \mathbf{G}(\mathbf{e}_i) \tag{14-23}$$

The element in the pth row and qth column of $\mathbf{G}(\mathbf{E}_L)$ enumerates the squared Euclidean distances associated with $\mathbf{E}_L$ and any L-step transition from state p to state q. We now need only sum over all possible $\mathbf{E}_L$ to obtain the final *error matrix* $\mathbf{G}$.

$$\mathbf{G} = \sum_{L=1}^{\infty} \sum_{\mathbf{E}_L} \mathbf{G}(\mathbf{E}_L) \tag{14-24}$$

The desired generating function follows immediately. Note that for an $n \times n$ matrix $\mathbf{A}$ and a column vector $\bar{\mathbf{1}}$ of n one's, $\bar{\mathbf{1}}^T \mathbf{A} \bar{\mathbf{1}}$ is the sum of all of the elements in $\mathbf{A}$.

$$T(D) = \frac{1}{2^\nu} \bar{\mathbf{1}}^T \mathbf{G} \bar{\mathbf{1}} \tag{14-25}$$

Unfortunately the summation in Eq. (14-24) is over a highly dependent set of error sequences. This dependence is best expressed and the summation best performed through the use of signal flow graphs. (See Section 11.2.3 for a detailed discussion.) In TCM systems the signal flow graphs are generally called **error-state diagrams**. The connections in these diagrams are determined by the convolutional encoder used in the TCM system. At this point we assume that the convolutional code is linear (though the overall TCM system need not be linear). The valid output sequences for the convolutional encoder thus form a commutative group. It follows that the difference between any pair of valid sequences of labels is also a valid sequence. The valid error sequences are thus identical to the valid sequences themselves. The labels $\mathbf{G}(\mathbf{e}_i)$ for the branches of the error-state diagram are thus determined by the signal labels on the corresponding branches of the encoder trellis diagram.

Example 14-1—An Error-State Diagram for a Two-State Encoder[4]

Consider the following simple two-state encoder that is used in conjunction with a QPSK constellation (also shown). The branches in the trellis diagram are labeled with the input bit and output label corresponding to the particular transition.

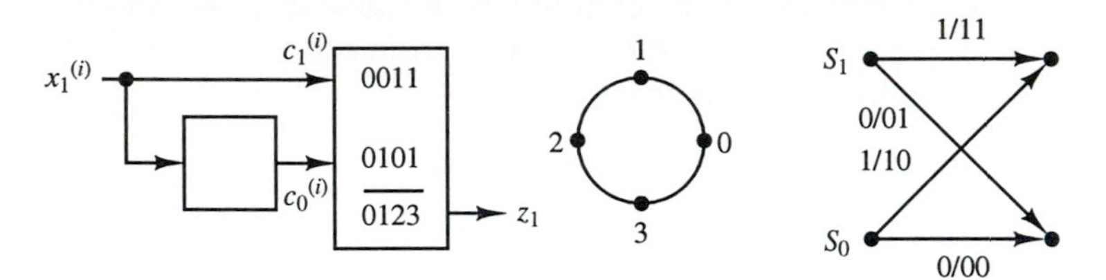

Since the error sequences $\mathbf{E}_L$ are identical to the valid sequences of signal labels, we can construct a signal flow graph that represents all possible error sequences. As with convolutional codes, we separate the starting and stopping state (S_0) into two separate states in the graph.

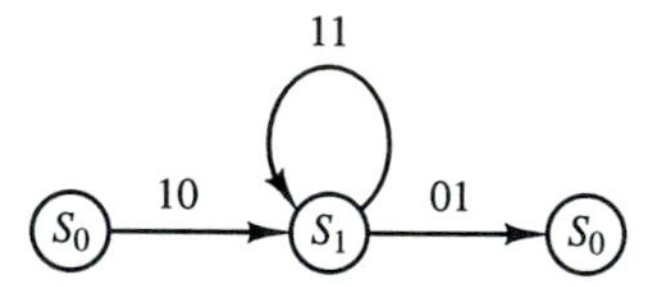

The matrices for the individual error blocks are constructed as follows.

$$\mathbf{G}(\mathbf{e}_i) = \mathbf{G}(e_1 e_2) = \frac{1}{2}\begin{bmatrix} D^{\|f(00) - f(00 \oplus e_1 e_2)\|^2} & D^{\|f(10) - f(10 \oplus e_1 e_2)\|^2} \\ D^{\|f(01) - f(01 \oplus e_1 e_2)\|^2} & D^{\|f(11) - f(11 \oplus e_1 e_2)\|^2} \end{bmatrix}$$

$$\mathbf{G}(01) = \frac{1}{2}\begin{bmatrix} D^2 & D^2 \\ D^2 & D^2 \end{bmatrix}, \quad \mathbf{G}(10) = \frac{1}{2}\begin{bmatrix} D^4 & D^4 \\ D^4 & D^4 \end{bmatrix}, \quad \mathbf{G}(11) = \frac{1}{2}\begin{bmatrix} D^2 & D^2 \\ D^2 & D^2 \end{bmatrix}$$

The error-state diagram is then modified.

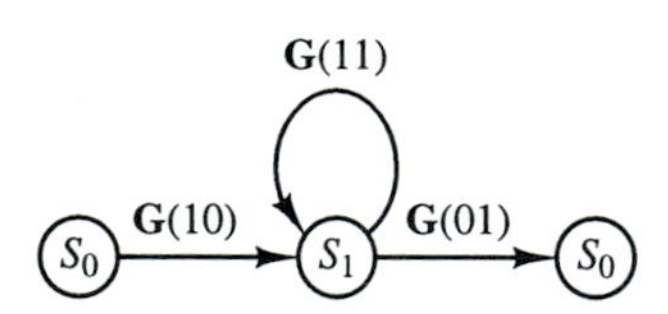

The error matrix $\mathbf{G}$ is the transfer function of the error-state diagram

$$\begin{aligned}
\mathbf{G} &= \mathbf{G}(10)[\mathbf{I}_2 - \mathbf{G}(11)]^{-1}\mathbf{G}(01) \\
&= \frac{1}{2}\begin{bmatrix} D^4 & D^4 \\ D^4 & D^4 \end{bmatrix} \cdot \begin{bmatrix} 1 - \frac{1}{2}D^2 & -\frac{1}{2}D^2 \\ -\frac{1}{2}D^2 & 1 - \frac{1}{2}D^2 \end{bmatrix}^{-1} \cdot \frac{1}{2}\begin{bmatrix} D^2 & D^2 \\ D^2 & D^2 \end{bmatrix} \\
&= \frac{D^6}{4(1 - D^2)}\begin{bmatrix} 1 & 1 \\ 1 & 1 \end{bmatrix} \cdot \begin{bmatrix} 1 - \frac{1}{2}D^2 & \frac{1}{2}D^2 \\ \frac{1}{2}D^2 & 1 - \frac{1}{2}D^2 \end{bmatrix} \cdot \begin{bmatrix} 1 & 1 \\ 1 & 1 \end{bmatrix} \\
&= \frac{D^6}{2(1 - D^2)}\begin{bmatrix} 1 & 1 \\ 1 & 1 \end{bmatrix}
\end{aligned}$$

[4] Adapted from [Big1, p. 103].

The generating function follows from Eq. (14-25).

$$T(D) = \left(\frac{1}{2}\right)[1 \quad 1] \cdot \left(\frac{D^6}{2(1 - D^2)}\begin{bmatrix} 1 & 1 \\ 1 & 1 \end{bmatrix}\right) \cdot \begin{bmatrix} 1 \\ 1 \end{bmatrix} = \frac{D^6}{(1 - D^2)}$$ ■

The above example treats one of the simplest trellis codes. Clearly the number of states in the error-state diagram as well as the size of the labeling matrices increases exponentially with the number of memory elements in the TCM encoder. The problem is greatly simplified if the branch labels can be replaced with scalar expressions. This is possible whenever the matrices $\mathbf{G}(\mathbf{e}_i)$ display a certain uniformity.

Consider the following. The error matrix element $[\mathbf{G}]_{p,q}$ denotes the contribution to the bound on error probability made by error sequences starting in the error-state diagram at state p and terminating at state q. $\mathbf{G}\bar{\mathbf{1}}$ is a column vector whose entries are the sums of the elements in each row of $\mathbf{G}$. The pth entry of $\mathbf{G}\bar{\mathbf{1}}$ then denotes the contribution to the bound on error probability made by all paths leaving state p and terminating at arbitrary states. Similarly the qth entry of the row vector $\bar{\mathbf{1}}^T\mathbf{G}$ denotes the contribution made by all paths terminating at state q. It follows that if all of the elements in $\mathbf{G}\bar{\mathbf{1}}$ are the same, then the probability of error is independent of the starting state of the transmitted sequence. The same holds for $\bar{\mathbf{1}}^T\mathbf{G}$ and the terminating state. These basic ideas are expressed in algebraic form through the use of eigenvectors and eigenvalues.

Let $\mathbf{A}$ be an $n \times n$ matrix. $\bar{\mathbf{1}}$ is by definition an eigenvector of $\mathbf{A}^T$ if $\bar{\mathbf{1}}^T\mathbf{A} = \alpha\bar{\mathbf{1}}^T$, where α is the corresponding eigenvalue. When this is true, $\mathbf{A}$ is said to be **column-uniform**. For another $n \times n$ matrix $\mathbf{B}$, $\bar{\mathbf{1}}$ is an eigenvector of $\mathbf{B}$ if $\mathbf{B}\bar{\mathbf{1}} = \beta\bar{\mathbf{1}}$, in which case $\mathbf{B}$ is said to be **row-uniform**. It is a simple matter to show that the sum or product of any pair of column- or row-uniform matrices is also column- or row-uniform. The case for row-uniformity of sums and products is demonstrated below.

$$\mathbf{B}_1\bar{\mathbf{1}} = \beta_1\bar{\mathbf{1}}, \quad \mathbf{B}_2\bar{\mathbf{1}} = \beta_2\bar{\mathbf{1}}$$

$$\Rightarrow (\mathbf{B}_1 + \mathbf{B}_2)\bar{\mathbf{1}} = (\beta_1 + \beta_2)\bar{\mathbf{1}} = \beta_3\bar{\mathbf{1}}$$

$$\Rightarrow (\mathbf{B}_1\mathbf{B}_2)\bar{\mathbf{1}} = \mathbf{B}_1(\mathbf{B}_2\bar{\mathbf{1}}) = \mathbf{B}_1\beta_2\bar{\mathbf{1}} = \beta_4\bar{\mathbf{1}}$$

Finally we note that if the $n \times n$ matrix $\mathbf{A}$ is either row- or column-uniform with eigenvalue α, then $\bar{\mathbf{1}}^T\mathbf{A}\bar{\mathbf{1}} = n\alpha$.

We can now put the pieces together as follows. If all of the $\{\mathbf{G}(\mathbf{e}_i)\}$ are row- or column-uniform, then so is the error matrix $\mathbf{G}$, for it is obtained by summing the products of various $\{\mathbf{G}(\mathbf{e}_i)\}$. The desired generating function (Eq. (14-25)) is then simply the sum of the elements in any row or column (respectively) of $\mathbf{G}$. A trellis code that satisfies these constraints is said to be **uniform**. It follows that if a code is uniform, then we need only label the branches of the error-state diagram with the sums of the elements in any row or column of the appropriate $\mathbf{G}(\mathbf{e}_i)$.

Example 14-2—Analysis of a Uniform Trellis Code

Each of the $\mathbf{G}(\mathbf{e}_i)$ for the code in Example 14-1 are both row- and column-uniform (trivially so—all of the elements in each matrix are the same). We can thus label the branches in the error-state diagram with the sums of the elements in any row or column of the corresponding $\mathbf{G}(\mathbf{e}_i)$.

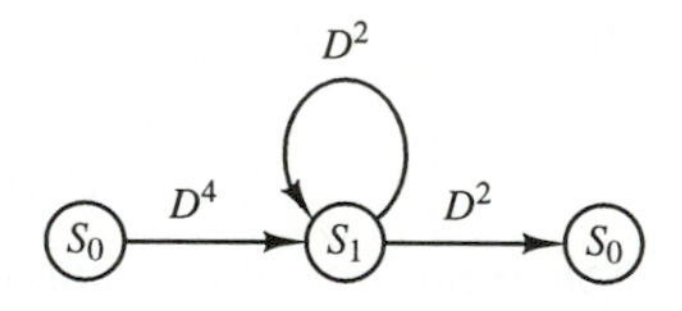

The generating function $T(D) = D^6/(1 - D^2)$ now follows by inspection. ■

It is possible to demonstrate that a given TCM system is uniform without having to construct the $\{\mathbf{G}(\mathbf{e}_i)\}$. Zehavi and Wolf [Zeh] presented a set of sufficient conditions for uniformity in 1987; they are deduced as follows.

Set the TCM encoder to the all-zero state. There are 2^m possible m-bit input blocks, and thus 2^m output labels. This set of output labels is denoted $\mathbf{C}_0$, and consists of half of the total 2^{m+1} possible output labels for the encoder. If we assume that the convolutional encoder is linear, then the set $\mathbf{C}_0$ is a commutative group (this follows from the fact that any linear combination of encoder inputs produces a linear combination of the corresponding outputs). Using Ungerboeck's design rules, the signals associated with $\mathbf{C}_0$ are assigned to half of the states in the encoder trellis; i.e., any transition out of one of these states is assigned a signal whose label is in $\mathbf{C}_0$. The 2^m labels that are not in $\mathbf{C}_0$ form a coset of $\mathbf{C}_0$. This coset can be represented by $\mathbf{C}_1 = \{\mathbf{c} \oplus \mathbf{c}' \mid \mathbf{c} \in \mathbf{C}_0\}$, where $\mathbf{c}'$ is a label not found in $\mathbf{C}_0$. The signals corresponding to labels in $\mathbf{C}_1$ are assigned to the remaining half of the trellis states.

The rows of the matrix $\mathbf{G}(\mathbf{e}_i)$ each correspond to a trellis state, so it follows that they are each associated with either $\mathbf{C}_0$ or $\mathbf{C}_1$. The rows of $\mathbf{G}(\mathbf{e}_i)$ will thus have equal sums (i.e., $\mathbf{G}(\mathbf{e}_i)$ will be row-uniform) if

$$\sum_{\mathbf{c}\in\mathbf{C}_0} D^{\|f(\mathbf{c})-f(\mathbf{c}\oplus\mathbf{e}_i)\|^2} = \sum_{\mathbf{c}\in\mathbf{C}_0} D^{\|f(\mathbf{c}\oplus\mathbf{c}')-f(\mathbf{c}\oplus\mathbf{c}'\oplus\mathbf{e}_i)\|^2} = \sum_{\mathbf{c}\in\mathbf{C}_1} D^{\|f(\mathbf{c})-f(\mathbf{c}\oplus\mathbf{e}_i)\|^2} \tag{14-26}$$

It follows from Eq. (14-26) that if the mapping $f(\mathbf{c}) \rightarrow f(\mathbf{c} \oplus \mathbf{c}'), \mathbf{c} \in \mathbf{C}_0$, is one-to-one and preserves distances, then the code is uniform. Such a mapping is called an **isometry**. As an example, consider the 8-PSK constellation below. Let $\mathbf{C}_0 = \{000, 010, 100, 110\}$ and define the mapping f to be $f(\mathbf{c}) \rightarrow f(\mathbf{c} \oplus 011)$. The image of the map is a reflection and rotation of the original constellation. The mapping is thus an isometry, and the distance between $f(\mathbf{c} \oplus \mathbf{e})$ and $\mathbf{f}(\mathbf{c} \oplus \mathbf{c}' \oplus \mathbf{e})$ is the same for any $\mathbf{c}$ and any $\mathbf{e}$. A trellis code based on a linear convolutional code and 8-PSK as labeled below is thus uniform.

010
011 001
100 000
101 111
110

$f(\mathbf{c} \oplus 011)$
reflection
+
rotation

001
000 010
111 011
110 100
101

Given uniformity, we can define a **weight profile** $W(\mathbf{e}_i)$ for the TCM system. $W(\mathbf{e}_i)$ is the sum of the elements in any row of $\mathbf{G}(\mathbf{e}_i)$.

$$W(\mathbf{e}_i) = \frac{1}{2^v} \sum_{\mathbf{c}\in\mathbf{C}_0} D^{\|f(\mathbf{c}) - f(\mathbf{c}\oplus\mathbf{e}_i)\|^2} \tag{14-27}$$

The generating function is then obtained by using the weight profiles as labels in the error-state diagram. The transfer function of the resulting graph is the desired generating function.

$$T(D) = \sum_{L=2}^{\infty} \sum_{\mathbf{E}_L} \prod_{\substack{\mathbf{e}_i \ni \\ \mathbf{E}_L = \mathbf{e}_1, \mathbf{e}_2, \ldots, \mathbf{e}_L}} W(\mathbf{e}_i) \qquad (14\text{-}28)$$

Note that the error-state diagram and thus the above generating function do not account for parallel transitions. They are an added factor in the overall bound, as shown in the next example.

Example 14-3—Node Error Probability Computation for Four-State Coded 8-PSK

Consider the four-state TCM encoder in Figure 14-10. It uses the same labeling for 8-PSK that is shown above in the discussion of isometries. The corresponding trellis is shown in Figure 14-12. The various distinct squared intersignal distances are as follows.

$$d^2(0,1) = d^2(000,001) = 0.586, \qquad d^2(0,3) = d^2(000,011) = 3.414$$

$$d^2(0,2) = d^2(000,010) = 2.000, \qquad d^2(0,4) = d^2(000,100) = 4.000$$

The set $\mathbf{C}_0$ consists of the labels {000, 010, 100, 110}. The individual weight profiles are computed using the expression

$$W(\mathbf{e}_i) = \frac{1}{4} \sum_{\mathbf{c} \in \{000, 010, 100, 110\}} D^{\|f(\mathbf{c}) - f(\mathbf{c} \oplus \mathbf{e}_i)\|^2}$$

$\mathbf{c}$	$\mathbf{c} \oplus 001$	$d^2[f(\mathbf{c}) - f(\mathbf{c} \oplus 001)]$	$\mathbf{c}$	$\mathbf{c} \oplus 010$	$d^2[f(\mathbf{c}) - f(\mathbf{c} \oplus 010)]$
000	001	0.586	000	010	2.000
010	011	0.586	010	000	2.000
100	101	0.586	100	110	2.000
110	111	0.586	110	100	2.000
$\Rightarrow W(001) = D^{0.586}$			$\Rightarrow W(010) = D^2$		

$\mathbf{c}$	$\mathbf{c} \oplus 011$	$d^2[f(\mathbf{c}) - f(\mathbf{c} \oplus 011)]$	$\mathbf{c}$	$\mathbf{c} \oplus 100$	$d^2[f(\mathbf{c}) - f(\mathbf{c} \oplus 100)]$
000	011	3.414	000	100	4.000
010	001	0.586	010	110	4.000
100	111	3.414	100	000	4.000
110	101	0.586	110	010	4.000
$\Rightarrow W(011) = (1/2)(D^{0.586} + D^{3.414})$			$\Rightarrow W(100) = D^4$		

$\mathbf{c}$	$\mathbf{c} \oplus 101$	$d^2[f(\mathbf{c}) - f(\mathbf{c} \oplus 101)]$	$\mathbf{c}$	$\mathbf{c} \oplus 110$	$d^2[f(\mathbf{c}) - f(\mathbf{c} \oplus 110)]$
000	101	3.414	000	110	2.000
010	111	3.414	010	100	2.000
100	001	3.414	100	010	2.000
110	011	3.414	110	000	2.000
$\Rightarrow W(101) = D^{3.414}$			$\Rightarrow W(110) = D^2$		

$\mathbf{c}$	$\mathbf{c}\oplus 111$	$d^2[f(\mathbf{c}) - f(\mathbf{c}\oplus 111)]$	$\mathbf{c}$	$\mathbf{c}\oplus 000$	$d^2[f(\mathbf{c}) - f(\mathbf{c}\oplus 000)]$
000	111	0.586	000	000	0.000
010	101	3.414	010	010	0.000
100	011	0.586	100	100	0.000
110	001	3.414	110	110	0.000
$\Rightarrow W(111) = (1/2)(D^{0.586} + D^{3.414})$			$\Rightarrow W(000) = D^0 = 1$		

The state diagram for the encoder has the following form (adapted from the trellis in Figure 14-12). The branches are labeled with the input information bits (x_0, x_1) and the encoder output label (c_0, c_1, c_2).

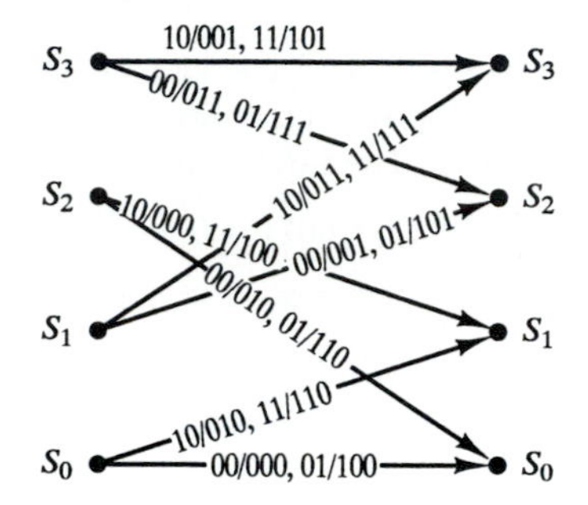

The following is a generic weighted, directed graph for a four-state encoder.

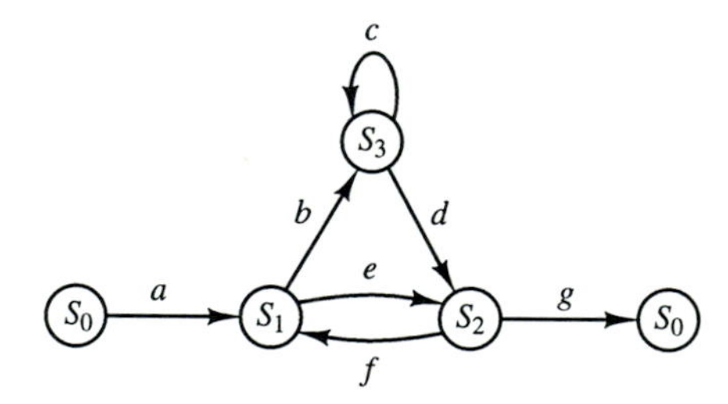

Using Mason's gain rule (Section 11.2.3) we obtain the following transfer function for the graph.

$$T = \frac{aeg(1 - c) + abdg}{(1 - c)(1 - ef) - bdf}$$

We now substitute the appropriate weight profiles for the graph labels,

$$a = W(010) + W(110) = 2D^2$$

$$b = W(011) + W(111) = D^{0.586} + D^{3.414}$$

$$c = W(001) + W(101) = D^{0.586} + D^{3.414}$$

$$d = W(011) + W(111) = D^{0.586} + D^{3.414}$$

$$e = W(001) + W(101) = D^{0.586} + D^{3.414}$$

$$f = W(000) + W(100) = 1 + D^4$$

$$g = W(010) + W(110) = 2D^2$$

and obtain the following generating function.

$$T(D) = \frac{4(D^{4.586} + D^{7.414})}{1 - 2D^{0.586} - 2D^{3.414} - D^{4.586} - D^{7.414}}$$

We see from the generating function that the squared minimum free distance for nonparallel paths is 4.586, and thus $d_{\text{free}_{\text{nonparallel}}} = 2.14$, as deduced earlier in the chapter. This generating function does not account for parallel paths, which have $d_{\text{free}_{\text{parallel}}} = 2.00$. The probability that a parallel path causes a node error is derived using Eq. (14-13). Combining this result with Eq. (14-18), the upper bound on node error rate is found to be

$$P_e \leq \underbrace{Q\left(\sqrt{\frac{2E_s}{N_0}}\right)}_{\text{parallel paths}} + \underbrace{Q\left(\sqrt{\frac{d_{\text{free}}^2 E_s}{2N_0}}\right) \cdot \exp\left(\frac{d_{\text{free}}^2 E_s}{4N_0}\right) \cdot T(D)\big|_{D=\exp(-E_s/4N_0)}}_{\text{nonparallel paths}} \quad \blacksquare$$

An augmented generating function $T(D,I)$ that accounts for information-bit errors can be constructed by modifying the approach discussed above. The resulting function has the form

$$T(D,I) = \sum_{L=1}^{\infty} \sum_{\mathbf{C}} P(\mathbf{C}) \sum_{\mathbf{E}_L} I^{b_{\mathbf{C},\mathbf{E}_L}} D^{\|f(\mathbf{C}) - f(\mathbf{C} \oplus \mathbf{E}_L)\|^2} \tag{14-29}$$

where $b_{\mathbf{C},\mathbf{E}_L}$ is the number of information bit errors caused by selecting the incorrect signal sequence $f(\mathbf{C} \oplus \mathbf{E}_L)$ instead of the correct sequence $f(\mathbf{C})$. For a nonuniform code, we modify each element of the matrices $\{\mathbf{G}(\mathbf{e}_i)\}$ as follows.

$$[\mathbf{G}(\mathbf{e}_i)]_{p,q} = \frac{1}{2^m} \sum_{\mathbf{c}_{p\to q}} I^{b_{\mathbf{c}_{p\to q},\mathbf{e}_i}} D^{\|f(\mathbf{c}_{p\to q}) - f(\mathbf{c}_{p\to q} \oplus \mathbf{e}_i)\|^2} \tag{14-30}$$

For uniform codes, the augmented weight profiles are as follows.

$$W(\mathbf{e}_i) = \frac{1}{2^\nu} \sum_{\mathbf{c} \in \mathbf{C}_0} I^{b_{\mathbf{c},\mathbf{e}_i}} D^{\|f(\mathbf{c}) - f(\mathbf{c} \oplus \mathbf{e}_i)\|^2} \tag{14-31}$$

Example 14-4—An Augmented Generating Function

In this example we derive an augmented generating function for the uniform, two-state encoder in Examples 14-1 and 14-2. The signal labels and the encoder trellis are repeated below.

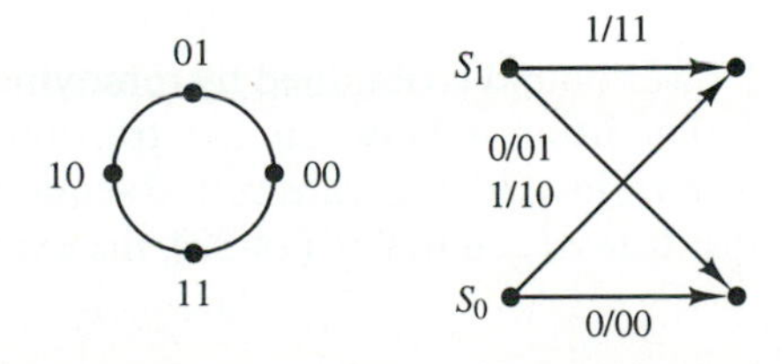

The augmented modified weight profiles are as follows.

$\mathbf{c}$	$\mathbf{c}\oplus 00$	$d^2[f(\mathbf{c}) - f(\mathbf{c}\oplus 00)]$	$\mathbf{c}$	$\mathbf{c}\oplus 01$	$d^2[f(\mathbf{c}) - f(\mathbf{c}\oplus 01)]$
00	00	0.000	00	01	2.000
10	10	0.000	10	11	2.000
$\Rightarrow W(00) = 1$			$\Rightarrow W(01) = D^2$		

$\mathbf{c}$	$\mathbf{c}\oplus 10$	$d^2[f(\mathbf{c}) - f(\mathbf{c}\oplus 10)]$	$\mathbf{c}$	$\mathbf{c}\oplus 11$	$d^2[f(\mathbf{c}) - f(\mathbf{c}\oplus 11)]$
00	10	4.000	00	11	2.000
10	00	4.000	10	01	2.000
$\Rightarrow W(10) = ID^4$			$\Rightarrow W(11) = ID^2$		

The error-state diagram now has the following form, and the augmented generating function follows immediately.

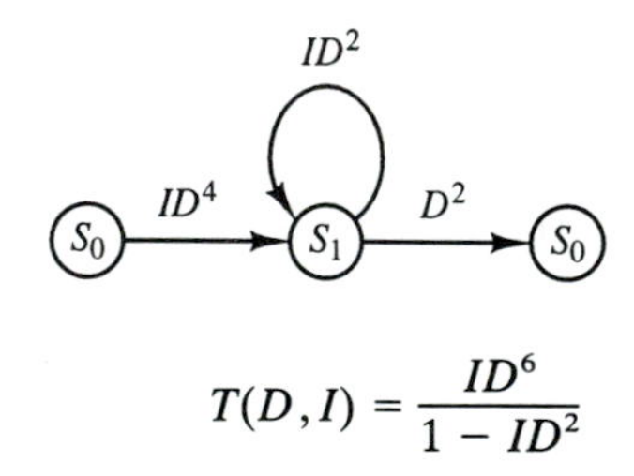

$$T(D,I) = \frac{ID^6}{1 - ID^2}$$ ■

14.4.4 A Lower Bound on Node Error Probability and Bit Error Rate for the AWGN Channel

The lower bound on node error probability is derived through the use of the "genie-assisted" receiver. The genie improves the receiver performance by offering it two signal sequences from which to choose: the transmitted sequence $f(\mathbf{C})$, and a sequence $f(\mathbf{C}')$ that is one of the sequences closest in Euclidean distance to the transmitted sequence. Such a receiver always performs at least as well as the real receiver, because the real receiver always has $f(\mathbf{C})$ and $f(\mathbf{C}')$ from which to choose, as well as all of the other valid sequences. The following lower bound on node error probability, conditioned on the transmission of $f(\mathbf{C})$, results.

$$P_{e|\mathbf{C}} \geq Q\left\{\frac{\|f(\mathbf{C}) - f(\mathbf{C}')\|}{\sqrt{2N_0/E_s}}\right\} \tag{14-32}$$

A general lower bound is obtained by removing the condition. Note, however, that in general the distance between the transmitted sequence and the closest incorrect sequence varies with the transmitted sequence; it is not always equal to d_{free}. If one simply substitutes d_{free} into Eq. (14-32), the expression will no longer be a lower bound. Biglieri et al. [Big1] take the following approach. Let $I(\mathbf{C})$ be an indicator function defined as follows.

$$I(\mathbf{C}) = \begin{cases} 1 & \text{if } \min_{\mathbf{C}'} \|f(\mathbf{C}) - f(\mathbf{C}')\| = d_{\text{free}} \\ 0 & \text{otherwise} \end{cases} \tag{14-33}$$

A general lower bound is then obtained as follows.

$$P_e \geq \sum_{\mathbf{C}} I(\mathbf{C})P(\mathbf{C})Q\left\{\sqrt{\frac{d_{\text{free}}^2 E_s}{2N_0}}\right\} = \Psi \cdot Q\left\{\sqrt{\frac{d_{\text{free}}^2 E_s}{2N_0}}\right\} \tag{14-34}$$

If the TCM system is uniform, then the distance structure of the code is independent of the transmitted sequence and $\Psi = 1$. Otherwise, a lower bound on Ψ is obtained as follows. Let Λ be the minimum length of the error sequence for a code that results in a signal sequence that is d_{free} from a transmitted sequence. Let $N_\Lambda(d_{\text{free}})$ be the number of signal sequences of length Λ that have a competing sequence at distance d_{free}. Since there are m information bits associated with each node in the trellis, it follows that there are $2^\nu \cdot 2^{m\Lambda}$ total paths of length Λ leaving the 2^ν states of the trellis at any given point in time. The ratio $N_\Lambda(d_{\text{free}})/2^\nu \cdot 2^{m\Lambda}$ is then the probability that a randomly selected trellis path has a path that diverges from it, accumulates a distance of d_{free}, and then remerges after exactly Λ branches. Given that $N_\Lambda(d_{\text{free}}) \geq 2$ (a competition at distance d_{free} requires two competitors), it follows that

$$\Psi \geq \frac{N_\Lambda(d_{\text{free}})}{2^\nu \cdot 2^{m\Lambda}} \geq \frac{1}{2^\nu \cdot 2^{m\Lambda - 1}} \tag{14-35}$$

Equation (14-34) can be modified to obtain a lower bound on bit error rate. We simply note that a node error results in a minimum of one information bit error, and that there are m information bits associated with each node.

$$P_b \geq \frac{\Psi}{m} \cdot Q\left\{\sqrt{\frac{d_{\text{free}}^2 E_s}{2N_0}}\right\} \tag{14-36}$$

Example 14-5—A Comparison of Upper and Lower Bounds

In this example we compare the upper and lower bounds for the node error probability for the TCM system in Example 14-3. The upper bound was found through a generating function analysis to be

$$P_e \leq Q\left(\sqrt{\frac{2E_s}{N_0}}\right) + Q\left(\sqrt{\frac{d_{\text{free}}^2 E_s}{2N_0}}\right) \cdot \exp\left(\frac{d_{\text{free}}^2 E_s}{4N_0}\right) \cdot T(D)\Big|_{D = \exp(-E_s/4N_0)}$$

Since this code is uniform, we have $\Psi = 1$, and the lower bound is obtained from Eq. (14-34). Note that the "genie" is offering a choice between the correct path and one that differs by only a single parallel transition.

$$P_e \geq Q\left\{\sqrt{\frac{2E_s}{N_0}}\right\}$$

The two bounds are graphed in the accompanying figure. Note that the bounds become quite tight with increasing signal-to-noise ratio. The performance curve for uncoded QPSK shows that, at high signal-to-noise ratios, the asymptotic coding gain of 3 dB is a good measure of the performance improvement. On the other hand, the performance improvement is substantially less than the asymptotic gain at low signal-to-noise ratios, for at that point the longer error paths are becoming significant contributors to the error probability.

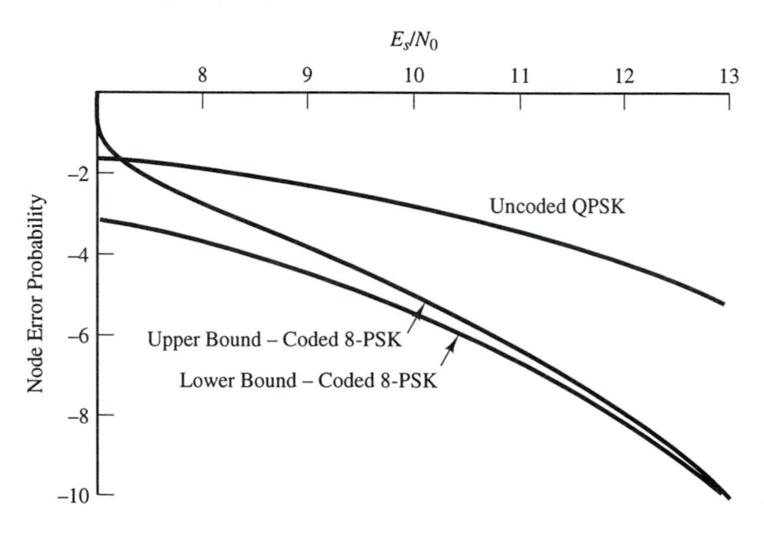

14.5 IMPLEMENTATION CONSIDERATIONS

In this section we consider two important considerations that may arise when TCM systems are designed and implemented. The question of **rotational invariance** arises when there is a possibility of significant phase offsets in the receiver. **Selective spectral efficiency** becomes an important issue when the condition of the communication channel for a given application changes significantly over time.

14.5.1 Rotational Invariance

In Section 11.4 we briefly discussed the problem of carrier phase offset in the receiver. The receiver generates an estimate of the carrier phase, then uses that estimate to demodulate the incoming signal. In many of the more powerful digital receivers, demodulator and decoder decisions are fed back to aid in the estimation of the carrier phase. In such systems an offset in the carrier phase estimate can lead to randomization of the estimate and a subsequent loss of receiver synchronization. Synchronization is only recovered when the random phase estimate results in a demodulated signal sequence that resembles a valid sequence. Since TCM systems use large signal constellations, the opportunities for invariance under various phase rotations would seem better than those for the binary system assumed in Section 11.4. Unfortunately this doesn't immediately follow without a great deal of effort.

The Ungerboeck encoder in Figure 14-10 uses an 8-PSK signal constellation in conjunction with a four-state convolutional code. Though 8-PSK is symmetric under any rotation that is an integer multiple of 45°, the signal sequences emerging from the encoder are not. A quick examination of the trellis in Figure 14-12 shows that the encoder is only invariant under 180° rotations. The phase offset estimate in the receiver will thus be in a random walk for carrier phase offsets between 22.5° and 127.5°; resynchronization can be expected to take some time [Ung4].

It is possible to achieve higher levels of rotational invariance through the use of higher-dimensional TCM systems. For example, several four- and eight-dimensional TCM systems based on two-dimensional MPSK have been developed that display the same rotational invariance as the MPSK constellations themselves. In these systems several consecutive two-dimensional MPSK signals are used to construct a single, higher-dimensional signal. Regrettably, space limitations preclude a treatment of multidimensional trellis codes in this text. The interested reader is referred to [Wei4] and [Pie] and the works referenced therein.

Though it is fairly difficult, a 90° rotational invariance can be achieved in TCM systems using two-dimensional, rectangular QAM. Wei has demonstrated the following conclusions concerning the design of such systems [Wei4].

- A nonlinear convolutional encoder must be used within the TCM encoder. This follows from the fact that a 90° rotation induces a nonlinear transformation on the signal labels.
- Rotationally invariant TCM systems may not provide the same coding gain or have the same complexity as comparable noninvariant systems.

There are, however, a few examples of rotationally invariant, two-dimensional systems whose performance and complexity match that of the best-known noninvariant codes. One such scheme, using 32-CROSS and an eight-state, nonlinear convolutional encoder, was found to have the same 4-dB coding gain as the best noninvariant code of the same complexity. This rotationally invariant system was later adopted in the CCITT V.32 Recommendation [IBM2]. V.32 is intended for 9.6 Kbps traffic over two-wire telephone lines and 14.4 Kbps traffic over four-wire circuits. The same nonlinear encoder was later adopted for use with 64-QAM and 128-CROSS in V.17 (14.4 Kbps FAX traffic over standard phone lines) and V.33 [McC], [Ung4].

Figures 14-20 and 14-21 show the encoder and 32-CROSS labeling recommended in V.32. Note that a differential encoder is placed at the input of the nonlinear convolutional encoder to allow for the resolution of phase ambiguities in the demodulated data at the receiver (see Section 11.4). The 32-CROSS constellation is shown here in its rotated form. The partitioning is the same as that shown in Figure 14-6. For example, the points surrounded by a square are the partition corresponding to a convolutional encoder output $c_0 c_1 c_2 = 111$. This is a rotated version of the partition D7 in Figure 14-6.

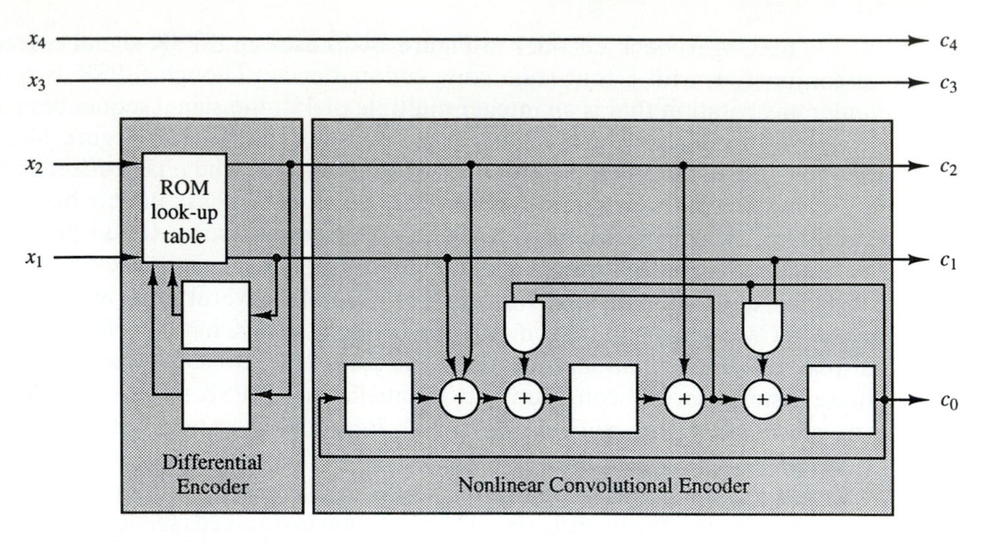

Figure 14-20. Encoder for the 90° Phase-Invariant TCM System Adopted for CCITT V.32

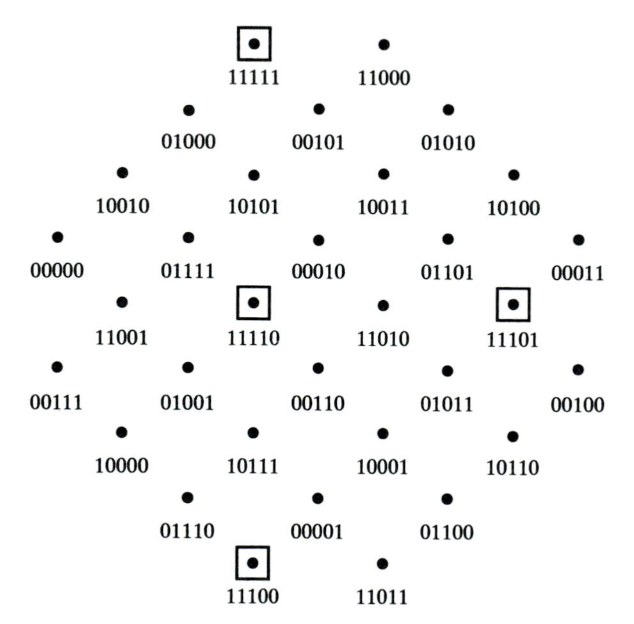

Figure 14-21. 32-CROSS Labeling ($c_0\, c_1\, c_2\, c_3\, c_4$) for V.32

14.5.2 TCM Systems with Adjustable Spectral Efficiency

The vast majority of radio channels experience some form of variation in noise and/or attenuation over time. In some cases there is a very rapid fluctuation in channel conditions (e.g., fading mobile communication channels), while in other cases the changes occur very slowly (e.g., geosynchronous satellite channels). It is frequently useful to design a communication system that adapts to the changing conditions, thereby always maximizing information throughput given some fixed reliability constraint.

TCM systems allow for reliable communication at high spectral efficiencies, but as channel conditions change, it may prove necessary to vary the number of bits/sec/Hz being transmitted to maintain reliability. Viterbi et al. [Vit4] have demonstrated that TCM systems with adjustable spectral efficiency can be readily constructed. Using a single convolutional code in conjunction with 4-, 8-, and 16-PSK, they designed an adaptive TCM system that provides 1, 2, and 3 bits/sec/Hz with asymptotic coding gains of 7.0 dB, 3.0 dB, and 5.3 dB, respectively. Though the convolutional code that they chose (the rate-1/2, constraint-length-7 code in Table 11-5) leads to an 8-PSK system that is demonstrably suboptimal to the corresponding system in Table 14-3, the efficiency of the adaptive system more than makes up for the loss. Figure 14-22 shows the "pragmatic" encoder proposed in [Vit4]. Note that Gray coding has been used to label the signal sets instead of the "natural" ordering used earlier in this chapter. The Gray code allows one to interpret the output of the convolutional encoder as selecting a phase within a sector. The Gray code mapping is as follows for 2^m-ary PSK.

$$00 \rightarrow 0 \text{ radians}$$

$$01 \rightarrow \pi/2^{m-1} \text{ radians}$$

$$11 \rightarrow 2\pi/2^{m-1} \text{ radians}$$

$$10 \rightarrow 3\pi/2^{m-1} \text{ radians}$$

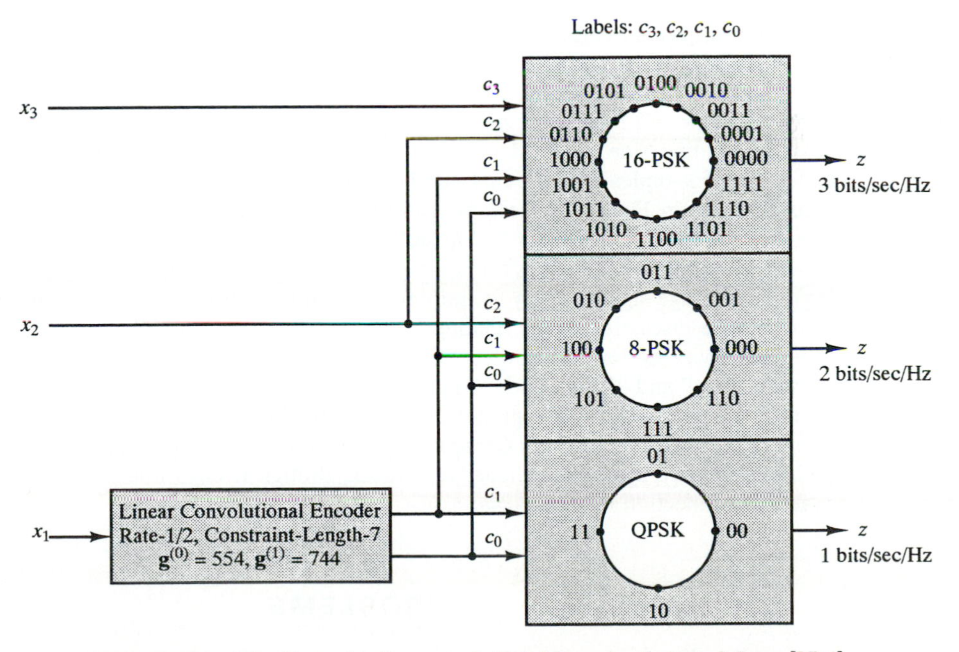

Figure 14-22. Viterbi et al.'s Pragmentic TCM Encoder (adapted from [Vit4], © 1989 IEEE)

Each 2^m-ary constellation is thus composed of 2^{m-2} sectors, and the Gray code maps the convolutional encoder output onto a phase in each sector. The constellation sector is selected by the $m - 2$ uncoded bits in lexicographic order.

SUGGESTIONS FOR FURTHER READING

Given the period of time that TCM has been with us, the quantity and depth of the literature on TCM is immense. In this chapter we have been able to introduce only the basic principles of design and analysis for Ungerboeck encoders, and then briefly to mention a few of the advanced topics that have had an obvious practical impact. The material that has been left out could fill several books. In fact, some of it has already filled one book. *Introduction to Trellis Coded Modulation*, by Biglieri, Divsalar, McLane, and Simon [Big1] was the first book devoted entirely to TCM systems. It includes a great deal of information on active research topics and is a good starting place for the graduate student interested in working in this area.

Some of the system-oriented issues that we have failed to cover in this chapter include multidimensional TCM systems. [Wei4] and [Pie] provide the best starting point for an investigation of this area. We have noted that multidimensional systems provide a degree of rotational invariance that is difficult to achieve in one- and two-dimensional systems.

The performance of TCM systems over fading channels has also not been considered here. [Big1] is almost certainly the best reference for this area, as the authors of [Big1] have played a leading role in the corresponding research. [Div2], [Div3], and [Div4] provide more concise presentations of some of the fading material in [Big1].

Another modification that we have not considered is the design of TCM systems using convolutional encoders that do not have rates of the form $k/(k + 1)$. Though few practical implementations have used these systems, they are of great interest to the research community. The subject is treated thoroughly in [For9] and [Big1]. Some of this work makes use of lattice theory, which brings us to the final, but greatest omission in this chapter.

In keeping with a promise made in Chapter 1, it was decided to keep the mathematics in this chapter to a minimum. As a result this chapter completely ignores what may be the most fascinating side of TCM research: the application of lattice theory. A lattice is a collection of n-tuples over the real numbers that forms an additive group. The rectangular constellations in Figure 14-3 can be considered as having been taken from Z^2, the two-dimensional lattice consisting of all n-tuples with integer coordinates. The theory of lattices provides mathematical analogs for, among other things, signal partitioning (lattice partitioning) and average signal energy vs. minimum Euclidean distance (lattice density). The best sources for a discussion of the relationship between lattices and TCM systems are a pair of seminal papers by Forney entitled "Coset Codes—Part 1: Introduction and Geometrical Classification" and "Coset Codes—Part 2: Binary Lattices and Related Codes" ([For9] and [For10], respectively). A comprehensive, theoretical discussion of lattices can be found in *Sphere Packings, Lattices, and Groups*, by Conway and Sloane [Con1]. This book is encyclopedic in its coverage of the theory of lattices, though it does not (explicitly) explore the practical connection to trellis codes.

PROBLEMS

1. Construct a partition tree for the following signal constellations and compute the intersignal distances at each level of the partition tree.

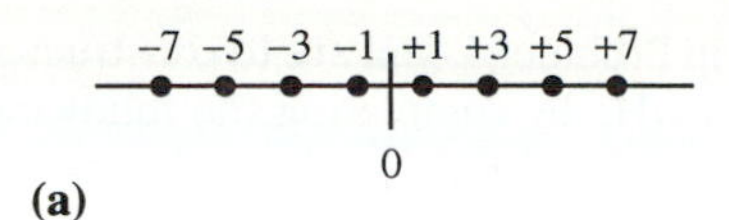

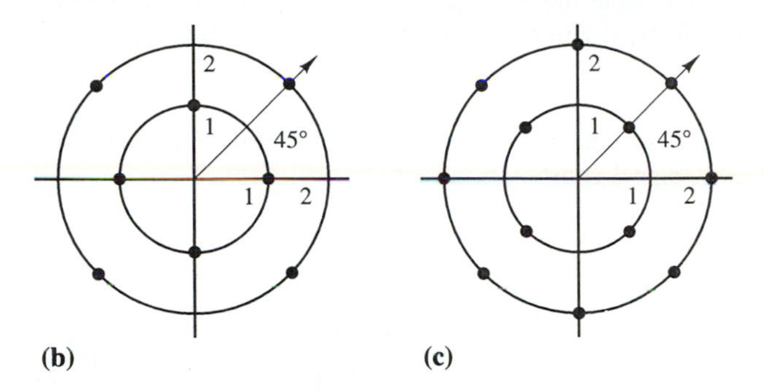

2. The signal constellation below was taken from a hexagonal lattice. The squared intersignal distances are as noted. Assuming that the constellation is centered on the axes so as to minimize the average signal energy, compute the average signal energy.

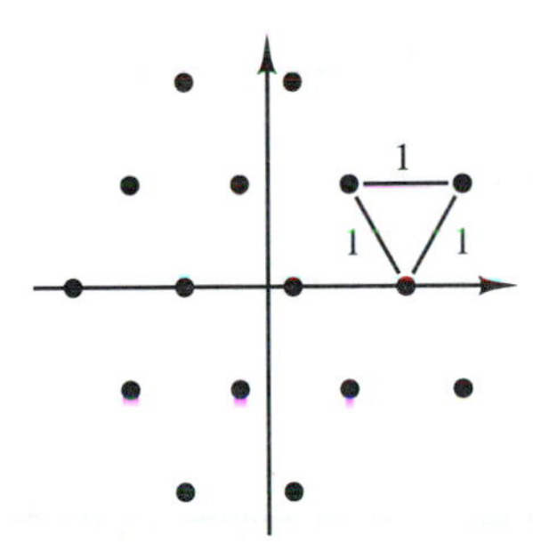

3. What advantages does the hexagonal constellation above have over a rectangular 16-QAM constellation with the same squared minimum distance? What disadvantages?
4. The following two-state encoder and 8-AM signal constellation is to be used to construct a TCM system that provides 2 bits/sec/Hz. Complete the steps listed below.

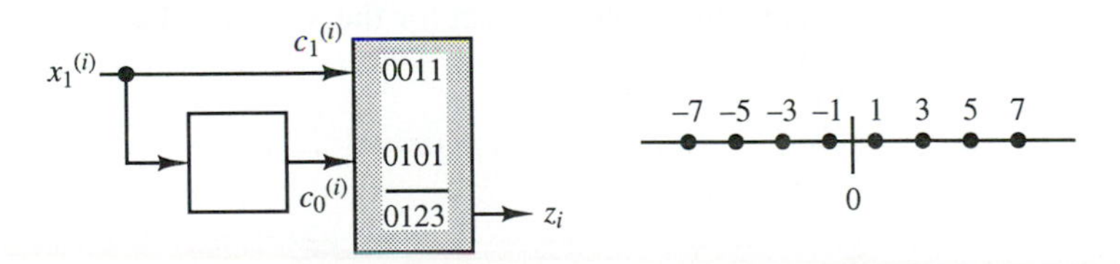

(a) Label the signals, and compute the squared intersignal distances and average signal energy.
(b) Determine the appropriate partitions for the signal constellation.
(c) Construct and label a trellis diagram for the system.
(d) Determine the squared minimum free distance and asymptotic coding gain for the system relative to an uncoded 4-AM system.

5. Using the two-state encoder in Problem 4, you are to construct a constant-envelope TCM system that provides 3 bits/sec/Hz by completing the following steps.
   (a) Select a signal constellation, label the signals, and compute the squared intersignal distances.
   (b) Determine the appropriate partitions for the signal constellation.
   (c) Construct and label a trellis diagram for the system.
   (d) Determine the squared minimum free distance and asymptotic coding gain for the system.

6. A TCM system is constructed using the following two-state encoder and asymmetric QPSK signal constellation [Sim].

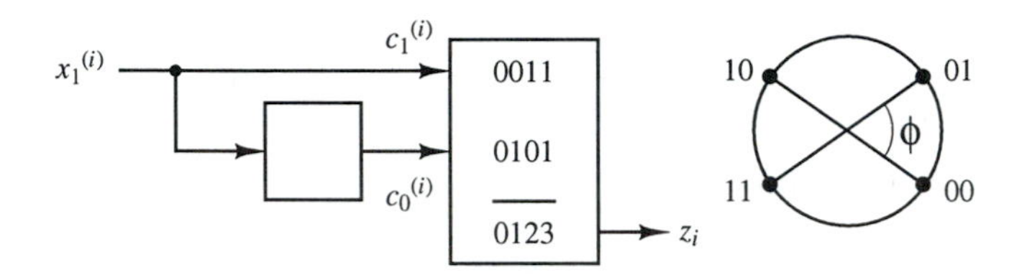

   (a) Determine the squared intersignal distances for the constellation.
   (b) Is this TCM system uniform? Why or why not?
   (c) Determine the squared minimum free distance of the system.
   (d) Determine the limit of the asymptotic coding gain as ϕ goes to π. Compare this result to the performance of the MTCM system with symmetric QPSK constellation discussed in the text.

7. Construct and label the trellis diagram for a two-state MTCM system using one-dimensional 4-AM. What is the asymptotic coding gain for this system?

8. Construct and label the trellis diagram for a two-state MTCM system using 8-PSK. What is the asymptotic coding gain for this system?

9. Construct and label the trellis diagram for a four-state MTCM system using 4-PSK and the convolutional encoder in Figure 14-10. What is the asymptotic coding gain for this system?

10. Construct and label the trellis diagram for the nonlinear convolutional encoder in Figure 14-20.

11. Construct the ROM-table for the differential encoder in Figure 14-20. (Section 11.4 may be of some use.)

12. Construct and label the trellis diagram for the V.32 TCM encoder described in Figures 14-20 and 14-21.

15

Error Control for Channels with Feedback

All of the error control systems discussed in the preceding fourteen chapters can be used on simplex channels (i.e., channels in which the information flow is in only one direction). Such error control systems are called **forward error correcting (FEC)** because of the one-way nature of the data traffic. If we allow for information flow in the other direction, however, a new world of design options becomes available. At its simplest, error control for channels with feedback takes the form of the pure **automatic-repeat-request (ARQ) protocol.**[1] In these protocols the transmitted data is encoded for error detection; detected errors at the receiver result in the generation of a retransmission request.

It is difficult to date the earliest introduction of ARQ protocols in digital communications. They were almost certainly an immediate consequence of the development of parity-check codes in the early days of digital computers. The next major development in the area came in 1960, when Wozencraft and Horstein [Woz3], [Woz4] described and analyzed a system that allowed for both error correction and error detection with retransmission requests. Their system, now known as a **type-I hybrid-ARQ protocol**, provided significantly improved performance over the pure ARQ protocols in many applications. Benice and Frey [Beni1] then systematized the three basic retransmission protocols (stop-and wait, go-back-N, and selective-repeat) and discussed their performance analysis in 1964.

In 1977 Sindhu [Sind] discussed a scheme that made use of the packets that cause retransmission requests (in the pure and type-I hybrid protocols they are simply discarded). Sindhu's idea was that such packets can be stored and later

[1] The "RQ" portion of "ARQ" comes from the Morse code designation for a repeat-request.

combined with additional copies of the packet, creating a single packet that is more reliable than any of its constituent packets. Since 1977 there have been innumerable systems proposed that involve some form of **packet combining**. These packet-combining systems can be loosely arranged into two categories: **code combining systems** and **diversity combining systems**.

In code combining systems the packets are concatenated to form noise-corrupted code words from increasingly longer and lower-rate codes. Code combining was first discussed in a 1985 paper by Chase [Cha], who coined the term. Subsequent code combining systems are exemplified by the work of Krishna, Morgera, and Odoul ([Kri], [Mor1], [Mor2]), Kallel and Haccoun [Kal], and Harvey and Wicker ([Har3], [Wic8]). One of the most popular code combining systems is the type-II hybrid-ARQ protocol invented by Lin and Yu [Lin2]. The type-II system allows for the combination of two packets, and is thus a **truncated** code combining system.

In diversity combining systems, the individual symbols from multiple, identical copies of a packet are combined to create a single packet with more reliable constituent symbols. Diversity combining systems are generally suboptimal with respect to code combining systems, but are simpler to implement. These schemes are represented by the work of Sindhu, Benelli ([Bene1], [Bene2]), Metzner ([Met1], [Met2]), and Wicker ([Wic6], [Wic9]).

This chapter focuses on the basic principles of retransmission request systems without any pretense of offering a comprehensive survey on the subject. It is hoped that this material will provide the interested reader with a good start in his or her exploration of the extensive literature on coding for channels with feedback.

We begin by considering the design and analysis of the pure ARQ protocols. At first we will assume that the feedback channel is noise free, but in the second section we eliminate this assumption. In the third section we examine the one-code and two-code type-I hybrid-ARQ protocols. A particular emphasis is placed on one-code hybrid-ARQ protocols that use Reed-Solomon codes. In the final section we take a quick look at type-II hybrid-ARQ protocols in particular and code combining systems in general. Example systems based on Reed-Solomon codes and convolutional codes are discussed.

15.1 PURE ARQ PROTOCOLS

In many applications the communication channel is error-free except for occasional bursts of noise of short duration. Such noise bursts may be caused, for example, by nearby power machinery, electrical storms, and single-event upsets in digital hardware. In such situations a simple error detecting ARQ protocol can provide a great deal of protection. The designer's objective is a system that will detect the error burst, discard the affected packet, and request a retransmission. The designer's hope is that the noise burst will have run its course before the retransmitted packet begins to make its way across the channel, and that multiple retransmission attempts will thus not be necessary.

In a pure ARQ protocol the message sequence is broken up into **packets** of length k. Each of these packets is encoded using a high-rate binary error detecting code **C** with length n. We assume here that the code **C** is linear with dimension k and is capable of detecting any error pattern that is not in itself a code word. There are thus $(2^n - 2^k)$ detectable error patterns.

15.1.1 Error Detection in Pure ARQ Protocols

The most frequently used error detecting codes are the CRC codes (see Section 5.3). CRC encoders and decoders are shift-register based and thus very easy to design and implement. They are also very flexible in the sense that a single encoder/decoder pair can be used with packets of widely varying length. On the other hand, the error detecting capabilities of most CRC codes are suboptimal. Kasami, Kløve, and Lin have examined the performance of several error detecting codes over a binary symmetric channel (BSC) with transition probability p. They first showed that there exist codes that satisfy the following upper bound on the undetected error probability P_e [Kas1].

$$P_e \leq 2^{-(n-k)}\{1 + (1-2p)^n - 2(1-p)^n\}, \qquad \text{for } 0 \leq p \leq \tfrac{1}{2} \tag{15-1}$$

They then identified four classes of codes that satisfy Eq. (15-1).

Class I: Even-weight linear codes of even length n and minimum distance $d_{\min}$ whose dual codes have minimum distance $d'_{\min}$ satisfying

$$d'_{\min} \geq \frac{n - \sqrt{(d_{\min}-1)(n-1)+1}}{2}.$$

This class includes extended Hamming codes and the extended binary Golay code.

Class II: Even-weight linear codes of odd length n and minimum distance $d_{\min}$ whose dual codes have minimum distance $d'_{\min}$ satisfying

$$d'_{\min} \geq \frac{n - \sqrt{n(d_{\min}-1)}}{2}$$

This class includes distance-6 primitive BCH codes of length $2^m - 1$ with $m \geq 5$ ($x + 1$ must be a factor of the generator polynomial).

Class III: Distance-8 extended primitive BCH codes of length $n = 2^m$ with m odd and $m \geq 5$.

Class IV: Distance-8 primitive BCH codes of length $n = 2^m - 1$ with m odd and $m \geq 5$.

Though these classes of codes do place constraints on packet length, their encoders and decoders have the same basic structure as those for the CRC codes.

15.1.2 The Retransmission Protocols and Their Performance

There are two basic parameters by which we can evaluate the performance of an ARQ protocol: reliability and throughput. In FEC systems we expressed reliability as bit or symbol error rate. In retransmission request systems it is often useful to express reliability in terms of $P(E)$, the accepted packet error rate.

Definition 15-1—Accepted Packet Error Rate

The accepted packet error rate $P(E)$ is the percentage of packets accepted by the receiver that contain one or more bit/symbol errors.

In some cases we may want to determine the bit error rate among accepted packets. The techniques discussed in Chapter 10 can be used to translate packet error rates into bit error rates.

The accepted packet error rate for a pure ARQ protocol can be computed quite easily. A packet is erroneously accepted if, on any transmission attempt, it arrives at the receiver containing an undetectable error pattern. Let the code **C** have probability P_e of undetected error. Let the probability of detected error (and thus the probability that a retransmission request is generated with each transmission) be P_r. $P(E)$ is computed by summing the probabilities of the various events that result in the acceptance of an erroneous packet. The packet can be accepted on the first transmission, the second (after a retransmission request), etc. We assume that the feedback channel is error-free, but we shall see in the next section that this does not affect the result of the analysis.

$$\begin{aligned} P(E) &= P_e + P_r P_e + P_r^2 P_e + P_r^3 P_e + \cdots + P_r^k P_e + \cdots \\ &= P_e \sum_{k=0}^{\infty} P_r^k = \frac{P_e}{1 - P_r} \end{aligned} \tag{15-2}$$

Note that since $P_r < 1$ in any functional system, we can assume that the series converges. Equation (15-2) shows that even though the probability of undetected error for the code remains fixed, the accepted word error rate increases with the probability of a retransmission request. This is because each additional retransmission provides the receiver with another opportunity to make a mistake.

The second performance measure, throughput, is defined as follows.

Definition 15-2—Throughput

The throughput (η) for an ARQ error control system is the average number of encoded data packets accepted by the receiver in the time it takes the transmitter to send a single k-bit data packet.

In the FEC systems the throughput is equal to the code rate $R = k/n$.

The throughput of a retransmission request protocol is a strong function of the number of times a packet has to be transmitted before it is accepted by the receiver.

Let T_r be the average number of times a packet must be transmitted before it is accepted. Given that we know the probability P_r that a retransmission request is generated, the derivation of T_r becomes an expected value problem.

$$\begin{aligned}
T_r &= (1 - P_r) + 2P_r(1 - P_r) + 3P_r^2(1 - P_r) + \cdots + kP_r^{k-1}(1 - P_r) + \cdots \\
&= (1 - P_r)\sum_{k=1}^{\infty} kP_r^{k-1} \\
&= (1 - P_r)\frac{\partial}{\partial P_r}\left(\sum_{k=0}^{\infty} P_r^k\right) \\
&= (1 - P_r)\frac{\partial}{\partial P_r}\left(\frac{1}{1 - P_r}\right) = \frac{1}{1 - P_r}
\end{aligned} \tag{15-3}$$

The throughput performance of a retransmission request system is also a strong function of how retransmission requests are handled by the transmitter and receiver. In selecting a retransmission protocol, the designer must strike a balance between the complexity of his or her design and the throughput performance of the resulting system. There are three basic retransmission protocols from which to choose: stop-and wait (SW-ARQ), go-back-*N* (GBN-ARQ), and selective repeat (SR-ARQ).

In SW-ARQ, the transmitter sends out a packet and waits for an acknowledgment. Once the receiver has processed the packet, it responds by sending an acknowledgment (ACK) if the packet was deemed error-free, or it sends a retransmission request (RQ) if the packet contained a detectable error pattern. This process is shown in Figure 15-1. Note that the transmitter is idle while waiting for the acknowledgment.

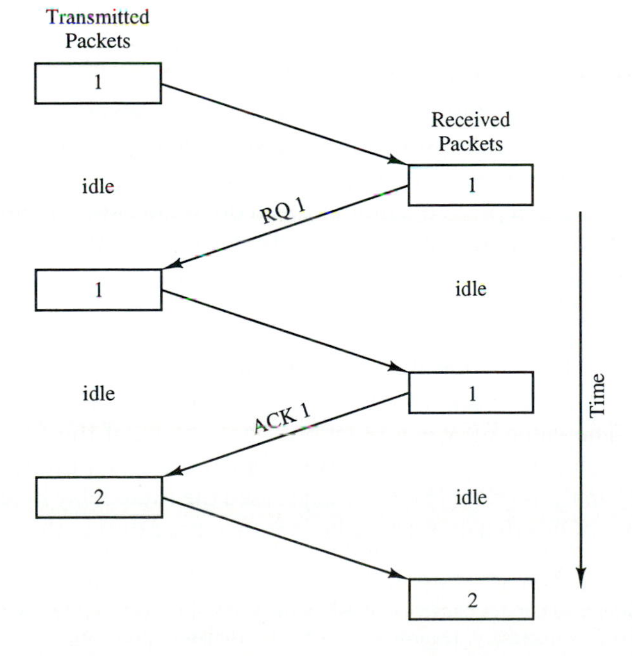

Figure 15-1. SW-ARQ Packet Traffic Diagram

This idle time is a function of the time required for the packet to reach the receiver (forward propagation delay = λ_1 seconds), the time required for the receiver to determine whether there are any errors in the received packet (receiver processing delay = δ seconds), and the time required for the receiver response (ACK or RQ) to reach the transmitter (feedback propagation delay = λ_2 seconds). If we assume that the bit transmission rate is D bits/second, we can express the idle time in terms of the number of bits Γ that could have been transmitted during the idle time.

$$\Gamma = D(\lambda_1 + \delta + \lambda_2) \quad \text{bits} \tag{15-4}$$

Suppose that T_r transmissions are necessary before the receiver accepts the transmitted packet. Each transmission involves the transmission of an n-bit packet followed by one idle period. The SW-ARQ system thus requires the transmission of $T_r(n + \Gamma)$ bits in order to move one k-bit packet of information from the transmitter to the receiver. The throughput for the SW-ARQ system is thus

$$\eta_{SW} = \frac{k}{T_r(n + \Gamma)} = R\left(\frac{1 - P_r}{1 + \Gamma/n}\right) \tag{15-5}$$

where R is the rate of the error detecting code used in the design.

The primary benefit of the SW-ARQ protocol is that there is no need for packet buffering at the transmitter or receiver.[2] In Figure 15-1 we see that packet #2 is not transmitted until packet #1 has been accepted by the receiver. The primary disadvantage of the SW-ARQ protocol is that it provides very poor throughput performance in applications in which Γ has any size whatsoever. SW-ARQ may thus prove useful in some computer applications (e.g., interprocessor transfer in multiprocessing systems), but most certainly not in satellite communications.

If we are willing to allow for some buffering in the transmitter, we can implement the go-back-N (GBN-ARQ) protocol shown in Figure 15-2. In this protocol the transmitter sends packets in a continuous stream. When the receiver detects an error in a received packet, it sends a retransmission request for that packet and waits for its second copy. All subsequent incoming packets are ignored until the second packet is received (it is assumed that packets cannot be sent on to the data sink unless they are in their proper order). By ignoring the packets that follow a retransmission request, receiver buffering is avoided. The transmitter, on the other hand, must respond by resending the requested packet *and all subsequent packets*. Buffering is thus necessary in the transmitter.

The term "go-back-N" derives from the fact that when a transmitter receives a retransmission request, it must go back into its buffer some N packets and restart transmission from there. The value of N is a function, once again, of the forward and feedback propagation delays (λ_1 and λ_2, respectively) and the receiver processing delay (δ). In the SW-ARQ analysis (Eq. (15-4)) we expressed the total delay in terms of Γ, the equivalent number of bits that could have been transmitted during this time.

[2] It is assumed that the source generates packets only when the transmitter is ready for them. In some applications source buffering is necessary, regardless of the retransmission protocol.

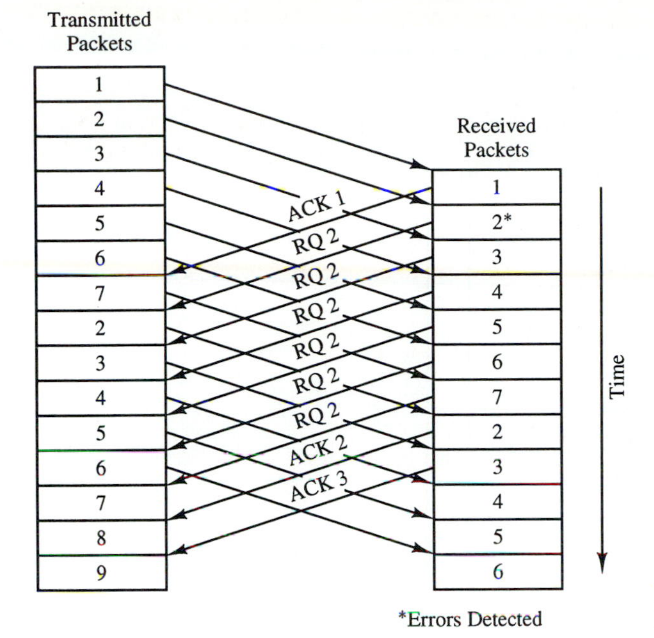

Figure 15-2. Go-Back-N Packet Traffic Diagram ($N = 6$)

N is simply the smallest integer number of packets that contain a total of at least Γ bits. Assuming a bit transmission rate D, we have

$$N = \left\lceil \frac{D(\lambda_1 + \delta + \lambda_2)}{n} \right\rceil = \left\lceil \frac{\Gamma}{n} \right\rceil \tag{15-6}$$

Every retransmission request causes the retransmission of a total of N packets: the requested packet and the $(N - 1)$ succeeding packets that are ignored by the receiver. The throughput for the GBN-ARQ protocol is thus

$$\eta_{GBN} = \left(\frac{k}{n}\right)\left(\frac{1}{1 + (T_r - 1)N}\right) = R\left(\frac{1 - P_r}{1 + P_r(N - 1)}\right) \tag{15-7}$$

Note that the parameter N in Eq. (15-7) is weighted by P_r (which we hope is small), whereas Γ/n in Eq. (15-5) has no such weighting. The GBN-ARQ protocol is thus not as sensitive to propagation delay as the SW-ARQ protocol. The reduced sensitivity is obtained at the expense of additional buffering in the transmitter.

If we allow for buffering in both the transmitter and the receiver, we can implement a selective-repeat (SR-ARQ) protocol, as shown in Figure 15-3.

Just as in the GBN-ARQ case, the transmitter sends a continuous stream of packets. In this case the transmitter responds to retransmission requests by sending the requested packet. It then returns to the point at which it stopped and resumes transmission of new packets. Each retransmission request results in the retransmission of only one packet. The throughput follows immediately.

$$\eta_{SR} = \left(\frac{k}{n}\right)\left(\frac{1}{T_r}\right) = R(1 - P_r) \tag{15-8}$$

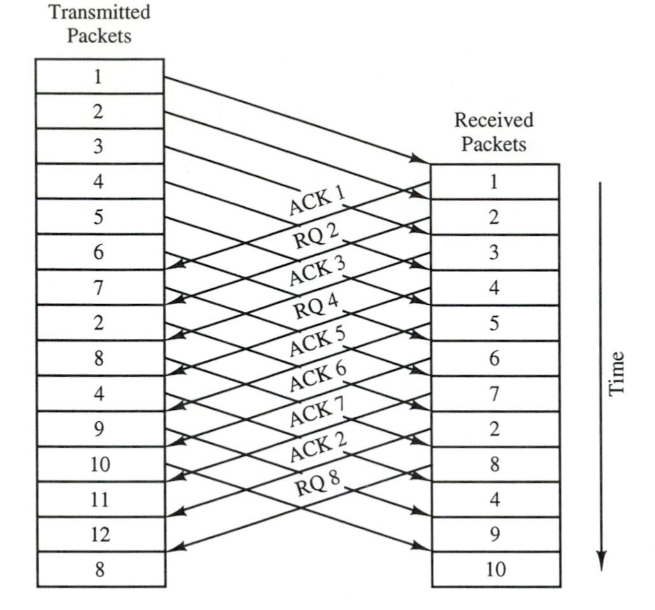

Figure 15-3. Selective-Repeat Packet Traffic Diagram

Example 15-1—A Satellite Channel

Suppose that a professor on sabbatical in Metz, France wants to communicate with a student in Atlanta, Georgia, by way of a 64-Kbps transponder in a satellite in geosynchronous orbit somewhere over the Atlantic. The round-trip propagation delay is approximately $\lambda_1 + \lambda_2 = 750$ ms. We assume (without comment) that the student's processing delay is negligible. The communication channel can be assumed to be binary symmetric with transition probability p. The (24, 12) extended Golay code, $\mathcal{G}_{24}$, is selected for error detection. $\mathcal{G}_{24}$ is in the first class of good error detecting codes discussed at the beginning of the chapter. It has the following weight distribution.

WEIGHT DISTRIBUTION FOR $\mathcal{G}_{24}$, (24, 12, 8)

i:	0	8	12	16	24
A_i:	1	759	2576	759	1

The probability of undetected error is thus (using Eq. (10-5))

$$P_e = 759p^8(1-p)^{16} + 2576p^{12}(1-p)^{12} + 759p^{16}(1-p)^8 + p^{24}$$

The probability of retransmission is obtained by subtracting P_e from the probability that a received word contains one or more bit errors.

$$P_r = 1 - (1-p)^{24} - P_e$$

The parameters Γ and N are computed as follows.

$$\Gamma = D(\lambda_1 + \delta + \lambda_2) = (64 \times 10^3 \text{ bits/s})(0.750 \text{ s}) = 48{,}000 \text{ bits}$$

$$N = \left\lceil \frac{48{,}000 \text{ bits}}{24 \text{ bits/packet}} \right\rceil = 2000 \text{ packets}$$

The throughput performance of the three retransmission protocols as a function of the channel transition probability p is shown in the accompanying figure.

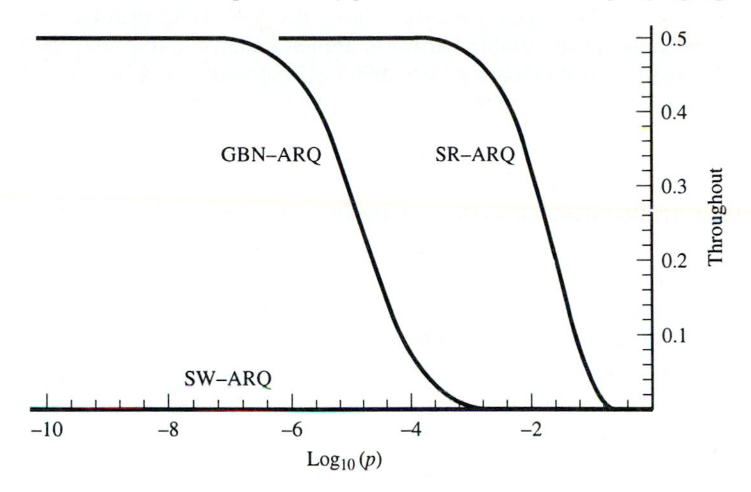

In this example we see that a substantial throughput penalty is paid between $p = 10^{-2}$ and 10^{-4} for the reduced complexity of the GBN-ARQ protocol. At lower values of p, however, SR-ARQ offers no improvement in performance. The SW-ARQ protocol is of no use at all crossover probabilities (the SW-ARQ performance curve is virtually coincident with the horizontal axis). This is because of the overwhelming amount of idle time associated with each transmission attempt. ■

Example 15-2—Terrestrial Line-of-Sight Communication

Now consider a 1-Mbps line-of-sight microwave link established between a pair of antennae on two mountain tops in the Alps. We assume a 25,000-foot round trip. Given that the speed of light in free space is 186,282 miles per second, we obtain the parameters $\Gamma = 26.8$ and $N = 2$. The extended binary Golay code is used once again for error detection. The throughput performance of the three protocols is shown in the accompanying graph.

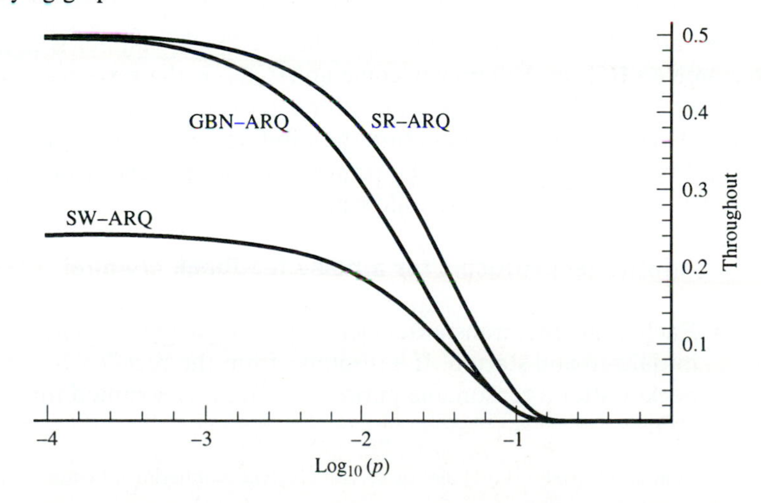

In this case we see that the added buffering in the SR-ARQ system does not provide a significant advantage over the GBN-ARQ system ($N = 2$). At lower values of p (higher signal-to-noise ratios), the SW-ARQ protocol may be the best choice because of the simple hardware implementation. Note that the SW-ARQ throughput performance reaches a throughput ceiling that is well below the code rate of 1/2. This ceiling is caused by the transmitter idle time during successful attempts. The ceiling can be raised by increasing the length of the packets (see Eq. (15-5)). ■

One of the primary design issues surrounding the SR-ARQ protocol is the size of the transmitter and receiver buffers. There are two basic approaches. The first approach is to select a buffer that is so large that the probability of overflow is negligible. The problems with this approach are obvious: memory is expensive and it consumes power. The second approach involves switching to a backup protocol when buffer overflow is imminent. Miller and Lin [Mil] investigated a scheme that uses GBN-ARQ as a backup protocol for SR-ARQ. In their (SR + GBN)-ARQ protocol, the transmitter automatically switches to GBN whenever the νth retransmission request is received for a given packet. The transmitter switches back to an SR-ARQ protocol once the troublesome packet is acknowledged. This scheme provides better performance than the standard GBN-ARQ protocol, but also limits the necessary size of the receiver buffer to $\nu(N - 1) + 1$ code words. The throughput is as follows [Mil].

$$\eta_{SR + GBN} = \left(\frac{k}{n}\right)\left[\frac{1 - P_r}{1 + P_r^{\nu+1}(N - 1)}\right] \tag{15-9}$$

Miller and Lin also examined a selective-repeat plus stutter scheme (SR-ST-ARQ) in which, after receiving the νth retransmission for a given packet, the transmitter sends copy after copy of the packet until it is accepted. This scheme is simpler than the (SR + GBN)-ARQ scheme, but less efficient [Mil].

15.2 NOISY FEEDBACK CHANNELS

There are additional concerns when the feedback channel is noisy: an ACK may become an RQ, an RQ may become an ACK, or the response may never reach its destination at all. Additional elements need to be integrated into the retransmission request system to handle these three possibilities. We begin by noting that all packets should be numbered so that the transmitter and receiver can keep track of them.[3] We can then implement the following.

Additional protocols for a noisy feedback channel

- Each time the transmitter sends out a packet, a timer for that packet is initialized and started. If a response from the receiver is not obtained for that packet after a reasonable period of time, it is assumed that the response is an RQ.

[3] Lin and Costello [Lin1] discuss several efficient numbering schemes.

- When the receiver sends an RQ back to the transmitter, the receiver initializes and starts a timer. If a new copy of the packet is not received after a reasonable period of time, the RQ is sent again.
- If the receiver receives a packet that has already been accepted, an ACK is sent to the transmitter and the packet is discarded.

Figure 15-4 shows the various things that can happen to a packet from the time the data is ready for encoding until it is purged from both the transmitter and receiver buffers. Once the data is ready for transmission, we enter the packet initiation state (state PI), in which the data is encoded. The packet is then transmitted to the receiver, which proceeds to check for errors (state ED). The decoder will either request a retransmission (transition to state RQ) or accept the packet (transition to state PA). There are self-loops about the states RQ and PA that allow for the conversion of RQs to ACKs and ACKs to RQs, respectively. Once the packet is accepted and the transmitter has received the acknowledgment, the process stops, and all copies of the packet are deleted from all system buffers.

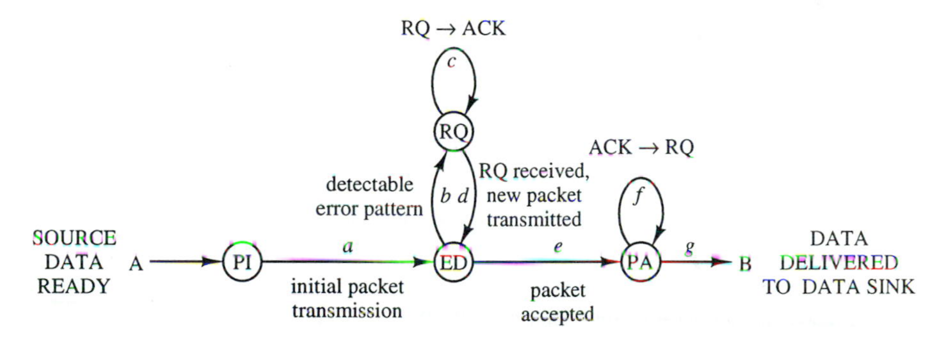

Figure 15-4. State Diagram for a Retransmission Request System with a Noisy Feedback Channel

In Section 11.2.3 we discussed Mason's formula for finding the generating function for a signal flow graph. The same techniques can be applied here by associating each branch of the diagram in Figure 15-4 with the appropriate expression. The generating function for the graph is obtained using the branch labels a, b, c, d, e, f, g.

FORWARD LOOPS (A→B)

K_1: A→PI→ED→PA→B $\quad F_1 = aeg$

LOOPS

L_1: ED→RQ→ED $\quad C_1 = bd$

L_2: RQ→RQ $\quad C_2 = c$

L_3: PA→PA $\quad C_3 = f$

NONTOUCHING PAIRS OF LOOPS

(L_1, L_3) $\qquad C_1 C_3 = bdf$

(L_2, L_3) $\qquad C_2 C_3 = cf$

There are no sets of three or more nontouching loops. The determinant is as follows.

$$\begin{aligned}\Delta &= 1 - \sum_{L_l} C_l + \sum_{(L_l, L_m)} C_l C_m - \sum_{(L_l, L_m, L_n)} C_l C_m C_n + \cdots \\ &= 1 - (C_1 + C_2 + C_3) + (C_1 C_3 + C_2 C_3) \\ &= 1 - bd - c - f + bdf + cf \\ &= (1 - bd - c)(1 - f)\end{aligned}$$

The single cofactor

$$\begin{aligned}\Delta_1 &= 1 - \sum_{(K_1, L_l)} C_l + \sum_{(K_1, L_l, L_m)} C_l C_m - \sum_{(K_1, L_l, L_m, L_n)} C_l C_m C_n + \cdots \\ &= 1 - C_2 = 1 - c\end{aligned}$$

The transfer function of the graph is then

$$T_{A \to B} = \frac{\sum_l F_l \Delta_l}{\Delta} = \frac{F_1 \Delta_1}{\Delta} = \frac{aeg(1 - c)}{(1 - c - bd)(1 - f)} \tag{15-10}$$

When we substitute the appropriate expressions for the labels in the graph in Figure 15-4, Eq. (15-10) provides reliability and throughput information. Figure 15-5 shows the labeling necessary for the computation of the probability of accepted packet error $P(E)$. The various labels are defined as follows.

P_r = the probability that a received packet contains a detectable error pattern and thus causes the generation of a retransmission request

P_e = the probability that a received packet contains an undetectable error pattern

P_c = the probability that a received packet is error-free (Note that $P_r + P_e + P_c = 1$)

$P_{RQ \to ACK}$ = the probability that noise on the feedback channel causes a retransmission request to become an acknowledgment

$P_{ACK \to RQ}$ = the probability that noise on the feedback channel causes an acknowledgment to become a retransmission request

Substituting into Eq. (15-10), we obtain the following interesting result.

$$P(E) = \frac{P_e(1 - P_{ACK \to RQ})(1 - P_{RQ \to ACK})}{[1 - P_{RQ \to ACK} - P_r(1 - P_{RQ \to ACK})](1 - P_{ACK \to RQ})} = \frac{P_e}{(1 - P_r)} \tag{15-11}$$

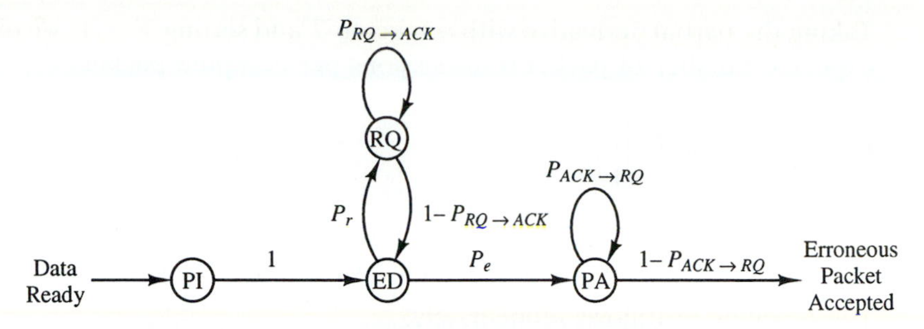

Figure 15-5. Graph Used for Determining Accepted Packet Error Rate

The accepted packet error rate is independent of the quality of the feedback channel! Unfortunately this is not the case for the throughput performance. Figure 15-6 shows the graph used to compute the throughput performance for the SW-ARQ protocol.

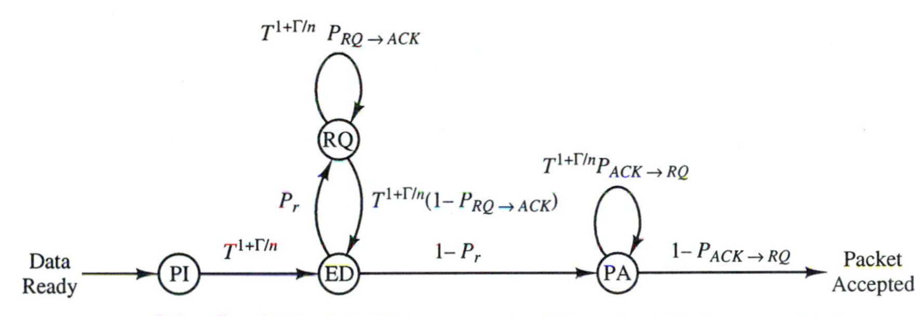

Figure 15-6. Graph Used for Determining the Throughput Performance for the SW-ARQ Protocol

In this graph we are using the same technique we used in Chapter 12 when computing a bound on the bit error rate for the Viterbi decoder. The variable T is simply a placeholder. The exponent for T is the number of n-bit packet transmissions associated with the state transition. For example, consider the RQ-RQ loop. In this case the receiver's request for the retransmission of a packet (say packet #1) has been converted into an ACK by noise on the feedback channel. The transmitter proceeds to send packet #2, which the receiver will ignore (it is waiting for packet #1 and will accept no other). The transmission of packet #2 and the subsequent idle time account for a time equivalent to that required for the transmission of $(1 + \Gamma/n)$ packets.

Substituting into Eq. (15-10), we obtain the following generator function.

$$G_{SW}(T) = \frac{T^{1+\Gamma/n}(1 - P_r)(1 - P_{ACK \to RQ})(1 - T^{1+\Gamma/n} P_{RQ \to ACK})}{[1 - T^{1+\Gamma/n} P_{RQ \to ACK} - T^{1+\Gamma/n} P_r(1 - P_{RQ \to ACK})](1 - T^{1+\Gamma/n} P_{ACK \to RQ})} \quad (15\text{-}12)$$

Taking the partial derivative with respect to T and setting $T = 1$, we obtain E_{SW}, the expected number of packet transmissions per accepted packet.

$$E_{SW} = \left[\frac{\partial}{\partial T} G_{SW}(T)\right]\Bigg|_{T=1}$$
$$= \frac{(1 + \Gamma/n)(1 - P_r P_{ACK\to RQ} - P_{RQ\to ACK} + P_r P_{RQ\to ACK})}{(1 - P_r)(1 - P_{ACK\to RQ})(1 - P_{RQ\to ACK})} \tag{15-13}$$

The throughput follows immediately.

$$\eta_{SW} = \left(\frac{k}{n}\right)\left(\frac{1}{E_{SW}}\right)$$
$$= R\left[\frac{(1 - P_r)(1 - P_{ACK\to RQ})(1 - P_{RQ\to ACK})}{(1 + \Gamma/n)(1 - P_r P_{ACK\to RQ} - P_{RQ\to ACK} + P_r P_{RQ\to ACK})}\right] \tag{15-14}$$

Note that Eq. (15-14) reduces to the noise-free feedback case in Eq. (15-5) when we set $P_{ACK\to RQ} = P_{RQ\to ACK} = 0$.

The impact of the noisy feedback channel is more easily seen if we assume that $P_{ACK\to RQ} = P_{RQ\to ACK}$. In this case Eq. (15-14) reduces to

$$\eta_{SW} = R\left[\frac{(1 - P_r)(1 - P_{ACK\to RQ})}{(1 + \Gamma/n)}\right] = (1 - P_{ACK\to RQ})\eta_{SW}^{\text{NOISE-FREE FEEDBACK}}$$

We now proceed to the GBN-ARQ protocol with a noisy feedback channel. Figure 15-7 contains the appropriate graph.

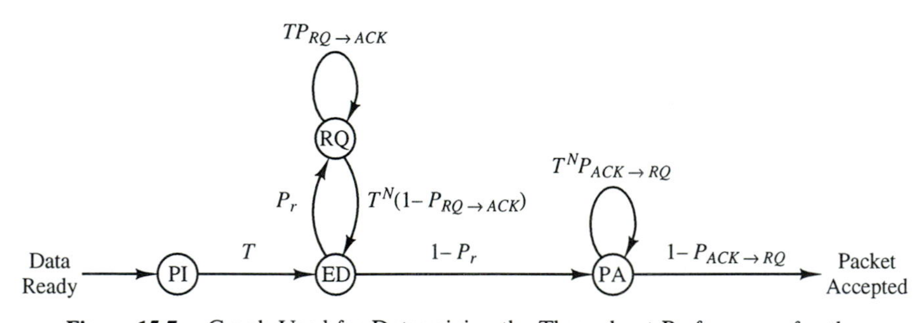

Figure 15-7. Graph Used for Determining the Throughput Performance for the GBN-ARQ Protocol

In this case the conversion of an RQ to an ACK results in the transmission of one additional packet, which is subsequently ignored by the receiver. The conversion of an ACK to an RQ, however, causes the transmitter to retransmit N packets that were not needed by the receiver.

Substituting into Eq. (15-10), we obtain the following generator function.

$$G_{GBN}(T) = \frac{T(1 - P_r)(1 - P_{ACK\to RQ})(1 - TP_{RQ\to ACK})}{[1 - TP_{RQ\to ACK} - T^N P_r(1 - P_{RQ\to ACK})](1 - T^N P_{ACK\to RQ})} \tag{15-15}$$

Taking the partial derivative with respect to T and setting $T = 1$ provides E_{GBN}, the expected number of packet transmissions required for each accepted packet.

$$E_{GBN} = \left[\frac{\partial}{\partial T} G_{GBN}(T)\right]\Bigg|_{T=1}$$

$$= \frac{\left[\begin{array}{c} 1 - P_{RQ\to ACK} - P_r P_{ACK\to RQ} + P_r P_{RQ\to ACK} \\ + (N-1)(P_{ACK\to RQ} + P_r - P_{ACK\to RQ} P_{RQ\to ACK} \\ - P_r P_{RQ\to ACK} - 2P_r P_{ACK\to RQ} + 2P_r P_{ACK\to RQ} P_{RQ\to ACK}) \end{array}\right]}{(1 - P_{ACK\to RQ})(1 - P_r)(1 - P_{RQ\to ACK})} \tag{15-16}$$

This somewhat unwieldy expression can then be translated into an expression for throughput.

$$\eta_{GBN} = \left(\frac{k}{n}\right)\left(\frac{1}{E_{GBN}}\right) \tag{15-17}$$

The reader may want to verify that Eq. (15-17) reduces to Eq. (15-7) when $P_{ACK\to RQ} = P_{RQ\to ACK} = 0$.

Finally we turn to the SR-ARQ protocol with noisy feedback. Figure 15-8 displays the relevant graph.

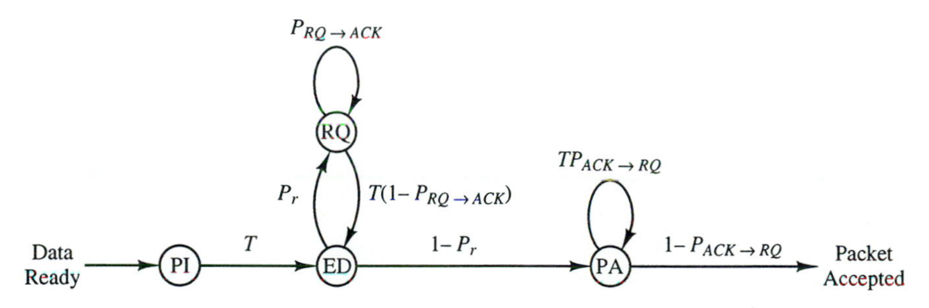

Figure 15-8. Graph Used for Determining the Throughput Performance of the SR-ARQ Protocol

In this case the only time that noise on the feedback causes the transmission of additional packets is when an acknowledgment is converted into a retransmission request (PA$\to$PA). Substituting into Eq. (15-10), we obtain the following generator function, which is then used to determine E_{SR}.

$$G_{SR}(T) = \frac{T(1 - P_r)(1 - P_{ACK\to RQ})(1 - P_{RQ\to ACK})}{(1 - P_{RQ\to ACK} - TP_r(1 - P_{RQ\to ACK}))(1 - TP_{ACK\to RQ})} \tag{15-18}$$

$$E_{SR} = \frac{1 - P_r P_{ACK\to RQ}}{(1 - P_r)(1 - P_{ACK\to RQ})} \tag{15-19}$$

The throughput follows immediately.

$$\eta_{SR} = \left(\frac{k}{n}\right)\left(\frac{1}{E_{SR}}\right) = R\frac{(1-P_r)(1-P_{ACK\to RQ})}{1-P_r P_{ACK\to RQ}} \tag{15-20}$$

It can be verified by observation that Eq. (15-20) reduces to Eq. (15-8) when $P_{ACK\to RQ} = P_{RQ\to ACK} = 0$.

Example 15-3—Noisy Feedback Channels

In Example 15-1 we considered the use of the three basic retransmission protocols on a satellite channel. At the time we assumed that the feedback channel was noise-free. We now reconsider the SR-ARQ protocol in this application with the addition of a noisy feedback channel. The curves in the accompanying drawings show the throughput as a function of the probability that an acknowledgment becomes a retransmission request. The individual curves are labeled with the forward channel transition probability. Note that the feedback channel has to be extremely noisy before the throughput of the SR-ARQ protocol is affected.

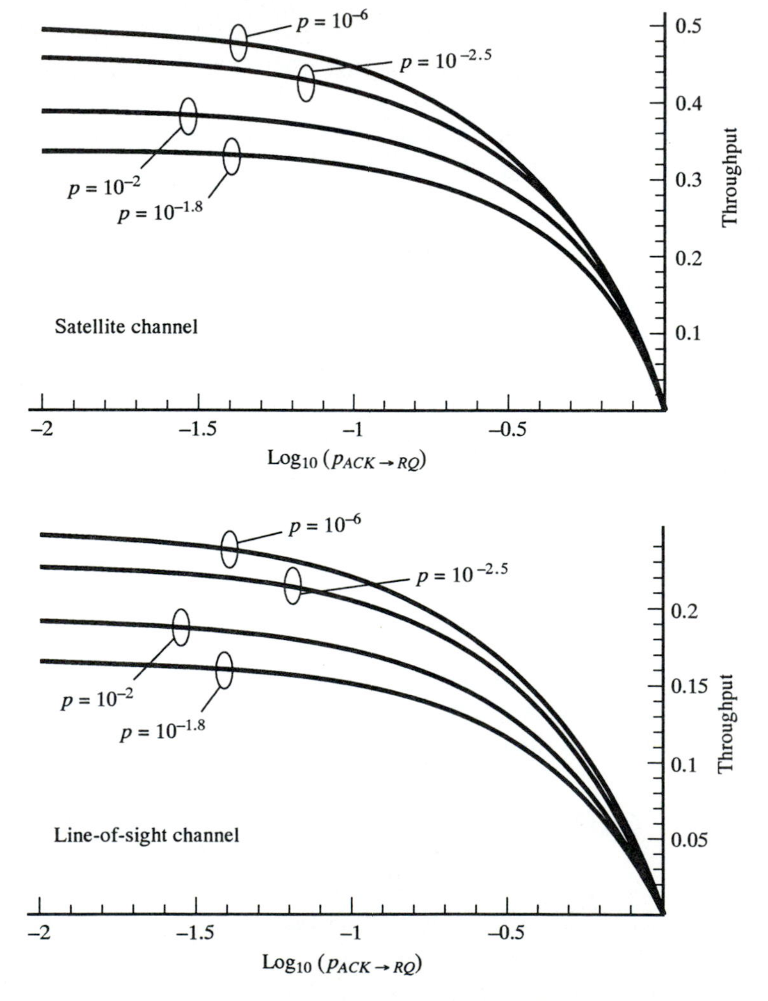

In Example 15-2 we examined a line-of-sight terrestrial link. If we allow for noise on the feedback channel, the SW-ARQ protocol provides the following performance. It is assumed that $P_{ACK \to RQ} = P_{RQ \to ACK}$. We see the exact same type of performance degradation that was shown in the case above. The feedback channel has to be quite noisy before performance is seriously affected. ■

15.3 TYPE-I HYBRID-ARQ PROTOCOLS

The various throughput graphs shown in the last section illustrate the primary problem associated with ARQ protocols: as the channel quality deteriorates, the increased frequency of retransmission requests has a severe impact on the throughput. The hybrid protocols counter this effect through the use of forward error correction in conjunction with error detection. The FEC portion of the protocol is designed to correct the error patterns most frequently caused by noise on the channel, while the error detection is used to detect the less frequently occurring patterns (which are generally those patterns most likely to cause decoder errors in FEC systems). The hybrid protocols can thus provide throughput similar to that of FEC systems, while offering reliability performance typical of ARQ protocols.

The type-I hybrid-ARQ protocol is the simplest of the hybrid protocols. Each packet is encoded for both error detection and error correction. These protocols can be implemented using either one-code or two-code systems. We first consider the two-code systems. The source is assumed to generate data packets of some fixed length k. The data is first encoded using a high-rate (k', k) error detecting code $\mathbf{C}_1$. CRC codes are frequently used for $\mathbf{C}_1$. The encoded data is then encoded once again using an (n, k') FEC code $\mathbf{C}_2$. When the packet arrives at the receiver, it is first decoded using the FEC decoder, as shown in Figure 15-9. The resulting k'-bit "message" is then sent to the error detecting decoder. If errors are detected, a retransmission request is sent back to the transmitter. Otherwise, the packet is accepted and the k-bit data packet passed along to the data sink.

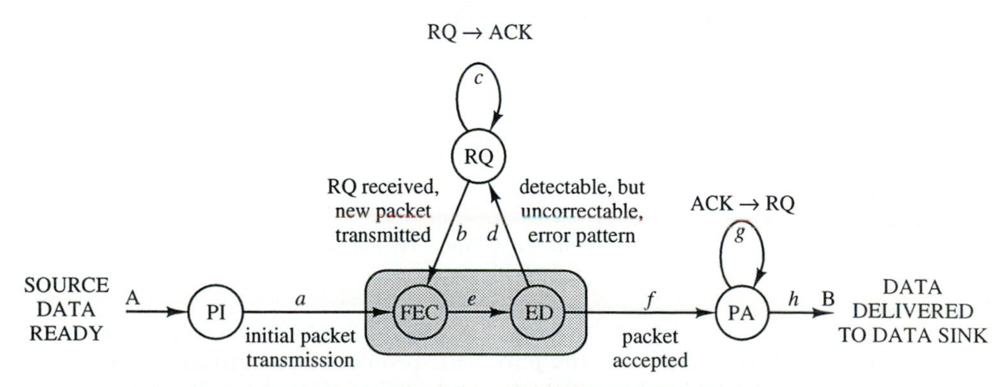

Figure 15-9. State Diagram for a Type-I Hybrid-ARQ Protocol Based on Two Codes

The transfer function for the graph in Figure 15-9 is a slightly modified version of Eq. (15-10).

$$T_{A \to B} = \frac{aefh(1 - c)}{(1 - bde - c)(1 - g)} \tag{15-21}$$

The probability of accepted word error is obtained by substituting the appropriate values into Eq. (15-21). Defining P_{DE} to be the probability of FEC decoder error, we obtain the following expression.

$$P(E) = \frac{P_{DE} P_e}{1 - P_{DE} P_r} \tag{15-22}$$

The throughput expressions for the type-I hybrid-ARQ protocol are the same as those for the pure ARQ cases with the exception that P_r is replaced by $P_{DE} P_r$. If the error detection codes in a pure and type-I hybrid-ARQ protocol are the same, the frequency of retransmissions in the latter is thus reduced by the factor P_{DE}.

Type-I hybrid-ARQ protocols can also be implemented using a single FEC code, as shown in Figure 15-10. The FEC decoder is modified to generate retransmission requests using one or both of the following two approaches.

- If the FEC code is not perfect and the decoder is a bounded-distance decoder, request retransmissions in the event of a decoder failure (this is an extremely powerful technique when applied to Reed-Solomon codes [Wic5], [Wic7]).
- If the FEC decoder is t-error-correcting, designate a retransmission threshold $t' < t$ such that a retransmission request is generated whenever the number of errors corrected exceeds t'.

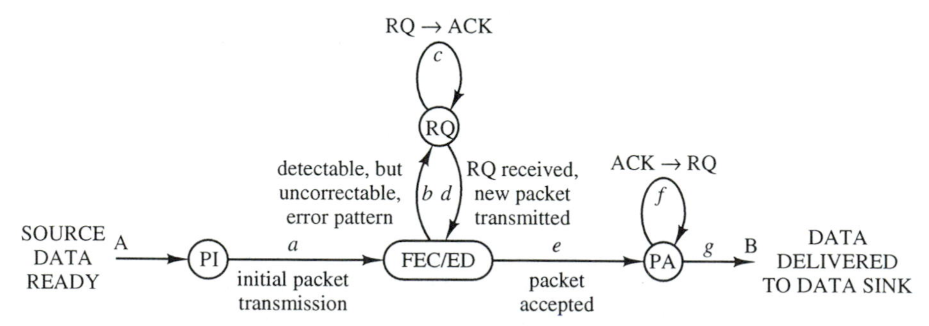

Figure 15-10. State Diagram for a Type-I Hybrid-ARQ Protocol Based on One Code

In both of the above techniques, the reliability of the decoded/accepted packets is increased with respect to the pure FEC protocol by a refusal to complete decoding of received packets for which the probability of decoder error is high.

We now consider the performance analysis of type-I protocols through the use of an example. Note that Figure 15-10 is the same as Figure 15-4 with the exception of the labeling on the decoding state. The analysis is thus essentially identical, with the exception that we must redefine the labels on the performance graphs as follows.

P_r = the probability that a received packet contains a detectable, but uncorrectable, error pattern and thus causes the generation of a retransmission request

P_e = the probability that a received packet causes an FEC decoder error

P_c = the probability that a received packet is error-free or contains a correctable error pattern (As before, $P_r + P_e + P_c = 1$)

Example 15-4—A Type-I Hybrid-ARQ Protocol Based on an Extended Golay Code

In this example we compare a type-I hybrid-ARQ protocol and a pure-ARQ protocol, both based on $\mathscr{G}_{24}$, the (24, 12, 8) extended binary Golay code. We adopt a selective-repeat retransmission protocol for both cases. In the pure ARQ protocol, $\mathscr{G}_{24}$ is used exclusively for error detection. In the type-I hybrid-ARQ protocol, we allow $\mathscr{G}_{24}$ to correct up to three errors per received word. Since $\mathscr{G}_{24}$ has minimum distance 8, it can also detect all four-error patterns. When such a pattern (or any other detectable but uncorrectable pattern) is detected, a retransmission request is generated. The communication channel is assumed to be binary symmetric with crossover probability p.

The performance of the pure ARQ protocol is discussed in detail in Examples 15-1 and 15-2. To analyze the type-I hybrid-ARQ protocol, we once again use the weight distribution for $\mathscr{G}_{24}$. A decoder error occurs whenever the received word is within distance three of a code word other than that which was transmitted. As in Chapter 10, we assume without loss of generality that the all-zero code word was transmitted. Equation (10-11) shows that the probability of being exactly distance k from a weight-j code word is

$$P_k^j = \sum_{r=0}^{k} \binom{j}{k-r} \binom{24-j}{r} p^{j-k+2r}(1-p)^{24-j+k-2r}$$

The probability that a code word is incorrectly accepted is then

$$P_e = \sum_{j=8}^{24} A_j \sum_{k=0}^{3} P_k^j$$

A received packet is correctly decoded and accepted if it contains three or fewer errors. We thus have

$$P_c = \sum_{j=0}^{3} \binom{24}{j} p^j (1-p)^{24-j}$$

The probability that a received word caused the generation of a retransmission request is then $P_r = 1 - P_c - P_e$.

Figure 15-11 shows the reliability performance for the pure ARQ and type-I hybrid-ARQ protocols, while Figure 15-12 compares their throughput performance. The type-I protocol offers substantially better throughput performance in the range $10^{-3} \leq p \leq 10^{-1}$, though at the expense of its reliability performance. Clearly the type-I protocol is the more useful for applications involving low SNRs. The type-I protocol throughput curve also shows an interesting phenomenon that is a function of the

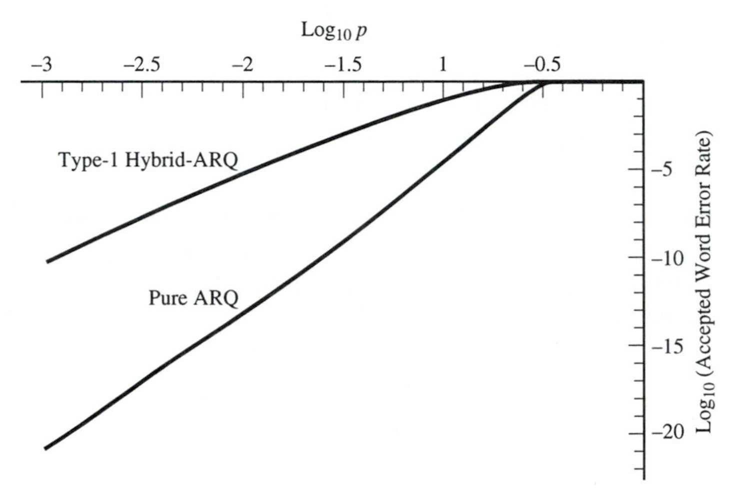

Figure 15-11. Comparison of Pure ARQ and Type-I Hybrid-ARQ Reliability Performance

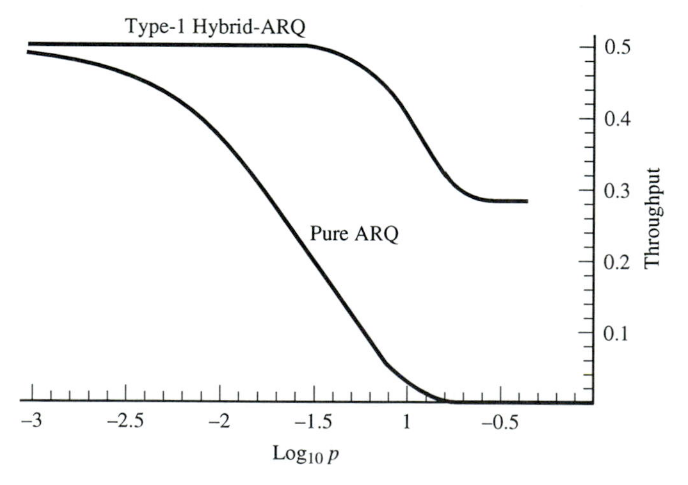

Figure 15-12. Comparison of Pure ARQ and Type-I Hybrid-ARQ Throughput Performance

geometry of the Golay codes. Though the extended binary Golay code is not perfect, it is still well packed. If the decoding spheres have radius three, then a random input to the decoder has a significant probability of falling into a decoding sphere; the throughput curve thus approaches a nonzero floor beneath which it cannot go. The throughput curve for the type-I protocol is meaningful only when considered in conjunction with the reliability curve, which shows what proportion of the throughput consists of correct packets. ■

The FEC code $\mathbf{C}_2$ should be chosen so that the effective channel seen by the error detecting decoder is relatively clean. As seen in the previous example, the type-I protocol suffers catastrophic throughput reduction once the channel deteriorates to a certain point. In many applications this type of "graceless" performance degradation is undesirable. One solution is to use a type-II hybrid-ARQ protocol or, more generally, a packet-combining protocol. This is the subject of the next section.

Before moving on, however, we note that the design of single-code type-I hybrid-ARQ protocols typified by the Golay protocol in Example 15-4 has been applied to a number of different block and convolutionally encoded FEC systems with great success. Drukarev and Costello [Dru2] were the first to discuss this approach in their development of type-I protocols based on the sequential decoding of convolutional codes. Yamamoto and Itoh [Yam] and Harvey and Wicker [Har2], [Har3] later examined systems based on the Viterbi decoder. Type-I systems based on the majority-logic decoding of both convolutional [Wic6] and block cyclic block codes [Ric1] have also been considered.

Perhaps the most powerful of the type-I single-code systems are those based on Reed-Solomon codes [Wic5], [Wic7]. Consider the case of an errors and erasures decoder for an RS code with minimum distance $d_{\min}$. Recall from Chapter 9 that such a decoder can correct all combinations of e errors and s erasures so long as $2e + s < d_{\min}$. Let d_e be the *effective diameter* of the type-I decoder. In the Reed-Solomon type-I protocol, FEC decoding is completed without a retransmission request if there are e errors and s erasures in the code word, such that $2e + s \leq d_e$. If the bound is not satisfied or a decoding failure occurs, then a retransmission is requested.

Figure 15-13 shows the reliability and throughput performance of an RS type-I hybrid-ARQ protocol over a Rayleigh fading channel. Note that the reliability performance improves substantially as the effective diameter is reduced. Unfortunately this improvement in reliability is obtained at the expense of throughput. A lower-rate code would provide better throughput at the lower signal-to-noise ratios.

It is interesting to compare the throughput performance of the Golay code in Figure 15-12 to that of the Reed-Solomon code in Figure 15-13. Unlike the Golay codes, Reed-Solomon codes are poorly packed. A random input to an RS decoder is far more likely to cause a decoder failure than a decoder error. The RS type-I hybrid-ARQ throughput curves thus go to zero asymptotically as the signal-to-noise ratio is reduced.

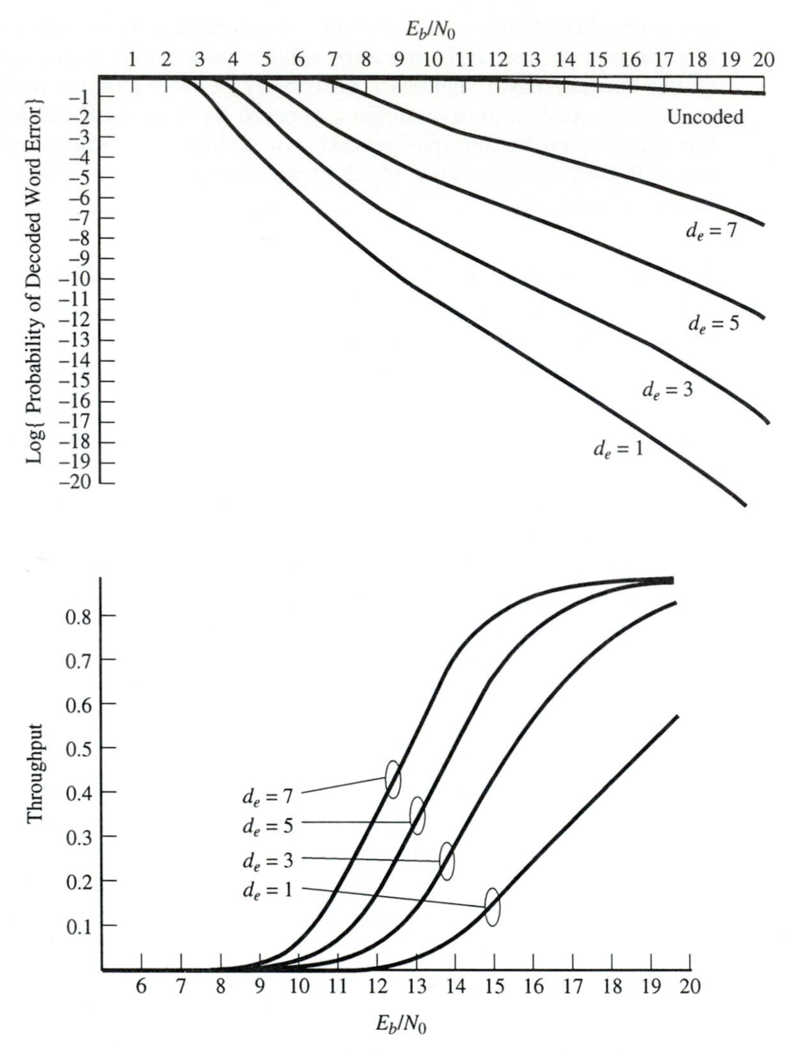

Figure 15-13. Word Error Rate and Throughput for a (64, 56) RS Code in a Type-I Hybrid-ARQ Protocol (Rayleigh Fading Channel) ([Wic7], © 1992 IEEE)

15-4 TYPE-II HYBRID-ARQ PROTOCOLS AND PACKET COMBINING

Type-II hybrid-ARQ protocols adapt to changing channel conditions through the use of **incremental redundancy**. Incremental redundancy was first suggested by Mandelbaum in 1974 [Man1]. The transmitter in these systems responds to retransmission requests by sending additional parity bits to the receiver. The receiver appends these bits to the received packet, allowing for increased error correction capability.

The type-II hybrid-ARQ protocol is the simplest form of incremental redundancy system. In this section we consider the type-II system originally suggested by Lin and Yu [Lin2] and modified by Wang and Lin [Wan]. We then proceed to a more general form that has proven quite powerful when used in conjunction with Reed-Solomon codes.

Wang and Lin's system uses two separate codes: a high-rate (n, k) error detecting code $\mathbf{C}_1$, and a $(2n, n)$ systematic invertible code $\mathbf{C}_2$. A k-bit message is first encoded using $\mathbf{C}_1$ to form an n-bit packet P_1. P_1 is then encoded using $\mathbf{C}_2$. The n parity bits (call them P_2) from the $\mathbf{C}_2$ code word are saved in a buffer, while the $\mathbf{C}_1$ code word P_1 is transmitted. Figure 15-14 shows the subsequent processing. The initial packet is checked for errors at the receiver (state ED_1). If it is found to contain errors, a retransmission request is sent back to the transmitter. The transmitter responds by sending P_2. Since $\mathbf{C}_2$ is invertible, the n bits used to create the $\mathbf{C}_2$ code word can be obtained by inverting the packet P_2. An inverted version of P_2 is created and checked for errors (ED_2). If the inverted version contains errors, P_2 is appended to P_1 to create a noise-corrupted $\mathbf{C}_2$ code word. After FEC decoding, the resulting message is checked once again for errors (ED_3). If there are still errors, the process continues, with the transmitter alternating transmission of P_1 and P_2 until one of the three error detection decoding operations is successfully passed.

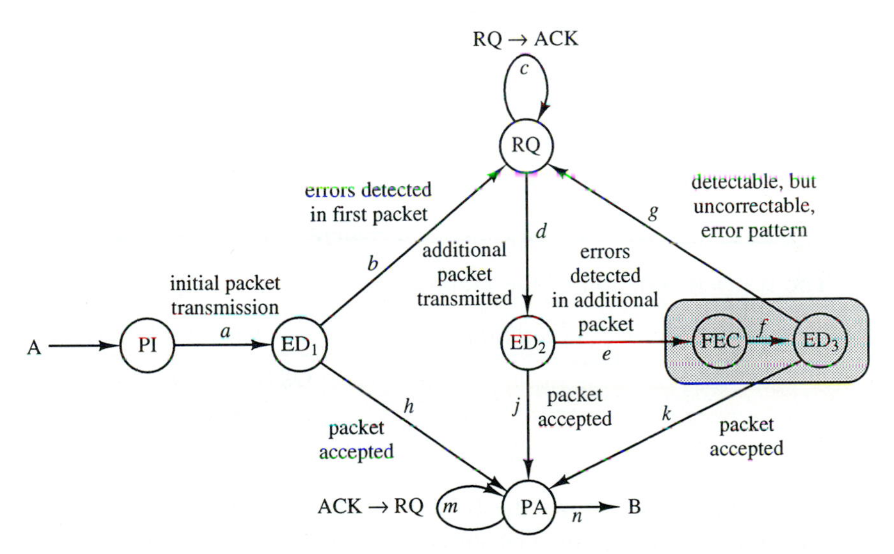

Figure 15-14. State Diagram for a Type-II Hybrid-ARQ Protocol (Wang-Lin Version)

As in the type-I system, the error detection role can be served by CRC codes. The half-rate systematic invertible code is not as easily selected. Lin and Yu [Lin2] suggested a class of codes based on shortened cyclic codes (and thus related to CRC codes). The inversion process is performed using shift-register-based circuits.

The transfer function for the graph in Figure 15-14 is found as follows.

FORWARD LOOPS (A→B)

K_1: A, PI, ED_1, PA, B $\quad F_1 = ahn$
K_2: A, PI, ED_1, RQ, ED_2, PA, B $\quad F_2 = abdjn$
K_3: A, PI, ED_1, RQ, ED_2, FEC, ED_3, PA, B $\quad F_3 = abdefkn$

LOOPS

L_1: RQ, ED_2, FEC, ED_3, RQ $\quad C_1 = defg$
L_2: RQ, RQ $\quad C_2 = c$
L_3: PA, PA $\quad C_3 = m$

NONTOUCHING PAIRS OF LOOPS

(L_1, L_3) $\quad C_1 C_3 = defgm$
(L_2, L_3) $\quad C_2 C_3 = cm$

There are no sets of three or more nontouching loops. We can now proceed with the computation of the determinant.

$$\begin{aligned}\Delta &= 1 - (C_1 + C_2 + C_3) + (C_1 C_3 + C_2 C_3)\\ &= 1 - defg - c - m + defgm + cm\\ &= (1 - defg - c)(1 - m)\end{aligned}$$

The cofactors are computed as follows.

$$\Delta_1 = 1 - C_1 - C_2 = 1 - defg - c, \qquad \Delta_2 = \Delta_3 = 1$$

The transfer function of the graph is then

$$\begin{aligned}T_{A\to B} &= \frac{ahn(1 - c - defg) + abdjn + abdefkn}{(1 - c - defg)(1 - m)}\\ &= \frac{an}{(1 - m)}\left[h + \frac{bd(j + efk)}{(1 - c - defg)}\right]\end{aligned} \tag{15-23}$$

Figure 15-15 shows the branch labels used to determine the reliability performance of the system. The superscripts on the labels reference the various probabilities to specific decoding operations. For example, $P_d^{(1)}$ is the probability of retransmission during the first decoding operation.

The corresponding probability of accepted packet error is as follows.

$$\begin{aligned}P(E) &= \frac{(1 - P_{ACK\to RQ})}{(1 - P_{ACK\to RQ})}\left[P_u^{(1)} + \frac{P_d^{(1)}(1 - P_{RQ\to ACK})(P_u^{(2)} + P_d^{(2)} P_{DE} P_u^{(3)})}{[1 - P_{RQ\to ACK} - (1 - P_{RQ\to ACK})P_d^{(2)} P_{DE} P_d^{(3)}]}\right]\\ &= P_u^{(1)} + \frac{P_d^{(1)}(P_u^{(2)} + P_d^{(2)} P_{DE} P_u^{(3)})}{(1 - P_d^{(2)} P_{DE} P_d^{(3)})}\end{aligned} \tag{15-24}$$

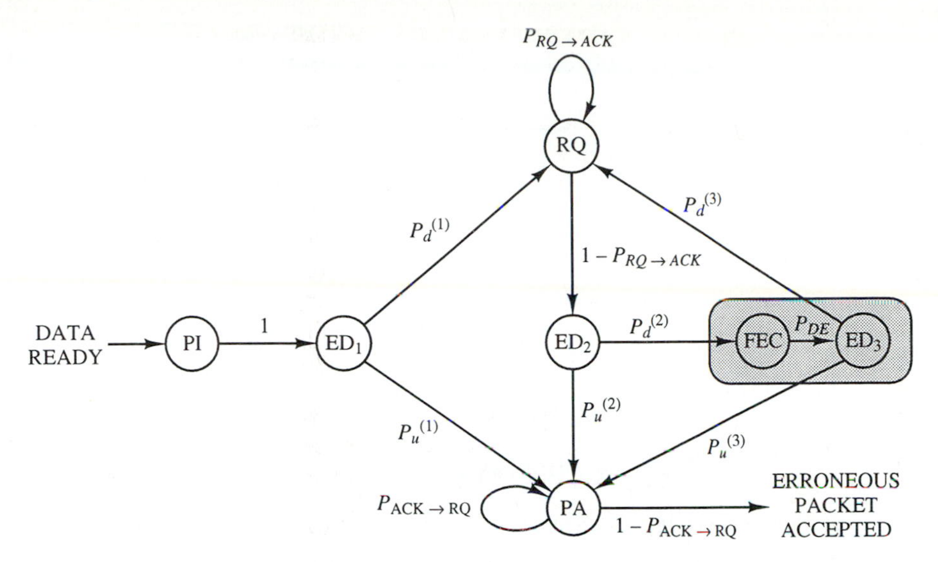

Figure 15-15. Diagram for Determining the Reliability Performance for a Type-II Hybrid-ARQ Protocol

The undetected error probabilities are often quite difficult to determine, for they require that the analyst completely characterize the output of the FEC decoder. A bound or approximation, however, is usually sufficient. When determining reliability performance for the entire system, an upper bound is easily obtained by assuming that FEC decoder errors cause a completely random output. The undetected error probability given an FEC decoder error is then the "coverage" λ of the error detecting code subtracted from one (see Chapter 5 for a discussion of error detection coverage). Lower bounds on throughput can be obtained by assuming that the error detecting codes detect all error patterns. This actually results in a very tight lower bound in most cases, for the probability of undetected error is much smaller than the probability of detected error.

The type-II protocol can be generalized in a number of ways. For example, the initial transmission and the parity transmission can both be implemented as type-I hybrid-ARQ protocols. This increases the flexibility of the system by allowing for some error correction on the first transmission [Wic8].

Perhaps the most powerful of the modified type-II protocols are those based on punctured Reed-Solomon codes. Mandelbaum was the first to note that the MDS nature of RS codes allows for a simple implementation in an incremental redundancy system. An RS-based incremental redundancy system was later investigated by Pursley for application to meteor-burst channels [Pur1], [Pur2], [Pur3]. Wicker and Bartz independently developed a similar scheme and provided an exact performance analysis in [Wic8]. In the following paragraphs, the Wicker/Bartz system is described. The interested reader can find the details of the performance analysis in [Wic8].

The basic structure of the codes in the RS type-II hybrid-ARQ protocol is shown in Figure 15-16. The figure applies not only to Reed-Solomon codes, but to all MDS codes used in this type-II protocol.

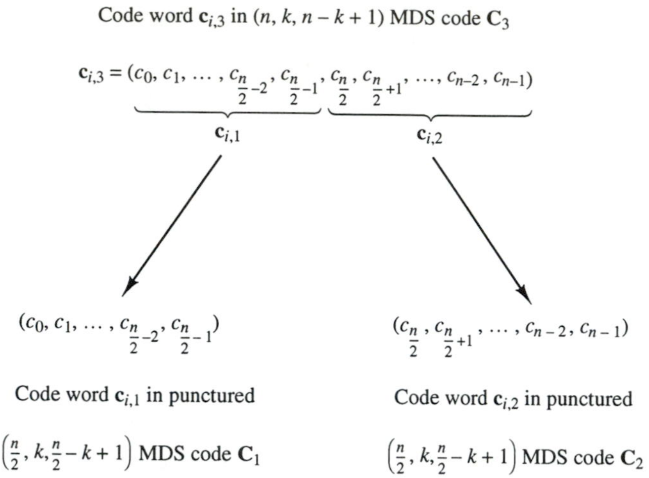

Code word $\mathbf{c}_{i,3}$ in $(n, k, n-k+1)$ MDS code $\mathbf{C}_3$

$$\mathbf{c}_{i,3} = (\underbrace{c_0, c_1, \ldots, c_{\frac{n}{2}-2}, c_{\frac{n}{2}-1}}_{\mathbf{c}_{i,1}}, \underbrace{c_{\frac{n}{2}}, c_{\frac{n}{2}+1}, \ldots, c_{n-2}, c_{n-1}}_{\mathbf{c}_{i,2}})$$

$(c_0, c_1, \ldots, c_{\frac{n}{2}-2}, c_{\frac{n}{2}-1})$

Code word $\mathbf{c}_{i,1}$ in punctured

$\left(\frac{n}{2}, k, \frac{n}{2}-k+1\right)$ MDS code $\mathbf{C}_1$

$(c_{\frac{n}{2}}, c_{\frac{n}{2}+1}, \ldots, c_{n-2}, c_{n-1})$

Code word $\mathbf{c}_{i,2}$ in punctured

$\left(\frac{n}{2}, k, \frac{n}{2}-k+1\right)$ MDS code $\mathbf{C}_2$

Figure 15-16. The Separation of an MDS Mother Code into Two Punctured Codes

A low-rate ($k < n/2$) "mother code" $\mathbf{C}_3$ is selected. Each code word in $\mathbf{C}_3$ is then split in half; all of the first halves forming an $(n/2, k)$ code $\mathbf{C}_1$ and the second halves forming another $(n/2, k)$ code $\mathbf{C}_2$. Since $\mathbf{C}_1$ and $\mathbf{C}_2$ are punctured versions of an MDS code, we know by Theorem 8-6 that $\mathbf{C}_1$ and $\mathbf{C}_2$ are themselves MDS. We further note that, given an errors and erasures decoder for $\mathbf{C}_3$, we can decode any code word from $\mathbf{C}_1$ or $\mathbf{C}_2$ by appending $n/2$ erasures to the end or beginning of the received word.

The complete RS type-II protocol is shown in Figure 15-17. The first transmission consists of a code word from $\mathbf{C}_1$. The encoding of $\mathbf{C}_1$ can be systematic, if desired (see Theorem 8-4). The $\mathbf{C}_1$ code words are used in the type-I hybrid-ARQ protocol described in the previous section. This allows for errors and erasures correction on the first transmission, making the application of the system more flexible. The transmitter responds to retransmission requests by sending the $\mathbf{C}_2$ code word associated with the $\mathbf{C}_1$ code word. In some cases it may be desirable to attempt to decode the $\mathbf{C}_2$ code word by itself, but in most cases involving MDS codes it is preferable to immediately concatenate the $\mathbf{C}_1$ and $\mathbf{C}_2$ code words to form a noise-corrupted $\mathbf{C}_3$ code word, as shown in Figure 15-17 [Wic8].

The state diagram in Figure 15-17 leads to the following transfer function.

$$T_{A\to B} = \frac{afj(1-c-de) + abdgj}{(1-c-de)(1-h)} = \frac{aj}{(1-h)}\left[f + \frac{bdg}{(1-c-de)}\right] \tag{15-25}$$

The labeling used in Figure 15-18 then provides the probability of accepted word error in Eq. (15-26).

$$\begin{aligned} P(E) &= \frac{(1-P_{ACK\to RQ})}{(1-P_{ACK\to RQ})}\left\{P_e^{(1)} + \frac{P_r^{(1)}P_e^{(2)}(1-P_{RQ\to ACK})}{[1-P_{RQ\to ACK} - P_r^{(2)}(1-P_{RQ\to ACK})]}\right\} \\ &= P_e^{(1)} + \frac{P_r^{(1)}P_e^{(2)}}{1-P_r^{(2)}} \end{aligned} \tag{15-26}$$

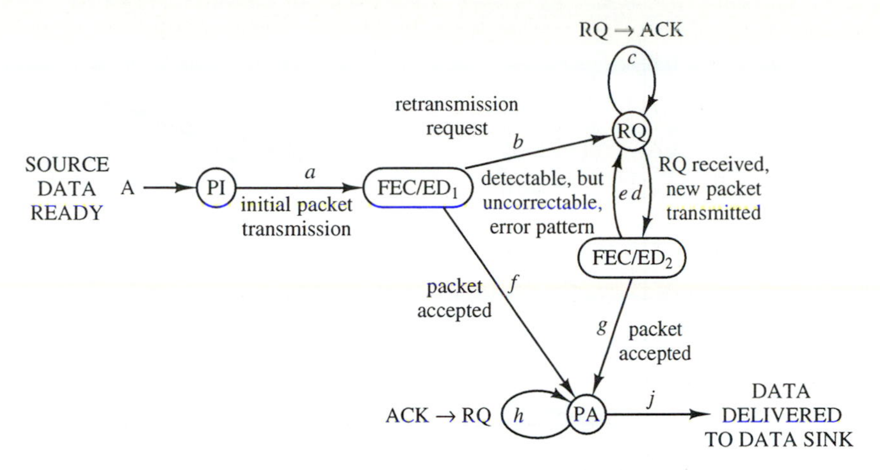

Figure 15-17. State Diagram for an RS Type-II Hybrid-ARQ Protocol

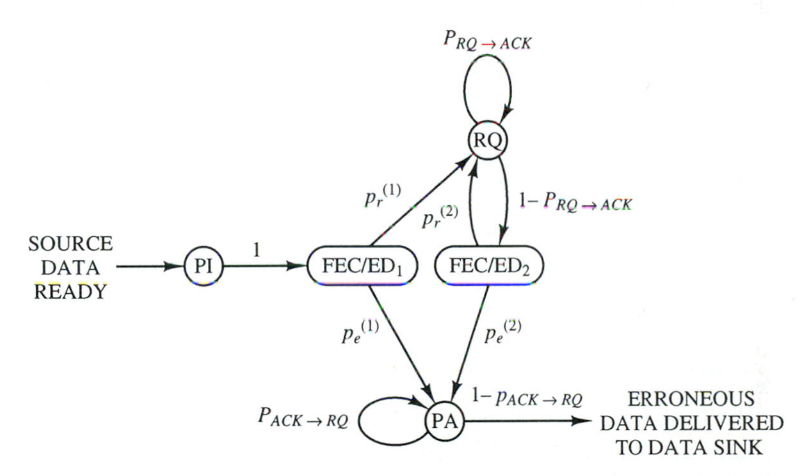

Figure 15-18. Graph Labeling Used for the Determination of Reliability Performance

Figure 15-19 compares the throughput performance of an RS-type-I and an RS-type-II hybrid-ARQ protocol. The parameters of the two protocols have been chosen so that their reliability performance is the same, allowing for a fair comparison. The type-I protocol is based on a punctured (16, 12) RS code. The type-II protocol uses a pair of punctured (16, 12) codes whose code words can be concatenated to form code words from a (32, 12) RS code. In this particular example the channel is assumed to be Rayleigh fading, and the decoder is able to decode erasures. Note that the type-II system allows for much more graceful performance degradation as the signal-to-noise ratio increases. When the throughput of the type-I protocol has dropped to zero, the type-II protocol is still able to offer slightly less than half its maximum possible throughput. This sort of performance is typical of the type-II

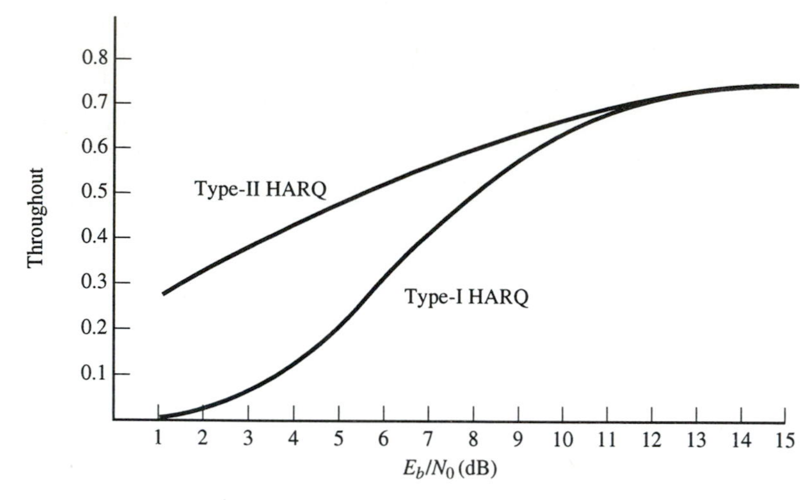

Figure 15-19. Throughput Performance for (16, 12)/(16, 12) MDS Type-II HARQ Protocol Compared to (16, 12) Type-I HARQ Protocol

systems. They are essentially two-rate adaptive coding systems. The high rate is used when the channel is good. The system is then able to automatically switch to the lower rate when the channel deteriorates.

Clearly if we can substantially improve performance by allowing for two code rates instead of one, we should benefit even more if we can offer a larger number of code rates. Packet combining systems can do just that. In a packet combining system each received packet is combined with its predecessors until the resulting combined packet can be reliably decoded.

Packet-combining systems can be separated into two distinct types. In a **code combining system** the individual transmissions are encoded at some code rate R. If the receiver has N packets that have caused retransmission requests, these packets are concatenated to form a single packet encoded at rate R/N. As N increases, the decoder eventually acquires sufficient power to reliably decode the packets under existing channel conditions. Code combining systems based on block [Cha], [Wic8] and convolutional codes [Har3], [Kal], [Hag2] have been investigated in the literature.

Diversity combining systems use multiple *identical* copies of a packet to create a single packet whose constituent symbols are more reliable than those of any of the individual copies. Diversity combining systems are typified by symbol voting schemes in hard-decision systems and by symbol averaging in soft-decision systems. The literature contains discussions of diversity combining systems based on majority logic decoding [Wic9] and Viterbi decoding [Har3] of convolutional codes, among others.

A simple averaging circuit is all that is needed to convert a soft-decision Viterbi decoder into a diversity combining decoder [Har2]. Let $\mathbf{X} = (x_0, x_1, \ldots, x_{N-1})$ be a transmitted packet and let $\mathbf{Y}^{(i)} = (y_0^{(i)}, y_1^{(i)}, y_2^{(i)}, \ldots, y_{N-1}^{(i)})$ be the ith received copy of that packet. Some L received copies $\mathbf{Y}^{(1)}, \mathbf{Y}^{(2)}, \ldots, \mathbf{Y}^{(L)}$ can be combined to form a single, more reliable packet by averaging the soft-decision values for the copies of each bit. A single packet $\mathbf{Z} = (z_0, z_1, \ldots, z_{N-1})$ is produced whose constituent coordinates have the form

$$z_j = \frac{1}{L}\sum_{i=1}^{L} y_j^{(i)} \tag{15-27}$$

Error detection in this system can be provided by residual redundancy in the convolutional code [Yam], [Har2] or by a CRC code [Ras3]. Figure 15-20 shows how averaged diversity combining can be used to improve the performance of a one-code convolutional type-I hybrid-ARQ protocol. Figure 15-20 has three sets of curves delineated by a parameter u. u is used in the one-code system to establish the allocation of the code's redundancy between error correction (lower values of u) and error detection (higher values of u). The details of this system can be found in [Har3]. It is sufficient here to note the substantial improvement in throughput performance provided by the introduction of averaged diversity combining.

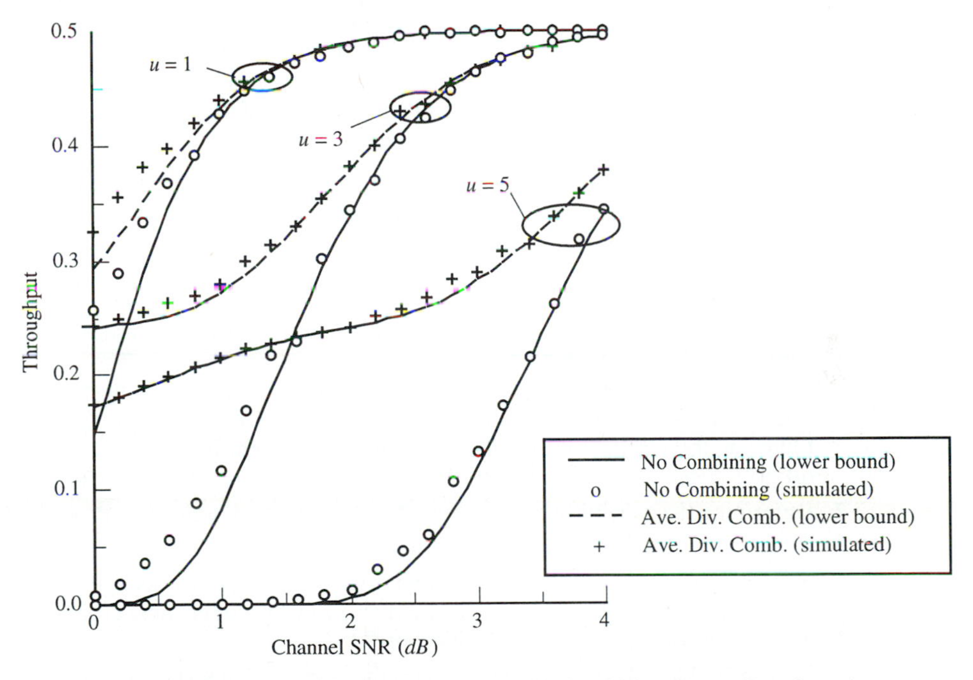

Figure 15-20. The Performance of a Diversity Combining System Based on the Viterbi Decoder

A generalized packet combining system is shown in Figure 15-21. Note that the various decoding operations are separated in the graph, and not treated as the same state. This is due to the fact that their performance parameters should be quite different. The probability of decoder error and retransmission request should both drop rapidly as the number of packets being combined increases. This is certainly the case for the Reed-Solomon-based systems that have been studied in the literature [Wic8], and only slightly less true for convolutionally coded systems [Har3] and TCM-based systems [Ras1], [Ras2].

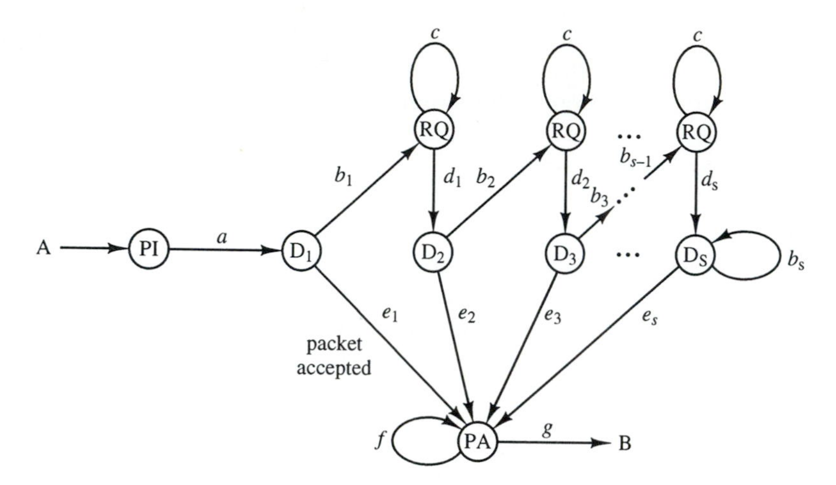

Figure 15-21. A Generalized Packet Combining System

PROBLEMS

1. A (7, 4) Hamming code is used in a pure ARQ protocol over a binary symmetric channel with crossover probability $p = 0.05$. Each packet consists of a single code word. The application has a 10,000-foot, free-space, line-of-sight channel. The transmitter has a 1×10^6 bit-per-second transmission rate, and the receiver processing delay is negligible. The feedback channel is assumed to be error-free. Determine the throughput for the following three protocols. (*Hint:* Equation (4-12) might be useful.)
   (a) SW-ARQ **(b)** GBN-ARQ **(c)** SR-ARQ
2. Determine the reliability performance for the system described in Problem 1.
3. Repeat Problem 1, but this time assume a 60,000-foot channel.
4. Repeat Problem 1, but this time assume that each packet consists of four code words. What is the probability of retransmission as a function of channel crossover probability when each packet consists of n code words?
5. Repeat Problem 1, but this time assume that the feedback channel has an error rate of $P_{ACK \to RQ} = P_{RQ \to ACK} = 0.2$.
6. A GBN-ARQ system has propagation delays λ_1 and λ_2 seconds, transmission rate D bits/second, and receiver processing delay δ. The channel is binary symmetric with crossover probability p. The packets have length N, and it is assumed that the probability of a retransmission request is $P_r = 1 - (1 - p)^n$. Determine the block length n that maximizes throughput in this system.

7. Assume that the system described in Problem 1 uses an SR-ARQ retransmission protocol and has a receiver buffer that can hold up to N code words. Using signal flow graph techniques, determine the probability of receiver buffer overflow as a function of N.
8. What proportion of all binary 23-tuples fall into the decoding spheres of a (23, 12) Golay decoder that is only allowed to correct error patterns of weight two or less (i.e., weight-three patterns cause retransmission requests)? Use this information to determine the accepted word error rate for a Golay-based type-I hybrid-ARQ protocol as the signal-to-noise ratio gets extremely low.
9. Repeat Problem 8, but this time use a (32, 28) Reed-Solomon code with a two-error correcting bounded-distance decoder (failures cause retransmission requests).

16

Applications

Though every application is unique, they display certain traits that allow one to loosely group them in a few overlapping categories. Within each category, certain types of error control systems are generally preferred. It would be misleading (and probably dangerous), however, to attempt to list these categories and their corresponding error control systems. In engineering one searches for *a* design solution, not *the* design solution. Bearing that in mind, we shall instead examine several applications in which error control coding has played a substantial role. In examining these applications, the principal considerations that led to the selection of the particular error control system are emphasized. The first application to be discussed here is the Compact Disc player, which uses a cross-interleaved pair of shortened Reed-Solomon codes. This is followed by an examination of several coding systems used in planetary satellite communications. These systems use convolutional codes by themselves, or convolutional codes concatenated with Reed-Solomon codes. We conclude with an examination of error control systems for computer storage. In Chapter 14 the design of bandwidth-limited systems (e.g., telephone modems) is discussed in great detail, so an example of such a system is not considered here.

The chapter begins with a discussion of interleavers, which figure prominently in all of the applications presented.

16.1 INTERLEAVERS

A **bursty channel** is defined as a channel over which errors tend to occur in bunches, or "bursts," as opposed to the random patterns associated with a Bernoulli-distributed process. Bursty channels usually contain some error-causing agent in the physical medium whose effective time constant exceeds the symbol transmission rate of the channel. For example, a scratch on a Compact Disc may obscure several consecutive bits on each of several adjacent tracks, thus causing multiple error bursts when the disc is played.

All of the block codes discussed in this text have been **random error correcting** (at least from the perspective of the code symbol alphabet). A random error correcting code can correct up to t symbol errors per code word, regardless of the placement of those errors. The performance analyses in Chapter 10 thus focused solely on error pattern weight. A problem arises with these codes whenever the channel encountered in the application is bursty. An error burst focuses several symbol errors within a small number of received code words, while the other code words may not be corrupted by any errors at all. The error correction capacity required by the words hit by bursts is thus wasted by the words that are not affected.

The performance of convolutional codes is also sensitive to bursty channels, though the reasons are much more subtle than in the block coding case. In the decoding of convolutional codes, error events occur whenever the received code word is closer to an incorrect code word than the code word that was transmitted. In Chapter 11 it was shown that linear convolutional code words can be viewed as paths through weighted directed graphs. Any pair of code words thus differs by one or more nonzero circuits through the graph, starting at the all-zero state. Since these circuits consist of several consecutive branches in the graph, the error patterns that cause one code word to look like another are bursty. A convolutional code may thus be able to correct an arbitrarily large number of well-spaced errors, while at the same time being unable to handle a short burst. This reasoning applies to TCM systems as well, with the additional note that parallel transitions introduce an increased sensitivity to random errors.

A great deal of effort has gone into the development of codes specifically designed for bursty channels. Such codes include the Fire codes and certain of the shortened cyclic codes. Fire codes are discussed later in this chapter. Here we focus on interleaving techniques that allow for the effective use of random error correcting codes over bursty channels. An **interleaver** is a device that mixes up the symbols from several code words so that the symbols from any given code word are well separated during transmission. When the code words are reconstructed by the deinterleaver, error bursts introduced by the channel are broken up and spread across several code words. The interleaver/deinterleaver pair thus creates an effectively random channel.

Block and **cross-interleave** are the most frequently used types of interleavers. The cross-interleave is sometimes called a **convolutional** or **periodic** interleaver for reasons that will become obvious. An $(n \times m)$ block interleaver and the corresponding deinterleaver are shown in Figure 16-1. The two circuits are identical, each consisting of n rows of m memory elements.

The coded data stream is read into the block interleaver rows in the order noted in the figure. The interleaver contents are then read out by columns. Any two adjacent symbols at the input are thus separated by $(n - 1)$ other symbols at the output. The row length m is frequently selected so that each row holds an entire code

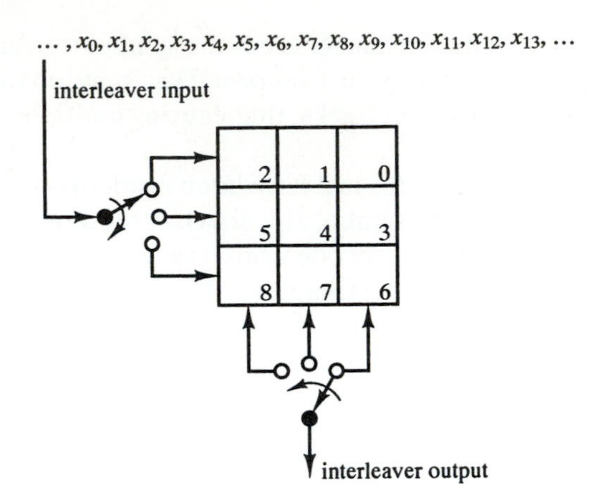

$\ldots, x_6, x_3, x_0, x_7, x_4, x_1, x_8, x_5, x_2, x_{15}, x_{12}, x_9, x_{16}, x_{13}, \ldots$

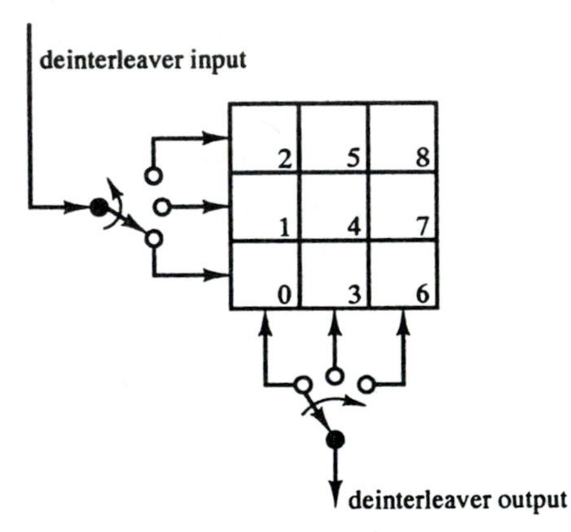

$\ldots, x_0, x_1, x_2, x_3, x_4, x_5, x_6, x_7, x_8, x_9, x_{10}, x_{11}, x_{12}, x_{13}, \ldots$

Figure 16-1. A 3×3 Block Interleaver and Deinterleaver

word. A burst of b errors causes a maximum of $\lceil b/n \rceil$ errors to occur in one or more code words.[1] It follows that if the code has the ability to correct random error patterns of weight t or less, one or more code words may be overwhelmed by a burst with length exceeding $(nt + 1)$.

The efficiency of an interleaver can be measured in a number of ways. Here we define the efficiency γ to be the ratio of the length of the smallest burst of errors that can cause the error correcting capability t of the code to be exceeded to the number of memory elements used in the interleaver. For an $(n \times m)$ block interleaver we thus have $\gamma = (nt + 1)/nm \approx t/m$.

A cross-interleave circuit is shown in Figure 16-2. The circuit is characterized by the index m, the number of delay lines. Each block D corresponds to a D-symbol

[1] As in Section 5.3.2, a burst stops and starts with an error, but may have correct symbols in the middle. We consider the worst case here, where a length-b burst causes b symbol errors.

delay. The design of this interleaver is due to Ramsey [Ram]. The input symbols are read onto the delay lines in the order shown in the figure. The output of the delay lines is read in the same order.

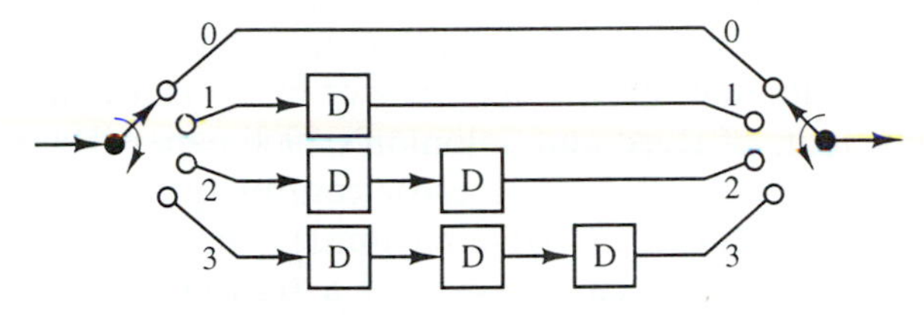

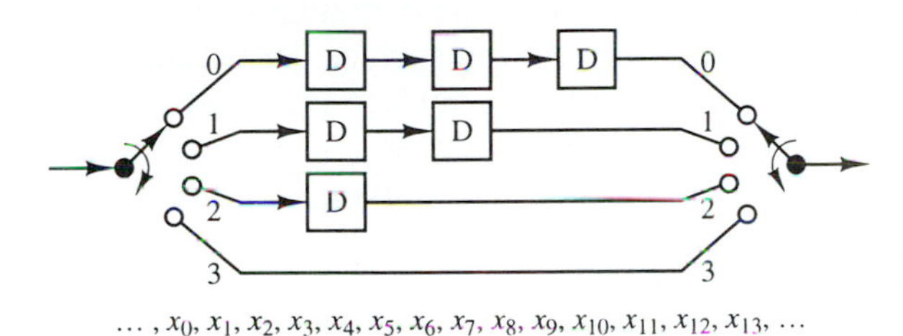

Figure 16-2. A Cross-Interleave Circuit and Corresponding Deinterleaver ($m = 4$, $D = 1$ symbol)

Consider a pair of consecutive input symbols x_0 and x_1. These two symbols are placed on adjacent delay lines, one with delay tD and the other with delay $(t + 1)D$. When x_0 reaches the output position of its delay line, x_1 will still be D delay elements short of the output of its own line. After x_0 is read, all of the m delay line outputs are read D times before x_1 is output. There are thus mD symbols separating adjacent code-word symbols at the output of the interleaver.

Suppose that m is chosen so that it equals or exceeds the length of the code words. Each symbol in a code word is thus placed on a different delay line. At the output of the interleaver, the code-word symbols will appear in order, with each symbol separated from its neighbor(s) in the code word by mD symbols from mD other code words. A length b burst of errors may thus cause $\lceil b/(mD + 1)\rceil$ errors in one or more code words. Given a t-error-correcting code, decoder errors are possible when the length of the error bursts equals or exceeds $(mD + 1)(t - 1) + 1$.

The efficiency of the cross-interleaver is

$$\gamma = \frac{(mD + 1)(t - 1) + 1}{[0 + 1 + 2 + \cdots + (m - 1)]D} = \frac{mD(t - 1) + t}{\left[\dfrac{m(m - 1)D}{2}\right]} \approx \frac{2t}{m - 1} \qquad \text{(16-1)}$$

The cross-interleaver is thus slightly more than twice as efficient as the block interleaver.

16.2 THE COMPACT DISC

It has been said that "without error correcting codes, digital audio would not be technically feasible" [Imm]. There is thus no more eloquent testimony to the power of error control coding than a comparison of the sound quality provided by an expensive turntable and that of a modestly priced Compact Disc (CD) player. The impact of the latter can be positively ethereal (depending upon one's selection of music, of course). It is unfortunate that the vast majority of listeners are unaware that they are receiving the benefits of Reed and Solomon's 1960 paper [Ree1]. In this section we briefly examine the Compact Disc error control system. A more detailed discussion of the entire player can be found in [Imm1], [Imm2], and [Car].

The channel in a CD playback system consists of a transmitting laser, a recorded disc, and a photodetector. Assuming that the player is working properly, the primary contributor to errors on this channel is the contamination of the surface of the disc (e.g., fingerprints and scratches). As the surface contamination affects an area that is usually quite large compared to the surface used to record a single bit, channel errors occur in bursts when the disc is played. As we shall see, the CD error control system handles bursts through cross-interleaving and through the burst error-correcting capability of Reed-Solomon codes.

Figure 16-3 shows the various stages through which music is processed on its way to being recorded on a disc. Each channel is sampled 44,100 times per second, allowing accurate reproduction of all frequencies up to 22 kHz. Each of the samples is then converted into digital form by a 16-bit analog-to-digital converter.

The output of the A/D converter forms a 1.41-Mbps data stream, which is passed directly to the CIRC encoder. The CIRC system will be discussed in detail in a moment, for now we note that the encoder has an effective rate of 3/4, and thus produces a 1.88-Mbps coded data stream.

Many CD players display timing and track information for the disc being played. This information is embedded on the disc by multiplexing a "control and display information" sequence with the coded data.

The multiplexed stream is then passed through a source encoder, whose job is to prepare the stream for effective recording on the optical disc. The CD system uses "eight-to-fourteen modulation" (EFM), which, as its name implies, converts eight-bit blocks of the input stream into fourteen-bit blocks. The EFM code is a **runlength-limited code**, which places upper and lower bounds on the length of sequences of bits having the same value (i.e., runlength). These limits relieve some of the design requirements for the receiver. The lower bound on runlength places an upper bound on the transition density within the recorded data, thus limiting the spectral occupancy of the signal. The upper bound on runlength insures that there will be enough transitions in the data to allow for proper operation of the bit synchronization loop in the receiver. Both bounds thus allow for simpler CD players (i.e., less expensive and more easily marketed).

A few additional synchronization and "merging" bits are added to the EFM encoder output, and the resulting bit stream is finally recorded as a series of pits on a laminated polyvinyl chloride disc.

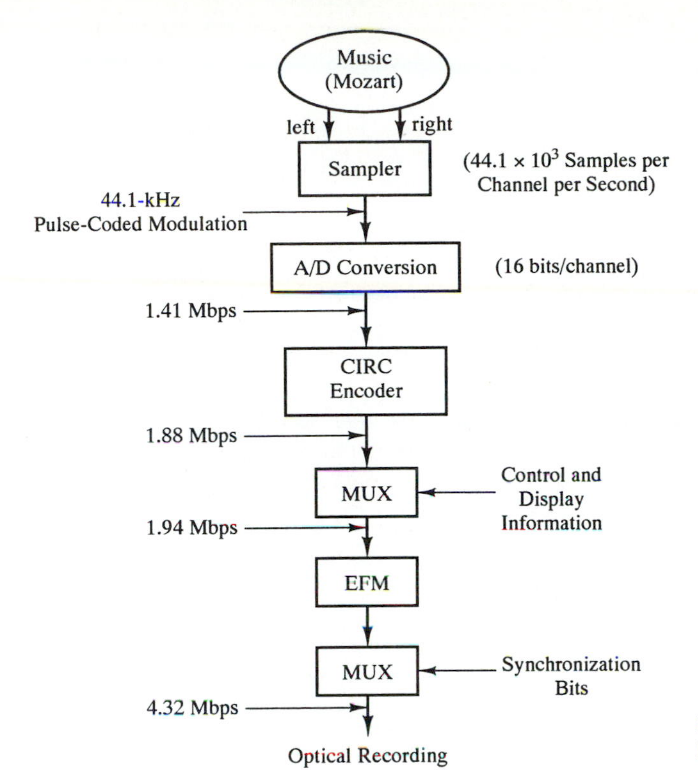

Figure 16-3. The CD Recording Process

The CD error control system is of primary interest here. The CIRC encoder in Figure 16-4 uses two shortened Reed-Solomon codes, $\mathbf{C}_1$ and $\mathbf{C}_2$. Both codes use 8-bit symbols from the code alphabet GF(256). This provides for a nice match with the 16-bit samples emerging from the A/D converter. The "natural" length of the RS code over GF(256) is 255, which would lead to 2040-bit code words and a relatively complicated decoder. It should be remembered that the decoder will reside in the retail player, and it is extremely important that its cost be minimized. The codes are thus shortened significantly (see Section 8.4 and Theorem 8-7): $\mathbf{C}_1$ is a (32, 28) code and $\mathbf{C}_2$ is a (28, 24) code. Both have redundancy 4 and minimum distance 5.

Each 16-bit sample is treated as a pair of symbols from GF(256). The samples are encoded 12 at a time by the $\mathbf{C}_2$ encoder to create a 28-symbol code word. The 28 symbols in each $\mathbf{C}_2$ code word are then passed through a cross-interleaver ($m = 28, D = 4$ symbols) before being encoded by the $\mathbf{C}_1$ encoder. The resulting 32-symbol $\mathbf{C}_1$ code words are then processed as shown in Figure 16-3.

The CIRC encoding process for the CD system is standard; no matter where you buy your CDs (standard size), they will play on any CD player. However, the CIRC decoding process is not standardized and can vary from player to player [Imm1]. This was done to allow manufacturers to experiment with various designs and to speed the player to market. The basic building blocks of the decoder are

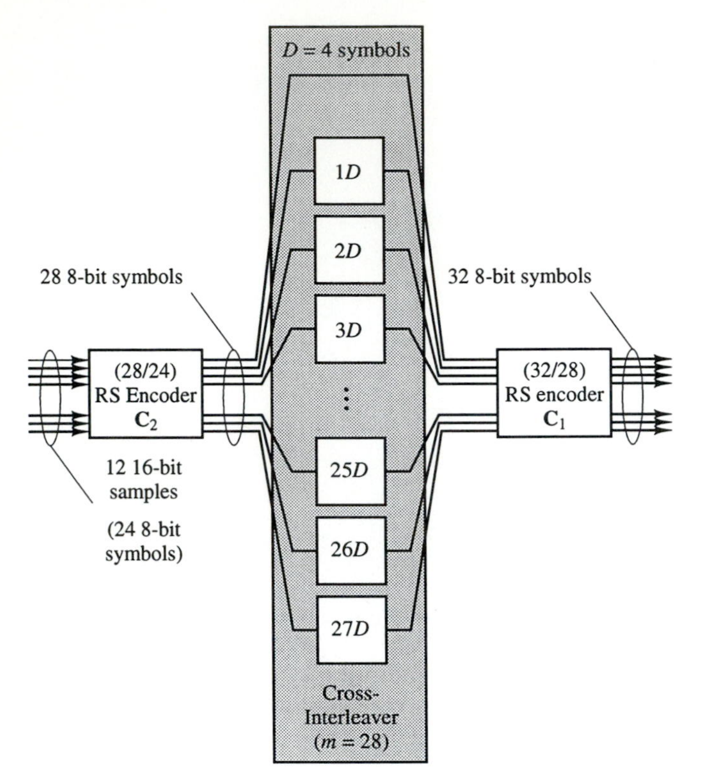

Figure 16-4. The Standard CIRC Encoder

shown in Figure 16-5. The $\mathbf{C}_1$ decoder is followed by a cross-deinterleaver and a $\mathbf{C}_2$ decoder. Since both codes have minimum distance 5, they can be used to correct all error patterns of weight 2 or less. In Chapters 10 and 15 it was noted that, even when the full error correcting capacity of an RS code is used, there remains a significant amount of error detecting capacity. This is clearly seen in Figures 10-7 and 10-8, which show that a high-weight error pattern is much more likely to cause a decoder failure (which is detectable) than a decoder error (which is not detectable). It was further seen in Chapter 15 that some of the error correction capacity of the RS code can be exchanged for an increase in error detection capacity and a substantial improvement in reliability performance. Most CIRC decoders take advantage of both of these principles. The $\mathbf{C}_1$ decoder is set to correct all single-error patterns. When the $\mathbf{C}_1$ decoder sees a higher-weight error pattern (double-error patterns and any pattern causing a decoder failure), the decoder outputs 28 erased symbols. The cross-deinterleaver spreads these erasures over 28 $\mathbf{C}_2$ code words. $\mathbf{C}_1$ may also be set to correct double-error patterns, or, at the other extreme, may simply be used to declare erasures (the least expensive implementation).

The $\mathbf{C}_2$ decoder can correct any combination of e errors and s erasures, where $2e + s < 5$. It is generally designed to decode erasures only (again, an inexpensive solution) owing to the small probability of a $\mathbf{C}_1$ decoder error. Whenever the number of erasures exceeds 4, the $\mathbf{C}_2$ decoder outputs 24 erasures, which corresponds to 12

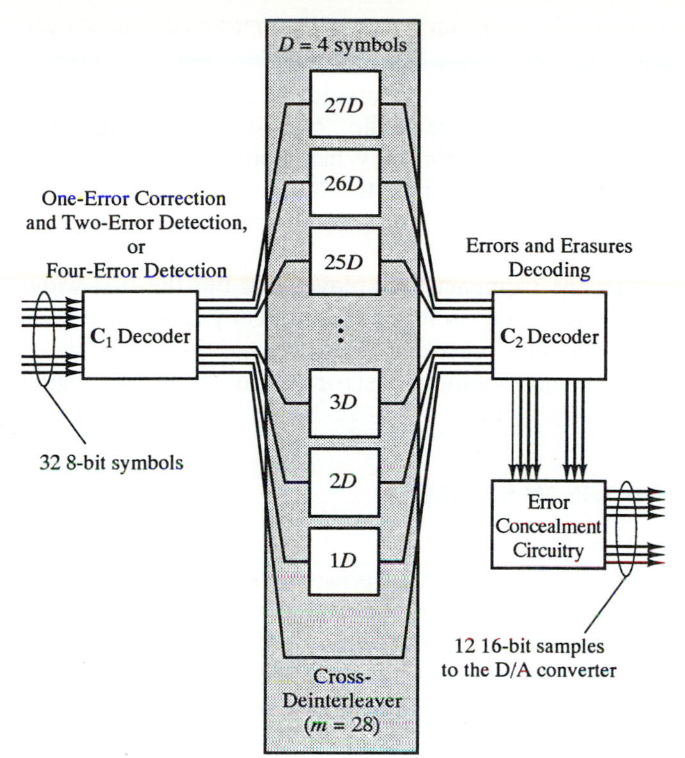

Figure 16-5. Basic Structure of a CIRC Decoder

erased music samples. The error concealment circuitry responds by muting these samples or by interpolating values through the use of correct samples adjacent to the erased samples. A number of additional interleaving and delay operations are included in the encoding and decoding operations in order to enhance the operation of the error concealment circuitry. For example, samples adjacent in time are further separated by additional interleavers to increase the impact of interpolation [Imm1].

Performance figures for the CIRC system in a typical CD player are provided in Table 16-1.

TABLE 16-1 CIRC performance in a typical CD player ([Imm1], © 1994 IEEE).

Longest completely correctable burst	≈4,000 data bits (2.5 mm track length)
Longest interpolatable burst (worst case)	≈12,300 data bits (7.7 mm track length)
Sample interpolation rate	One every 10 hours at BER = 10^{-4}, 1000 samples per minute at BER = 10^{-3}
Undetected error rate (click in output)	<1 every 750 hours at BER = 10^{-3}, negligible at BER $\leq 10^{-4}$
Implementation	1 LSI chip and 2048 bytes RAM

It should be noted that all of the interleaving done in the CIRC system is done on a symbol-by-symbol basis. The bits comprising a given symbol thus remain contiguous throughout the entire encoding, transmission, and decoding process. This allows the CIRC system to take advantage of the "burst-trapping" capability of Reed-Solomon codes. The channel itself is binary. When a burst of errors occurs, it may affect many consecutive received bits. Since the RS decoder does not care whether a symbol error has been caused by 1 bit error or 8, we obtain some significant burst-error correcting capability.

As a final application note, the Compact Disc player was but the first widespread commercial application of RS codes in digital audio. They have since been used in the DAT™ recorder, the Digital Compact Cassette™, and the D1-D2 video recorders. [Imm2] provides an excellent treatment of error control systems for digital recorders of various types.

16.3 CONCATENATED CODES AND THE NASA DEEP SPACE STANDARD

Satellite downlinks are generally characterized as power-limited channels.[2] Onboard batteries and solar cells are heavy and thus contribute significantly to launch costs. This is a particularly troublesome issue with deep space probes. A communication-channel bit error rate of 10^{-5} is desired for many applications. There is thus a need for strong error control codes that operate efficiently at extremely low signal-to-noise ratios. Convolutional codes have been particularly successful in these applications, beginning with the convolutional encoder/sequential decoder used in the *Pioneer* missions [Mas2]. A NASA Planetary Standard was later developed for other missions, including a portion of the *Voyager* mission. The *Voyager* mission allowed for the use of the two convolutional codes listed in Tables 16-2 and 16-3. Both offer the maximal d_{free} for their respective rates and constraint length.

TABLE 16-2

$\mathbf{C}_1$: Rate 1/2, Constraint Length 7, $d_{\text{free}} = 10$
$\mathbf{g}^{(0)}(D) = 1 + D + D^3 + D^4 + D^6$
$\mathbf{g}^{(1)}(D) = 1 + D^3 + D^4 + D^5 + D^6$

TABLE 16-3

$\mathbf{C}_2$: Rate 1/3, Constraint Length 7, $d_{\text{free}} = 15$
$\mathbf{g}^{(0)}(D) = 1 + D + D^3 + D^4 + D^6$
$\mathbf{g}^{(1)}(D) = 1 + D^3 + D^4 + D^5 + D^6$
$\mathbf{g}^{(2)}(D) = 1 + D^2 + D^4 + D^5 + D^6$

[2] Geosynchronous systems are becoming an exception as bandwidth crowding becomes an increasingly serious problem. TCM is now being considered for some applications.

With 3-bit soft-decision decoding, $\mathbf{C}_1$ and $\mathbf{C}_2$ provide 5.1 dB and 5.6 dB of coding gain, respectively, at a decoded bit error rate of 10^{-5} [Lin1]. $\mathbf{C}_1$ has since become an "industry standard" [Vit4] and has been used in applications from military satellite communications to commercial cellular telephony. Figure 16-6 shows the bit-error-rate performance of $\mathbf{C}_1$ as compared to an uncoded system and a (64, 32) Reed-Solomon error control system. The modulation format for all three curves is coherently demodulated BPSK. The $\mathbf{C}_1$ performance curve is an upper bound that remains tight until the SNR becomes very small (≤ 2.5 dB). The Reed-Solomon curve treats decoder failures as errors. Note that at low channel SNRs (≤ 4.5 dB), the convolutional code outperforms the Reed-Solomon code. It is only at higher SNRs that the Reed-Solomon code provides better performance. This is an example of a general tendency shown by convolutional and Reed-Solomon codes. The convolutional codes are good choices for applications requiring a moderate amount of reliability over poor channels. This is particularly true when soft-decision decoding is used in conjunction with channel side information. Reed-Solomon codes are a good choice for applications requiring extremely high levels of reliability on channels that are only moderately noisy.

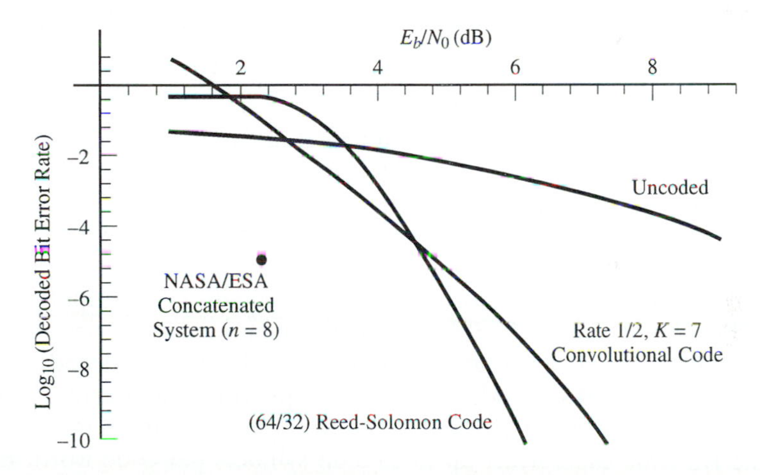

Figure 16-6. Bit-Error-Rate Performance for a Rate-1/2, $K = 7$ Convolutional Code, a (64, 32) Reed-Solomon Code, the NASA/ESA Concatenated System, and an Uncoded BPSK Channel

When the channel is extremely poor, it is often necessary to use more than one error control code. In the previous section, a pair of shortened RS codes were used in tandem to improve the fidelity of a digital audio system. Such a system is generally referred to as a **concatenated** coding system. Given the bursty nature of the digital audio channel, a pair of RS codes was a good choice. In deep space satellite applications the primary problem is random noise. Figure 16-7 shows the NASA/European Space Agency standard adopted in 1987 for deep space missions. The rate-1/2 Planetary Standard convolutional code is joined by a (255, 223) RS code in a concatenated system.

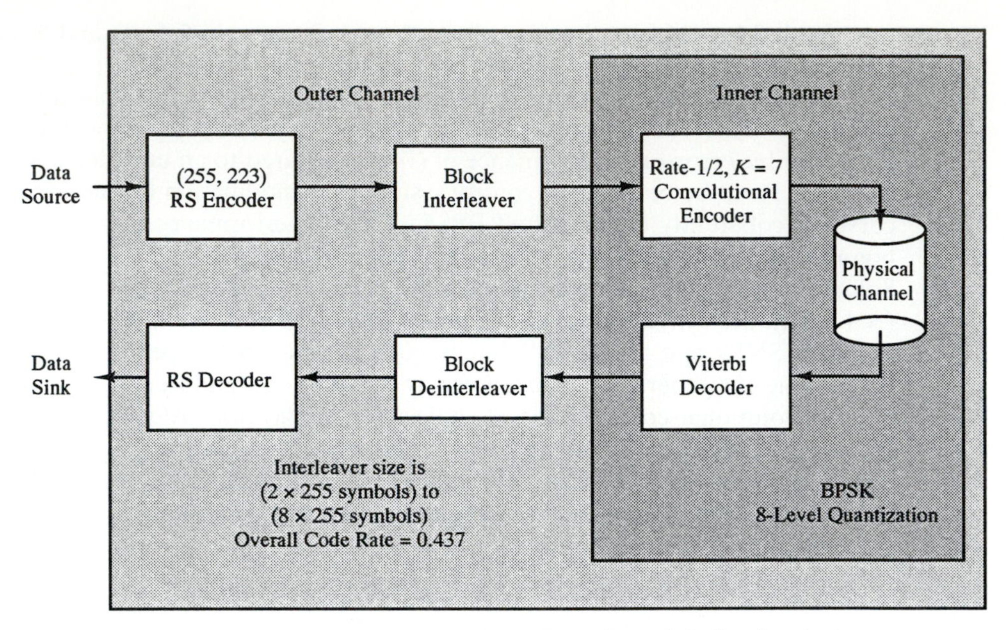

Figure 16-7. NASA/ESA Deep Space Error Control Coding Standard

The Viterbi decoder in this system is designed to improve the effective channel quality to the point where the RS code can be used efficiently. The RS code also provides some additional advantages in this scheme. When the Viterbi decoder suffers a decoder error, a code word is output that usually differs from the transmitted word over a few consecutive trellis branches. It follows that, though the input to the Viterbi decoder is corrupted by random noise, the output of the decoder tends to have burst errors. The RS code, with its inherent burst-error correcting capability, is thus a natural choice to "clean up" after the Viterbi decoder.

A single RS encoder/decoder pair provides burst-error "trapping" that is a function of the length (in bits) of the RS code symbols. In the NASA/ESA system, the RS code can correct up to 16 8-bit symbols per code word. Eight-bit symbols were chosen because the smallest nonzero path through the $K = 7$ convolutional code traverses 7 branches. The most frequently occurring error bursts at the output of the Viterbi decoder will thus have length 7 or 8, and can be trapped by a single RS symbol (in the best case).

If the Viterbi decoder has output bursts whose lengths exceed 128 bits, the RS decoder will probably fail or, less likely, a decoder error will occur. In order to allow for the correction of longer bursts, a symbol interleaver is placed between the two encoders. Current NASA missions use block interleavers that hold from $n = 2$ to $n = 8$ Reed-Solomon code words. For example, the NASA *Galileo* mission to Jupiter used $n = 2$, while ESA's *Giotto* mission to Halley's Comet used $n = 8$. The impact of interleaver size on the performance of the NASA/ESA system is discussed in detail in [Hag1]. Hagenauer et al. conducted simulation studies that showed that the NASA/ESA system with interleaver sizes $n = 2$, 4, and 8 provides a decoder BER of 10^{-5} at an E_b/N_0 of 2.6 dB, 2.45 dB, and 2.35 dB respectively. The uncoded BPSK system requires an E_b/N_0 of 9.6 dB to provide the same BER performance. The coding gain with respect to an uncoded system is thus 7.0 to 7.25 dB, depending on the size of the interleaver. The concatenated system thus provides an increase of 2 dB in coding gain over the rate-1/2 Planetary Standard convolutional code by itself (see Figure 16-6).

16.4 COMPUTER APPLICATIONS

The various computer applications for error control can be loosely grouped into four categories: memory (random access memory and read-only memory), disk storage, tape storage, and interprocessor communication. Each has its unique characteristics that indicate the use of certain types of codes. [Rao] provides the best detailed reference for this area, while [Pat1] and [Lin1, Ch. 16] provide nice summaries. In this section we will take a look at single-error-correcting/double-error-detecting (SEC-DED) codes for computer memory, Fire codes for magnetic disks, and the Reed-Solomon based system for the IBM 3850 tape mass storage system.

16.4.1 Computer Memory

The earliest memory storage units were mechanical relays. The contents of these storage devices could actually be verified through visual inspection. Simple parity-check codes provided error detection in these systems, though they remain legendary for their lack of reliability. The relays were eventually replaced by ferrite cores in which a magnetic field could be induced and its orientation read. These core memories had to be "refreshed" periodically and their contents verified to see if any errors had crept in since the last refresh cycle. In 1961 IBM introduced the 7030 computer, their first to have core memory with error correction capability. The IBM 7030 used a Hamming code to provide single-error correction and double-error detection [Lin1]. Subsequent core memory systems used variations of this code, as did later systems using semiconductor memory.

Semiconductor memory is currently the predominant choice for random access and read-only memory applications that require rapid access. Read/write errors in semiconductor memory are generally labeled as either "hard" or "soft" errors. Hard errors are device failures that are relatively permanent, while soft errors are transient, and are sometimes called "single event upsets" (SEUs). SEUs are frequently caused by radiation; atomic particles leave an ion trail, or "tunnel," as they pass through a device substrate. In many device technologies, this can cause a change in the state of the affected memory cells. SEUs can also be caused by electronic transients induced by lightning, adjacent electrical machinery, and the ionic scintillation generated by thermonuclear weapons. Note that some of these error sources can also cause hard failures.

Computers manipulate data in logical groups that contain one or more bytes, each byte consisting of 8 bits. It is thus convenient and natural for the error control system to operate on the same logical groups. The simpler semiconductor memory error control systems use a variant of the SEC-DED code originally developed for core memory. A form of spatial interleaving is achieved by distributing the bits comprising a given byte on 8 separate memory devices (chips or cards). A localized SEU thus appears at the error correcting decoder as a series of bytes or words containing a single bit error.

The extended Hamming codes have the following parameters. Note that m is a positive integer.

EXTENDED HAMMING CODE PARAMETERS

Length $= 2^m$
Dimension $= 2^m - m - 1$
Minimum distance $= 4$

A typical parity-check matrix is shown in Example 4-8. If a systematic encoding is assumed, the parity-check matrix defines a set of parity equations. These equations can be readily implemented in hardware and used to generate the redundant bits quite rapidly. Each row in the matrix denotes an addition operation, with the weight of the row being one more than the number of data and parity bits being summed in the encoding operation (one of the 1s denotes the parity bit being computed). There is thus a direct connection between the number of 1s in the parity-check matrix for the code and the complexity of the corresponding encoder and decoder circuits.

The parity-check matrix in Eq. (4-8) results in a circuit that is more complex than is necessary. Hsiao [Hsi] developed a set of shortened Hamming codes that provide the same minimum distance as the extended Hamming codes, but result in a simpler encoder and decoder. The basic Hamming code has the following parameters.

HAMMING CODE PARAMETERS

Length $= 2^m - 1$
Dimension $= 2^m - m - 1$
Minimum distance $= 3$

The parity-check matrix (see Section 4.5) has as columns all $(2^m - 1)$ distinct, binary m-tuples. A Hamming code is shortened by removing some j columns, resulting in a $(2^m - j - 1, 2^m - m - j - 1)$ code whose minimum distance is at least 3. Hsiao suggested the following requirements for the shortened codes (from [Lin1]).

1. Every column in the shortened parity-check matrix must have an odd number of 1s.
2. The total number of 1s in the matrix should be minimized.
3. The number of 1s per row should be approximately the same.

The first requirement ensures that the resulting shortened code has minimum distance 4. Consider the following. The syndrome $\mathbf{S} = \mathbf{r}\mathbf{H}^T$ corresponding to any single-error pattern will be one of the columns of the parity-check matrix, and will thus have odd weight. The syndrome corresponding to a double-error pattern will be the sum of two distinct, odd-weight columns, and will thus be nonzero and have even weight. All double-error patterns are thus detectable, so the shortened code must have minimum distance of at least 4.

The second and third requirements ensure, respectively, that the number of summands for each parity computation is minimal, and that the number of computations necessary for each parity bit are approximately the same. Complexity is thus kept to a minimum. The parity-check matrix for a (22, 16) SEC-DED code is shown below. It operates on two information bytes at a time and is optimal with respect to the above three requirements.

$$\mathbf{H} = \begin{bmatrix} 1&0&0&0&0&0&1&0&0&1&1&0&0&1&0&0&1&1&1&1&0&0 \\ 0&1&0&0&0&0&0&0&1&1&1&1&1&0&1&0&0&0&1&0&1&0 \\ 0&0&1&0&0&0&1&1&1&0&1&1&1&0&0&1&1&0&0&0&0&0 \\ 0&0&0&1&0&0&1&1&1&0&0&0&0&1&1&1&0&1&0&0&0&1 \\ 0&0&0&0&1&0&0&0&0&1&0&0&1&1&1&1&0&0&0&1&1&1 \\ 0&0&0&0&0&1&0&1&0&0&0&1&0&0&0&0&1&1&1&1&1&1 \end{bmatrix}$$

The IBM model 370 series of computers uses a (72, 64) SEC-DED Hsiao code. The parity-check matrix for this code has 216 ones, with an average of 27 ones per row. Suppose that 3-input XOR gates are used to perform the parity computation. The same three-level circuit containing a total of 13 gates is sufficient for the computation of each of the parity bits, as well as the computation of the syndrome.

16.4.2 Magnetic Disk Storage

Magnetic disks are frequently used for mass storage applications. As with the Compact Disc™, these storage devices involve moving media and/or record/read heads. Though not as prone to radiation-induced SEUs as semiconductor memory, disks do suffer from burst errors caused by surface contamination and material defects.

The early disk drives, such as the IBM 3330 (1971), IBM 3340 (1973), and IBM 3350 (1976), used variants of the Fire code, allowing for recording densities as high as 253 bits per millimeter. Fire described these codes in a report published in 1959 [Fir]. Fire codes are capable of correcting a single burst in a variable-length code word and are constructed as follows.

Let $g_1(x)$ be an irreducible, binary polynomial with degree m and period κ. The **period** of a polynomial $f(x)$ is defined as the smallest positive integer α such that $f(x)$ divides $(x^\alpha - 1)$. Note that the period of binary primitive polynomials of degree m is $2^m - 1$ by Definition 2-16. For irreducible polynomials it is only necessary that the period of the polynomial divide $2^m - 1$. Let b be a positive integer such that $b \leq m$

and κ does not divide $(2b - 1)$. A Fire code capable of correcting bursts of length less than or equal to b is then defined by the following generator polynomial.

$$g(x) = (x^{2b-1} + 1)g_1(x) \tag{16-2}$$

Note that the constraints on $(2b - 1)$ and κ, and the irreducibility of $g_1(x)$, ensure that the two factors $(x^{2b-1} + 1)$ and $g_1(x)$ are relatively prime. If $g(x)$ is to be a valid generator polynomial, then it must divide $(x^n + 1)$, where n is the length of the code. The smallest such n is the least common multiple of $(2b - 1)$ and κ.

The decoding operation for Fire codes is an interesting example of "error trapping." The general form of the decoding circuit is shown in Figure 16-8. Note that there are two key components: an error pattern linear feedback shift register (LFSR) and an error location LFSR. These two LFSRs use the techniques discussed in Section 5.2 to compute the syndromes of the received word with respect to the two factors of the generator polynomial in Eq. (16-2). Recall that the syndrome of a received polynomial $r(x)$ given a binary generator polynomial $g(x)$ is computed as follows.

$$S(x) \equiv r(x)\ modulo\ g(x) \tag{16-3}$$

Equivalently, we say that the syndrome of $r(x)$ with respect to $g(x)$ is the remainder for the division operation $r(x)/g(x)$.

Once $r(x)$ has been completely fed into the two LFSRs, the error location LFSR contains the syndrome $S_1(x)$ of $r(x)$ with respect to $g_1(x)$, while the error pattern LFSR contains the syndrome $S_2(x)$ of $r(x)$ with respect to $(x^{2b-1} + 1)$. Suppose that the received word has been corrupted by an error pattern $E(x)$. Since the syndrome is only a function of the error pattern and not of the transmitted code word, we have

$$\begin{aligned} S_1(x) &\equiv E(x)\ modulo\ g_1(x) \\ S_2(x) &\equiv E(x)\ modulo\ (x^{2b-1} + 1) \end{aligned} \tag{16-4}$$

If we assume that the error pattern is a burst of length b or less, we can reexpress the error pattern as $E(x) = x^p E_b(x)$, where $E_b(x)$ has degree less than b. $E_b(x)$ is said to be the burst pattern, while p is the burst location (the received word coordinate at which the burst begins). We now have

$$\begin{aligned} S_1(x) &\equiv x^p E_b(x)\ modulo\ g_1(x) \\ S_2(x) &\equiv x^p E_b(x)\ modulo\ (x^{2b-1} + 1) \end{aligned} \tag{16-5}$$

Theorem 5-3 showed that the syndromes for cyclically shifted versions of a received word are obtained through repeated shifts of the contents of the shift register, once the original syndrome has been calculated. We could thus have created a single shift register for computing Fire code syndromes with respect to $g(x)$, and trapped the error burst through $(n - p)$ shifts of the LFSR after the syndrome computation. The decoder in Figure 16-8 is able to find the burst pattern and location much sooner by taking advantage of the structure of the generator polynomial.

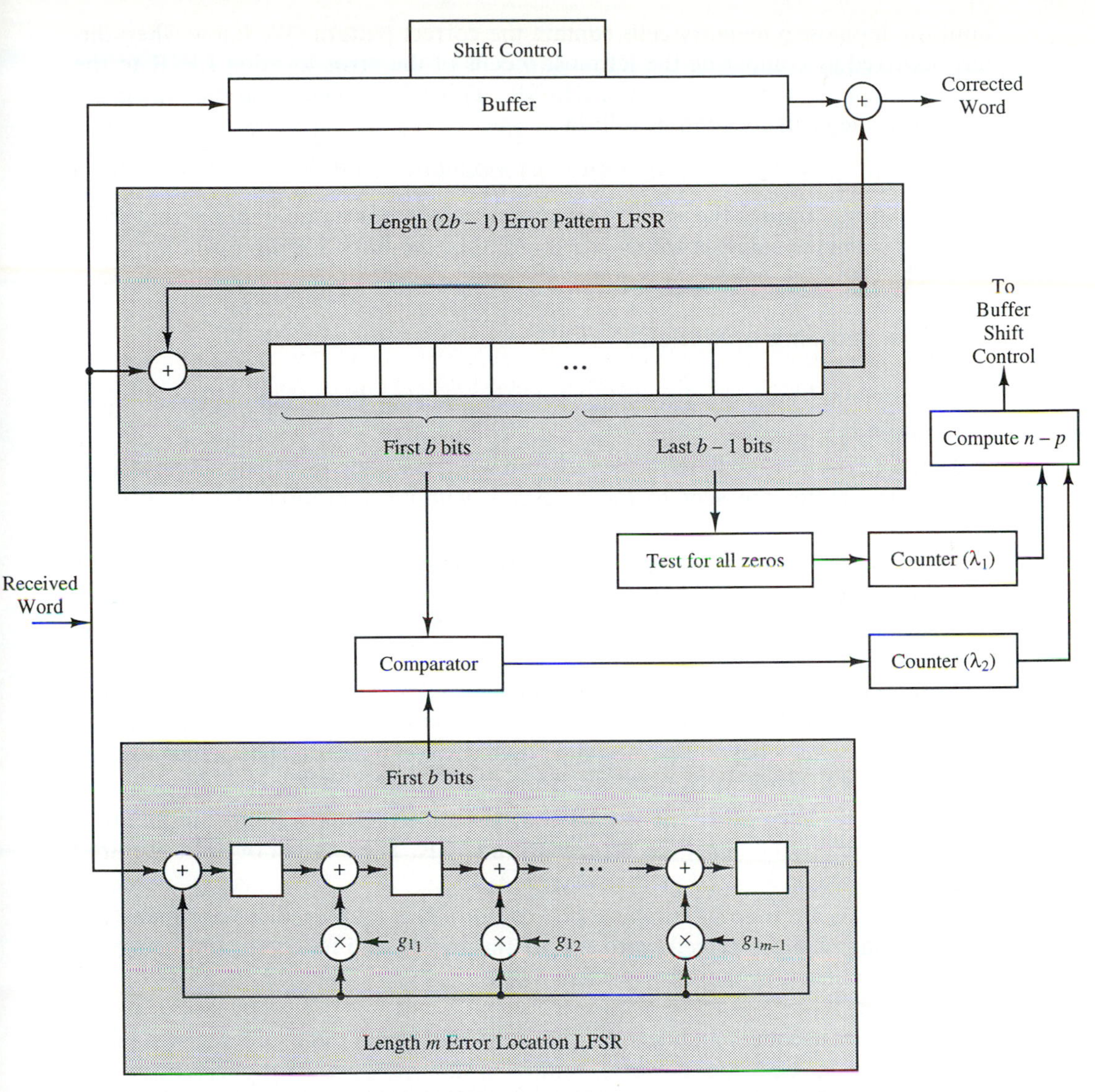

Figure 16-8. A Decoding Circuit for Fire Codes

The polynomial $(x^{2b-1} + 1)$ defines a circulating shift register. After some λ_1 additional shifts it is thus possible to move the error burst into the b leftmost memory cells of the error pattern LFSR. We will know this has occurred whenever the $(b - 1)$ rightmost cells contain zeros. λ_1 is related to p by the expression

$$\lambda_1 = (n - p) \ modulo \ (2b - 1) \tag{16-6}$$

Once the error pattern has been identified, we can then shift the error location LFSR until the leftmost b memory cells contain the correct pattern. We know when this has occurred by comparing the leftmost b cells of the error location LFSR to the pattern trapped in the error pattern LFSR. This takes some λ_2 shifts, which are related to the error location as follows.

$$\lambda_2 = (n - p) \textit{ modulo } \kappa \tag{16-7}$$

It can be shown through the application of some interesting number theory that there is always a unique solution for $(n - p)$, and thus the burst location p.

The decoding operation is thus performed as follows.

Fire decoding operation

1. $r(x)$ is fed into the two shift registers, causing the computation of $S_1(x)$ and $S_2(x)$.
2. If $S_1(x) = S_2(x) = 0$, then the received word is a code word, and the contents of the buffer are sent to the data sink without alteration and decoding is stopped.
3. If one syndrome is nonzero and the other is zero, the received word contains an uncorrectable, though detectable, error burst. The contents of the buffer are then marked as unreliable (perhaps erased) and sent to the data sink. Decoding is stopped.
4. If both $S_1(x)$ and $S_2(x)$ are nonzero, then the error pattern LFSR is shifted until its rightmost $(b - 1)$ bits are zero. This is guaranteed to require no more than $(2b - 2)$ shifts beyond the original syndrome computation. The actual number of shifts required is counted and stored as the variable λ_1.
5. The error location LFSR is then shifted until its leftmost b bits are identical to the leftmost b bits in the error pattern LFSR. The number of shifts required is counted and stored as the variable λ_2.
6. λ_1 and λ_2 are used to compute $(n - p)$, which is then used to coordinate the shifting of the received word out of the buffer and its correction using the trapped burst in the error pattern LFSR.

The Fire encoders and decoders are very fast, given their shift-register-based implementation. They are, however, limited in their random-error-correcting capability. More recent magnetic disk drives (e.g., IBM 3370 (1979), IBM 3375 (1980), and IBM 3380E (1985)) have used modified Reed-Solomon codes [Pat1]. These codes allow for recording densities as high as 638 bits per millimeter.

16.4.3 Magnetic Tape Storage

Magnetic tape systems suffer from error-causing mechanisms similar to those encountered in magnetic and optical disk systems (e.g., material defects and surface contamination). In addition, the flexibility of the magnetic tape can create its own unique set of problems (e.g. bit insertion due to tape stretching). In most digital systems the data on the tape is organized in parallel tracks that run the length of the tape. For example, the IBM 3420 series magnetic tape storage system uses 9 data tracks on half-inch tape, as shown in Figure 16-9. Given this organization and the direction in which the tape moves, contaminants tend to corrupt one track or a small number of adjacent tracks during the reading process. In almost all error control systems for magnetic tape, the code words are thus defined vertically (across the tracks), providing a form of interleaving. The loss of one or two tracks then requires the correction of only one or two symbols per code word.

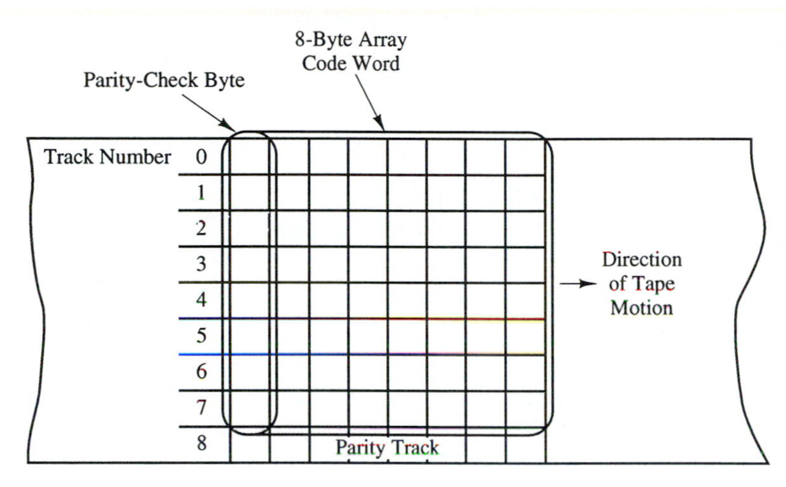

Figure 16-9. Data format for the IBM 3420 Magnetic Tape Storage System

The IBM 3420 system is particularly interesting in that, during recording, the code-word bytes are arranged vertically, allowing for "on the fly" computation and recording of the parity-check byte. Since the code words thus run horizontally, it would seem that no track interleaving is provided, but this is not the case. During decoding the bytes are read vertically, with 8 bits on each track contributing a byte to the received word. The details of this code and its decoding are described by the code's inventors, Patel and Hong, in [Pat2]. [Lin1] also provides a nice description. This code can correct errors caused by the loss of a single track, and, given an external error-pointing mechanism that declares track erasures, it can also correct double-track errors. The resulting reliability allows for recording densities of 250 bits per millimeter [Pat1].

In 1975 IBM introduced the 3850 Mass Storage System (MSS), an enormous tape-based storage facility. It contains up to 4720 tape cartridges stored in hexagonal compartments, each cartridge being 3.5 inches long and 1.9 inches in diameter [Pat1]. Each cartridge contains 64 feet of 2.7-inch-wide magnetic tape. The MSS has a mechanical system that selects cartridges for reading, then replaces them when they are no longer needed (much like an old juke box). The total capacity of the system is 50.4 MBytes, rendering it quite obsolete by current standards, but the IBM 3850 remains an interesting system to study because of its coding scheme.

The IBM 3850 uses a rotary read-write head. The tape moves past the head in a helical turn, as shown in Figure 16-10. Owing to the relative positions of the tape and head, the data is recorded in slanted "stripes," as shown in Figure 16-11. Each stripe is divided into 20 segments, which are in turn divided into 15 sections of 16 bytes each [Pat3]. The sections also contain a parity bit (odd) and several synchronization bits.

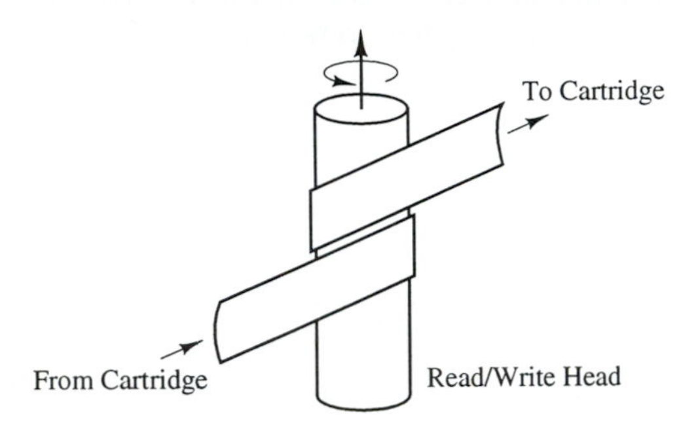

Figure 16-10. Rotating Read-Write Head in IBM 3850 MSS

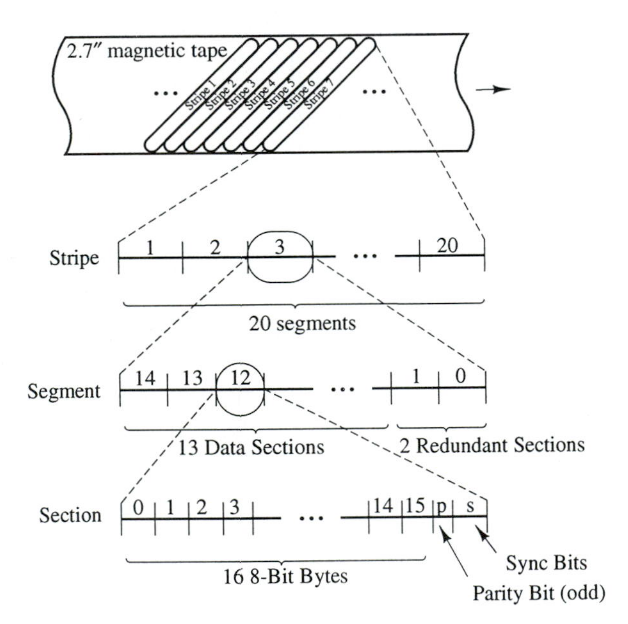

Figure 16-11. Data Format for IBM 3850 MSS

The primary organizational block for the IBM 3850 MSS is the segment. Each segment consists of 16 interleaved code words from an 8-ary (15, 13) BCH code. The 15 8-bit symbols in each code word are distributed across the 15 sections of the segment. The first section contains the first symbols for the 16 code words, the second section contains the second symbols, and so on.

The code is defined as follows. $f(x) = x^8 + x^5 + x^3 + x + 1$ is primitive and can thus be used to construct $GF(2^8)$. Let α be a root of $f(x)$. From Chapter 2 we know that α has order 255, and that all of the nonzero elements of $GF(2^8)$ can be expressed as powers of α. Let $\beta = \alpha^{17}$. Since $17 \cdot 15 = 255$, it follows that the order of β must be 15. It follows that $GF(2^8)$ contains the subfield $GF(2^4) = \{0, \beta, \beta^2, \beta^3, \ldots, \beta^{15}\}$. The IBM 3850 MSS code is defined by the generator polynomial $g(x) = (x + 1)(x + \beta)$. This code has minimum distance 3, and can thus correct single-byte errors. Given an external error pointer, the code can correct double-byte erasures.

Since the code is cyclic, simple shift-register-based circuits can be used for both encoding and decoding. The interested reader is referred to [Pat1] and [Pat3].

Appendix A

Binary Primitive Polynomials

This appendix contains a program for generating primitive polynomials. This is followed by a list of all of the binary primitive polynomials of degrees 2 through 14 and a list of all weight-three and weight-five binary primitive polynomials of degrees 15 through 32. The polynomials are listed as binary numbers (e.g., 100101 $\leftrightarrow x^5 + x^2 + 1$) in increasing order.

A.1 C PROGRAM FOR GENERATING PRIMITIVE POLYNOMIALS[1]

```
#define N 32

#include <stdio.h>

int i,n,xx; char d[N];

void see () {
    for (i=n; i>=0; i--) printf("%c",d[i]+'0');
    printf("\n");
}

char s[N],t,j,f; unsigned long c,max;
void visit () {
    for (i=0;i<n;i++) s[i]=1; c=0;
```

[1] This program was written by Haluk Aydinoglu and is printed here with his permission.

```
        do {c++;
            for (i=t=0;i<n;i++)
                t = (t^(s[i]&d[i]));
            for (i=0;i<n-1;i++)
                s[i]=s[i+1];
        s[n-1]=t;
            for (i=f=0; i<n; i++)
                if (!s[i]) {f=1; break;}
        } while (f);
        if (c==max) see ();
    }
    void gp (1) char 1; {
            if (!1) visit ();
        else {
            d[1] = 0;gp (1-1);
            d[1] = 1;gp (1-1);
        }
    }
    void gc (1,rw) char 1, rw; {char q;
        if (rw == 2) {visit(); return;}
        for (q=1;q>=rw-2;q--) {
            d[q]=1;gc(q-1,rw-1);d[q]=0;
        }
    }

    void main () {
        printf("\n");
         for (n=21; n<= 25; n++) {printf("%d\n",n);/*this line determines the interval*/
        for (xx=max=1;xx <= n;++) max = 2*max;max-;
        d[n]=d[0]=1; gc (n-1,3);/*if you want to generate all polynomials*/
                        /*then use gp(n-1)*/

            printf("\n");
        }
    }
```

A.2 ALL PRIMITIVE POLYNOMIALS OF DEGREES 2 THROUGH 14

2
111

3
1011
1101

4
10011
11001

5
100101
101001
101111
110111
111011
111101

6
1000011
1011011
1100001
1100111
1101101
1110011

7
10000011
10001001
10001111
10010001
10011101
10100111
10101011
10111001
10111111
11000001
11001011
11010011
11010101

11100101
11101111
11110001
11110111
11111101

8

100011101
100101011
100101101
101001101
101011111
101100011
101100101
101101001
101110001
110000111
110001101
110101001
111000011
111001111
111100111
111110101

9

1000010001
1000011011
1000100001
1000101101
1000110011
1001011001
1001011111
1001101001
1001101111
1001110111
1001111101
1010000111
1010010101
1010100011
1010100101
1010101111
1010110111
1010111101
1011001111
1011010001
1011011011
1011110101
1011111001
1100010011
1100010101
1100011111
1100100011
1100110001
1100111011
1101001111
1101011011
1101100001
1101101011
1101101101
1101110011
1101111111
1110000101
1110001111
1110110101
1110111001
1111000111
1111001011
1111001101
1111010101
1111011001
1111100011
1111101001
1111111011

10

10000001001
10000011011
10000100111
10000101101
10001100101
10001101111
10010000001
10010001011
10011000101
10011010111
10011100111
10011110011
10011111111
10100001101
10100011001
10100100011
10100110001
10100111101
10101000011
10101010111
10101101011
10110000101
10110001111
10110010111
10110100001
10111000111
10111100101
10111110111
10111111011
11000010011
11000010101
11000100101
11000110111
11001000011
11001001111
11001011011
11001111001
11001111111
11010001001
11010110101
11011000001
11011010011
11011011111
11011111101
11100010111
11100011101
11100100001
11100111001
11101000111
11101001101
11101010101
11101011001
11101100011
11101111101
11110001101
11110010011
11110110001
11111011011
11111110011
11111111001

11

100000000101
100000010111
100000101011
100000101101
100001000111
100001100011
100001100101
100001110001
100001111011
100010001101
100010010101
100010011111
100010101001
100010110001
100011001111
100011010001
100011100001
100011100111
100011101011
100011110101
100100001101
100100010011
100100100101
100100101001
100100111011
100100111101
100101000101
100101001001
100101010001
100101011011
100101110011
100101110101
100101111111
100110000011
100110001111
100110101011
100110101101
100110111001
100111000111
100111011001
100111100101
100111110111
101000000001
101000000111
101000010011
101000010101
101000101001
101001001001
101001100001
101001101101
101001111001
101001111111
101010000101
101010010001
101010011101
101010100111
101010101011
101010110011
101010110101

101011010101
101011011111
101011101001
101011101111
101011110001
101011111011
101100000011
101100001001
101100010001
101100110011
101100111111
101101000001
101101001011
101101011001
101101011111
101101100101
101101101111
101101111101
101110000111
101110001011
101110010011
101110010101
101110101111
101110110111
101110111101
101111001001
101111011011
101111011101
101111100111
101111101101
110000001011
110000001101
110000011001
110000011111
110001010111
110001100001
110001101011
110001110011
110010000101
110010001001
110010010111
110010011011
110010011101
110010110011
110010111111
110011000111
110011001101
110011010011
110011010101
110011100011
110011101001
110011110111
110100000011
110100001111
110100011101
110100100111
110100101101
110101000001
110101000111
110101010101
110101011001
110101100011
110101101111
110101110001
110110010011
110110011111
110110101001
110110111011
110110111101
110111001001
110111010111
110111011011
110111100001
110111100111
110111110101
111000000101
111000011101
111000100001
111000100111
111000101011
111000110011
111000111001
111001000111
111001001011
111001010101
111001011111
111001110001
111001111011
111001111101
111010000001
111010010011
111010011111
111010100011
111010111011
111011001111
111011011101
111011110011
111011111001
111100001011
111100011001
111100110001
111100110111
111101011101
111101101011
111101101101
111101110101
111110000011
111110010001
111110010111
111110011011
111110100111
111110101101
111110110101
111111001101
111111010011
111111100101
111111101001

12

1000001010011
1000001101001
1000001111011
1000001111101
1000010011001
1000011010001
1000011101011
1000100000111
1000100011111
1000100100011
1000100111011
1000101001111
1000101010111
1000101100001
1000101101011
1000110000101
1000110110011
1000111011001
1000111011111
1001000001101
1001000110111
1001000111101
1001001100111
1001001110011
1001001111111
1001010111001
1001011000001
1001011001011
1001100001111
1001100011101
1001100100001
1001100111001
1001100111111
1001101001101
1001101110001
1001110011001
1001110100011
1001110101001
1010000000111
1010000110001
1010000110111
1010001001111
1010001011101
1010001100111
1010001110101
1010010100111
1010010101101
1010011010011
1010100001111
1010100011101
1010101001101
1010110010011
1010111000101
1010111010111
1010111011101
1010111101011
1011000001001
1011001000111
1011001010101
1011001011001
1011010100101
1011010111101
1011100010101
1011100011001
1011101000011
1011101000101
1011101110101
1011110001001
1011110101101
1011110110011
1011110111111
1011111000001
1100001010111
1100001011101
1100010010001
1100010010111
1100010111001

1100011101111
1100100011011
1100100110101
1100101000001
1100101100101
1100101111011
1100110001011
1100110110001
1100110111101
1100111001001
1100111001111
1100111100111
1101000011011
1101000101011
1101000110011
1101001101001
1101010001011
1101011010001
1101011100001
1101011110101
1101100001011
1101100010011
1101100011111
1101101010111
1101110010001
1101110100111
1101110111111
1101111000001
1101111010011
1110000000101
1110000010001
1110000010111
1110000100111
1110001001101
1110010000111
1110010011111
1110010100101
1110010111011
1110011000101
1110011001001
1110011001111
1110011110011
1110100000111
1110100100011
1110101000011
1110101010001
1110101011011
1110101110101
1110110000101
1110110001001
1111000010101
1111000011001
1111000101111
1111001000101
1111001010001
1111001100111
1111001110011
1111010001111
1111011100011
1111100010001
1111100011011
1111100100111
1111101110001
1111110011001
1111110111011
1111110111101
1111111001001

13

10000000011011
10000000100111
10000000110101
10000001010011
10000001100101
10000001101111
10000010001011
10000010001101
10000010011111
10000010100101
10000010101111
10000010111011
10000010111101
10000011000011
10000011001001
10000011100001
10000011110011
10000100001101
10000100010101
10000100101001
10000100101111
10000100111011
10000101000011
10000101100111
10000101101011
10000101111001
10000110001001
10000110010111
10000110011101
10000110111111
10000111000001
10000111000111
10000111001101
10000111011111
10000111100011
10000111110001
10000111111011
10001000011001
10001000100101
10001000110111
10001000111101
10001001000011
10001001011011
10001001011101
10001001111001
10001001111111
10001010001001
10001010010111
10001010011011
10001010110011
10001010111111
10001011001101
10001011101111
10001011110111
10001011111011
10001100000101
10001100100111
10001100101011
10001101000111
10001101010101
10001101011001
10001101101111
10001101110001
10001101111101
10001110000111
10001110001101
10001110010101
10001110100011
10001110101001
10001110110001
10001110110111
10001110111011
10001111100001
10001111101101
10001111111001
10010000001011
10010000010011
10010000011111
10010000100101
10010000101001
10010000111101
10010001010001
10010001010111
10010001100001
10010001101101
10010001111111
10010010000011
10010010011011
10010010011101
10010010110101
10010010111111
10010011000001
10010011000111
10010011001011
10010011100011
10010100001001
10010100010111
10010100011101
10010100100001
10010100101101
10010100111001
10010101010011
10010101010101
10010101100011
10010101110001
10010101110111
10010110000111
10010110001011
10010110010101
10010110011001
10010110011111
10010110101111
10010110111101
10010111000101
10010111001111
10010111010111
10010111101011
10011000000011
10011000000101
10011000010001
10011000101101
10011000111111
10011001001011
10011001010011
10011001011001
10011001101001
10011001110111

10011001111011
10011010000111
10011010010011
10011010011001
10011010110001
10011010110111
10011010111101
10011011000011
10011011101011
10011011110101
10011100010011
10011100101001
10011100111011
10011101001111
10011101010111
10011101011101
10011101101011
10011101110011
10011101111001
10011110000011
10011110010001
10011110100001
10011110111001
10011111000111
10011111001011
10011111011111
10011111101111
10011111110001
10100000000111
10100000011001
10100000011111
10100000100011
10100000110001
10100000111011
10100000111101
10100001000101
10100001100111
10100001110101
10100010000101
10100010101011
10100010101101
10100010111111
10100011001101
10100011010101
10100011011111
10100011100011
10100011101001
10100011111011
10100100001001
10100100001111
10100100010001
10100100011011
10100100101011
10100100110101
10100100111111
10100101000001
10100101001011
10100101010101
10100101110111
10100101111101
10100110000001
10100110010011
10100110011111
10100110101111
10100110110111
10100110111101
10100111000011
10100111010111
10100111110011
10100111110101
10101000000011
10101000001111
10101000011101
10101000100001
10101000110011
10101000110101
10101001001101
10101001101001
10101001101111
10101001110001
10101001111011
10101001111101
10101010100101
10101010101001
10101010110001
10101011000101
10101011010111
10101011011011
10101011101011
10101011110011
10101100000001
10101100010101
10101100100011
10101100100101
10101100101111
10101100110111
10101101000011
10101101001001
10101101101101
10101101111111
10101110000101
10101110010111
10101110011011
10101110101101
10101110110011
10101111011001
10101111100101
10101111111101
10110000001111
10110000100001
10110000101011
10110000101101
10110000111111
10110001000001
10110001001101
10110001110001
10110010001011
10110010001101
10110010010101
10110010100011
10110010101111
10110010111101
10110011000101
10110011010001
10110011010111
10110011100001
10110011100111
10110011101011
10110100001101
10110100011001
10110100101001
10110100101111
10110100110111
10110100111011
10110101000101
10110101011011
10110101100111
10110101110101
10110110001001
10110110001111
10110110100111
10110110101011
10110110110101
10110111100011
10110111110001
10110111111101
10111000000111
10111000010011
10111000010101
10111000101001
10111001001001
10111001001111
10111001011011
10111001011101
10111001100001
10111001101011
10111010001111
10111010010001
10111010010111
10111010011101
10111010101011
10111010110011
10111010111001
10111011011111
10111011111011
10111011111101
10111100000101
10111100001001
10111100010001
10111100010111
10111100111111
10111101000001
10111101001011
10111101001101
10111101011001
10111101011111
10111101100101
10111101101001
10111110010101
10111110100101
10111110101111
10111110110001
10111111001111
10111111011101
10111111100111
10111111101101
10111111110101
10111111111111
11000000000111
11000000010101
11000000011001
11000000101111
11000001001001
11000001001111
11000001100111
11000001111001

11000001111111
11000010010001
11000010100001
11000010110101
11000010111111
11000011000001
11000011010011
11000011011001
11000011100101
11000011101111
11000100000101
11000100001111
11000100110101
11000101000111
11000101001101
11000101011111
11000101100011
11000101110001
11000101111011
11000110100011
11000110101001
11000110110111
11000111000101
11000111001001
11000111011011
11000111100001
11000111101011
11000111101101
11000111110011
11000111111111
11001000001001
11001000001111
11001000011101
11001000100111
11001000111001
11001001001011
11001001010011
11001001011001
11001001100101
11001010000001
11001010010011
11001010011001
11001010011111
11001010101001
11001010110111
11001010111011
11001011000011
11001011010111
11001011011011
11001011100111
11001100000111
11001100010101
11001100101111
11001101010001
11001101011101
11001101110101
11001110010111
11001110011011
11001110101011
11001110111001
11001111000001
11001111000111
11001111010101
11001111100011
11001111100101
11001111110111
11001111111011
11010000001001
11010000011011
11010000100111
11010001000001
11010001001101
11010001011111
11010001101001
11010001110111
11010001111011
11010010000111
11010010010011
11010010011001
11010010100101
11010010111101
11010011001001
11010011011011
11010011100111
11010011111001
11010100001101
11010100011111
11010100100101
11010100110001
11010100110111
11010101000101
11010101001111
11010101011101
11010101101101
11010101110011
11010101111111
11010110011101
11010110100001
11010110111001
11010111001101
11010111010101
11010111011001
11010111100011
11010111101001
11010111101111
11011000000001
11011000001011
11011000011111
11011000100101
11011000101111
11011000111011
11011001001001
11011001010001
11011001011011
11011001110011
11011001110101
11011010010001
11011010011011
11011010011101
11011010101101
11011011001011
11011011010011
11011011010101
11011011100011
11011011101111
11011100000101
11011100001111
11011100011011
11011100100001
11011100101101
11011100111001
11011101000001
11011101000111
11011101010011
11011101110001
11011101110111
11011110001011
11011110010101
11011110011001
11011110100011
11011111000101
11011111001111
11011111010001
11011111010111
11011111011101
11011111100001
11011111110011
11100000000011
11100000000101
11100000010111
11100000011101
11100000100111
11100000110011
11100001001011
11100001011001
11100001101001
11100001110001
11100010100011
11100010110001
11100010111011
11100011001001
11100011001111
11100011100001
11100011110011
11100011111001
11100100000001
11100100000111
11100100001011
11100100010011
11100100110001
11100101001111
11100101100111
11100101101101
11100110000011
11100110000101
11100110010111
11100110100001
11100110100111
11100110101101
11100111001011
11100111001101
11100111010011
11100111101111
11100111110111
11100111111101
11101000000111
11101000101001
11101000101111
11101000111101
11101001010001
11101001011101
11101001100001
11101001100111
11101001110011
11101001110101
11101010001001

11101010111001
11101010111111
11101011001101
11101011010011
11101011010101
11101011011111
11101011100101
11101011101001
11101011111011
11101100010001
11101100101011
11101100101101
11101100110101
11101100111111
11101101010011
11101101011001
11101101100011
11101101100101
11101101101111
11101101110001
11101101110111
11101110001011
11101110011001
11101110100101
11101110101001
11101110110111
11101110111011
11101111010001
11101111100111
11101111110011
11101111111111
11110000001101
11110000010011
11110000010101
11110000011111
11110000100011
11110000100101
11110000111011
11110001001111
11110001011101
11110001101101
11110010000011
11110010001111
11110010011101
11110010100111
11110010101011
11110010111001
11110011000111
11110011101001
11110111111011
11110011111101
11110100000011
11110100010111
11110100011011
11110100100001
11110100101101
11110100110011
11110100110101
11110101000001
11110101001101
11110101100101
11110101101001
11110101111101
11110110000001
11110110010101
11110110110001
11110110110111
11110111000011
11110111010001
11110111011011
11110111100111
11110111101011
11110111111001
11111000000101
11111000001001
11111000001111
11111000011011
11111000101011
11111000111111
11111001000001
11111001010011
11111001100101
11111001101001
11111010001011
11111010100011
11111010111101
11111011000101
11111011010111
11111011011101
11111011100001
11111011111001
11111100001101
11111100011001
11111100011111
11111100100101
11111100110111
11111100111101
11111101000011
11111101000101
11111101001001
11111101010001
11111101010111
11111101100001
11111110000011
11111110001001
11111110010001
11111110101011
11111110110101
11111111100011
11111111110111
11111111111101

14

100000000101011
100000000111001
100000001010011
100000001011111
100000001111011
100000010101001
100000010101111
100000010111011
100000010111101
100000011001111
100000011101011
100000011110011
100000100001101
100000100010011
100000100111011
100000101000011
100000110011011
100000110011101
100000110100111
100000110101101
100000110110101
100000111010101
100000111011001
100000111110001
100001000001101
100001001010111
100001001100001
100001001111111
100001010000101
100001010011101
100001011000111
100001011001011
100001011001101
100001011100011
100001011101001
100001011101111
100001100001001
100001100100001
100001100111111
100001101111101
100001110000111
100001110010101
100001110101111
100001111001001
100001111101011
100001111101101
100010000001011
100010001000011
100010001110011
100010011010011
100010011010101
100010011011111
100010011100011
100010011111011
100010100101011
100010100111001
100010101011001
100010101101111
100010110011001
100010110011111
100010110100101
100010110110111
100010111000101
100010111010111
100010111100111
100010111110011
100010111111111
100011000001111
100011000011101
100011000100111
100011000110101
100011001000111
100011001011001
100011001100011
100011001110001
100011001111011
100011010100101
100011011000101
100011011001111
100011011011011
100011100110001
100011101001111
100011101111111

100011110100111
100011111000001
100011111100101
100011111101001
100011111101111
100100000010011
100100000011001
100100000111011
100100001000011
100100001011011
100100001100001
100100001100111
100100001111111
100100010000011
100100011100011
100100100011101
100100100101101
100100100111001
100100101101001
100100101110111
100100110000111
100100110101001
100100111001111
100100111100001
100101000110101
100101001100101
100101001111011
100101010000001
100101010000111
100101010001011
100101010100011
100101011001001
100101011011101
100101011111001
100101100110111
100101101001001
100101101011011
100101101101011
100101101101101
100101101110011
100101101111001
100101110000101
100101110010111
100101110100001
100101110101011
100101110111001
100101111001011
100101111001101
100101111010101
100101111011111
100101111110001
100101111111011
100110000001001
100110000011101
100110000111111
100110001000111
100110001001011
100110001010101
100110001111101
100110010001101
100110010100101
100110010111101
100110011000101
100110011010001
100110011010111
100110011101101
100110100011111
100110100110001
100110101000101
100110101001111
100110101010001
100110101110101
100110110001111
100110110101101
100110110110011
100110111000001
100110111111011
100111000000001
100111000010101
100111000110111
100111001000011
100111001001001
100111001010001
100111001101011
100111001101101
100111010000011
100111010011011
100111011010011
100111011100101
100111011101001
100111011101111
100111100011101
100111101010101
100111101011111
100111101101001
100111101101111
100111110011111
100111110101001
100111110110111
100111110111011
100111111101011
101000000000111
101000000011111
101000000100101
101000001011101
101000010100001
101000011010101
101000011101001
101000100101101
101000101010011
101000101011001
101000101011111
101000101111101
101000110001011
101000110011001
101000110100101
101000110110001
101000111010001
101000111011011
101000111101011
101000111101101
101000111110101
101000111111111
101001000000101
101001000001111
101001000101101
101001001100101
101001010001011
101001010011001
101001010100011
101001010110001
101001011000011
101001011000101
101001011010001
101001011100111
101001100100101
101001100101001
101001101011011
101001110011011
101001110011101
101001110100111
101001110111001
101001110111111
101001111000111
101001111110001
101010000000011
101010000001111
101010000010111
101010000110101
101010000111001
101010001000111
101010001011111
101010001111011
101010010001101
101010010011111
101010010111011
101010011011101
101010011100001
101010100011001
101010100101111
101010100110111
101010101111001
101010101111111
101010110000011
101010110000101
101010110010001
101010110101101
101010111000001
101010111010011
101010111101001
101010111110111
101011000001101
101011000010101
101011000101001
101011000101111
101011000110001
101011010011011
101011010101011
101011010101101
101011010110011
101011011000001
101011011000111
101011011001101
101011011110111
101011100101011
101011100101101
101011100111111
101011101011001
101011101100011
101011101111101
101011110010011
101011110101111
101011110111101
101011111000011
101011111000101
101011111010111

101100000000011
101100000100001
101100000110011
101100000110101
101100000111111
101100001000001
101100010010101
101100010011001
101100010100011
101100010101111
101100011011101
101100011100111
101100011101011
101100011111111
101100101001111
101100101111111
101100110001111
101100110010111
101100110011101
101100110100001
101100110100111
101100110110101
101100111011111
101100111101001
101101000011111
101101000100101
101101000111011
101101000111101
101101001000011
101101001000101
101101001001001
101101001110101
101101010110101
101101011000001
101101011010011
101101011010101
101101011011001
101101100001111
101101100010111
101101100101011
101101100111001
101101101100011
101101101101001
101101101110111
101101110011001
101101111000011
101101111000101
101101111100001
101101111101011
101110000000111
101110000001011
101110000011001
101110000110001
101110001001001
101110001011101
101110001111001
101110010000011
101110010100001
101110010111111
101110011000001
101110011001011
101110011001101
101110011100101
101110100000101
101110100011101
101110100110011
101110101101111
101110110001101
101110110010101
101110110100011
101110110101001
101110110111011
101110111001111
101111000000011
101111000001111
101111000011011
101111000101101
101111001000111
101111001111101
101111010000001
101111010011001
101111011010111
101111011011011
101111011100111
101111011110101
101111100000111
101111100011001
101111100100011
101111100111101
101111101000011
101111101000101
101111101100001
101111101101011
101111101110101
101111110010111
101111111011111
110000000001101
110000000010101
110000000101111
110000000111101
110000001100111
110000001101011
110000010001001
110000010011101
110000010101011
110000010110011
110000010111001
110000011010101
110000100001001
110000100010001
110000100101101
110000100111001
110000101000001
110000101111101
110000110011111
110000110100101
110000110101111
110000110111011
110000111101101
110000111110101
110001000110011
110001001100011
110001001101111
110001001111101
110001010001101
110001010011111
110001010100101
110001010101001
110001010110111
110001011010111
110001011011011
110001011011101
110001011110011
110001100100011
110001100110001
110001100110111
110001101011011
110001101101101
110001101110101
110001110001001
110001110010001
110001110100001
110001110111111
110001111101111
110010000001001
110010000010111
110010001000001
110010001000111
110010010001011
110010010011111
110010011100111
110010011110101
110010100000001
110010100001011
110010100110111
110010100111011
110010101000101
110010101011011
110010110010001
110010110101101
110010110111001
110010110111111
110010111010101
110010111101111
110010111110111
110011000000111
110011000001101
110011000100011
110011000111011
110011001011101
110011001101011
110011010000011
110011010010111
110011010100111
110011010101011
110011010110101
110011011011001
110011011011111
110011100010001
110011100011011
110011101001011
110011101011111
110011101101001
110011101110111
110011110000001
110011110100011
110011111010001
110100000010001
110100000011101
110100001010011
110100001110111
110100010010011
110100010011111
110100010100101
110100010101001
110100011000101

110100011111111
110100100011001
110100100101111
110100101101011
110100101110011
110100110001111
110100110011101
110100110100001
110100110101011
110100111101001
110101000000001
110101000101111
110101001010001
110101001011011
110101001100111
110101001101011
110101001101101
110101001110101
110101010000011
110101010010111
110101010110011
110101010110101
110101010111111
110101011001011
110101011101001
110101011101111
110101100000011
110101100010111
110101100101011
110101100110011
110101100111001
110101101000111
110101101001011
110101101101001
110101101111101
110101110000001
110101110001101
110101110110111
110101111000101
110101111010111
110101111100001
110101111101101
110101111111001
110110000101111
110110000111101
110110001010111
110110001110011
110110010110101
110110010111001
110110011000001
110110011000111
110110011100101
110110100001001
110110100010111
110110100101011
110110101010011
110110101100011
110110101100101
110110101101001
110110110100011
110110110110001
110110110111011
110110110111101
110110111000101
110111000001001
110111000100111
110111000101101
110111000110011
110111001000001
110111001010011
110111001111011
110111010000001
110111010000111
110111010010101
110111010101111
110111011000011
110111011011011
110111011011101
110111011111001
110111100000001
110111100010101
110111100101001
110111100110001
110111100111011
110111101001111
110111101010111
110111110010001
110111111000111
110111111011001
110111111011111
110111111101001
111000000000101
111000000011101
111000000110011
111000001111101
111000010101001
111000010111011
111000011001001
111000011001111
111000011100001
111000100010011
111000100010101
111000100011001
111000100110001
111000100110111
111000100111101
111000101101011
111000110011011
111000110100001
111000110110101
111000111011111
111000111100101
111000111110111
111000111111011
111001000011111
111001000110001
111001000111011
111001001001111
111001001100111
111001010110011
111001011000001
111001011001101
111001011011111
111001011100101
111001011110001
111001011110111
111001100000011
111001100001001
111001100100111
111001100101011
111001110001101
111001110010011
111001110100101
111001110111101
111001111010001
111001111111111
111010000010011
111010000010101
111010001001111
111010001011011
111010001101011
111010001101101
111010010101011
111010010110011
111010011001101
111010011011111
111010011101001
111010011111101
111010100011011
111010100100001
111010101110111
111010101111011
111010110011001
111010110100011
111010110111101
111010111001111
111010111010001
111010111101011
111010111110101
111011000111001
111011001000111
111011001010011
111011001010101
111011001011111
111011001100011
111011001101001
111011010011111
111011010100011
111011010101111
111011011010001
111011011101011
111011011111001
111011100001011
111011101001001
111011101010111
111011101101101
111011101110011
111011110100111
111011110110101
111011111000111
111011111010011
111011111010101
111100000010101
111100000100101
111100000110001
111100000111101
111100001101101
111100011001011
111100011001101
111100011011001
111100011101111
111100100010111
111100100100111
111100101001101
111100101011001
111100101101111

111100101110001
111100101111011
111100110000001
111100110000111
111100110110001
111100111001001
111100111010111
111100111011101
111101000000011
111101000011011
111101000101011
111101000110101
111101000111111
111101001001011
111101001010101
111101010000001
111101010001101
111101010110111
111101010111011
111101011000011
111101011100001
111101011110101
111101100100011
111101101001111
111101101010001
111101101011101
111101101111001
111101110001111
111101110100001
111101110101011
111101110111001
111101111010011
111101111100011
111101111110001
111110000000101
111110000100111
111110000101101
111110001011001
111110010001011
111110010010011
111110010010101
111110010100011
111110010110111
111110011000011
111110011010001
111110011111001
111110011111111
111110100000001
111110100010101
111110100110111
111110101000101
111110101110011
111110101111001
111110110010001
111110110010111
111110110100111
111110110110011
111110111000111
111110111001101
111110111101001
111110111111011
111110111111101
111111000001101
111111000011001
111111000101111
111111001100001
111111001110101
111111001111111
111111010011101
111111010101011
111111011010011
111111011100011
111111011100101
111111100001001
111111100100001
111111100111111
111111101001101
111111101010101
111111101110001
111111110001011
111111110001101
111111110011111
111111111000101
111111111010001
111111111100111

A.3 ALL WEIGHT-THREE AND WEIGHT-FIVE PRIMITIVE POLYNOMIALS OF DEGREES 15 THROUGH 32

15
1100000000000001
1000100000000001
1000000100000001
1000000010000001
1000000000010001
1000000000000011
1110100000000001
1110000100000001
1110000000000011
1101000000001001
1101000000000101
1100100100000001
1100100001000001
1100100000000011
1100010000100001
1100010000000101
1100001100000001
1100001000001001
1100001000000101
1100000100100001
1100000100010001
1100000010000101
1100000001001001
1100000000100101
1100000000010011
1100000000000111
1011010000000001
1010110000000001
1010100010000001
1010010100000001
1010010010000001
1010010000001001
1010010000000101
1010010000000011
1010001001000001
1010000110000001
1010000100000011
1010000010010001
1010000010001001
1010000001000011
1010000000100101
1010000000100011
1010000000001011
1001100000100001
1001100000001001
1001001000010001
1001001000000011
1001000101000001
1001000100001001
1001000100000101
1001000010010001
1001000010001001
1001000001100001
1001000001000011
1001000000100101
1001000000011001
1001000000001011
1000110000100001
1000100100001001
1000100100000101
1000100010000011
1000100001100001
1000100001001001
1000011010000001
1000011000010001

1000011000001001
1000010110000001
1000010100100001
1000010010100001
1000010010000011
1000010000110001
1000010000100011
1000010000011001
1000001010001001
1000001001000101
1000001000010011
1000000110100001
1000000110000101
1000000101100001
1000000100100101
1000000100010101
1000000011000011
1000000010100101
1000000010010011
1000000010000111
1000000000110101
1000000000101101
1000000000010111

16

11010000000010001
11001010000000001
11001000000000011
11000010000010001
11000001001000001
11000001000010001
11000000010000101
11000000000010101
11000000000010011
10110100000000001
10110000000100001
10101000010000001
10101000000000011
10100100010000001
10100001010000001
10100001000010001
10100001000000011
10100000100001001
10011100000000001
10011000010000001
10010100001000001
10010001001000001
10010000100000101
10010000001010001
10001010000001001
10001001010000001
10001001001000001
10001000010000101
10001000010000011
10001000001000011
10001000000001011
10000110000100001
10000101100000001
10000101010000001
10000100001100001
10000100000001101
10000011001000001
10000010011000001
10000010010010001
10000010010001001
10000010010000011
10000010000101001
10000001010100001
10000001010010001
10000001010000101
10000001000100101
10000001000011001
10000001000010101
10000000110100001
10000000001010011
10000000000111001
10000000000101101

17

100100000000000001
100001000000000001
100000100000000001
100000000001000001
100000000000100001
100000000000001001
111100000000000001
111000000010000001
110100001000000001
110100000001000001
110100000000010001
110100000000000011
110011000000000001
110010000000100001
110010000000000101
110001010000000001
110001000100000001
110001000000010001
110001000000000011
110000100010000001
110000100000000011
110000011000000001
110000010100000001
110000010000100001
110000001000100001
110000001000010001
110000000101000001
110000000100001001
110000000001000011
110000000000110001
110000000000101001
110000000000100011
110000000000001101
110000000000001011
101101000000000001
101100010000000001
101100001000000001
101100000100000001
101100000001000001
101100000000100001
101100000000000011
101010100000000001
101010000010000001
101010000000100001
101001001000000001
101001000000000101
101000110000000001
101000100100000001
101000100000100001
101000100000000101
101000010010000001
101000010000000101
101000001001000001
101000001000100001
101000000110000001
101000000100100001
101000000010000101
101000000001000101
101000000000110001
101000000000100101
101000000000010011
100110010000000001
100110001000000001
100110000010000001
100101100000000001
100101000100000001
100101000000100001
100101000000001001

10010100000000011
10010010000010000 1
10010001000001000 1
10010001000000100 1
10010000110000000 1
10010000100001000 1
10010000100000001 1
10010000001001000 1
10010000001000100 1
10010000000010100 1
10001100010000000 1
10001100000010000 1
10001100000001000 1
10001100000000010 1
10001100000000001 1
10001010000100000 1
10001001010000000 1
10001001000100000 1
10001001000000100 1
10001000110000000 1
10001000100010000 1
10001000100001000 1
10001000010001000 1
10001000010000100 1
10001000010000001 1
10001000001000100 1
10001000000110000 1
10001000000011000 1
10001000000010001 1
10001000000000101 1
10000110100000000 1
10000110001000000 1
10000110000001000 1
10000100100000010 1
10000100011000000 1
10000100010001000 1
10000100010000010 1
10000100010000001 1
10000100001000001 1
10000100000100100 1
10000100000100010 1
10000100000011000 1
10000100000010100 1
10000100000001010 1
10000100000001001 1
10000100000000110 1
10000011100000000 1
10000011001000000 1
10000010100100000 1

10000010100000001 1
10000010010100000 1
10000010010000010 1
10000010001001000 1
10000010000101000 1
10000010000000110 1
10000010000000101 1
10000001100010000 1
10000001100000010 1
10000001001100000 1
10000001001000010 1
10000001000110000 1
10000001000100001 1
10000001000001100 1
10000001000001010 1
10000001000000011 1
10000000110001000 1
10000000110000100 1
10000000101001000 1
10000000101000001 1
10000000100100010 1
10000000100011000 1
10000000100010100 1
10000000100010001 1
10000000100000110 1
10000000011100000 1
10000000011000001 1
10000000010110000 1
10000000010010010 1
10000000010001100 1
10000000010000110 1
10000000010000101 1
10000000001100010 1
10000000001010001 1
10000000001001100 1
10000000001000110 1
10000000000110100 1
10000000000101010 1
10000000000011001 1
10000000000010110 1
10000000000000111 1

18

1000000100000000001
1000000000010000001
1110010000000000001
1110000010000000001
1110000000000100001
1101000001000000001

1100100001000000001
1100010000000010001
1100010000000000011
1100000010000000101
1100000000101000001
1100000000000100011
1011001000000000001
1010010000001000001
1010010000000001001
1010000100010000001
1010000100000010001
1010000010010000001
1010000001000001001
1010000000100100001
1010000000100000011
1010000000010001001
1010000000000110001
1001100000100000001
1001010000001000001
1001010000000010001
1001000100100000001
1001000100001000001
1001000100000000101
1001000001100000001
1001000001000100001
1001000001000010001
1001000001000000101
1001000000000100101
1000110000100000001
1000110000000100001
1000110000000000101
1000101000000100001
1000100110000000001
1000100001000001001
1000100000010000101
1000100000000101001
1000100000000100011
1000011001000000001
1000010100100000001
1000010100000100001
1000010010100000001
1000010010000000101
1000010001001000001
1000010001000001001
1000010000010100001
1000010000001010001
1000010000000110001
1000010000000000111
1000001010000000011

1000001001000100001
1000001000010001001
1000001000000101001
1000001000000100101
1000000100100000101
1000000100010000101
1000000011000001001
1000000010100100001
1000000010010100001
1000000010010001001
1000000010000110001
1000000010000011001
1000000001001100001
1000000001000010011
1000000001000001011
1000000000110010001
1000000000100000111
1000000000001001101
1000000000000100111

19

1110010000000000001
1110001000000000001
1110000000000100001
1101000001000000001
1101000000000100001
1100101000000000001
1100100100000000001
1100100001000000001
1100100000000001001
1100011000000000001
1100010000000000011
1100001010000000001
1100001000010000001
1100001000000001001
1100001000000000011
1100000101000000001
1100000100000001001
11000000010001000001
1100000001000000101
11000000001001000001
11000000001000010001
11000000000110000001
11000000000100000101
11000000000010100001
11000000000010001001
11000000000001000011
1100000000000100101
1100000000000100011
1011000000001000001
10101000000100000001
10100100010000000001
1010010000000001001
1010010000000000011
10100010000100000001
10100010000000100001
10100001100000000001
10100001010000000001
10100001000000010001
10100000100001000001
10100000100000000011
10100000010001000001
10100000010000001001
10100000001000000011
10100000000010001001
1010000000001100001
1010000000000110001
1001101000000000001
1001100001000000001
1001100000001000001
1001010000001000001
1001010000000100001
1001001000000001001
1001000100100000001
1001000100000100001
1001000100000000101
1001000100000000011
1001000001100000001
10010000010001000001
10010000010000010001
10010000001010000001
10010000001000100001
10010000001000000101
10010000000101000001
10010000000100100001
10010000000011000001
10010000000010000011
10010000000001010001
10010000000001001001
10010000000001000011
10010000000000100101
10010000000000010011
10001100010000000001
10001100000000010001
10001100000000000101
10001010010000000001
10001010001000000001
10001010000100000001
1000101000000001001
10001001001000000001
10001000011000000001
10001000010100000001
10001000010001000001
10001000010000000011
10001000001010000001
10001000001000001001
10001000000010000101
1000100000000110001
1000011100000000001
1000011010000000001
1000011000000000101
1000010110000000001
1000010101000000001
10000101000000100001
10000101000000000011
10000100110000000001
1000010010000001001
1000010001000001001
1000010000101000001
1000010000011000001
1000010000010100001
1000010000010001001
1000010000001000101
1000010000000101001
1000010000000001011
1000010000000000111
10000011000000100001
10000011000000001001
10000010101000000001
10000010100100000001
10000010100000001001
10000010011000000001
10000010010000000011
10000010001000010001
10000010001000001001
10000010001000000101
10000010001000000011
10000010000100000101
10000010000000101001
10000010000000011001
10000010000000001101
10000001100000000011
10000001010000100001

10000001010000010001
10000001010000001001
10000000101000010001
10000000100101000001
10000000100001010001
10000000100001000101
10000000100001000011
10000000100000010101
10000000011001000001
10000000011000010001
10000000011000001001
10000000010101000001
10000000010010010001
10000000010010001001
10000000010001010001
10000000001100100001
10000000001010100001
10000000001010000101
10000000001010000011
10000000001001010001
10000000001000110001
10000000001000100101
10000000001000011001
10000000001000010011
10000000001000001011
10000000000110100001
10000000000110000101
10000000000101100001
10000000000101000011
10000000000011100001
10000000000010010011
10000000000001100011
10000000000001011001
10000000000001010011
10000000000001000111
10000000000000100111

20

100100000000000000001
100000000000000001001
110010100000000000001
110010000000000000101
110001000100000000001
110001000010000000001
110001000000000010001
110000100001000000001
110000100000000010001
110000001000100000001
110000000100010000001
110000000001100000001
110000000001000000101
110000000000100010001
110000000000011000001
110000000000000011001
101001100000000000001
101000010010000000001
101000010000010000001
101000000110000000001
101000000100100000001
101000000100000100001
101000000100000000011
101000000001000001001
101000000000000010011
100110000001000000001
100110000000000000011
100101100000000000001
100101000100000000001
100101000001000000001
100100100000100000001
100100100000000001001
100100010000000010001
100100001010000000001
100100000101000000001
100100000100000000101
100100000001100000001
100100000001010000001
100100000001000010001
100100000000001010001
100100000000001001001
100011000100000000001
100011000000000100001
100010100000000001001
100010010001000000001
100010010000000100001
100010001000010000001
100010001000000000011
100010000100000001001
100010000011000000001
100010000001100000001
100010000000010001001
100010000000001000011
100010000000000100011
100001100000100000001
100001001000100000001
100001000100010000001
100001000010010000001
100001000001000000101
100001000000110000001
100001000000010010001
100010000000000110001
100000110010000000001
100000110000000000011
100000011000010000001
100000011000000100001
100000010100000001001
100000010010000100001
100000010001100000001
100000010001000100001
100000010001000000011
100000010000110000001
100000010000100010001
100000010000010000101
100000001100100000001
100000001100010000001
100000001100000010001
100000001100000001001
100000001100000000011
100000001001100000001
100000001001000000101
100000001000100100001
100000001000100000011
100000001000001100001
100000001000001001001
100000000110000010001
100000000101000001001
100000000100010010001
100000000100001000011
100000000100000101001
100000000100000011001
100000000011000000101
100000000010100001001
100000000010011000001
100000000010010000101
100000000010000100011
100000000001000110001
100000000001000101001
100000000001000100011
100000000000001101001
100000000000001100101
100000000000001010011

21

1010000000000000000001
1000000000000000000101

22
11000000000000000000001
10000000000000000000011

23
100001000000000000000001
100000000100000000000001
100000000000001000000001
100000000000000000100001

24
1000000000000000010000111
1110000100000000000000001

25
10010000000000000000000001
10000001000000000000000001
10000000000000000010000001
10000000000000000000001001

26
100000000000000000001000111
111000100000000000000000001

27
1000000000000000000000010111
1110100000000000000000000001

28
10000000000000000000000001001
10010000000000000000000000001

29
100000000000000000000000000101
101000000000000000000000000001

30
1000000100000000000000000000111
1110000000000000000000010000001

31
10000000000000000000000000001001
10010000000000000000000000000001

32
100000000010000000000000000000111
111000000000000000000010000000001

Appendix B

Add-One Tables and Vector Space Representations for Galois Fields of Size 2^m

This appendix contains add-one tables (also known as Zech's logarithms) for GF(8) through GF(64). A program is then provided that can generate vector-space representations for any binary extension field. Sample outputs are provided for GF(8) through GF(1024).

B.1 ADD-ONE TABLES FOR GF(8) THROUGH GF(64)

GF(8)

$p(x) = x^3 + x + 1 \Rightarrow \alpha^3 + \alpha + 1 = 0$

$\log_\alpha x$	$\log_\alpha (x + 1)$	$\log_\alpha x$	$\log_\alpha (x + 1)$
0	*	4	5
1	3	5	4
2	6	6	2
3	1		

GF(16)

$p(x) = x^4 + x + 1 \Rightarrow \alpha^4 + \alpha + 1 = 0$

$\log_\alpha x$	$\log_\alpha (x + 1)$	$\log_\alpha x$	$\log_\alpha (x + 1)$
0	*	8	2
1	4	9	7
2	8	10	5
3	14	11	12
4	1	12	11
5	10	13	6
6	13	14	3
7	9		

GF(32)

$p(x) = x^5 + x^2 + 1 \Rightarrow \alpha^5 + \alpha^2 + 1 = 0$

$\log_\alpha x$	$\log_\alpha (x+1)$	$\log_\alpha x$	$\log_\alpha (x+1)$
0	*	16	9
1	18	17	30
2	5	18	1
3	29	19	11
4	10	20	8
5	2	21	25
6	27	22	7
7	22	23	12
8	20	24	15
9	16	25	21
10	4	26	28
11	19	27	6
12	23	28	26
13	14	29	3
14	13	30	17
15	24		

GF(64)

$p(x) = x^6 + x + 1 \Rightarrow \alpha^6 + \alpha + 1 = 0$

$\log_\alpha x$	$\log_\alpha (x+1)$	$\log_\alpha x$	$\log_\alpha (x+1)$
0	*	32	3
1	6	33	16
2	12	34	31
3	32	35	13
4	24	36	54
5	62	37	44
6	1	38	49
7	26	39	43
8	48	40	55
9	45	41	28
10	61	42	21
11	25	43	39
12	2	44	37
13	35	45	9
14	52	46	30
15	23	47	17
16	33	48	8
17	47	49	38
18	27	50	22
19	56	51	53
20	59	52	14
21	42	53	51
22	50	54	36
23	15	55	40
24	4	56	19
25	11	57	58
26	7	58	57
27	18	59	20
28	41	60	29
29	60	61	10
30	46	62	5
31	34		

B.2 C PROGRAM FOR THE GENERATION OF GALOIS FIELD REPRESENTATIONS[1]

```
/* generate table for Galois Fields */

#include <stdio.h>
#include <math.h>

/* currently works for fields GF(2**3) through GF(2**10) */
/* for larger fields [up to GF(2**31)], calculate the variable "reduce" */
/*     from the primitive polynomial as described in comments in createfield() below */
/*     and add another "else if" clause at beginning of createfield function */

/*****************************************************************************/
#define M 10                              /*size of field = 2 to the Mth, M must be <= 31 */
#define     maxGFsize     1024            /* must be >= two to the Mth power (array sizes)*/
/*****************************************************************************/

void printbin(FILE* prtfile, int n);
void createfield(void);

/* global variables */
unsigned long alphapow[maxGFsize];        /* contains integers whose binary rep's indicate
polynomial form */
unsigned long logalpha[maxGFsize];        /* log base alpha of given element */
unsigned long twotothe[31];                     /* powers of 2 */
main ()
{
      unsigned long i;
      FILE *outfile;
      twotothe[0]=1;
      for (i=1;i<=M;i++)
            twotothe[i]=2*twotothe[i-1];
      if (maxGFsize < twotothe[M]) {
            fprintf(stderr, "Error: maxGFsize must be >= two to the Mth power \n");
            exit (1);
      }
      createfield();
      outfile = fopen("gfgen.out", "w");
      fprintf(outfile, "GF(2**%d)\n",M);
/*    fprintf(outfile,"logalpha\n"); */

/* The following routine, which is commented out, prints the field sorted */
```

[1] This program was written by Thomas Tapp, of the Georgia Tech Research Institute, and is printed here with his permission.

```
/* by vector space coordinates */
/*      for (i=0;i<twotothe[M];i++) {
                printbin(outfile,i);
                fprintf(outfile,"\t%ld\n",logalpha[i]);
        } */
/*      fprintf(outfile,"alphapow\n"); */
        for (i=0;i<twotothe[M]-1;i++) {
                fprintf(outfile, "%ld\t",i);
                printbin(outfile,alphapow[i]);
                fprintf(outfile,"\n");
        }
        fclose(outfile);
}
void createfield(void) {
        unsigned long reduce;
        unsigned long temp;
        int i,j;
        unsigned long n;                        /* power of alpha */
        if (M==3) reduce=3;
        else if (M==4) reduce=3;
        else if (M==5) reduce=5;
        else if (M==6) reduce=3;
        else if (M==7) reduce=9;
        else if (M==8) reduce=29;
        else if (M==9) reduce=17;
        else if (M==10) reduce=9;
        else {
                fprintf(stderr, "Error : Primitive Polynomial not known for
                GF(2**%x)\n",M);
                exit(1);
        }
/*      reduce is calculated from the primitive polynomial as follows:
                subtract the highest order term (x to the Mth)
                and evaluate remaining polynomial at x=2
                e.g., GF(2**5) generated by p(x)=1 + x**2 + x**5
                get rid of x**5 term and evaluate 1 + x**2 at x=2 which gives reduce=5
 */
        alphapow[0] = 1;                /* alpha^0 = 1 */
        for (n=1;n<twotothe[M]-1;n++) {
                temp = alphapow[n-1];
                temp = (temp<<1);    /* multiply by alpha */
                if (temp & twotothe[M])         /* contains alpha**M term */
                        alphapow[n] = (temp & ~twotothe[M]) ^ reduce;
                else
                        alphapow[n] = temp;    /* if no alpha**M term, store as is */
  }
  /* create table to go in opposite direction */
  logalpha[0] = -1; /* special case, actually log(0)=-inf */
```

```
    for(n=0;n<twotothe[M]-1;n++)
        logalpha[alphapow[n]]=n;
}

void printbin(FILE* prtfile, int n) {
    int i;
    for (i=0;i<M;i++)
        fprintf(prtfile, "%1i",!!(n&twotothe[i]));
}
/***** end of program *****/
```

B.3 GALOIS FIELD VECTOR-SPACE REPRESENTATIONS FOR GF(8) THROUGH GF(1024)

GF(8)

$p(x) = x^3 + x + 1$

$\Rightarrow \alpha^3 + \alpha + 1 = 0$

$x = a + b\alpha + c\alpha^2$

x	*abc*
0	100
1	010
2	001
3	110
4	011
5	111
6	101

GF(16)

$p(x) = x^4 + x + 1$

0	1000
1	0100
2	0010
3	0001
4	1100
5	0110
6	0011
7	1101
8	1010
9	0101
10	1110
11	0111
12	1111
13	1011
14	1001

GF(32)

$p(x) = x^5 + x^2 + 1$

0	10000
1	01000
2	00100
3	00010
4	00001
5	10100
6	01010
7	00101
8	10110
9	01011
10	10001
11	11100
12	01110
13	00111
14	10111
15	11111
16	11011
17	11001
18	11000
19	01100
20	00110
21	00011
22	10101
23	11110
24	01111
25	10011
26	11101
27	11010
28	01101
29	10010
30	01001

GF(64)

$p(x) = x^6 + x + 1$

0	100000
1	010000
2	001000
3	000100
4	000010
5	000001
6	110000
7	011000
8	001100
9	000110
10	000011
11	110001
12	101000
13	010100
14	001010
15	000101
16	110010
17	011001
18	111100
19	011110
20	001111
21	110111
22	101011
23	100101
24	100010
25	010001
26	111000
27	011100
28	001110
29	000111
30	110011
31	101001
32	100100
33	010010
34	001001
35	110100
36	011010
37	001101
38	110110
39	011011
40	111101
41	101110
42	010111
43	111011
44	101101
45	100110
46	010011
47	111001
48	101100
49	010110
50	001011
51	110101
52	101010
53	010101
54	111010
55	011101

56	111110
57	011111
58	111111
59	101111
60	100111
61	100011
62	100001

GF(128)

$p(x) = x^7 + x^3 + 1$

0	1000000
1	0100000
2	0010000
3	0001000
4	0000100
5	0000010
6	0000001
7	1001000
8	0100100
9	0010010
10	0001001
11	1001100
12	0100110
13	0010011
14	1000001
15	1101000
16	0110100
17	0011010
18	0001101
19	1001110
20	0100111
21	1011011
22	1100101
23	1111010
24	0111101
25	1010110
26	0101011
27	1011101
28	1100110
29	0110011
30	1010001
31	1100000
32	0110000
33	0011000
34	0001100
35	0000110
36	0000011
37	1001001
38	1101100
39	0110110
40	0011011
41	1000101
42	1101010
43	0110101
44	1010010
45	0101001
46	1011100
47	0101110
48	0010111
49	1000011
50	1101001
51	1111100
52	0111110
53	0011111
54	1000111
55	1101011
56	1111101
57	1110110
58	0111011
59	1010101
60	1100010
61	0110001
62	1010000
63	0101000
64	0010100
65	0001010
66	0000101
67	1001010
68	0100101
69	1011010
70	0101101
71	1011110
72	0101111
73	1011111
74	1100111
75	1111011
76	1110101
77	1110010
78	0111001
79	1010100
80	0101010
81	0010101
82	1000010
83	0100001
84	1011000
85	0101100
86	0010110
87	0001011
88	1001101
89	1101110
90	0110111
91	1010011
92	1100001
93	1111000
94	0111100
95	0011110
96	0001111
97	1001111
98	1101111
99	1111111
100	1110111
101	1110011
102	1110001
103	1110000
104	0111000
105	0011100
106	0001110
107	0000111
108	1001011
109	1101101
110	1111110
111	0111111
112	1010111
113	1100011
114	1111001
115	1110100
116	0111010
117	0011101
118	1000110
119	0100011
120	1011001
121	1100100
122	0110010
123	0011001
124	1000100
125	0100010
126	0010001

GF(256)

$p(x) = x^8 + x^4 + x^3 + x^2 + 1$

0	10000000
1	01000000
2	00100000
3	00010000
4	00001000
5	00000100
6	00000010
7	00000001
8	10111000
9	01011100
10	00101110
11	00010111
12	10110011
13	11100001
14	11001000
15	01100100
16	00110010
17	00011001
18	10110100
19	01011010
20	00101101
21	10101110
22	01010111
23	10010011
24	11110001
25	11000000
26	01100000
27	00110000
28	00011000
29	00001100
30	00000110
31	00000011
32	10111001
33	11100100
34	01110010
35	00111001
36	10100100
37	01010010
38	00101001
39	10101100
40	01010110
41	00101011
42	10101101
43	11101110
44	01110111
45	10000011
46	11111001
47	11000100
48	01100010
49	00110001
50	10100000
51	01010000
52	00101000
53	00010100

54	00001010
55	00000101
56	10111010
57	01011101
58	10010110
59	01001011
60	10011101
61	11110110
62	01111011
63	10000101
64	11111010
65	01111101
66	10000110
67	01000011
68	10011001
69	11110100
70	01111010
71	00111101
72	10100110
73	01010011
74	10010001
75	11110000
76	01111000
77	00111100
78	00011110
79	00001111
80	10111111
81	11100111
82	11001011
83	11011101
84	11010110
85	01101011
86	10001101
87	11111110
88	01111111
89	10000111
90	11111011
91	11000101
92	11011010
93	01101101
94	10001110
95	01000111
96	10011011
97	11110101
98	11000010
99	01100001
100	10001000
101	01000100
102	00100010
103	00010001
104	10110000
105	01011000
106	00101100
107	00010110
108	00001011
109	10111101
110	11100110
111	01110011
112	10000001
113	11111000
114	01111100
115	00111110
116	00011111
117	10110111
118	11100011
119	11001001
120	11011100
121	01101110
122	00110111
123	10100011
124	11101001
125	11001100
126	01100110
127	00110011
128	10100001
129	11101000
130	01110100
131	00111010
132	00011101
133	10110110
134	01011011
135	10010101
136	11110010
137	01111001
138	10000100
139	01000010
140	00100001
141	10101000
142	01010100
143	00101010
144	00010101
145	10110010
146	01011001
147	10010100
148	01001010
149	00100101
150	10101010
151	01010101
152	10010010
153	01001001
154	10011100
155	01001110
156	00100111
157	10101011
158	11101101
159	11001110
160	01100111
161	10001011
162	11111101
163	11000110
164	01100011
165	10001001
166	11111100
167	01111110
168	00111111
169	10100111
170	11101011
171	11001101
172	11011110
173	01101111
174	10001111
175	11111111
176	11000111
177	11011011
178	11010101
179	11010010
180	01101001
181	10001100
182	01000110
183	00100011
184	10101001
185	11101100
186	01110110
187	00111011
188	10100101
189	11101010
190	01110101
191	10000010
192	01000001
193	10011000
194	01001100
195	00100110
196	00010011
197	10110001
198	11100000
199	01110000
200	00111000
201	00011100
202	00001110
203	00000111
204	10111011
205	11100101
206	11001010
207	01100101
208	10001010
209	01000101
210	10011010
211	01001101
212	10011110
213	01001111
214	10011111
215	11110111
216	11000011
217	11011001
218	11010100
219	01101010
220	00110101
221	10100010
222	01010001
223	10010000
224	01001000
225	00100100
226	00010010
227	00001001
228	10111100
229	01011110
230	00101111
231	10101111
232	11101111
233	11001111
234	11011111
235	11010111
236	11010011
237	11010001
238	11010000
239	01101000
240	00110100
241	00011010
242	00001101
243	10111110
244	01011111
245	10010111
246	11110011
247	11000001
248	11011000
249	01101100

250	00110110
251	00011011
252	10110101
253	11100010
254	01110001

GF(512)

$p(x) = x^9 + x^4 + 1$

0	100000000
1	010000000
2	001000000
3	000100000
4	000010000
5	000001000
6	000000100
7	000000010
8	000000001
9	100010000
10	010001000
11	001000100
12	000100010
13	000010001
14	100011000
15	010001100
16	001000110
17	000100011
18	100000001
19	110010000
20	011001000
21	001100100
22	000110010
23	000011001
24	100011100
25	010001110
26	001000111
27	100110011
28	110001001
29	111010100
30	011101010
31	001110101
32	100101010
33	010010101
34	101011010
35	010101101
36	101000110
37	010100011
38	101000001
39	110110000
40	011011000
41	001101100
42	000110110
43	000011011
44	100011101
45	110011110
46	011001111
47	101110111
48	110101011
49	111000101
50	111110010
51	011111001
52	101101100
53	010110110
54	001011011
55	100111101
56	110001110
57	011000111
58	101110011
59	110101001
60	111000100
61	011100010
62	001110001
63	100101000
64	010010100
65	001001010
66	000100101
67	100000010
68	010000001
69	101010000
70	010101000
71	001010100
72	000101010
73	000010101
74	100011010
75	010001101
76	101010110
77	010101011
78	101000101
79	110110010
80	011011001
81	101111100
82	010111110
83	001011111
84	100111111
85	110001111
86	111010111
87	111111011
88	111101101
89	111100110
90	011110011
91	101101001
92	110100100
93	011010010
94	001101001
95	100100100
96	010010010
97	001001001
98	100110100
99	010011010
100	001001101
101	100110110
102	010011011
103	101011101
104	110111110
105	011011111
106	101111111
107	110101111
108	111000111
109	111110011
110	111101001
111	111100100
112	011110010
113	001111001
114	100101100
115	010010110
116	001001011
117	100110101
118	110001010
119	011000101
120	101110010
121	010111001
122	101001100
123	010100110
124	001010011
125	100111001
126	110001100
127	011000110
128	001100011
129	100100001
130	110000000
131	011000000
132	001100000
133	000110000
134	000011000
135	000001100
136	000000110
137	000000011
138	100010001
139	110011000
140	011001100
141	001100110
142	000110011
143	100001001
144	110010100
145	011001010
146	001100101
147	100100010
148	010010001
149	101011000
150	010101100
151	001010110
152	000101011
153	100000101
154	110010010
155	011001001
156	101110100
157	010111010
158	001011101
159	100111110
160	010011111
161	101011111
162	110111111
163	111001111
164	111110111
165	111101011
166	111100101
167	111100010
168	011110001
169	101101000
170	010110100
171	001011010
172	000101101
173	100000110
174	010000011
175	101010001
176	110111000
177	011011100
178	001101110
179	000110111
180	100001011
181	110010101
182	111011010
183	011101101
184	101100110
185	010110011
186	101001001

187 110110100
188 011011010
189 001101101
190 100100110
191 010010011
192 101011001
193 110111100
194 011011110
195 001101111
196 100100111
197 110000011
198 111010001
199 111111000
200 011111100
201 001111110
202 000111111
203 100001111
204 110010111
205 111011011
206 111111101
207 111101110
208 011110111
209 101101011
210 110100101
211 111000010
212 011100001
213 101100000
214 010110000
215 001011000
216 000101100
217 000010110
218 000001011
219 100010101
220 110011010
221 011001101
222 101110110
223 010111011
224 101001101
225 110110110
226 011011011
227 101111101
228 110101110
229 011010111
230 101111011
231 110101101
232 111000110
233 011100011
234 101100001
235 110100000
236 011010000
237 001101000
238 000110100
239 000011010
240 000001101
241 100010110
242 010001011
243 101010101
244 110111010
245 011011101
246 101111110
247 010111111
248 101001111
249 110110111
250 111001011
251 111110101
252 111101010
253 011110101
254 101101010
255 010110101
256 101001010
257 010100101
258 101000010
259 010100001
260 101000000
261 010100000
262 001010000
263 000101000
264 000010100
265 000001010
266 000000101
267 100010010
268 010001001
269 101010100
270 010101010
271 001010101
272 100111010
273 010011101
274 101011110
275 010101111
276 101000111
277 110110011
278 111001001
279 111110100
280 011111010
281 001111101
282 100101110
283 010010111
284 101011011
285 110111101
286 111001110
287 011100111
288 101100011
289 110100001
290 111000000
291 011100000
292 001110000
293 000111000
294 000011100
295 000001110
296 000000111
297 100010011
298 110011001
299 111011100
300 011101110
301 001110111
302 100101011
303 110000101
304 111010010
305 011101001
306 101100100
307 010110010
308 001011001
309 100111100
310 010011110
311 001001111
312 100110111
313 110001011
314 111010101
315 111111010
316 011111101
317 101101110
318 010110111
319 101001011
320 110110101
321 111001010
322 011100101
323 101100010
324 010110001
325 101001000
326 010100100
327 001010010
328 000101001
329 100000100
330 010000010
331 001000001
332 100110000
333 010011000
334 001001100
335 000100110
336 000010011
337 100011001
338 110011100
339 011001110
340 001100111
341 100100011
342 110000001
343 111010000
344 011101000
345 001110100
346 000111010
347 000011101
348 100011110
349 010001111
350 101010111
351 110111011
352 111001101
353 111110110
354 011111011
355 101101101
356 110100110
357 011010011
358 101111001
359 110101100
360 011010110
361 001101011
362 100100101
363 110000010
364 011000001
365 101110000
366 010111000
367 001011100
368 000101110
369 000010111
370 100011011
371 110011101
372 111011110
373 011101111
374 101100111
375 110100011
376 111000001
377 111110000
378 011111000
379 001111100
380 000111110
381 000011111
382 100011111

383	110011111
384	111011111
385	111111111
386	111101111
387	111100111
388	111100011
389	111100001
390	111100000
391	011110000
392	001111000
393	000111100
394	000011110
395	000001111
396	100010111
397	110011011
398	111011101
399	111111110
400	011111111
401	101101111
402	110100111
403	111000011
404	111110001
405	111101000
406	011110100
407	001111010
408	000111101
409	100001110
410	010000111
411	101010011
412	110111001
413	111001100
414	011100110
415	001110011
416	100101001
417	110000100
418	011000010
419	001100001
420	100100000
421	010010000
422	001001000
423	000100100
424	000010010
425	000001001
426	100010100
427	010001010
428	001000101
429	100110010
430	010011001
431	101011100
432	010101110
433	001010111
434	100111011
435	110001101
436	111010110
437	011101011
438	101100101
439	110100010
440	011010001
441	101111000
442	010111100
443	001011110
444	000101111
445	100000111
446	110010011
447	111011001
448	111111100
449	011111110
450	001111111
451	100101111
452	110000111
453	111010011
454	111111001
455	111101100
456	011110110
457	001111011
458	100101101
459	110000110
460	011000011
461	101110001
462	110101000
463	011010100
464	001101010
465	000110101
466	100001010
467	010000101
468	101010010
469	010101001
470	101000100
471	010100010
472	001010001
473	100111000
474	010011100
475	001001110
476	000100111
477	100000011
478	110010001
479	111011000
480	011101100
481	001110110
482	000111011
483	100001101
484	110010110
485	011001011
486	101110101
487	110101010
488	011010101
489	101111010
490	010111101
491	101001110
492	010100111
493	101000011
494	110110001
495	111001000
496	011100100
497	001110010
498	000111001
499	100001100
500	010000110
501	001000011
502	100110001
503	110001000
504	011000100
505	001100010
506	000110001
507	100001000
508	010000100
509	001000010
510	000100001

GF(1024)

$p(x) = x^{10} + x^3 + 1$

0	1000000000
1	0100000000
2	0010000000
3	0001000000
4	0000100000
5	0000010000
6	0000001000
7	0000000100
8	0000000010
9	0000000001
10	1001000000
11	0100100000
12	0010010000
13	0001001000
14	0000100100
15	0000010010
16	0000001001
17	1001000100
18	0100100010
19	0010010001
20	1000001000
21	0100000100
22	0010000010
23	0001000001
24	1001100000
25	0100110000
26	0010011000
27	0001001100
28	0000100110
29	0000010011
30	1001001001
31	1101100100
32	0110110010
33	0011011001
34	1000101100
35	0100010110
36	0010001011
37	1000000101
38	1101000010
39	0110100001
40	1010010000
41	0101001000
42	0010100100
43	0001010010
44	0000101001
45	1001010100
46	0100101010
47	0010010101
48	1000001010
49	0100000101
50	1011000010
51	0101100001
52	1011110000
53	0101111000
54	0010111100
55	0001011110
56	0000101111
57	1001010111
58	1101101011
59	1111110101
60	1110111010
61	0111011101
62	1010101110
63	0101010111

64	1011101011	113	1010001110	162	1001000010	211	0110100010
65	1100110101	114	0101000111	163	0100100001	212	0011010001
66	1111011010	115	1011100011	164	1011010000	213	1000101000
67	0111101101	116	1100110001	165	0101101000	214	0100010100
68	1010110110	117	1111011000	166	0010110100	215	0010001010
69	0101011011	118	0111101100	167	0001011010	216	0001000101
70	1011101101	119	0011110110	168	0000101101	217	1001100010
71	1100110110	120	0001111011	169	1001010110	218	0100110001
72	0110011011	121	1001111101	170	0100101011	219	1011011000
73	1010001101	122	1101111110	171	1011010101	220	0101101100
74	1100000110	123	0110111111	172	1100101010	221	0010110110
75	0110000011	124	1010011111	173	0110010101	222	0001011011
76	1010000001	125	1100001111	174	1010001010	223	1001101101
77	1100000000	126	1111000111	175	0101000101	224	1101110110
78	0110000000	127	1110100011	176	1011100010	225	0110111011
79	0011000000	128	1110010001	177	0101110001	226	1010011101
80	0001100000	129	1110001000	178	1011111000	227	1100001110
81	0000110000	130	0111000100	179	0101111100	228	0110000111
82	0000011000	131	0011100010	180	0010111110	229	1010000011
83	0000001100	132	0001110001	181	0001011111	230	1100000001
84	0000000110	133	1001111000	182	1001101111	231	1111000000
85	0000000011	134	0100111100	183	1101110111	232	0111100000
86	1001000001	135	0010011110	184	1111111011	233	0011110000
87	1101100000	136	0001001111	185	1110111101	234	0001111000
88	0110110000	137	1001100111	186	1110011110	235	0000111100
89	0011011000	138	1101110011	187	0111001111	236	0000011110
90	0001101100	139	1111111001	188	1010100111	237	0000001111
91	0000110110	140	1110111100	189	1100010011	238	1001000111
92	0000011011	141	0111011110	190	1111001001	239	1101100011
93	1001001101	142	0011101111	191	1110100100	240	1111110001
94	1101100110	143	1000110111	192	0111010010	241	1110111000
95	0110110011	144	1101011011	193	0011101001	242	0111011100
96	1010011001	145	1111101101	194	1000110100	243	0011101110
97	1100001100	146	1110110110	195	0100011010	244	0001110111
98	0110000110	147	0111011011	196	0010001101	245	1001111011
99	0011000011	148	1010101101	197	1000000110	246	1101111101
100	1000100001	149	1100010110	198	0100000011	247	1111111110
101	1101010000	150	0110001011	199	1011000001	248	0111111111
102	0110101000	151	1010000101	200	1100100000	249	1010111111
103	0011010100	152	1100000010	201	0110010000	250	1100011111
104	0001101010	153	0110000001	202	0011001000	251	1111001111
105	0000110101	154	1010000000	203	0001100100	252	1110100111
106	1001011010	155	0101000000	204	0000110010	253	1110010011
107	0100101101	156	0010100000	205	0000011001	254	1110001001
108	1011010110	157	0001010000	206	1001001100	255	1110000100
109	0101101011	158	0000101000	207	0100100110	256	0111000010
110	1011110101	159	0000010100	208	0010010011	257	0011100001
111	1100111010	160	0000001010	209	1000001001	258	1000110000
112	0110011101	161	0000000101	210	1101000100	259	0100011000

260 0010001100
261 0001000110
262 0000100011
263 1001010001
264 1101101000
265 0110110100
266 0011011010
267 0001101101
268 1001110110
269 0100111011
270 1011011101
271 1100101110
272 0110010111
273 1010001011
274 1100000101
275 1111000010
276 0111100001
277 1010110000
278 0101011000
279 0010101100
280 0001010110
281 0000101011
282 1001010101
283 1101101010
284 0110110101
285 1010011010
286 0101001101
287 1011100110
288 0101110011
289 1011111001
290 1100111100
291 0110011110
292 0011001111
293 1000100111
294 1101010011
295 1111101001
296 1110110100
297 0111011010
298 0011101101
299 1000110110
300 0100011011
301 1011001101
302 1100100110
303 0110010011
304 1010001001
305 1100000100
306 0110000010
307 0011000001
308 1000100000
309 0100010000
310 0010001000
311 0001000100
312 0000100010
313 0000010001
314 1001001000
315 0100100100
316 0010010010
317 0001001001
318 1001100100
319 0100110010
320 0010011001
321 1000001100
322 0100000110
323 0010000011
324 1000000001
325 1101000000
326 0110100000
327 0011010000
328 0001101000
329 0000110100
330 0000011010
331 0000001101
332 1001000110
333 0100100011
334 1011010001
335 1100101000
336 0110010100
337 0011001010
338 0001100101
339 1001110010
340 0100111001
341 1011011100
342 0101101110
343 0010110111
344 1000011011
345 1101001101
346 1111100110
347 0111110011
348 1010111001
349 1100011100
350 0110001110
351 0011000111
352 1000100011
353 1101010001
354 1111101000
355 0111110100
356 0011111010
357 0001111101
358 1001111110
359 0100111111
360 1011011111
361 1100101111
362 1111010111
363 1110101011
364 1110010101
365 1110001010
366 0111000101
367 1010100010
368 0101010001
369 1011101000
370 0101110100
371 0010111010
372 0001011101
373 1001101110
374 0100110111
375 1011011011
376 1100101101
377 1111010110
378 0111101011
379 1010110101
380 1100011010
381 0110001101
382 1010000110
383 0101000011
384 1011100001
385 1100110000
386 0110011000
387 0011001100
388 0001100110
389 0000110011
390 1001011001
391 1101101100
392 0110110110
393 0011011011
394 1000101101
395 1101010110
396 0110101011
397 1010010101
398 1100001010
399 0110000101
400 1010000010
401 0101000001
402 1011100000
403 0101110000
404 0010111000
405 0001011100
406 0000101110
407 0000010111
408 1001001011
409 1101100101
410 1111110010
411 0111111001
412 1010111100
413 0101011110
414 0010101111
415 1000010111
416 1101001011
417 1111100101
418 1110110010
419 0111011001
420 1010101100
421 0101010110
422 0010101011
423 1000010101
424 1101001010
425 0110100101
426 1010010010
427 0101001001
428 1011100100
429 0101110010
430 0010111001
431 1000011100
432 0100001110
433 0010000111
434 1000000011
435 1101000001
436 1111100000
437 0111110000
438 0011111000
439 0001111100
440 0000111110
441 0000011111
442 1001001111
443 1101100111
444 1111110011
445 1110111001
446 1110011100
447 0111001110
448 0011100111
449 1000110011
450 1101011001
451 1111101100
452 0111110110
453 0011111011
454 1000111101
455 1101011110

456	0110101111	505	1110010110	554	0001100010	603	0111011111
457	1010010111	506	0111001011	555	0000110001	604	1010101111
458	1100001011	507	1010100101	556	1001011000	605	1100010111
459	1111000101	508	1100010010	557	0100101100	606	1111001011
460	1110100010	509	0110001001	558	0010010110	607	1110100101
461	0111010001	510	1010000100	559	0001001011	608	1110010010
462	1010101000	511	0101000010	560	1001100101	609	0111001001
463	0101010100	512	0010100001	561	1101110010	610	1010100100
464	0010101010	513	1000010000	562	0110111001	611	0101010010
465	0001010101	514	0100001000	563	1010011100	612	0010101001
466	1001101010	515	0010000100	564	0101001110	613	1000010100
467	0100110101	516	0001000010	565	0010100111	614	0100001010
468	1011011010	517	0000100001	566	1000010011	615	0010000101
469	0101101101	518	1001010000	567	1101001001	616	1000000010
470	1011110110	519	0100101000	568	1111100100	617	0100000001
471	0101111011	520	0010010100	569	0111110010	618	1011000000
472	1011111101	521	0001001010	570	0011111001	619	0101100000
473	1100111110	522	0000100101	571	1000111100	620	0010110000
474	0110011111	523	1001010010	572	0100011110	621	0001011000
475	1010001111	524	0100101001	573	0010001111	622	0000101100
476	1100000111	525	1011010100	574	1000000111	623	0000010110
477	1111000011	526	0101101010	575	1101000011	624	0000001011
478	1110100001	527	0010110101	576	1111100001	625	1001000101
479	1110010000	528	1000011010	577	1110110000	626	1101100010
480	0111001000	529	0100001101	578	0111011000	627	0110110001
481	0011100100	530	1011000110	579	0011101100	628	1010011000
482	0001110010	531	0101100011	580	0001110110	629	0101001100
483	0000111001	532	1011110001	581	0000111011	630	0010100110
484	1001011100	533	1100111000	582	1001011101	631	0001010011
485	0100101110	534	0110011100	583	1101101110	632	1001101001
486	0010010111	535	0011001110	584	0110110111	633	1101110100
487	1000001011	536	0001100111	585	1010011011	634	0110111010
488	1101000101	537	1001110011	586	1100001101	635	0011011101
489	1111100010	538	1101111001	587	1111000110	636	1000101110
490	0111110001	539	1111111100	588	0111100011	637	0100010111
491	1010111000	540	0111111110	589	1010110001	638	1011001011
492	0101011100	541	0011111111	590	1100011000	639	1100100101
493	0010101110	542	1000111111	591	0110001100	640	1111010010
494	0001010111	543	1101011111	592	0011000110	641	0111101001
495	1001101011	544	1111101111	593	0001100011	642	1010110100
496	1101110101	545	1110110111	594	1001110001	643	0101011010
497	1111111010	546	1110011011	595	1101111000	644	0010101101
498	0111111101	547	1110001101	596	0110111100	645	1000010110
499	1010111110	548	1110000110	597	0011011110	646	0100001011
500	0101011111	549	0111000011	598	0001101111	647	1011000101
501	1011101111	550	1010100001	599	1001110111	648	1100100010
502	1100110111	551	1100010000	600	1101111011	649	0110010001
503	1111011011	552	0110001000	601	1111111101	650	1010001000
504	1110101101	553	0011000100	602	1110111110	651	0101000100

652 0010100010
653 0001010001
654 1001101000
655 0100110100
656 0010011010
657 0001001101
658 1001100110
659 0100110011
660 1011011001
661 1100101100
662 0110010110
663 0011001011
664 1000100101
665 1101010010
666 0110101001
667 1010010100
668 0101001010
669 0010100101
670 1000010010
671 0100001001
672 1011000100
673 0101100010
674 0010110001
675 1000011000
676 0100001100
677 0010000110
678 0001000011
679 1001100001
680 1101110000
681 0110111000
682 0011011100
683 0001101110
684 0000110111
685 1001011011
686 1101101101
687 1111110110
688 0111111011
689 1010111101
690 1100011110
691 0110001111
692 1010000111
693 1100000011
694 1111000001
695 1110100000
696 0111010000
697 0011101000
698 0001110100
699 0000111010
700 0000011101
701 1001001110
702 0100100111
703 1011010011
704 1100101001
705 1111010100
706 0111101010
707 0011110101
708 1000111010
709 0100011101
710 1011001110
711 0101100111
712 1011110011
713 1100111001
714 1111011100
715 0111101110
716 0011110111
717 1000111011
718 1101011101
719 1111101110
720 0111110111
721 1010111011
722 1100011101
723 1111001110
724 0111100111
725 1010110011
726 1100011001
727 1111001100
728 0111100110
729 0011110011
730 1000111001
731 1101011100
732 0110101110
733 0011010111
734 1000101011
735 1101010101
736 1111101010
737 0111110101
738 1010111010
739 0101011101
740 1011101110
741 0101110111
742 1011111011
743 1100111101
744 1111011110
745 0111101111
746 1010110111
747 1100011011
748 1111001101
749 1110100110
750 0111010011
751 1010101001
752 1100010100
753 0110001010
754 0011000101
755 1000100010
756 0100010001
757 1011001000
758 0101100100
759 0010110010
760 0001011001
761 1001101100
762 0100110110
763 0010011011
764 1000001101
765 1101000110
766 0110100011
767 1010010001
768 1100001000
769 0110000100
770 0011000010
771 0001100001
772 1001110000
773 0100111000
774 0010011100
775 0001001110
776 0000100111
777 1001010011
778 1101101001
779 1111110100
780 0111111010
781 0011111101
782 1000111110
783 0100011111
784 1011001111
785 1100100111
786 1111010011
787 1110101001
788 1110010100
789 0111001010
790 0011100101
791 1000110010
792 0100011001
793 1011001100
794 0101100110
795 0010110011
796 1000011001
797 1101001100
798 0110100110
799 0011010011
800 1000101001
801 1101010100
802 0110101010
803 0011010101
804 1000101010
805 0100010101
806 1011001010
807 0101100101
808 1011110010
809 0101111001
810 1011111100
811 0101111110
812 0010111111
813 1000011111
814 1101001111
815 1111100111
816 1110110011
817 1110011001
818 1110001100
819 0111000110
820 0011100011
821 1000110001
822 1101011000
823 0110101100
824 0011010110
825 0001101011
826 1001110101
827 1101111010
828 0110111101
829 1010011110
830 0101001111
831 1011100111
832 1100110011
833 1111011001
834 1110101100
835 0111010110
836 0011101011
837 1000110101
838 1101011010
839 0110101101
840 1010010110
841 0101001011
842 1011100101
843 1100110010
844 0110011001
845 1010001100
846 0101000110
847 0010100011

848 1000010001
849 1101001000
850 0110100100
851 0011010010
852 0001101001
853 1001110100
854 0100111010
855 0010011101
856 1000001110
857 0100000111
858 1011000011
859 1100100001
860 1111010000
861 0111101000
862 0011110100
863 0001111010
864 0000111101
865 1001011110
866 0100101111
867 1011010111
868 1100101011
869 1111010101
870 1110101010
871 0111010101
872 1010101010
873 0101010101
874 1011101010
875 0101110101
876 1011111010
877 0101111101
878 1011111110
879 0101111111
880 1011111111
881 1100111111
882 1111011111
883 1110101111
884 1110010111
885 1110001011
886 1110000101
887 1110000010
888 0111000001
889 1010100000
890 0101010000
891 0010101000
892 0001010100
893 0000101010
894 0000010101
895 1001001010
896 0100100101
897 1011010010
898 0101101001
899 1011110100
900 0101111010
901 0010111101
902 1000011110
903 0100001111
904 1011000111
905 1100100011
906 1111010001
907 1110101000
908 0111010100
909 0011101010
910 0001110101
911 1001111010
912 0100111101
913 1011011110
914 0101101111
915 1011110111
916 1100111011
917 1111011101
918 1110101110
919 0111010111
920 1010101011
921 1100010101
922 1111001010
923 0111100101
924 1010110010
925 0101011001
926 1011101100
927 0101110110
928 0010111011
929 1000011101
930 1101001110
931 0110100111
932 1010010011
933 1100001001
934 1111000100
935 0111100010
936 0011110001
937 1000111000
938 0100011100
939 0010001110
940 0001000111
941 1001100011
942 1101110001
943 1111111000
944 0111111100
945 0011111110
946 0001111111
947 1001111111
948 1101111111
949 1111111111
950 1110111111
951 1110011111
952 1110001111
953 1110000111
954 1110000011
955 1110000001
956 1110000000
957 0111000000
958 0011100000
959 0001110000
960 0000111000
961 0000011100
962 0000001110
963 0000000111
964 1001000011
965 1101100001
966 1111110000
967 0111111000
968 0011111100
969 0001111110
970 0000111111
971 1001011111
972 1101101111
973 1111110111
974 1110111011
975 1110011101
976 1110001110
977 0111000111
978 1010100011
979 1100010001
980 1111001000
981 0111100100
982 0011110010
983 0001111001
984 1001111100
985 0100111110
986 0010011111
987 1000001111
988 1101000111
989 1111100011
990 1110110001
991 1110011000
992 0111001100
993 0011100110
994 0001110011
995 1001111001
996 1101111100
997 0110111110
998 0011011111
999 1000101111
1000 101010111
1001 1111101011
1002 1110110101
1003 1110011010
1004 0111001101
1005 1010100110
1006 0101010011
1007 1011101001
1008 1100110100
1009 0110011010
1010 0011001101
1011 1000100110
1012 0100010011
1013 1011001001
1014 1100100100
1015 0110010010
1016 0011001001
1017 1000100100
1018 0100010010
1019 0010001001
1020 1000000100
1021 0100000010
1022 0010000001

—Appendix C—

Cyclotomic Cosets *modulo* $2^m - 1$

The following is a list of the cyclotomic cosets *modulo* $2^m - 1$ with respect to GF(2). The coset containing the integer x is as follows.

$$\{x \; mod \, (2^m - 1), x \cdot 2 \; mod \, (2^m - 1), x \cdot 2^2 \; mod \, (2^m - 1), \ldots, x \cdot 2^{d-1} \; mod \, (2^m - 1)\}$$

The cardinality d of the coset must be a divisor of m. In coding theory we are primarily interested in cyclotomic cosets *modulo* $(2^m - 1)$ because they partition the $(2^m - 1)$ powers of a primitive element in GF(2^m) into distinct sets of conjugate elements. By Theorem 3-4 these sets correspond to the binary minimal polynomials for the elements in GF(2^m). For example, let α be primitive (i.e., order 7) in GF(8). Under the heading "*modulo* 7" we see {0}, {1, 2, 4} and {3, 5, 6}—one cyclotomic coset of size 1 and two of size 3. It follows that $m_0(x) = (x + 1)$, $m_1(x) = (x + \alpha)(x + \alpha^2)(x + \alpha^4)$, and $m_3(x) = (x^3 + \alpha)(x + \alpha^5)(x + \alpha^6)$ are the three distinct minimal polynomials for elements in GF(8).

***modulo* 3**

{0},
{1, 2}

***modulo* 7**

{0},
{1, 2, 4}, {3, 5, 6}

***modulo* 15**

{0},
{5, 10},
{1, 2, 4, 8}, {3, 6, 9, 12}, {7, 11, 13, 14}

***modulo* 31**

{0},
{1, 2, 4, 8, 16}, {3, 6, 12, 17, 24}, {5, 9, 10, 18, 20}, {7, 14, 19, 25, 28}, {11, 13, 21, 22, 26},
{15, 23, 27, 29, 30}

***modulo* 63**

{0},
{21, 42},
{9, 18, 36}, {27, 45, 54},
{1, 2, 4, 8, 16, 32}, {3, 6, 12, 24, 33, 48}, {5, 10, 17, 20, 34, 40},
{7, 14, 28, 35, 49, 56}, {11, 22, 25, 37, 44, 50}, {13, 19, 26, 38, 41, 52},
{15, 30, 39, 51, 57, 60}, {23, 29, 43, 46, 53, 58}, {31, 47, 55, 59, 61, 62}

***modulo* 127**

{0},
{1, 2, 4, 8, 16, 32, 64}, {3, 6, 12, 24, 48, 65, 96}, {5, 10, 20, 33, 40, 66, 80},
{7, 14, 28, 56, 67, 97, 112}, {9, 17, 18, 34, 36, 68, 72}, {11, 22, 44, 49, 69, 88, 98},
{13, 26, 35, 52, 70, 81, 104}, {15, 30, 60, 71, 99, 113, 120}, {19, 25, 38, 50, 73, 76, 100},
{21, 37, 41, 42, 74, 82, 84}, {23, 46, 57, 75, 92, 101, 114}, {27, 51, 54, 77, 89, 102, 108},
{29, 39, 58, 78, 83, 105, 116}, {31, 62, 79, 103, 115, 121, 124}, {43, 45, 53, 85, 86, 90, 106},
{47, 61, 87, 94, 107, 117, 122}, {55, 59, 91, 93, 109, 110, 118},
{63, 95, 111, 119, 123, 125, 126}

***modulo* 255**

{0},
{85, 170},
{17, 34, 68, 136}, {51, 102, 153, 204}, {119, 187, 221, 238},
{1, 2, 4, 8, 16, 32, 64, 128}, {3, 6, 12, 24, 48, 96, 129, 192}, {5, 10, 20, 40, 65, 80, 130, 160},
{7, 14, 28, 56, 112, 131, 193, 224}, {9, 18, 33, 36, 66, 72, 132, 144},
{11, 22, 44, 88, 97, 133, 176, 194}, {13, 26, 52, 67, 104, 134, 161, 208},
{15, 30, 60, 120, 135, 195, 225, 240}, {19, 38, 49, 76, 98, 137, 152, 196},
{21, 42, 69, 81, 84, 138, 162, 168}, {23, 46, 92, 113, 139, 184, 197, 226},
{25, 35, 50, 70, 100, 140, 145, 200}, {27, 54, 99, 108, 141, 177, 198, 216},
{29, 58, 71, 116, 142, 163, 209, 232}, {31, 62, 124, 143, 199, 227, 241, 248},
{37, 41, 73, 74, 82, 146, 148, 164}, {39, 57, 78, 114, 147, 156, 201, 228},
{43, 86, 89, 101, 149, 172, 178, 202}, {45, 75, 90, 105, 150, 165, 180, 210},

{47, 94, 121, 151, 188, 203, 229, 242}, {53, 77, 83, 106, 154, 166, 169, 212},
{55, 110, 115, 155, 185, 205, 220, 230}, {59, 103, 118, 157, 179, 206, 217, 236},
{61, 79, 122, 158, 167, 211, 233, 244}, {63, 126, 159, 207, 231, 243, 249, 252},
{87, 93, 117, 171, 174, 186, 213, 234}, {91, 107, 109, 173, 181, 182, 214, 218},
{95, 125, 175, 190, 215, 235, 245, 250}, {111, 123, 183, 189, 219, 222, 237, 246},
{127, 191, 223, 239, 247, 251, 253, 254}

modulo 511

{0},
{73, 146, 292}, {219, 365, 438},
{1, 2, 4, 8, 16, 32, 64, 128, 256}, {3, 6, 12, 24, 48, 96, 192, 257, 384},
{5, 10, 20, 40, 80, 129, 160, 258, 320}, {7, 14, 28, 56, 112, 224, 259, 385, 448},
{9, 18, 36, 65, 72, 130, 144, 260, 288}, {11, 22, 44, 88, 176, 193, 261, 352, 386},
{13, 26, 52, 104, 131, 208, 262, 321, 416}, {15, 30, 60, 120, 240, 263, 387, 449, 480},
{17, 33, 34, 66, 68, 132, 136, 264, 272}, {19, 38, 76, 97, 152, 194, 265, 304, 388},
{21, 42, 84, 133, 161, 168, 266, 322, 336}, {23, 46, 92, 184, 225, 267, 368, 389, 450},
{25, 50, 67, 100, 134, 200, 268, 289, 400}, {27, 54, 108, 195, 216, 269, 353, 390, 432},
{29, 58, 116, 135, 232, 270, 323, 417, 464}, {31, 62, 124, 248, 271, 391, 451, 481, 496},
{35, 49, 70, 98, 140, 196, 273, 280, 392}, {37, 74, 81, 137, 148, 162, 274, 296, 324},
{39, 78, 113, 156, 226, 275, 312, 393, 452}, {41, 69, 82, 138, 145, 164, 276, 290, 328},
{43, 86, 172, 177, 197, 277, 344, 354, 394}, {45, 90, 139, 180, 209, 278, 325, 360, 418},
{47, 94, 188, 241, 279, 376, 395, 453, 482}, {51, 99, 102, 198, 204, 281, 305, 396, 408},
{53, 106, 141, 163, 212, 282, 326, 337, 424}, {55, 110, 220, 227, 283, 369, 397, 440, 454},
{57, 71, 114, 142, 228, 284, 291, 401, 456}, {59, 118, 199, 236, 285, 355, 398, 433, 472},
{61, 122, 143, 244, 286, 327, 419, 465, 488}, {63, 126, 252, 287, 399, 455, 483, 497, 504},
{75, 89, 150, 178, 201, 293, 300, 356, 402}, {77, 105, 147, 154, 210, 294, 308, 329, 420},
{79, 121, 158, 242, 295, 316, 403, 457, 484}, {83, 101, 153, 166, 202, 297, 306, 332, 404},
{85, 149, 165, 169, 170, 298, 330, 338, 340}, {87, 174, 185, 229, 299, 348, 370, 405, 458},
{91, 182, 203, 217, 301, 357, 364, 406, 434}, {93, 151, 186, 233, 302, 331, 372, 421, 466},
{95, 190, 249, 303, 380, 407, 459, 485, 498}, {103, 115, 206, 230, 307, 313, 409, 412, 460},
{107, 179, 205, 214, 309, 345, 358, 410, 428}, {109, 155, 211, 218, 310, 333, 361, 422, 436},
{111, 222, 243, 311, 377, 411, 444, 461, 486}, {117, 157, 167, 234, 314, 334, 339, 425, 468},
{119, 231, 238, 315, 371, 413, 441, 462, 476}, {123, 207, 246, 317, 359, 414, 435, 473, 492},
{125, 159, 250, 318, 335, 423, 467, 489, 500}, {127, 254, 319, 415, 463, 487, 499, 505, 508},
{171, 173, 181, 213, 341, 342, 346, 362, 426}, {175, 189, 245, 343, 350, 378, 427, 469, 490},
{183, 221, 235, 347, 366, 373, 429, 442, 470}, {187, 215, 237, 349, 363, 374, 430, 437, 474},
{191, 253, 351, 382, 431, 471, 491, 501, 506}, {223, 251, 367, 381, 439, 446, 475, 493, 502},
{239, 247, 375, 379, 443, 445, 477, 478, 494}, {255, 383, 447, 479, 495, 503, 507, 509, 510}

modulo 1023

{0},
{341, 682},
{33, 66, 132, 264, 528}, {99, 198, 396, 561, 792}, {165, 297, 330, 594, 660},
{231, 462, 627, 825, 924}, {363, 429, 693, 726, 858}, {495, 759, 891, 957, 990},

{1, 2, 4, 8, 16, 32, 64, 128, 256, 512},
{3, 6, 12, 24, 48, 96, 192, 384, 513, 768},
{5, 10, 20, 40, 80, 160, 257, 320, 514, 640},
{7, 14, 28, 56, 112, 224, 448, 515, 769, 896},
{9, 18, 36, 72, 129, 144, 258, 288, 516, 576},
{11, 22, 44, 88, 176, 352, 385, 517, 704, 770},
{13, 26, 52, 104, 208, 259, 416, 518, 641, 832},
{15, 30, 60, 120, 240, 480, 519, 771, 897, 960},
{17, 34, 65, 68, 130, 136, 260, 272, 520, 544},
{19, 38, 76, 152, 193, 304, 386, 521, 608, 772},
{21, 42, 84, 168, 261, 321, 336, 522, 642, 672},
{23, 46, 92, 184, 368, 449, 523, 736, 773, 898},
{25, 50, 100, 131, 200, 262, 400, 524, 577, 800},
{27, 54, 108, 216, 387, 432, 525, 705, 774, 864},
{29, 58, 116, 232, 263, 464, 526, 643, 833, 928},
{31, 62, 124, 248, 496, 527, 775, 899, 961, 992},
{35, 70, 97, 140, 194, 280, 388, 529, 560, 776},
{37, 74, 148, 161, 265, 296, 322, 530, 592, 644},
{39, 78, 156, 225, 312, 450, 531, 624, 777, 900},
{41, 82, 133, 164, 266, 289, 328, 532, 578, 656},
{43, 86, 172, 344, 353, 389, 533, 688, 706, 778},
{45, 90, 180, 267, 360, 417, 534, 645, 720, 834},
{47, 94, 188, 376, 481, 535, 752, 779, 901, 962},
{49, 67, 98, 134, 196, 268, 392, 536, 545, 784},
{51, 102, 195, 204, 390, 408, 537, 609, 780, 816},
{53, 106, 212, 269, 323, 424, 538, 646, 673, 848},
{55, 110, 220, 440, 451, 539, 737, 781, 880, 902},
{57, 114, 135, 228, 270, 456, 540, 579, 801, 912},
{59, 118, 236, 391, 472, 541, 707, 782, 865, 944},
{61, 122, 244, 271, 488, 542, 647, 835, 929, 976},
{63, 126, 252, 504, 543, 783, 903, 963, 993, 1008},
{69, 81, 138, 162, 273, 276, 324, 546, 552, 648},
{71, 113, 142, 226, 284, 452, 547, 568, 785, 904},
{73, 137, 145, 146, 274, 290, 292, 548, 580, 584},
{75, 150, 177, 300, 354, 393, 549, 600, 708, 786},
{77, 154, 209, 275, 308, 418, 550, 616, 649, 836},
{79, 158, 241, 316, 482, 551, 632, 787, 905, 964},
{83, 166, 197, 305, 332, 394, 553, 610, 664, 788},
{85, 170, 277, 325, 337, 340, 554, 650, 674, 680},
{87, 174, 348, 369, 453, 555, 696, 738, 789, 906},
{89, 139, 178, 278, 356, 401, 556, 581, 712, 802},
{91, 182, 364, 395, 433, 557, 709, 728, 790, 866},
{93, 186, 279, 372, 465, 558, 651, 744, 837, 930},
{95, 190, 380, 497, 559, 760, 791, 907, 965, 994},
{101, 163, 202, 281, 326, 404, 562, 593, 652, 808},

$\{103, 206, 227, 412, 454, 563, 625, 793, 824, 908\}$,
$\{105, 141, 210, 282, 291, 420, 564, 582, 657, 840\}$,
$\{107, 214, 355, 397, 428, 565, 689, 710, 794, 856\}$,
$\{109, 218, 283, 419, 436, 566, 653, 721, 838, 872\}$,
$\{111, 222, 444, 483, 567, 753, 795, 888, 909, 966\}$,
$\{115, 199, 230, 398, 460, 569, 611, 796, 817, 920\}$,
$\{117, 234, 285, 327, 468, 570, 654, 675, 849, 936\}$,
$\{119, 238, 455, 476, 571, 739, 797, 881, 910, 952\}$,
$\{121, 143, 242, 286, 484, 572, 583, 803, 913, 968\}$,
$\{123, 246, 399, 492, 573, 711, 798, 867, 945, 984\}$,
$\{125, 250, 287, 500, 574, 655, 839, 931, 977, 1000\}$,
$\{127, 254, 508, 575, 799, 911, 967, 995, 1009, 1016\}$,
$\{147, 153, 201, 294, 306, 402, 585, 588, 612, 804\}$,
$\{149, 169, 293, 298, 329, 338, 586, 596, 658, 676\}$,
$\{151, 185, 302, 370, 457, 587, 604, 740, 805, 914\}$,
$\{155, 217, 310, 403, 434, 589, 620, 713, 806, 868\}$,
$\{157, 233, 295, 314, 466, 590, 628, 659, 841, 932\}$,
$\{159, 249, 318, 498, 591, 636, 807, 915, 969, 996\}$,
$\{167, 229, 313, 334, 458, 595, 626, 668, 809, 916\}$,
$\{171, 342, 345, 357, 405, 597, 684, 690, 714, 810\}$,
$\{173, 299, 346, 361, 421, 598, 661, 692, 722, 842\}$,
$\{175, 350, 377, 485, 599, 700, 754, 811, 917, 970\}$,
$\{179, 203, 358, 406, 409, 601, 613, 716, 812, 818\}$,
$\{181, 301, 331, 362, 425, 602, 662, 677, 724, 850\}$,
$\{183, 366, 441, 459, 603, 732, 741, 813, 882, 918\}$,
$\{187, 374, 407, 473, 605, 715, 748, 814, 869, 946\}$,
$\{189, 303, 378, 489, 606, 663, 756, 843, 933, 978\}$,
$\{191, 382, 505, 607, 764, 815, 919, 971, 997, 1010\}$,
$\{205, 211, 307, 410, 422, 614, 617, 665, 820, 844\}$,
$\{207, 243, 414, 486, 615, 633, 819, 828, 921, 972\}$,
$\{213, 309, 333, 339, 426, 618, 666, 678, 681, 852\}$,
$\{215, 371, 430, 461, 619, 697, 742, 821, 860, 922\}$,
$\{219, 411, 435, 438, 621, 717, 729, 822, 870, 876\}$,
$\{221, 311, 442, 467, 622, 667, 745, 845, 884, 934\}$,
$\{223, 446, 499, 623, 761, 823, 892, 923, 973, 998\}$,
$\{235, 359, 413, 470, 629, 691, 718, 826, 857, 940\}$,
$\{237, 315, 423, 474, 630, 669, 723, 846, 873, 948\}$,
$\{239, 478, 487, 631, 755, 827, 889, 925, 956, 974\}$,
$\{245, 317, 335, 490, 634, 670, 679, 851, 937, 980\}$,
$\{247, 463, 494, 635, 743, 829, 883, 926, 953, 988\}$,
$\{251, 415, 502, 637, 719, 830, 871, 947, 985, 1004\}$,
$\{253, 319, 506, 638, 671, 847, 935, 979, 1001, 1012\}$,
$\{255, 510, 639, 831, 927, 975, 999, 1011, 1017, 1020\}$,
$\{343, 349, 373, 469, 683, 686, 698, 746, 853, 938\}$,

$\{347, 365, 427, 437, 685, 694, 725, 730, 854, 874\}$,
$\{351, 381, 501, 687, 702, 762, 855, 939, 981, 1002\}$,
$\{367, 445, 491, 695, 734, 757, 859, 890, 941, 982\}$,
$\{375, 471, 477, 699, 747, 750, 861, 885, 942, 954\}$,
$\{379, 431, 493, 701, 727, 758, 862, 875, 949, 986\}$,
$\{383, 509, 703, 766, 863, 943, 983, 1003, 1013, 1018\}$,
$\{439, 443, 475, 731, 733, 749, 877, 878, 886, 950\}$,
$\{447, 507, 735, 765, 879, 894, 951, 987, 1005, 1014\}$,
$\{479, 503, 751, 763, 887, 893, 955, 958, 989, 1006\}$,
$\{511, 767, 895, 959, 991, 1007, 1015, 1019, 1021, 1022\}$

Appendix D[1]

Minimal Polynomials of Elements in GF(2^m)

The following is a list of the binary minimal polynomials for all elements in binary extension fields from GF(2^2) through GF(2^{10}). The polynomials are denoted here by the exponent of one root and the powers of the polynomials' nonzero terms. For example, the entry "3 (0, 2, 3)" under the heading "GF(8)" corresponds to $m(x) = 1 + x^2 + x^3$, which has as roots the conjugate elements $\{\alpha^3, \alpha^6, \alpha^5\}$, where α is primitive in GF(8). To conserve space, the minimal polynomials are enumerated using the root with the smallest exponent. A complete listing of the exponents of the elements in the various conjugacy classes (i.e., the cyclotomic cosets) can be found in Appendix C.

GF(4)

1	(0, 1, 2)		

GF(8)

1	(0, 1, 3)	3	(0, 2, 3)

GF(16)

1	(0, 1, 4)	3	(0, 1, 2, 3, 4)
5	(0, 1, 2)	7	(0, 3, 4)

GF(32)

1	(0, 2, 5)	3	(0, 2, 3, 4, 5)
5	(0, 1, 2, 4, 5)	7	(0, 1, 2, 3, 5)
11	(0, 1, 3, 4, 5)	15	(0, 3, 5)

[1] The material in this appendix is taken, with permission, from S. Lin and D. J. Costello, Jr., *Error Control Coding: Fundamentals and Applications*, pp. 579–582, Englewood Cliffs, NJ: Prentice Hall, 1983.

GF(64)

1 (0, 1, 6)
3 (0, 1, 2, 4, 6)
5 (0, 1, 2, 5, 6)
7 (0, 3, 6)
9 (0, 2, 3)
11 (0, 2, 3, 5, 6)
13 (0, 1, 3, 4, 6)
15 (0, 2, 4, 5, 6)
21 (0, 1, 2)
23 (0, 1, 4, 5, 6)
27 (0, 1, 3)
31 (0, 5, 6)

GF(128)

1 (0, 3, 7)
3 (0, 1, 2, 3, 7)
5 (0, 2, 3, 4, 7)
7 (0, 1, 2, 4, 5, 6, 7)
9 (0, 1, 2, 3, 4, 5, 7)
11 (0, 2, 4, 6, 7)
13 (0, 1, 7)
15 (0, 1, 2, 3, 5, 6, 7)
19 (0, 1, 2, 6, 7)
21 (0, 2, 5, 6, 7)
23 (0, 6, 7)
27 (0, 1, 4, 6, 7)
29 (0, 1, 3, 5, 7)
31 (0, 4, 5, 6, 7)
43 (0, 1, 2, 5, 7)
47 (0, 3, 4, 5, 7)
55 (0, 2, 3, 4, 5, 6, 7)
63 (0, 4, 7)

GF(256)

1 (0, 2, 3, 4, 8)
3 (0, 1, 2, 4, 5, 6, 8)
5 (0, 1, 4, 5, 6, 7, 8)
7 (0, 3, 5, 6, 8)
9 (0, 2, 3, 4, 5, 7, 8)
11 (0, 1, 2, 5, 6, 7, 8)
13 (0, 1, 3, 5, 8)
15 (0, 1, 2, 4, 6, 7, 8)
17 (0, 1, 4)
19 (0, 2, 5, 6, 8)
21 (0, 1, 3, 7, 8)
23 (0, 1, 5, 6, 8)
25 (0, 1, 3, 4, 8)
27 (0, 1, 2, 3, 4, 5, 8)
29 (0, 2, 3, 7, 8)
31 (0, 2, 3, 5, 8)
37 (0, 1, 2, 3, 4, 6, 8)
39 (0, 3, 4, 5, 6, 7, 8)
43 (0, 1, 6, 7, 8)
45 (0, 3, 4, 5, 8)
47 (0, 3, 5, 7, 8)
51 (0, 1, 2, 3, 4)
53 (0, 1, 2, 7, 8)
55 (0, 4, 5, 7, 8)
59 (0, 2, 3, 6, 8)
61 (0, 1, 2, 3, 6, 7, 8)
63 (0, 2, 3, 4, 6, 7, 8)
85 (0, 1, 2)
87 (0, 1, 5, 7, 8)
91 (0, 2, 4, 5, 6, 7, 8)
95 (0, 1, 2, 3, 4, 7, 8)
111 (0, 1, 3, 4, 5, 6, 8)
119 (0, 3, 4)
127 (0, 4, 5, 6, 8)

GF(512)

1 (0, 4, 9)
3 (0, 3, 4, 6, 9)
5 (0, 4, 5, 8, 9)
7 (0, 3, 4, 7, 9)
9 (0, 1, 4, 8, 9)
11 (0, 2, 3, 5, 9)
13 (0, 1, 2, 4, 5, 6, 9)
15 (0, 5, 6, 8, 9)
17 (0, 1, 3, 4, 6, 7, 9)
19 (0, 2, 7, 8, 9)

21	(0, 1, 2, 4, 9)	23	(0, 3, 5, 6, 7, 8, 9)
25	(0, 1, 5, 6, 7, 8, 9)	27	(0, 1, 2, 3, 7, 8, 9)
29	(0, 1, 3, 5, 6, 8, 9)	31	(0, 1, 3, 4, 9)
35	(0, 8, 9)	37	(0, 1, 2, 3, 5, 6, 9)
39	(0, 2, 3, 6, 7, 8, 9)	41	(0, 1, 4, 5, 6, 8, 9)
43	(0, 1, 3, 6, 7, 8, 9)	45	(0, 2, 3, 4, 5, 6, 9)
47	(0, 1, 3, 4, 6, 8, 9)	51	(0, 2, 4, 6, 7, 8, 9)
53	(0, 2, 4, 7, 9)	55	(0, 2, 3, 4, 5, 7, 9)
57	(0, 2, 4, 5, 6, 7, 9)	59	(0, 1, 2, 3, 6, 7, 9)
61	(0, 1, 2, 3, 4, 6, 9)	63	(0, 2, 5, 6, 9)
73	(0, 1, 3)	75	(0, 1, 3, 4, 5, 6, 7, 8, 9)
77	(0, 3, 6, 8, 9)	79	(0, 1, 2, 6, 7, 8, 9)
83	(0, 2, 4, 8, 9)	85	(0, 1, 2, 4, 6, 7, 9)
87	(0, 2, 5, 7, 9)	91	(0, 1, 3, 6, 8)
93	(0, 3, 4, 5, 6, 7, 9)	95	(0, 3, 4, 5, 7, 8, 9)
103	(0, 1, 2, 3, 5, 7, 9)	107	(0, 1, 5, 7, 9)
109	(0, 1, 2, 3, 4, 5, 6, 8, 9)	111	(0, 1, 2, 3, 4, 8, 9)
117	(0, 1, 2, 3, 6, 8, 9)	119	(0, 1, 9)
123	(0, 1, 2, 7, 9)	125	(0, 4, 6, 7, 9)
127	(0, 3, 5, 6, 9)	171	(0, 2, 4, 5, 7, 8, 9)
175	(0, 5, 7, 8, 9)	183	(0, 1, 3, 5, 8, 9)
187	(0, 3, 4, 6, 7, 8, 9)	191	(0, 1, 4, 5, 9)
219	(0, 2, 3)	223	(0, 1, 5, 8, 9)
239	(0, 2, 3, 5, 6, 8, 9)	255	(0, 5, 9)

GF(1024)

1	(0, 3, 10)	3	(0, 1, 2, 3, 10)
5	(0, 2, 3, 8, 10)	7	(0, 3, 4, 5, 6, 7, 8, 9, 10)
9	(0, 1, 2, 3, 5, 7, 10)	11	(0, 2, 4, 5, 10)
13	(0, 1, 2, 3, 5, 6, 10)	15	(0, 1, 3, 5, 7, 8, 10)
17	(0, 2, 3, 6, 8, 9, 10)	19	(0, 1, 3, 4, 5, 6, 7, 8, 10)
21	(0, 1, 3, 5, 6, 7, 8, 9, 10)	23	(0, 1, 3, 4, 10)
25	(0, 1, 5, 8, 10)	27	(0, 1, 3, 4, 5, 6, 8, 9, 10)
29	(0, 4, 5, 8, 10)	31	(0, 1, 5, 9, 10)
33	(0, 2, 3, 4, 5)	35	(0, 1, 4, 9, 10)
37	(0, 1, 5, 6, 8, 9, 10)	39	(0, 1, 2, 6, 10)
41	(0, 2, 5, 6, 7, 8, 10)	43	(0, 3, 4, 8, 10)
45	(0, 4, 5, 9, 10)	47	(0, 1, 2, 3, 4, 5, 6, 9, 10)
49	(0, 2, 4, 6, 8, 9, 10)	51	(0, 1, 2, 5, 6, 8, 10)
53	(0, 1, 2, 3, 7, 8, 10)	55	(0, 1, 3, 5, 8, 9, 10)
57	(0, 4, 6, 9, 10)	59	(0, 3, 4, 5, 8, 9, 10)
61	(0, 1, 4, 5, 6, 7, 8, 9, 10)	63	(0, 2, 3, 5, 7, 9, 10)
69	(0, 6, 7, 8, 10)	71	(0, 1, 4, 6, 7, 9, 10)
73	(0, 1, 2, 6, 8, 9, 10)	75	(0, 1, 2, 3, 4, 8, 10)
77	(0, 1, 3, 8, 10)	79	(0, 1, 2, 5, 6, 7, 10)

83 (0, 1, 4, 7, 8, 9, 10)
87 (0, 3, 6, 7, 10)
91 (0, 2, 4, 5, 7, 9, 10)
95 (0, 2, 5, 6, 10)
101 (0, 2, 3, 5, 10)
105 (0, 1, 2, 7, 8, 9, 10)
109 (0, 1, 2, 5, 10)
115 (0, 1, 2, 4, 5, 6, 7, 8, 10)
119 (0, 1, 3, 4, 6, 9, 10)
123 (0, 4, 8, 9, 10)
127 (0, 1, 2, 3, 4, 5, 6, 7, 10)
149 (0, 2, 4, 9, 10)
155 (0, 3, 5, 7, 10)
159 (0, 1, 2, 4, 5, 6, 7, 9, 10)
167 (0, 1, 4, 5, 6, 7, 10)
173 (0, 1, 2, 3, 4, 6, 7, 9, 10)
179 (0, 3, 7, 9, 10)
183 (0, 1, 2, 3, 8, 9, 10)
189 (0, 1, 5, 6, 10)
205 (0, 1, 3, 7, 10)
213 (0, 1, 3, 4, 7, 8, 10)
219 (0, 3, 4, 5, 7, 8, 10)
223 (0, 2, 5, 9, 10)
235 (0, 1, 2, 3, 6, 9, 10)
239 (0, 1, 2, 4, 6, 8, 10)
247 (0, 1, 6, 9, 10)
253 (0, 5, 6, 8, 10)
341 (0, 1, 2)
347 (0, 1, 6, 8, 10)
363 (0, 2, 5)
375 (0, 2, 3, 4, 10)
383 (0, 2, 7, 8, 10)
447 (0, 3, 5, 7, 8, 9, 10)
495 (0, 1, 2, 3, 5)

85 (0, 1, 2, 6, 7, 8, 10)
89 (0, 1, 2, 4, 6, 7, 10)
93 (0, 1, 2, 3, 4, 5, 6, 7, 8, 9, 10)
99 (0, 1, 2, 4, 5)
103 (0, 2, 3, 4, 5, 6, 8, 9, 10)
107 (0, 3, 4, 5, 6, 9, 10)
111 (0, 1, 4, 6, 10)
117 (0, 3, 4, 7, 10)
121 (0, 1, 2, 5, 7, 9, 10)
125 (0, 6, 7, 9, 10)
147 (0, 2, 3, 5, 6, 7, 10)
151 (0, 5, 8, 9, 10)
157 (0, 1, 3, 5, 6, 8, 10)
165 (0, 3, 5)
171 (0, 2, 3, 6, 7, 9, 10)
175 (0, 2, 3, 7, 8, 10)
181 (0, 1, 3, 4, 6, 7, 8, 9, 10)
187 (0, 2, 7, 9, 10)
191 (0, 4, 5, 7, 8, 9, 10)
207 (0, 2, 4, 5, 8, 9, 10)
215 (0, 5, 7, 8, 10)
221 (0, 3, 4, 6, 8, 9, 10)
231 (0, 1, 3, 4, 5)
237 (0, 2, 6, 7, 8, 9, 10)
245 (0, 2, 6, 7, 10)
251 (0, 2, 3, 4, 5, 6, 7, 9, 10)
255 (0, 7, 8, 9, 10)
343 (0, 2, 3, 4, 8, 9, 10)
351 (0, 1, 2, 3, 4, 5, 7, 9, 10)
367 (0, 2, 3, 4, 5, 8, 10)
379 (0, 1, 2, 4, 5, 9, 10)
439 (0, 1, 2, 4, 8, 9, 10)
479 (0, 1, 2, 4, 7, 8, 10)
511 (0, 7, 10)

Appendix E[1]

Generator Polynomials of Binary BCH Codes of Lengths 7 Through 255

The following is a list of all of the generator polynomials for all binary, narrow-sense, primitive BCH codes of lengths 7 through 255. The polynomials are ordered by the length n, dimension k, and error correcting capability t of the corresponding BCH code. The polynomials themselves are written using an octal representation. To read an entry, the octal expression is first expanded into binary form. The ones in the binary form correspond to the nonzero terms in the generator polynomial, while the position of the ones determines the terms' degrees. For example, the two-error correcting (31, 21) BCH code has the following generator polynomial.

$$3551 \Rightarrow 011, 101, 101, 001 \Rightarrow g(x) = x^{10} + x^9 + x^8 + x^6 + x^5 + x^3 + 1$$

n	k	t	generator polynomial in octal form
7	4	1	13
15	11	1	23
15	7	2	721
15	5	3	2467
31	26	1	45
31	21	2	3551
31	16	3	107657
31	11	5	5423325
31	6	7	313365047

[1] The material in this appendix is taken, with permission, from S. Lin and D. J. Costello, Jr., *Error Control Coding: Fundamentals and Applications*, pp. 583–586, Englewood Cliffs, NJ: Prentice Hall, 1983.

n	k	t	generator polynomial in octal form
63	57	1	103
63	51	2	12471
63	45	3	1701317
63	39	4	166623567
63	36	5	1033500423
63	30	6	157464165547
63	24	7	17323260404441
63	18	10	1363026512351725
63	16	11	6331141367235453
63	10	13	472622305527250155
63	7	15	5231045543503271737
127	120	1	211
127	113	2	41567
127	106	3	11554743
127	99	4	3447023271
127	92	5	624730022327
127	85	6	130704476322273
127	78	7	26230002166130115
127	71	9	6255010713253127753
127	64	10	1206534025570773100045
127	57	11	335265252505705053517721
127	50	13	54446512523314012421501421
127	43	14	17721772213651227521220574343
127	36	15	3146074666522075044764574721735
127	29	21	403114461367670603667530141176155
127	22	23	123376070404722522435445626637647043
127	15	27	22057042445604554770523013762217604353
127	8	31	7047264052751030651476224271567733130217
255	247	1	435
255	239	2	267543
255	231	3	156720665
255	223	4	75626641375
255	215	5	23157564726421
255	207	6	16176560567636227
255	199	7	7633031270420722341
255	191	8	2663470176115333714567
255	187	9	52755313540001322236351
255	179	10	22624710717340432416300455
255	171	11	15416214212342356077061630637
255	163	12	7500415510075602551574724514601
255	155	13	3757513005407665015722506464677633

n	k	t	generator polynomial in octal form
255	147	14	16421301735371655253041653054410117111
255	139	15	461401732060175561570722730247453567445
255	131	18	215713331471510151261250277442142024165471
255	123	19	120614052242066003717210326516141226272506267
255	115	21	60526665572100247263636404600276352556313472737
255	107	22	22205772322066256312417300235347420176574750154441
255	99	23	10656667253473174222741416201574332252411076432303431
255	91	25	6750265030327444172723631724732511075550762707243445611
255	87	26	110136763414743236435231634307172046206722545273311721317
255	79	27	66700035637657500020270344207366174621015326711766541342355
255	71	29	2402471052064432151555417211233116320544425036255764322 1706035
255	63	30	10754475055163544325315217357707003666111726455267613656702543301
255	55	31	7315425203501100133015275306032054325414326755010557044426035473617
255	47	42	2533542017062646563033041377406233175123334145446045005066024552543173
255	45	43	15202056055234161131101346376423701563670024470762373033202157025051541
255	37	45	5136330255067007414177447245437530420735706174323432347644354737403044003
255	29	47	3025715536673071465527064012361377115342242324201174114060254657410403565037
255	21	55	1256215257060332656001773153607612103227341405653074542521153121614466513473725
255	13	59	464173200505256454442657371425006600433067744547656140317467721357026134460500547
255	9	63	157260252174724632010310432553551346141623672120440745451127661155477055616775160570

General Bibliography

Books

[Ber1] E. R. BERLEKAMP, *Algebraic Coding Theory*, New York: McGraw-Hill, 1968 (rev. ed., Laguna Hills, Aegean Park Press, 1984).

[Ber2] E. R. BERLEKAMP (ed.), *Key Papers in the Development of Coding Theory*, New York: IEEE Press, 1974.

[Bha] V. K. BHARGAVA, D. HACCOUN, R. MATYAS, and P. NUSPL, *Digital Communications by Satellite*, New York: John Wiley & Sons, 1981.

[Big1] E. BIGLIERI, D. DIVSALAR, P. J. MCLANE, M. K. SIMON, *Introduction to Trellis-Coded Modulation with Applications*, New York: Macmillan Publishing Company, 1991.

[Bla1] R. E. BLAHUT, *Information Theory*, Reading, MA: Addison-Wesley, 1988.

[Bla2] R. E. BLAHUT, *Theory and Practice of Error Control Codes*, Reading, MA: Addison-Wesley, 1984.

[Bla3] R. E. BLAHUT, *Digital Transmission of Information*, Reading, MA: Addison-Wesley, 1990.

[Blak] I. F. BLAKE and R. C. MULLIN, *The Mathematical Theory of Coding*, New York: Academic Press, 1975.

[Cla] G. CLARK and J. CAIN, *Error-Correction Coding for Digital Communications*, New York: Plenum Press, 1981.

[Con1] J. H. CONWAY and N. J. A. SLOANE, *Sphere Packings, Lattices, and Groups*, New York: Springer-Verlag, 1988.

[Cov] T. COVER and J. THOMAS, *Elements of Information Theory*, New York: Wiley Interscience, 1991.

[For1] G. D. FORNEY, JR., *Concatenated Codes*, Cambridge, MA: MIT Press, 1966.

[Gag1] R. M. GAGLIARDI, *Introduction to Communications Engineering*, 2d ed., New York: John Wiley & Sons, 1988.

[Gag2] R. M. GAGLIARDI, *Satellite Communications*, New York: John Wiley & Sons, 1985.

[Gal1] R. G. GALLAGER, *Information Theory and Reliable Communication*, New York: John Wiley & Sons, 1968.

[Gol1] S. W. GOLOMB, *Shift Register Sequences*, San Francisco: Holden-Day, 1967 (Rev. ed., Laguna Hills: Aegean Park Press, 1982).

[Hag1] J. HAGENAUER, E. OFFER, and L. PAPKE, "Matching Viterbi Decoders and Reed-Solomon Decoders in a Concantenated System," in *Reed Solomon Codes and Their Applications*, Wicker and Bhargava (ed.), New York: IEEE Press, 1994.

[Hir] K. HIRADE, "Mobile Radio Communications," in *Advanced Digital Communications*, Chap. 10, K. Feher (ed.), Englewood Cliffs, NJ: Prentice Hall, 1987.

[Hun] T. W. HUNGERFORD, *Algebra*, New York: Springer-Verlag, 1974.

[Imm1] K. A. S. IMMINK, "RS Codes and the Compact Disc," in *Reed Solomon Codes and Their Applications*, Wicker and Bhargava (ed.), New York: IEEE Press, 1994.

[Imm2] K. A. S. IMMINK, *Coding Techniques for Digital Recorders*, Englewood Cliffs, NJ: Prentice Hall International (U.K.), 1991.

[Knu] D. E. KNUTH, *The Art of Computer Programming*, 2d ed., Vol. 2, Reading, MA: Addison-Wesley, 1981.

[Lee1] W. C. Y. LEE, *Mobile Communications Engineering*, New York: McGraw-Hill, 1982.

[Lee2] W. C. Y. LEE, *Mobile Communications Design Fundamentals*, Indianapolis: Howard Sams and Co., 1986.

[Lid] R. LIDL and H. NIEDERREITER, *Finite Fields*, Reading, MA: Addison-Wesley, 1983.

[Lin1] S. LIN and D. J. COSTELLO JR., *Error Control Coding: Fundamentals and Applications*, Englewood Cliffs, NJ: Prentice Hall, 1983.

[Lind] W. C. LINDSEY and M. K. SIMON, *Telecommunication Systems Engineering*, Englewood Cliffs, NJ: Prentice Hall, 1973.

[Liu] C. L. LIU, *Introduction to Combinatorial Mathematics*, New York: McGraw Hill, 1968.

[Mac] F. J. MACWILLIAMS and N. J. A. SLOANE, *The Theory of Error Correcting Codes*, Amsterdam: North-Holland, 1977.

[Man] H. B. MANN (ed.), *Error Correcting Codes*, New York: John Wiley & Sons, 1968.

[Mas] S. MASON and H. ZIMMERMANN, *Electronic Circuits*, *Signals*, *and Systems*, New York: Wiley, 1960.

[Mas1] J. L. MASSEY, *Threshold Decoding*, Cambridge, MA: MIT Press, 1963.

[Mas2] J. L. MASSEY, "Deep Space Communications and Coding: A Match Made in Heaven," in *Advanced Methods for Satellite and Deep Sapce Communications*, J. Hagenauer (ed.), Lecture Notes in Control and Information Sciences, Vol. 182, Berlin: Springer-Verlag, 1992.

[McC] K. R. MCCONNELL, D. BODSON, and R. SCHAPHORST, *FAX: Digital Facsimile Technology and Applications*, Boston: Artech House, 1992.

[McE1] R. J. MCELIECE, *Finite Fields for Computer Scientists and Engineers*, Boston: Kluwer Academic Publishers, 1987.

[Mic] A. M. MICHELSON and A. H. LEVESQUE, *Error Control Techniques for Digital Communication*, New York: John Wiley & Sons, 1985.

[Pap] A. PAPOULIS, *Probability*, *Random Variables, and Stochastic Processes*, 2d ed., New York: McGraw-Hill, 1984.

[Pat1] A. M. PATEL, "Signal and Error-Control Coding," in *Magnetic Recording, Volume II: Computer Data Storage*, New York: McGraw-Hill, 1988.

[Pet1] W. W. PETERSON and E. J. WELDON, JR., *Error Correcting Codes*, 2d ed., Cambridge, MA: MIT Press, 1972.

[Pro] J. G. PROAKIS, *Digital Communications*, 2d ed., New York: McGraw-Hill, 1989.

[Rao] T. R. N. RAO and E. FUJIWARA, *Error-Control Coding for Computer Systems*, Englewood Cliffs, NJ: Prentice Hall, 1989.

[Rio] J. RIORDAN, *An Introduction to Combinatorial Analysis*, New York: John Wiley & Sons, 1958.

[Rys] H. J. RYSER, *Combinatorial Mathematics*, Carus Mathematical Monograph No. 14, The Mathematical Association of America, 1963.

[Skl] B. SKLAR, *Digital Communications, Fundamentals and Applications*, Englewood Cliffs, NJ: Prentice Hall, 1988.

[Tan] A. S. TANENBAUM, *Computer Networks*, 2d ed., Englewood Cliffs, NJ: Prentice Hall, 1988.

[VanL] J. H. VAN LINT, *Introduction to Coding Theory*, New York: Springer, 1982.

[VanS] S. A. VANSTONE and P.C. VAN OORSCHOT, *An Introduction to Error Correcting Codes with Applications*, Boston: Kluwer Academic Publishers, 1989.

[Vit1] A. J. VITERBI and J. K. OMURA, *Principles of Digital Communication and Coding*, New York: McGraw-Hill, 1979.

[Wel] E. J. WELDON, JR., "Some Results on Majority Logic Decoding, *Error Correcting Codes* (H. B. Mann ed.), New York: John Wiley & Sons, 1968.

[Woz1] J. M. WOZENCRAFT and I. M. JACOBS, *Principles of Communication Engineering*, New York: John Wiley & Sons, 1965.

[Woz2] J. M. WOZENCRAFT and B. REIFFEN, *Sequential Decoding*, Cambridge, MA: MIT Press, 1961.

Journal articles, conference papers, technical reports

[Bene1] G. BENELLI, "An ARQ Scheme with Memory and Soft Error Detectors," *IEEE Transactions on Communications*, Vol. COM-33, pp. 285–288, March 1985.

[Bene2] G. BENELLI, "An ARQ Scheme with Memory and Integrated Modulation," *IEEE Transactions on Communications*, Vol. COM-35, pp. 689–697, July 1987.

[Beni1] R. J. BENICE and A. H. FREY, "An Analysis of Retransmission Systems," *IEEE Transactions on Communications Technology*, Vol. COM-12, pp. 135–145, December 1964.

[Beni2] R. J. BENICE and A. H. FREY, "Comparisons of Error Control Techniques," *IEEE Transactions on Communications Technology*, Vol. COM-12, pp. 146–154, December 1964.

[Ber3] E. R. BERLEKAMP, "The Technology of Error-Correcting Codes," *Proceedings of the IEEE*, Vol. 68, pp. 564–593, May 1980.

[Ber4] E. R. BERLEKAMP, R. PEILE, and S. POPE, "The Application of Error Control to Communications," *IEEE Communications Magazine*, Vol. 25, pp. 44–57, April 1987.

[Ber5] E. R. BERLEKAMP, R. J. MCELIECE, H. C. A. VAN TILBORG, "On the Inherent Intractability of Certain Coding Problems," *IEEE Transactions on Information Theory*, Vol. IT-24, pp. 384–386, May 1978.

[Ber6] E. BERLEKAMP, "Nonbinary BCH Decoding," presented at the 1967 International Symposium on Information Theory, San Remo, Italy.

[Big2] E. BIGLIERI, "High-Level Modulation and Coding for Nonlinear Satellite Channels," *IEEE Transactions on Communications*, Vol. COM-32, No. 5, pp. 616–626, May 1984.

[Bla4] R. E. BLAHUT, "Transform Techniques for Error Control Codes," *IBM Journal of Research and Development*, Vol. 23, pp. 299–315, 1979.

[Bos1] R. C. BOSE and D. K. RAY-CHAUDHURI, "On a Class of Error Correcting Binary Group Codes," *Information and Control*, Vol. 3, pp. 68–79, March 1960.

[Bos2] R. C. BOSE and D. K. RAY-CHAUDHURI, "Further Results on Error Correcting Binary Group Codes," *Information and Control*, Vol. 3, pp. 279–290, September 1960.

[Bru] H. BRUNEEL and M. MOENECLAEY, "On the Throughput Performance of Some Continuous ARQ Strategies with Repeated Transmissions," *IEEE Transactions on Communications*, Vol. COM-34, pp. 244–249, March 1986.

[Cai] J. B. CAIN, G. C. CLARK, JR., and J. M. GEIST, "Punctured Convolutional Codes of Rate $(n-1)/n$ and Simplified Maximum Likelihood Decoding," *IEEE Transactions on Information Theory*, IT-25, pp. 97–100, January 1979.

[Cal1] A. R. CALDERBANK and J. MAZO, "A New Description of Trellis Codes," *IEEE Transactions on Information Theory*, Vol. IT-30, pp. 784–791, November 1984.

[Cal2] A. R. CALDERBANK and N. J. A. SLOANE, "New Trellis Codes Based on Lattices and Cosets," *IEEE Transactions on Information Theory*, Vol. IT-33, pp. 177–195, March 1987.

[Cal3] R. CALDERBANK, J. E. MAZO, and V. K. WEI, "Asymptotic Upper Bounds on the Minimum Distance of Trellis Codes," *IEEE Transactions on Communications*, Vol. COM-33, No. 4, pp. 305–309, April 1985.

[Car] M. G. CARASSO, J. B. H. PEEK, and J. P. SINJOU, "The Compact Disc Digital Audio System," *Philips Technical Review*, Vol. 40, No. 6, pp. 151–156, 1982.

[Cas] G. CASTAGNOLI, J. GANZ, and P. GRABER, "Optimum Cyclic Redundancy-Check Codes with 16-Bit Redundancy," *IEEE Transactions on Communications*, Vol. 38, No. 1, pp. 111–114, January 1990.

[Cha] D. CHASE, "Code Combining—A Maximum-Likelihood Decoding Approach for Combining an Arbitrary Number of Noisy Packets," *IEEE Transactions on Communications*, Vol. COM-33, pp. 385–393, May 1985.

[Che] P. R. CHEVILLAT and D. J. COSTELLO, JR., "An Analysis of Sequential Decoding for Specific Time-Invariant Convolutional Codes," *IEEE Transactions on Information Theory*, Vol. IT-24, pp. 443–451, July 1978.

[Chi] R. T. CHIEN, "Cyclic Decoding Procedure for the Bose-Chaudhuri-Hocquenghem Codes," *IEEE Transactions on Information Theory*, Vol. IT-10, pp. 357–363, October 1964.

[Com] R. A. COMROE and D. J. COSTELLO JR., "ARQ Schemes for Data Transmission in Mobile Radio Systems," *IEEE Journal on Selected Areas in Communications*, Vol. SAC-2, No. 4, pp. 472–481, July 1984.

[Con2] J. H. CONWAY and N. J. A. SLOANE, "Fast Quantizing and Decoding Algorithms for Lattice Quantizers and Decoders," *IEEE Transactions on Information Theory*, Vol. IT-28, pp. 227–232, March 1982.

[Cor] K. G. CORNETT and S. B. WICKER, "Bit Error Rate Estimation Techniques for Digital Mobile Radios," *Proceedings of the 41st Annual IEEE Vehicular Technology Conference*, St. Louis, Missouri, May 19–20, 1991.

[Cos] D. J. COSTELLO, JR., "Free Distance Bounds for Convolutional Codes," *IEEE Transactions on Information Theory*, Vol. IT-20, No. 3, pp. 356–365, May 1974.

[Cus] E. L. CUSACK, "Error Control for QAM Signalling," *Electronics Letters*, Vol. 20, pp. 62–63, 1984.

[Dav] F. DAVARIAN, "Mobile Digital Communications via Tone Calibration," *IEEE Transactions on Vehicular Technology*, Vol. VT-36, No. 2, pp. 55–62, May 1987.

[Den] R. H. DENG, "Hybrid-ARQ Scheme Using TCM and Code Combining," *Electronics Letters*, Vol. 27, No. 10, pp. 866–868, May 1991.

[Dif] W. DIFFIE and M. E. HELLMAN, "New Directions in Cryptography," *IEEE Transactions in Information Theory*, Vol. IT-22, pp. 644–654, November 1976.

[Div1] D. DIVSALAR and M. K. SIMON, "Multiple Trellis Coded Modulation (MTCM)," *IEEE Transactions on Communications*, Vol. COM-36, No. 4, pp. 410–419, April 1988.

[Div2] D. DIVSALAR and M. K. SIMON, "The Performance of Trellis Coded Multilevel DPSK on a Fading Satellite Mobile Channel," *IEEE Transactions on Vehicular Technology*, Vol. VT-37, No. 2, pp. 78–91, May 1988.

[Div3] D. DIVSALAR and M. K. SIMON, "The Design of Trellis Coded MPSK for Fading Channels: Performance Criteria," *IEEE Transactions on Communications*, Vol. COM-36, No. 9, pp. 1004–1012, September 1988.

[Div4] D. DIVSALAR and M. K. SIMON, "The Design of Trellis Coded MPSK for Fading Channels: Set Partitioning for Optimum Code Design," *IEEE Transactions on Communications*, Vol. COM-36, No. 9, pp. 1013–1021, September 1988.

[Dru1] A. DRUKAREV and D. J. COSTELLO, JR., "A Comparison of Block and Convolutional Codes in ARQ Error Control Systems," *IEEE Transactions on Communications*, Vol. COM-30, pp. 2449–2455, November 1982.

[Dru2] A. DRUKAREV and D. J. COSTELLO, JR., "Hybrid ARQ Error Control Using Sequential Decoding," *IEEE Transactions on Information Theory*, Vol. IT-29, pp. 521–535, July 1983.

[Du] J. DU, M. KASAHARA, and T. NAMEKAWA, "Separable Codes on Type-II Hybrid-ARQ Systems," *IEEE Transactions on Communications*, Vol. 36, pp. 1089–1097, October 1988.

[Eli] P. ELIAS, "Coding for Noisy Channels," *IRE Conv. Record*, Part 4, pp. 37–47, 1955.

[Elia] M. ELIA, "Algebraic Decoding of the (23, 12, 7) Golay Code," *IEEE Transactions on Information Theory*, IT-33, Vol. 1, pp. 150–151, January 1987.

[Fan] R. M. FANO, "A Heuristic Discussion of Probabilistic Decoding," *IEEE Transactions on Information Theory*, IT-9, pp. 64–74, April 1963.

[Fir] P. FIRE, "A Class of Multiple-Error Correcting Binary Codes for Non-Independent Errors," Sylvania Report Number RSL-E-2, Sylvania Electronic Defense Laboratory, Reconnaissance Systems Division, Mountain View, California, March 1959.

[For2] G. D. FORNEY, JR., "The Viterbi Algorithm," *Proceedings of the IEEE*, Vol. 61, pp. 268–278, March 1973.

[For3] G. D. FORNEY, JR., "Convolutional Codes I: Algebraic Structure," *IEEE Transactions on Information Theory*, IT-16, pp. 720–738, November 1970.

[For4] G. D. FORNEY, JR., "Convolutional Codes II: Maximum Likelihood Decoding," *Information and Control*, Vol. 25, pp. 222–266, July 1974.

[For5] G. D. FORNEY, JR., "Convolutional Codes III: Sequential Decoding," *Information and Control*, Vol. 25, pp. 267–297, July 1974.

[For6] G. D. FORNEY, R. G. GALLAGER, G. R. LANG, F. M. LONGSTAFF, and S. U. QURESHI, "Efficient Modulation for Band-Limited Channels," *IEEE Journal on Selected Areas in Communications*, Vol. SAC-2, No. 5, September 1984.

[For7] G. D. FORNEY, "On Decoding BCH Codes," *IEEE Transactions on Information Theory*, Vol. IT-11, pp. 549–557, October 1965.

[For8] G. D. FORNEY and E. K. BOWER, "A High Speed Sequential Decoder: Prototype Design and Test," *IEEE Transactions on Communications Technology*, Vol. COM-19, pp. 821–835, October 1971.

[For9] G. D. FORNEY, JR., "Coset Codes—Part 1: Introduction and Geometrical Classification," *IEEE Transactions on Information Theory*, IT-34, pp. 1123–1151, September 1988.

[For10] G. D. FORNEY, JR., "Coset Codes—Part 2: Binary Lattices and Related Codes," *IEEE Transactions on Information Theory*, IT-34, pp. 1152–1187, September 1988.

[Gil] E. N. GILBERT, "A Comparison of Signalling Alphabets," *Bell System Technical Journal*, Vol. 31, pp. 504–522, 1952.

[Gol2] M. J. E. GOLAY, "Notes on Digital Coding" *Proceedings of the IRE,* Vol. 37, p. 657, June 1949.

[Gop] V. D. GOPPA, "Codes on Algebraic Curves," *Soviet Math. Doklady*, Vol. 24, pp. 170–172, 1981.

[Gor1] D. GORENSTEIN and N. ZIERLER, "A Class of Error Correcting Codes in p^m Symbols," *Journal of the Society of Industrial and Applied Mathematics*, Vol. 9, pp. 207–214, June 1961.

[Gor2] W. C. GORE, "Transmitting Binary Symbols with Reed-Solomon Codes," *Proceedings of the Princeton Conference on Information Science and Systems*, Princeton, New Jersey, pp. 495–497, 1973.

[Gre] R. R. GREEN, "A Serial Orthogonal Decoder," *Space Programs Summary*, No. 37–39, Vol. IV, pp. 247–251, June 1966.

[Had] J. HADAMARD, "Résolution d'une Question Relative aux Déterminants," Bulletin on Science and Mathematics, Vol. 17, pp. 240–248, 1893.

[Hag2] J. HAGENAUER and E. LUTZ, "Forward Error Correction Coding for Fading Compensation in Mobile Satellite Channels," *IEEE Journal on Selected Areas in Communications*, Vol. SAC-5, No. 2, pp. 215–225, February 1987.

[Hag3] J. HAGENAUER, "Rate Compatible Punctured Convolutional Codes and Their Applications," *IEEE Transactions on Communications*, Vol. COM-36, pp. 389–400, April 1988.

[Ham] R. W. HAMMING, "Error Detecting and Error Correcting Codes," *Bell System Technical Journal*, Vol. 29, pp. 147–160, 1950.

[Har] C. R. P. HARTMANN, J. B. DUCEY, and L. D. RUDOLPH, "On the Structure of Generalized Finite Geometry Codes," *IEEE Transactions on Information Theory*, IT-20, Vol. 2, pp. 240–252, March 1974.

[Har1] B. A. HARVEY and S. B. WICKER, "Error-Trapping Viterbi Decoding in a Type-I Hybrid-ARQ Protocol," *Proceedings of the 1990 IEEE International Conference on Communications*, Atlanta, Georgia, pp. 332.5.1–332.5.5, April 16–19, 1990.

[Har2] B. A. HARVEY and S. B. WICKER, "Décodage de Viterbi avec piégeage d'erreurs pour les protocoles ARQ-hybride de type I," *Canadian Electrical and Computer Engineering Journal*, Vol. 16, No. 1, pp. 5–12, January 1991.

[Har3] B. A. HARVEY and S. B. WICKER, "Packet Combining Systems Based on the Viterbi Decoder," *IEEE Transactions on Communications*, April 1994.

[Hel] J. A. HELLER and I. M. JACOBS, "Viterbi Decoding for Satellite and Space Communication," *IEEE Transactions on Communications Technology*, COM-19, pp. 835–848, October 1971.

[Hem] F. HEMMATI and D. J. COSTELLO, JR., "Truncation Error Probability in Viterbi Decoding," *IEEE Transactions on Communications*, Vol. COM-25, No. 5, pp. 530–532, May 1977.

[Hoc] A. HOCQUENGHEM, "Codes Correcteurs d'Erreurs," *Chiffres*, Vol. 2, pp. 147–156, 1959.

[Hsi] M. Y. HSIAO, "A Class of Optimal Minimum Odd-Weight-Column SEC-DED Codes," *IBM Journal of Research and Development*, Vol. 14, July 1970.

[IBM1] IBM EUROPE, "Trellis-Coded Modulation Schemes for Use in Data Modems Transmitting 3–7 Bits per Modulation Interval," CCITT SG XVII Contribution COM XVII, No. D114, April 1983.

[IBM2] IBM EUROPE, "Trellis-Coded Modulation Schemes with 8-state Symmetric Encoder and 90° Symmetry for Use in Data Modems Transmitting 3–7 Bits per Modulation Interval," CCITT SG XVII Contribution COM XVII, No. D180, October 1983.

[Jac] I. M. JACOBS and E. R. BERLEKAMP, "A Lower Bound to the Distribution of Computation for Sequential Decoding," *IEEE Transactions on Information Theory*, Vol. IT-13, pp. 167–174, April 1967.

[Jel1] F. JELINEK, "A Fast Sequential Decoding Algorithm Using a Stack," *IBM Journal of Research and Development*, Vol. 13, pp. 675–685, November 1969.

[Jel2] F. JELINEK, "An Upper Bound on Moments of Sequential Decoding Effort," *IEEE Transactions on Information Theory*, Volume IT-15, pp. 140–149, January 1969.

[Joh] R. JOHANNESSON, "Robustly Optimal Rate One-Half Binary Convolutional Codes," *IEEE Transactions on Information Theory*, Vol. IT-21, pp. 464–468, July 1975.

[Kal1] S. KALLEL and D. HACCOUN, "Sequential Decoding with ARQ and Code Combining: A Robust Hybrid FEC/ARQ System," *IEEE Transactions on Communications*, Vol. 36, pp. 773–780, July 1988.

[Kal2] S. KALLEL and D. HACCOUN, "Sequential Decoding with an Efficient Partial Retransmission ARQ Strategy," *IEEE Transactions on Communications*, Vol. 39, No. 2, pp. 208–213, February 1991.

[Kas1] T. KASAMI, T. KLØVE, and S. LIN, "Linear Block Codes for Error Detection," *IEEE Transactions on Information Theory*, Vol. IT-29, pp. 131–137, January 1983.

[Kas2] T. KASAMI and S. LIN, "On the Probability of Undetected Error for the Maximum Distance Separable Codes," *IEEE Transactions on Communications*, Vol. COM-32, No. 9, pp. 998–1006, September 1984.

[Kas3] T. KASAMI, S. LIN, and W. W. PETERSON, "Some Results on Weight Distributions of BCH Codes," *IEEE Transactions on Information Theory*, Volume IT-12, No. 2, p. 274, April 1966.

[Kas4] T. KASAMI, "A Decoding Procedure for Multiple-Error-Correction Cyclic Codes," *IEEE Transactions on Information Theory*, Vol. IT-10, No. 2, pp. 134–139, April 1964.

[Kha] R. E. KHAN, S.A. GRONEMEYER, J. BURHFIEL, and R. C. KUNZELMAN, "Advances in Packet Radio Technology," *Proceedings of the IEEE*, Vol. 66, No. 11, pp. 1468–1496, November 1978.

[Klø] T. KLØVE and M. MILLER, "The Detection of Errors After Error-Correction Decoding," *IEEE Transactions on Communications*, Vol. COM-32, pp. 511–517, May 1984.

[Kri] H. KRISHNA and S. MORGERA, "A New Error Control Scheme for Hybrid-ARQ Systems," *IEEE Transactions on Communications*, Vol. COM-35, pp. 981–990, October 1987.

[Lae] R. P. LAESER, "Engineering *Voyager 2*'s Encounter with Uranus," *Scientific American*, Vol. 255, No. 5, pp. 36–45, November 1986.

[Lau] C. LAU and C. LEUNG, "Performance Analysis of a Memory-ARQ Scheme with Soft Decision Detectors," *IEEE Transactions on Communications*, Vol. COM-34, pp. 827–832, August 1986.

[Lee3] W. C. Y. LEE, "Narrowbanding in Cellular Mobile Systems," *Telephony*, pp. 44–46, December 1, 1986.

[Lee4] W. C. Y. LEE, "Elements of Cellular Mobile Radio Systems," *IEEE Transactions on Vehicular Technology*, Vol. VT-35, No. 2, pp. 48–56, May 1986.

[Lee5] A. LEE and P. J. MCLANE, "Convolutionally Interleaved PSK and DPSK Trellis Codes for Shadowed Fast Fading Mobile Satellite Channels," *IEEE Transactions on Vehicular Technology*, No. 1, pp. 37–47, February 1990.

[Lin2] S. LIN and P. S. YU, "A Hybrid-ARQ Scheme with Parity Retransmission for Error Control of Satellite Channels," *IEEE Transactions on Communications*, Vol. COM-30, pp. 1701–1719, July 1982.

[Lin3] S. LIN, D. J. COSTELLO, JR., and M. J. MILLER, "Automatic-Repeat-Request Error Control Schemes," *IEEE Communications Magazine*, Vol. 22, pp. 5–17, December 1984.

[Lu] D. L. LU and J. F. CHANG, "Analysis of ARQ Protocols via Signal Flow Graphs," *IEEE Transactions on Communications*, Vol. COM-37, No. 3, pp. 245–251, March 1989.

[Lug1] L. LUGAND and D. J. COSTELLO, JR., "A Comparison of Three Hybrid-ARQ Schemes Using Convolutional Codes on a Non-Stationary Channel," *Proceedings of GLOBECOM 1982*, pp. C.8.4.1-C8.4.5.

[Lug2] L. R. LUGAND and D. J. COSTELLO, JR., "Parity Retransmission Hybrid-ARQ Using Convolutional Codes on a Nonstationary Channel," *IEEE Transactions on Communications*, Vol. 37, pp. 755–765, July 1989.

[Man1] D. M. MANDELBAUM, "An Adaptive Feedback Coding Scheme Using Incremental Redundancy," *IEEE Transactions on Information Theory*, pp. 388–389, May 1974.

[Man2] D. M. MANDELBAUM, "On Forward Error Correction with Adaptive Decoding," *IEEE Transactions on Information Theory*, pp. 230–233, March 1975.

[Mas3] J. L. MASSEY and M. K. SAIN, "Inverses of Linear Sequential Circuits," *IEEE Transactions on Computers*, Vol. C-17, pp. 330–337, April 1968.

[Mas4] J. L. MASSEY, "Shift Register Synthesis and BCH Decoding," *IEEE Transactions on Information Theory*, Vol. IT-15, No. 1, pp. 122–127, January 1969.

[Mas5] J. L. MASSEY, "Variable-Length Codes and the Fano Metric," *IEEE Transactions on Information Theory*, Vol. IT-18, No. 1, pp. 196–198, January 1972.

[Mas6] J. L. MASSEY and D. J. COSTELLO, JR., "Nonsystematic Convolutional Codes for Sequential Decoding in Space Applications," *IEEE Transactions on Communications*, Vol. COM-19, pp. 806–813, October 1971.

[Mas7] J. L. MASSEY, "Coding and Modulation in Digital Communications," *Proceedings of the 1974 International Zurich Seminar on Digital Communications*, Zurich, Switzerland, pp. E2(1)–E2(4), March 1974.

[McE2] R. McEliece and L. Swanson, "On the Decoder Error Probability for Reed-Solomon Codes," *IEEE Transactions on Information Theory*, Vol. IT-32, No. 5, pp. 701–703, September 1986.

[McE3] R. J. McEliece, E. R. Rodemich, H. C. Rumsey, Jr., and L. R. Welch, "New Upper Bounds on the Rate of a Code Using the Delsarte-MacWilliams Inequalities," *IEEE Transactions on Information Theory*, Vol. 23, pp. 157–166, 1977.

[McG] J. P. McGeehan and A. J. Bateman, "Phase-Locked Transparent Tone-in-Band (TTIB): A New Spectrum Configuration Particularly Suited to the Transmission of Data over SSB Mobile Radio Networks," *IEEE Transactions on Communications*, Vol. COM-32, No. 1, pp. 81–87, January 1984.

[Meg1] J. E. Meggitt, "Error Correcting Codes and Their Implementation for Data Transmission Systems," *IRE Transactions on Information Theory*, Vol. IT-7, pp. 234–244, October 1961.

[Meg2] J. E. Meggitt, "Error Correcting Codes for Correcting Bursts of Errors," *IBM Journal of Research and Development*, Vol. 4, pp. 329–334, July 1960.

[Mer] P. Merkey and E. C. Posner, "Optimum Cyclic Redundancy Codes for Noisy Channels," *IEEE Transactions on Information Theory*, Vol. IT-30, No. 6, pp. 865–867, November 1984.

[Met1] J. Metzner, "Improvements in Block-Retransmission Schemes," *IEEE Transactions on Communications*, pp. 524–532, February 1979.

[Met2] J. Metzner and D. Chang, "Efficient Selective-Repeat ARQ Strategies for Very Noisy and Fluctuating Channels," *IEEE Transactions on Communications*, Vol. COM-33, pp. 409–416, May 1985.

[Mil] M. J. Miller and S. Lin, "The Analysis of Some Selective-Repeat-ARQ Schemes with Finite Receiver Buffer," *IEEE Transactions on Communications*, Vol. COM-29, pp. 1307–1315, September 1981.

[Min] G. J. Minty, "A Comment on the Shortest Route Problem," *Operations Research*, Vol. 5, p. 724, October 1957.

[Mor1] S. Morgera and V. Oduol, "Soft-Decision Decoding Applied to the Generalized Type-II Hybrid-ARQ Scheme," *Proceedings of GLOBECOM 1988*, pp. 21.1.1–21.1.5.

[Mor2] S. Morgera and V. Oduol, "Soft-Decision Decoding Applied to the Generalized Type-II Hybrid-ARQ Scheme," *IEEE Transactions on Communications*, Vol. 37, pp. 393–396, April 1989.

[Mul] D. E. Muller, "Application of Boolean Algebra to Switching Circuit Design," *IEEE Transactions on Computers*, Vol. 3, pp. 6–12, September 1954.

[Omu] J. K. Omura, "On the Viterbi Decoding Algorithm," *IEEE Transactions on Information Theory*, Vol. IT-15, pp. 177–179, January 1969.

[Pal] R. E. A. C. Paley, "On Orthogonal Matrices," *Journal of Mathematics and Physics*, Vol. 12, pp. 311–320, 1933.

[Pat2] A. M. Patel and S. J. Hong, "Optimal Rectangular Code for High Density Magnetic Tapes," *IBM Journal of Research and Development*, Vol. 18, pp. 579–588, November 1974.

[Pat3] A. M. Patel, "Error Recovery Scheme for the IBM 3850 Mass Storage System," *IBM Journal of Research and Development*, Vol. 24, 1980.

[Pet2] W. W. Peterson, "Encoding and Error-Correction Procedures for the Bose-Chaudhuri Codes," *IRE Transactions on Information Theory*, Vol. IT-6, pp. 459–470, September 1960.

[Pie] S. S. PIETROBON, R. H. DENG, A. LAFANECHÉRE, G. UNGERBOECK, and D. J. COSTELLO, JR., "Trellis Coded Multidimensional Phase Modulation," *IEEE Transactions on Information Theory*, Vol. IT-36, No. 1, January 1990.

[Pos] E. POSNER, "Combinatorial Structures in Planetary Reconnaissance," in H. B. Mann (ed.), *Error Correcting Codes*, New York: John Wiley & Sons, 1969, pp. 15–46.

[Pot1] G. J. POTTIE and D. P. TAYLOR, "Multilevel Codes Based on Partitioning," *IEEE Transactions on Information Theory*, Vol. IT-35, No. 1, pp. 87–98, January 1989.

[Pur1] M. B. PURSLEY and S. D. SANDBERG, "Variable Rate Coding for Meteor-Burst Communications," *IEEE Transactions on Communications*, Vol. 37, No. 11, pp. 1105–1112, November 1989.

[Pur2] M. B. PURSLEY and S.D. SANDBERG, "Delay and Throughput for Three Transmission Schemes in Packet Radio Networks," *IEEE Transactions on Communications*, Vol. 37, No. 12, pp. 1264–1274, December 1989.

[Pur3] M. B. PURSLEY and S. D. SANDBERG, "Incremental Redundancy Transmission for Meteor-Burst Communications," *IEEE Transactions on Communications*, Vol. 39, No. 5, pp. 689–702, May 1991.

[Pra1] E. PRANGE, "Cyclic Error-Correcting Codes in Two Symbols," *Air Force Cambridge Research Center*-TN-57-103, Cambridge, MA: September 1957.

[Pra2] E. PRANGE, "Some Cyclic Error-Correcting Codes with Simple Decoding Algorithms," *Air Force Cambridge Research Center*-TN-58-156, Cambridge, MA: April 1958.

[Pra3] E. PRANGE, "The Use of Coset Equivalence in the Analysis and Decoding of Group Codes," *Air Force Cambridge Research Center*-TR-59-164, Cambridge, MA: 1959.

[Ram] J. L. RAMSEY, "Realization of Optimum Interleavers," *IEEE Transactions on Information Theory*, Vol. IT-16, pp. 338–345, May 1970.

[Ras1] L. K. RASMUSSEN, "Trellis Coded, Adaptive Rate Hybrid-ARQ Protocols over AWGN and Slowly Fading Rician Channels," Ph.D. Dissertation, Georgia Institute of Technology, August 1993.

[Ras2] L. RASMUSSEN and S. B. WICKER, "Trellis Coded Hybrid-ARQ Protocols over AWGN and Slowly Fading Rician Channels," *IEEE Transactions on Information Theory*, March 1994.

[Ras3] L. RASMUSSEN and S. B. WICKER, "Trellis Coded, Type-I Hybrid-ARQ Protocols Based on CRC Error-Detecting Codes," *IEEE Transactions on Communications*, to appear.

[Ree1] I. S. REED, "A Class of Multiple-Error-Correcting Codes and a Decoding Scheme, *IEEE Transactions on Information Theory*, Vol. 4, pp. 38–49, September 1954.

[Ree2] I. S. REED and G. SOLOMON, "Polynomial Codes over Certain Finite Fields," *SIAM Journal on Applied Mathematics*, Vol. 8, pp. 300–304, 1960.

[Ree3] I. S. REED, X. YIN, and T-K TRUONG, "Algebraic Decoding of the (32, 16, 8) Quadratic Residue Code," *IEEE Transactions on Information Theory*, Vol. 36, No. 4, pp. 876–880, July 1990.

[Ree4] I. S. REED, T-K TRUONG, X. CHEN, and X. YIN, "Algebraic Decoding of the (41, 21, 9) Quadratic Residue Code," *IEEE Transactions on Information Theory*, Vol. 38, No. 3, pp. 974–986, May 1992.

[Ree5] I. S. REED, R. A. SCHOLTZ, T-K TRUONG, and L. R. WELCH, "The Fast Decoding of Reed-Solomon Codes Using Fermat-Theoretic Transforms and Continued Fractions," *IEEE Transactions on Information Theory*, Vol. IT-24, pp. 100–106, January 1978.

[Ric1] M. D. RICE and S. B. WICKER, "Modified Majority-Logic Decoding of Cyclic Codes in Hybrid-ARQ Systems," *IEEE Transactions on Communications*, Vol. 40, No. 9, pp. 1413–1417, September 1992.

[Ric2] M. D. RICE and S. B. WICKER, "Adaptive Error Control for Slowly Varying Channels," *IEEE Transactions on Communications*, March 1994.

[Ric3] M. D. RICE and WICKER S.B., "A Sequential Testing Scheme for Adaptive Error Control on Slowly Varying Channels," *IEEE Transactions on Communications*, April 1994.

[Rob] J. P. ROBINSON and A. J. BERNSTEIN, "A Class of Binary Recurrent Codes with Limited Error Propagation," *IEEE Transactions on Information Theory*, Vol. IT-13, No. 1, pp. 106–113, January 1967.

[Sas] A. R. K. SASTRY, "Improving Automatic Repeat Request (ARQ) Performance on Satellite Channels Under High Error Rate Conditions," *IEEE Transactions on Communications*, Vol. COM-23, pp. 436–439, April 1975.

[Sch] C. SCHLEGEL and D. J. COSTELLO, JR., "Bandwidth Efficient Coding for Fading Channels: Code Construction and Performance Analysis," *IEEE Journal on Selected Areas in Communications*, Vol. 7, No. 9, pp. 1356–1368, December 1989.

[Sha1] C. E. SHANNON, "A Mathematical Theory of Communication," *Bell System Technical Journal*, Vol. 27, pp. 379–423 and pp. 623–656, 1948.

[Sha2] C. E. SHANNON, "Communication Theory of Secrecy Systems, *Bell System Technical Journal*, Vol. 28, pp. 656–715, 1949.

[Shac] N. SHACHAM, "Adaptive Link Level Protocol for Packet Radio Channels," *Proceedings of IEEE INFOCOM*, pp. 626–635, April 1986.

[Sim] M. K. SIMON and D. DIVSALAR, "Combined Trellis Coding with Asymmetric MPSK Modulation," Jet Propulsion Laboratory Publication 85-24, May 1, 1985.

[Sind] P. SINDHU, "Retransmission Error Control with Memory," *IEEE Transactions on Communications*, Vol. COM-25, pp. 473–479, May 1977.

[Sing] R. C. SINGLETON, "Maximum Distance Q-Nary Codes," *IEEE Transactions on Information Theory*, Vol. IT-10, pp. 116–118, 1964.

[Sug] Y. SUGIYAMA, Y. KASAHARA, S. HIRASAWA, and T. NAMEKAWA, "A Method for Solving Key Equation for Goppa Codes," *Information and Control*, Vol. 27, pp. 87–99, 1975.

[Tie] A. TIETÄVÄINEN, "On the Nonexistence of Perfect Codes over Finite Fields," *SIAM Journal of Applied Mathematics,* Vol. 24, pp. 88–96, 1973.

[Tsf] M. A. TSFASMAN, S. G. VLADUT, and TH. ZINK, "Modular Curves, Shimura Curves, and Goppa Codes, Better than the Varsharmov-Gilbert Bound," *Mathematische Nachrichten*, Vol. 104, pp. 13–28, 1982.

[Ung1] G. UNGERBOECK and I. CSAJKA, "On Improving Data Link Performance by Increasing the Channel Alphabet and Introducing Sequence Coding," *1976 International Symposium on Information Theory*, Ronneby, Sweden, June 1976.

[Ung2] G. UNGERBOECK, "Channel Coding with Multilevel/Phase Signals," *IEEE Transactions on Information Theory*, Vol. IT-28, No. 1, pp. 55–67, January 1982.

[Ung3] G. UNGERBOECK, "Trellis-Coded Modulation with Redundant Signal Sets, Part I: Introduction," *IEEE Communications Magazine*, Vol. 25, No. 2, pp. 5–11, February 1987.

[Ung4] G. UNGERBOECK, "Trellis-Coded Modulation with Redundant Signal Sets, Part II: State of the Art," *IEEE Communications Magazine*, Vol. 25, No. 2, pp. 12–21, February 1987.

[Var] R. R. VARSHARMOV, "Estimate of the Number of Signals in Error Correcting Codes," *Soviet Math. Doklady*, Vol. 117, pp. 739–741, 1957.

[Vau] R. G. VAUGHN, "Signals in Mobile Communications: A Review," *IEEE Transactions on Vehicular Technology*, Vol. VT-35, No. 4, pp. 133–145, November 1986.

[Vit2] A. J. VITERBI, "Error Bounds for Convolutional Codes and an Asymptotically Optimum Decoding Algorithm," *IEEE Transactions on Information Theory*, Vol. IT-13, pp. 260–269, April 1967.

[Vit3] A. J. Viterbi, "Convolutional Codes and Their Performance in Communication Systems," *IEEE Transactions on Communications Technology*, Vol. COM-19, No. 5, pp. 751–772, October 1971.

[Vit4] A. J. VITERBI, JACK K. WOLF, EPHRAIM ZEHAVI, and ROBERTO PADOVANI, "A Pragmatic Approach to Trellis-Coded Modulation," *IEEE Communications Magazine*, pp. 11–19, July 1989.

[Wan] Y. WANG and S. LIN, "A Modified Selective-Repeat Type-II Hybrid-ARQ System and Its Performance Analysis," *IEEE Transactions on Communications*, Vol. COM-31, pp. 593–607, May 1983.

[Wei1] L.-F. Wei, "Rotationally Invariant Convolutional Channel Coding with Expanded Signal Space—Part 1: 180°," *IEEE Journal on Selected Areas in Communications*, Vol. SAC-2, No. 5, pp. 659–671, September 1984.

[Wei2] L.-F. WEI, "Rotationally Invariant Convolutional Channel Coding with Expanded Signal Space—Part 2: Nonlinear Codes," *IEEE Journal on Selected Areas in Communications*, Vol. SAC-2, No. 5, pp. 672–686, September 1984.

[Wei3] L.-F. WEI, "Trellis-Coded Modulation with Multidimensional Constellations," *IEEE Transactions on Information Theory*, Vol. IT-33, No. 4, pp. 483–501, July 1987.

[Wei4] L.-F. WEI, "Rotationally Invariant Trellis Coded Modulation with Multidimensional MPSK," *IEEE Journal on Selected Areas in Communications*, Vol. SAC-7, No. 9, pp. 1281–1295, December 1989.

[Wel] L. R. WELCH, "Computation of Finite Fourier Series," *Space Programs Summary*, No. 37-37, Vol. IV, pp. 295–297, February 1966.

[Wic1] S. B. WICKER, "A Geometric Approach to Error Correcting Codes," doctoral dissertation, May 1987, University of Southern California.

[Wic2] S. B. WICKER and L. R. WELCH, "The Geometry of MDS Codes," *Proceedings of the 1987 IEEE Pacific Rim Conference for Communications, Computers, and Signal Processing*, Victoria, British Columbia, pp. 89–92, June 4, 1987.

[Wic3] S. B. WICKER, "Cyclic Codes and Rational Curves," *1988 International Symposium on Information Theory*, Kobe, Japan, Paper Number MC7-2, June 19, 1988.

[Wic4] S. B. WICKER, "Adaptive Rate Coding Using Reed-Solomon Codes in a Hybrid-ARQ Protocol," *Proceedings of the 1989 International Symposium on Signals, Systems, and Electronics*, Nürnberg, West Germany, pp. 642–645, September 18, 1989.

[Wic5] S. B. WICKER, "High Reliability Data Transfer over the Land Mobile Radio Channel," *IEEE Transactions on Vehicular Technology*, Vol. 39, No. 1, pp. 48–55, February 1990.

[Wic6] S. B. WICKER, "Modified Majority-Logic Decoders for Use in Convolutionally Encoded Hybrid-ARQ Systems," *IEEE Transactions on Communications*, Vol. 38, No. 3, pp. 263–266, March 1990.

[Wic7] S. B. WICKER, "Reed-Solomon Error Control Coding for Data Transmission over Rayleigh Fading Channels with Feedback," *IEEE Transactions on Vehicular Technology*, Vol. 41, No. 2, pp. 124–133, May 1992.

[Wic8] S. B. WICKER and M. D. BARTZ, "Type-II Hybrid-ARQ Protocols Using Punctured MDS Codes," *IEEE Transactions on Communications*, April 1994.

[Wic9] S. B. WICKER, "Adaptive Rate Error Control Through the Use of Diversity Combining and Majority-Logic Decoding in a Hybrid-ARQ Protocol," *IEEE Transactions on Communications*, Vol. 39, No. 3, pp. 380–385, March 1991.

[Wil1] S. G. WILSON, H. A. SLEEPER, P. J. SCHOTTLER, and M. T. LYONS, "Rate-3/4 Convolutional Coding of 16-PSK: Code Design and Performance Study," *IEEE Transactions on Communications*, Vol. COM-32, pp. 1308–1315, December 1984.

[Wit] K. A. WITZKE and C. LEUNG, "A Comparison of Some Error Detecting CRC Code Standards," *IEEE Transactions on Communications*, Vol. COM-33, No. 9, pp. 996–998, September 1985.

[Wol] J. K. WOLF, "Decoding of Bose-Chaudhuri-Hocquenghem Codes and Prony's Method for Curve Fitting," *IEEE Transactions on Information Theory*, Vol. IT-13, p. 608, 1967.

[Woz3] J. M. WOZENCRAFT and M. HORSTEIN, "Coding for Two-Way Channels," Technical Report 383, Research Laboratory of Electronics, M.I.T., January 3, 1961.

[Woz4] J. M. WOZENCRAFT and M. HORSTEIN, "Digitalised Communication over Two-Way Channels," in *The Fourth London Symposium on Information Theory*, London, England, August 29–September 3, 1960.

[Yam] H. YAMAMOTO and K. ITOH, "Viterbi Decoding Algorithm for Convolutional Codes with Repeat Request," *IEEE Transactions on Information Theory*, Vol. IT-26, pp. 540–547, September 1980.

[Zeh] E. ZEHAVI and J. K. WOLF, "On the Performance Evaluation of Trellis Codes," *IEEE Transactions on Information Theory*, Vol. IT-33, No. 2, pp. 196–202, March 1987.

[Zig] K. ZIGANGIROV, "Some Sequential Decoding Procedures," *Probl. Peredachi Inf.*, Vol. 2, pp. 13–25, 1966 (Russian).

Index

A

B

D

E

K

L

M

N

O

P

T

U